Plant Adaptation Strategies in Changing Environment

Vertika Shukla • Sanjeev Kumar
Narendra Kumar
Editors

Plant Adaptation Strategies in Changing Environment

Springer

Editors
Vertika Shukla
School of Environmental Sciences
Babasaheb Bhimrao Ambedkar University
Lucknow, UP, India

Sanjeev Kumar
School of Environmental Sciences
Babasaheb Bhimrao Ambedkar University
Lucknow, UP, India

Narendra Kumar
Department of Environmental Science,
School of Environmental Sciences
Babasaheb Bhimrao Ambedkar University
Lucknow, UP, India

ISBN 978-981-10-6743-3 ISBN 978-981-10-6744-0 (eBook)
https://doi.org/10.1007/978-981-10-6744-0

Library of Congress Control Number: 2017961561

Printed on acid-free paper

This Springer imprint is published by Springer Nature
The registered company is Springer Nature Singapore Pte Ltd.
The registered company address is: 152 Beach Road, #21-01/04 Gateway East, Singapore 189721, Singapore

Preface

Plants are sessile organisms, and to cope up with disturbed/altered ambient conditions, they have evolved diverse adaptation strategies to support their growth and development. The ability of various plant groups to tolerate the extremes posed by natural (arid/cold) conditions and/or chemically rich environments involves both morphological and physiological adaptation as well as changes in ecological behaviour to sustain in relatively protected niches within an extreme environment. These adaptations are manifested as changes in enzyme activity, cellular organization, metabolomics and ultimately genetic alterations expressed as mutations.

The present compilation addresses various stress conditions to which plants are exposed to and how they resist or avoid the particular stress condition/s. Plants endure temperature fluctuations and varying solar irradiance and also tolerate anthropogenic emissions. Lichens, pioneers of plant succession, are among the most ubiquitous organisms which are widely distributed in polar regions and cold deserts. They are highly mechanized to freezing and desiccation. Resurrection strategy in plants presents an interesting aspect for research, which requires thorough investigation at the cellular and genetic levels.

Similar to lichens, algae are also present globally, occurring in a range of habitats. Algae are found growing abundantly in extreme habitats indicating their adaptation to harsh environments. Lower plant groups not only adapt to natural alterations but also have been potential sequesters of anthropogenic emissions especially. Pteridophytes are widely utilized as hyperaccumulators of metals. Fly ash-induced metabolic adaptation in different fern species provides an insight into the metabolic adaptations adopted by ferns to tolerate higher levels of metals. Further utility of these species in phytoremediation of toxic metals from fly ash as well as ecorestoration of fly ash landfills is an added advantage.

In view of the increasing urbanization and industrialization, metallic emissions are posing a risk at spatio-temporal scale, but crops have adapted to the high levels by various detoxification processes.

A number of physiological and biochemical changes occur under heavy metal (HM) toxicity including alteration in water uptake and transport, root and shoot growth and generation and scavenging of reactive oxygen species (ROS) and changes in HM-complexing ligands for sequestration of these HMs into vacuoles to reduce the HM concentration in the cytoplasm. Physiological and biochemical responses of plants to HM toxicity have been discussed in detail along with shedding light on uptake and transport mechanisms of HMs in brief.

Agriculture productivity is greatly affected by the salinity of the soil. Crop growth and yield are severely decreased by salt stress. Proteomic-based techniques have emerged as a powerful tool to reveal the molecular mechanisms of salt stress responses. Several salt-responsive proteins have been identified for crops by using these techniques. It will be possible to change salt-sensitive crops to salt-tolerant crops in the near future by using these proteins and their corresponding genes. In the past few years, great progress has been made towards understanding plant salt stress responses and tolerance mechanisms, but still many challenges lie ahead. Both molecular breeding and advanced biotechnology methods should help scientists to develop crops with enhanced salt tolerance. Apart from the proteomics approach, usage of plant-based biochar increases plant growth and biomass under salt stress. Biochar-treated soil increases the photosynthesis, nutrient uptake and modified gas exchange characteristics in plants; the details about omics approaches and biochar have also been well discussed in the chapters.

Plants which are adapted to toxic chemicals (hyperaccumulating plants) show an important part in the terrestrial food web and are significant in both agricultural and natural ecosystems. These plants are frequently utilized for phytoremediation of contaminated sites. The chapters provide an understanding about the applicability and mechanism of the phytoremediation process.

Plants biosynthesize and store a variety of metabolites especially carbon-based secondary compounds that protect plants from stress conditions by acting as antioxidants, cryoprotectants and heat protectants. These organic compounds are chemically diverse, comprising sugars (sucrose, glucose, xylose, galactose, mannose, inositol, fructose, glycerol, mannitol, sorbitol), fatty acids and amino acids and their derivatives. Use of hyphenated instrumentation techniques like LC-MS/MS, GC-MS, etc. provides the high-throughput estimation of metabolic profile changes in valuable plant resources. Variation in sugar, amino acid and fatty acids in plant species exposed to different environmental conditions has been discussed in detail.

Lichens are known to synthesize more than 1000 metabolites. The range of chemicals synthesized underlines their ecological importance. Lichens are potential natural resources of a wide range of chemicals having a potential biological role in sustainability in a diverse range of habitats and extreme conditions. The chapter reviews the progress made in the past 30 years in the field of lichenology to isolate and characterize bioactive constituents.

The overview on plant adaptation to outer space environment and survival capability of some plants underlines the importance of lithopanspermia process, i.e. the transport of living organisms by meteorites in the Solar System between the planets of the Earth type is an interesting aspect for studies on existence of extraterrestrial life in the universe and origin of life on Earth.

Overall the book provides comprehensive information on the strategies adopted by various plant groups to sustain changing environmental condition whether natural or man-made. Ultimately it is the strategy based on 'survival of the fittest' which has resulted in the evolution of various plant groups. This book aims to serve as a valuable knowledge resource especially for beginners, students and researchers from different academic backgrounds of basic research and applied sciences.

Lucknow, UP, India

Vertika Shukla
Sanjeev Kumar
Narendra Kumar

Contents

1 **Adaptation of Lichens to Extreme Conditions** 1
Richard A. Armstrong

2 **Adaptive Mechanisms of Desiccation Tolerance in Resurrection Plants** 29
Farah Deeba and Vivek Pandey

3 **Dehydration and Freezing Resistance of Lichenized Fungi** 77
H. Harańczyk, A. Casanova-Katny, M. Olech, and K. Strzałka

4 **Adaptation in Algae to Environmental Stress and Ecological Conditions** 103
Sanjeeva Nayaka, Kiran Toppo, and Sushma Verma

5 **Biochemical and Molecular Responses in Higher Plants Under Salt Stress** 117
Neerja Srivastava

6 **Biochar Mitigates Salinity Stress in Plants** 153
Anju Patel, Puja Khare, and D.D. Patra

7 **Fly Ash-Induced Metabolic Adaptations in Three Ferns** 183
Alka Kumari

8 **The Multiple Properties of Some of the Lichenized Ascomycetes: Biological Activity and Active Metabolites** 201
Valery M. Dembitsky

9 **Metabolic Profiling and Its Plausible Environmental Significance in a Common Himalayan Lichen** 235
Vertika Shukla, Ankita Asati, Devendra K. Patel, Manoj Semwal, Narendra Kumar, and Dalip K. Upreti

10 Heavy Metal Tolerance in Crop Plants: Physiological and Biochemical Aspects ... 253
Reshu Chauhan, Surabhi Awasthi, Amit Pal Singh, Sudhakar Srivastava, Veena Pande, Rudra Deo Tripathi, and Amit Kumar

11 Plant Adaptation to Recalcitrant Chemicals ... 269
Hina Khatoon, Apourv Pant, and J.P.N. Rai

12 Plant Growth Under Stress Conditions: Boon or Bane ... 291
Poonam, Shamshad Ahmad, Narendra Kumar, Paromita Chakraborty, and Richa Kothari

13 Adaptation Strategies of Plants Against Common Inorganic Pollutants and Metals ... 315
Dhananjay Kumar, Sanjeev Kumar, Vertika Shukla, and Narendra Kumar

14 Impacts of Climate Change on Agriculture: Adaptation, Mitigation, and Environmental Policy ... 329
Ravindra Prasad, Sanjeev Kumar, Anuj Kumar Yadav, Shailender Kumar, Mahesh Kumar, Ritu Singh, and Narendra Kumar

15 Nitric Oxide (NO) and Physio-biochemical Adaptation in Plants Against Stress ... 347
Arun Kumar Maurya and Anita Rani

Editors and Contributors

About the Editors

Vertika Shukla received her M.Sc. in organic chemistry and Ph.D. (2003) on *Chemical Study of Macrolichens of Garhwal Himalayas* from H.N.B. Garhwal University, Srinagar (Garhwal). She currently teaches chemistry to undergraduate applied geology students at Babasaheb Bhimrao Ambedkar (Central) University, Lucknow. Dr. Shukla has completed three postdoc fellowships awarded by the Department of Science and Technology (DST, New Delhi) at CSIR National Botanical Research Institute and Babasaheb Bhimrao Ambedkar (Central) University. She has published 64 scientific articles, mostly in respected international SCI journals. She has also coauthored and coedited three books published by Springer. Currently, she is the chief editor of the journal *Cryptogam Biodiversity and Assessment*, the official journal of Indian Lichenological Society.

Dr. Shukla has more than 10 years of research and teaching experience. She has carried out extensive research on air pollution and climate change in Garhwal Himalayas. She has contributed much on secondary metabolite chemistry, accumulation and spatio-temporal behaviour of persistent organic pollutants (POPs) and metals in high-altitude ecosystems.

Sanjeev Kumar is an assistant professor at the School for Environmental Science, Babasaheb Bhimrao Ambedkar University, Lucknow. He received his Ph.D. degree in environmental sciences from Babasaheb Bhimrao Ambedkar University, Lucknow, in 2015. He has published several research articles in national and international journals and has several book chapters in national and international books to his credit. He has been awarded a "Fellow of Society of Plant Research" and has also received the "Significant Contribution in Academic Activities" award from Babasaheb Bhimrao Ambedkar University, Lucknow. He is joint secretary of the Prof. H.S. Srivastava Foundation for Science and Society and the Society for Science of Climate Change and Sustainable Environment and is coeditor of the multilingual magazine *KAHAAR*. Dr. Sanjeev Kumar has also worked as assistant editor for international journals such as *Functional Plant Biology*, *Physiology and Molecular Biology of Plants* and the *Journal of Climate Change and Environmental Sustainability*.

Narendra Kumar obtained his M.Sc. and Ph.D. degrees in environmental science from the Babasaheb Bhimrao Ambedkar University, Lucknow. Dr. Kumar is an active researcher and academician with almost 15 years of postgraduate teaching and research experience. Dr. Kumar started his academic career as a project fellow at National Botanical Research Institute (NBRI-CSIR), Lucknow, U.P., India. He is currently working as an assistant professor at the Department of Environmental Science, Babasaheb Bhimrao Ambedkar University (a Central University), Lucknow. Dr. Kumar has successfully guided more than 50 M.Sc. environmental science students for their dissertation work and produced one Ph.D.

Dr. Kumar has published more than 20 international and national research papers as well as many articles in the leading newspapers. He has published five book chapters with reputed national and international publishers.

Contributors

Shamshad Ahmad Department of Environmental Sciences, Babasaheb Bhimrao Ambedkar University, Lucknow, UP, India

Richard A. Armstrong Department of Vision Sciences, Aston University, Birmingham, UK

Ankita Asati Analytical Chemistry Division, CSIR-Indian Institute of Toxicology Research, Lucknow, India

Surabhi Awasthi CSIR-National Botanical Research Institute, Lucknow, India

Institute of Environment & Sustainable Development, Banaras Hindu University, Varanasi, India

A. Casanova-Katny Veterinary school, Faculty of Natural Resources, Catholic University of Temuco, Temuco, Chile

Paromita Chakraborty Department of Civil Engineering, SRM Research Institute, SRM University, Kattankulathur, Tamil Nadu, India

Water Science Laboratory, University of Nebraska-Lincoln, Lincoln, NE, USA

Reshu Chauhan CSIR-National Botanical Research Institute, Lucknow, India

Department of Biotechnology, Kumaun University, Bhimtal, Bhimtal, Nainital, Uttarakhand, India

Farah Deeba Biotechnology Department, CSIR-Central Institute of Medicinal & Aromatic Plants, Lucknow, India

Valery M. Dembitsky National Scientific Center of Marine Biology, Vladivostok, Russia

H. Harańczyk Institute of Physics, Jagiellonian University, Cracow, Poland

Puja Khare Agronomy and Soil Science Division, CSIR-CIMAP, Lucknow, India

Hina Khatoon Department of Environmental Science, G.B. Pant University of Agriculture and Technology, Pantnagar, Uttarakhand, India

Richa Kothari Department of Environmental Sciences, Babasaheb Bhimrao Ambedkar University, Lucknow, UP, India

Amit Kumar Cytogenetic and Plant Breeding Lab, Department of Botany, University of Lucknow, Lucknow, India

Dhananjay Kumar Department of Environmental Science, Babasaheb Bhimrao Ambedkar (Central) University, Lucknow, UP, India

Mahesh Kumar Department of Environmental Science, Babasaheb Bhimrao Ambedkar University, Lucknow, UP, India

Shailender Kumar Department of Environmental Studies, University of Delhi, New Delhi, Delhi, India

Alka Kumari Department of Botany, University of Lucknow, Lucknow, UP, India

Arun Kumar Maurya Department of Botany, Multanimal Modi PG College, Modinagar Ghaziabad, UP, India

Sanjeeva Nayaka Algology Section, CSIR-National Botanical Research Institute, Lucknow, India

M. Olech Institute of Botany, Jagiellonian University, Cracow, Poland

Institute of Biochemistry and Biophysics, Polish Academy of Sciences, Warsaw, Poland

Veena Pande Department of Biotechnology, Kumaun University, Bhimtal, Nainital, Uttarakhand, India

Vivek Pandey Plant Ecology & Environmental Sciences Division, CSIR-National Botanical Research Institute, Lucknow, India

Apourv Pant Department of Environmental Science, G.B. Pant University of Agriculture and Technology, Pantnagar, Uttarakhand, India

Anju Patel School of Earth Sciences, Banasthali Vidyapeeth, Banasthali, Rajasthan, India

Devendra K. Patel Analytical Chemistry Division, CSIR-Indian Institute of Toxicology Research, Lucknow, India

D.D. Patra Bidhan Chandra Krishi Viswa Vidyalaya, Nadia, West Bengal, India

Poonam Department of Environmental Sciences, Babasaheb Bhimrao Ambedkar University, Lucknow, UP, India

Ravindra Prasad Department of Environmental Studies, University of Delhi, New Delhi, Delhi, India

J.P.N. Rai Department of Environmental Science, G.B. Pant University of Agriculture and Technology, Pantnagar, Uttarakhand, India

Anita Rani Department of Botany, Dyal Singh College, Delhi University, Delhi, India

Manoj Semwal ICT Department, CSIR-Central Institute of Medicinal & Aromatic Plants, Lucknow, India

Amit Pal Singh CSIR-National Botanical Research Institute, Lucknow, India

Ritu Singh Department of Environmental Science, School of Earth Sciences, Central University of Rajasthan, Ajmer, Rajasthan, India

Neerja Srivastava Department of Biochemistry, IBSBT, CSJMU, Kanpur, UP, India

Sudhakar Srivastava Institute of Environment & Sustainable Development, Banaras Hindu University, Varanasi, India

K. Strzałka Malopolska Centre of Biotechnology, Jagiellonian University, Cracow, Poland

Faculty of Biochemistry, Biophysics and Biotechnology, Jagiellonian University, Cracow, Poland

Kiran Toppo Algology Section, CSIR-National Botanical Research Institute, Lucknow, India

Rudra Deo Tripathi CSIR-National Botanical Research Institute, Lucknow, India

Dalip K. Upreti Lichenology Laboratory, CSIR-National Botanical Research Institute, Lucknow, India

Sushma Verma Algology Section, CSIR-National Botanical Research Institute, Lucknow, India

Anuj Kumar Yadav Department of Zoology & Environmental Sciences, Gurukula Kangri Vishwavidyalaya, Haridwar, Uttarakhand, India

Adaptation of Lichens to Extreme Conditions

1

Richard A. Armstrong

Abstract

Lichens exhibit the classic features of stress-tolerant organisms, viz. slow growth rates, considerable longevity, low demand for nutrients, and the presence of specific adaptations to survive in the most inhospitable environments on Earth. The ability of lichens to tolerate the extremes posed by deserts, polar regions, and chemically rich environments involves both morphological and physiological adaptation and changes in ecological behaviour so that species adapt to relatively protected niches within an extreme environment. This chapter discusses those aspects of the lichen symbiosis relevant to survival in extreme conditions and then describes the adaptation of lichens to (1) wet forests, (2) deserts, (3) the Arctic, (4) alpine regions, (5) Antarctica, (6) chemically rich environments, and (7) extraterrestrial environments such as outer space and Mars. It is evident that the lichen symbiosis is more tolerant to hostile conditions than its symbionts, morphological and physiological adaptations are intimately associated, and convergent evolution has resulted in similar changes in different environments.

Keywords

Lichen · Extreme environments · Adaptation · Arctic/alpine regions · Antarctica · Deserts · Chemically rich environments · Extraterrestrial environment

R.A. Armstrong (✉)
Department of Vision Sciences, Aston University, Birmingham, UK
e-mail: R.A.Armstrong@aston.ac.uk

V. Shukla et al. (eds.), *Plant Adaptation Strategies in Changing Environment*,
https://doi.org/10.1007/978-981-10-6744-0_1

Abbreviations

ABC transporter	ATP-binding casette transporter
EBF	European BIOPAN Facility
GSH	Glutathione
GST	Glutathione S-transferase
ISS	International Space Station
nPS	Net photosynthesis
ROS	Reactive oxygen species
S/V	Surface/volume ratio
SOD	Superoxide dismutase
THEMIS	Thermal Emission Imaging System
UV	Ultraviolet

1.1 Introduction

Lichens occur in every major ecosystem on Earth from the poles to the tropics. Although many species are found in the tropics, they only become a significant component of the vegetation in more extreme environments such as the hot arid and semiarid deserts and the cold polar regions (Mattick 1954). Lichens are influenced by many environmental factors including the long-term effects of climate, local changes in microclimate, and a variety of factors associated with the substratum such as type of rock, bark, or soil, substrate chemistry, and the degree of nutrient enrichment by birds, salinity, or pollution (Armstrong 1974, 2015; James et al. 1977). Plants in general have been divided into three groups according to whether a ruderal, competitive, or stress-tolerant life-cycle strategy predominates (Grime 1979). In this scheme, lichens are considered to be stress-tolerant organisms and therefore to occur in communities in which competition is unlikely to play a significant role. Such organisms would be characterized by slow growth rates, considerable longevity, low demands for nutrients, and the presence of specific adaptations to survive in stressful conditions (Grime 1979). The ability of lichens to tolerate extreme conditions involves both morphological and physiological adaptation as well as changes in ecological behaviour so that they adapt to relatively protected niches within an extreme environment. This chapter discusses those aspects of the lichen symbiosis relevant to survival in extreme conditions and then describes the adaptation of lichens to (1) wet forest, (2) desert, (3) the Arctic, (4) alpine regions, (5) Antarctica, (6) chemically rich environments, and (7) extraterrestrial environments such as outer space and Mars.

1.2 The Lichen Symbiosis

1.2.1 Morphology

A lichen is composite, symbiotic organism comprising two or more constituents, the most frequent being an alga and a fungus. In cross section (Fig. 1.1), a typical lichen is composed mainly of fungal tissue, but embedded in the upper cortical layers are eukaryotic algal cells. Some lichens are also associated with cyanobacteria (blue-green algae) found in special structures called cephalodia, and lichens may also associate with other microorganisms. Hence, a lichen can be considered as an 'ecosystem' in which several constituent organisms collaborate (Farrar 1976). The algal partner carries out photosynthesis and supplies the fungus with carbohydrate, but there is little experimental evidence that the fungus supplies nutrients directly to the alga (Smith and Douglas 1987). Nevertheless, a thick upper cortical layer and associated hyphae may protect the algal cells from rapid water loss and intense radiation but at the expense of slower uptake of materials (Rundel 1988). There are three major types of lichen growth form (Fig. 1.2), viz. the crustose type composed of a thin crust that is tightly attached to the substratum, the foliose type that comprises a series of radially arranged leaflike lobes, and the fruticose type in which the lichen thallus is attached to the substratum at a single point and forms a complex branched structure. Most habitats in which lichens are frequent have a mixture of all three growth forms. These growth forms have very different surface to volume (S/V) ratios, which is maximal in fruticose lichens and an important factor

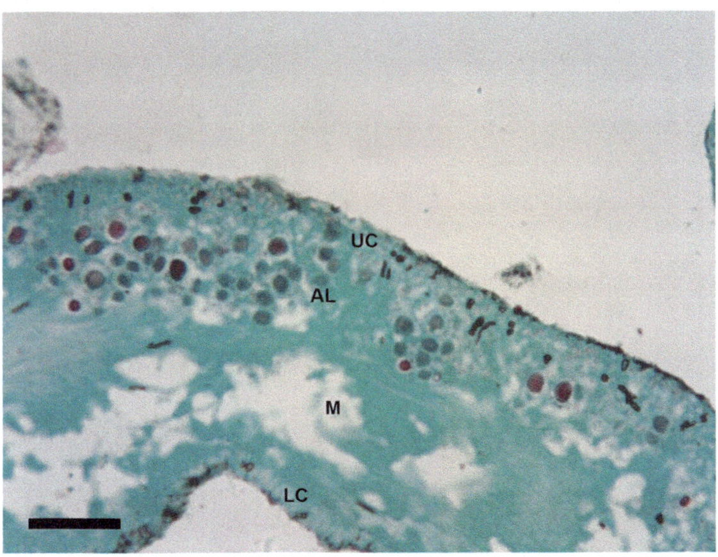

Fig. 1.1 Vertical section through a thallus of the foliose lichen *Xanthoria parietina* (L.) Th. Fr. showing the various layers: *UC* Upper cortex, *A* algal layer, *M* medulla, *LC* lower cortex. Some desert lichens subvert this structure and have a ventral rather than a dorsal algal layer

Fig. 1.2 The three main growth forms of lichens: (**a**) the crustose lichen *Caloplaca marina* (Wedd.) Zahlbr. Ex Du Riez, (**b**) the foliose lichen *Parmelia sulcata* T. Tayl., and (**c**) the fruticose lichen *Ramalina subfarinacea* (Nyl. ex Cromb.) Nyl. The proportions of these three growth forms often vary with type of extreme environment

determining water uptake and evaporative loss. Various morphological adaptations in lichens result in an increase in this ratio by changing growth form, branching pattern, or thallus thickness (Larson 1984). Hence, the development of a prothallus by many crustose species, often observable as a ring of fungal tissue at the thallus margin, may be particularly important in the uptake and accumulation of water. Additional growth forms include the 'placodioid' type in which a largely crustose thallus has distinct marginal lobes as in some species of *Buellia* and *Caloplaca*, a 'powdery' type ('leprose'), as in *Lepraria*, and a 'gelatinous' type in which algal cells are generally distributed throughout the thallus, as in *Collema* and *Leptogium.* Many lichens exclusively produce fungal spores, and reproduction involves 'lichenization' which results when a germinating spore contacts a suitable alga. However, many species have also evolved vegetative 'diaspores' which disperse the fungal and algal symbionts together (Bailey 1976). The most common methods of vegetative reproduction include isidia (Armstrong 1981), soredia (Armstrong 1991), and thallus fragments (Armstrong 1990a).

1.2.2 Growth and Physiology

The foliose and crustose types of lichen grow radially over the substratum rather like a fungus on an agar plate, but growth rates can be very slow. Foliose species have rates of radial extension between 2 and 5 mm per year (Armstrong and Bradwell 2011), but many crustose lichens grow much more slowly with rates of less than 0.5 mm year (Armstrong 2005, Armstrong and Bradwell 2010). Some species grow so slowly that larger thalli growing in the Arctic may live to be over 5000 years old and thus may be some of the oldest living organisms on Earth (Beschel 1961). The slow growth of lichens is not attributable to slower than normal physiological processes but to the fact that they may spend a considerable portion of their time in a dehydrated state and metabolically inactive. Moreover, when the lichen is wetted, there is a loss of carbon due to respiration (Smith and Molesworth 1973). After wetting, photosynthesis begins to replace the carbon lost, but the lichen has to remain wet for a sufficient period in the light to make good the carbon losses and then to make new carbon for growth (Armstrong 1976). Hence, frequent rain showers combined with rapid rates of drying in the sun may continually deplete carbon sources with little available for growth. As a result of slow growth, however, a lichen may make relatively little demand on the environment for nutrients, thus enabling the organism to grow in potentially nutrient-poor habitats.

A large group of lichens contain the green alga *Trebouxia* as the algal partner in which carbohydrate is released from the alga as the polyol ribitol and then converted into arabitol and mannitol by the fungus (Richardson et al. 1968). In some foliose and fruticose lichens, transfer of carbohydrate occurs within an extracellular envelope of hydrophobic proteinaceous material creating an 'apoplastic continuum', but this system is unlikely to be present in crustose lichens where the mechanism of transfer is less clear (Honegger 1978). Hence, algal cells in *Rhizocarpon lecanorinum* Anders are not penetrated by haustoria, a mechanism which can extract

Table 1.1 Mean concentrations (μg mg^{-1} extracted tissue) of the poyols ribitol, arabitol, and mannitol in the foliose lichen *Xanthoparmelia conspersa* (Ehrh ex Ach) Hale and crustose lichen *Rhizocarpon geographicum* (L.) DC

Species	Region	Ribitol	Arabitol	Mannitol
X. conspersa	Marginal lobes	7.26	19.78	19.66
R. geographicum	Prothallus	1.16	6.58	4.25
	Areolae	8.73	22/73	11.20

Data from Armstrong and Smith (2009)

nutrients in lichens with a lesser degree of thallus organization (Clayden 1998). In *Trebouxia*-containing lichens, arabitol and mannitol are used as carbohydrate reserves, arabitol as a respiratory reserve, while mannitol having a more protective function (Farrar 1973). In Table 1.1, for example, the concentrations of arabitol and mannitol are similar in the foliose lichen *Xanthoparmelia conspersa* (Ehrh ex Duby) Hale, but the level of arabitol is greater than mannitol in *Rhizocarpon geographicum* (L.) DC., especially in the prothallus. In addition, the arabitol to ribitol ratio is similar in the areolae of *R. geographicum* and in *X. conspersa*, but in the more vulnerable marginal prothallus of the former, the arabitol and mannitol to ribitol ratio is greater suggesting a greater proportion of the photosynthate is stored as arabitol and mannitol (Armstrong and Smith 1987). Hence, lichens may maintain pools of polyols as a stress-tolerant mechanism, and the relative level of these polyols in different parts of the thallus is a useful indicator of the stress response in lichens (Farrar 1973, 1976).

A number of other stress-tolerant mechanisms are likely to be present in lichens including an antioxidant system (Li and Wei 2016). Hence, Kranner et al. (2005) investigated the effect of desiccation and irradiation on the level of reactive oxygen species (ROS) in *Cladonia vulcani* Savicz. Antioxidant and photoprotective mechanisms were more effective by an order of magnitude compared with the isolated symbionts. In addition, it was concluded that the alga and fungus could induce increased regulation of the protective systems in the other evidence that the lichen symbiosis is more stress resistant than either of its partners (Table 1.2).

1.3 Adaptations of Lichens to Extreme Conditions

1.3.1 Wet Forest

There are two types of wet forest environment on Earth. First, the temperate wet forest, also known as temperate rainforest, which occupies regions along the west coast of North America from California to Alaska. This region is characterized by higher temperatures than normal for boreal forest environments, relatively limited seasonal variation, and high levels of humidity (Odum 1971). Rainfall is high, usually between 30 and 150 in a year, with periods of fog compensating for lower rainfall in the more southern regions of the forest. Humidity is usually very high throughout the forests with a precipitation/evaporation ratio which is very favourable for plant growth. In southern regions, fog may account for two to three times

Table 1.2 Summary of morphological and physiological adaptations of lichens to extreme environments

Environment	Type	Adaptations
Wet forest	Morphological	Often shade adapted: thinner thalli, less fertile, and less pigmented. Tomentum may be present. Upper or lower surface impermeable to water
	Physiological	Can tolerate high temperatures when wet. Some species show little depression of nPS at thallus saturation. May use lichen substances as defensive allelochemicals
Desert	Morphological	Thin crustose growth form. Light-coloured thalli in some areas, darkly pigmented when very arid
		Increasing cortical thickness in some species. Inversion of structure with ventral algal layer. May contain photoprotective pigments (yellow/orange species). Increase of surface area in foggy regions with fruticose, pendulous, or tufted growth form
	Physiological	nPS may be optimal at high temperatures and recorded at very low thallus water potentials
The Arctic	Morphological	Crustose or fruticose. Dark pigmentation with light-coloured forms in more protected niches. Fruticose species in dense columnar mats
	Physiological	Takes advantage of short favourable periods
		Tolerant of freezing: strong recovery of nPS and respiration after storage at low temperature
		Increasing levels of soluble carbohydrates especially trehalose
Alpine regions	Morphological	Greater area of black prothallus in some species at higher altitudes
	Physiological	Accumulate UV-absorbing phenolic usnic acid. Storage of pools of polyols. Takes advantage of short favourable periods. Changes in protein and fatty acid composition with altitude
Antarctic	Morphological	Crustose species dominant and fruticose species also present. Develop stipitate and pulvinate growth. Increase in cortical thickness. Dark pigmentation with yellow-green forms in cracks and fissures. Endolithic growth form in dry valleys
	Physiological	Similar to Arctic. Photoprotective pigment melanin present in some species
Chemically rich	Morphological	Increased surface hydrophobicity with pores to provide 'breathable' surface
		Increased surface 'roughness'
	Physiological	Tolerance of high pH. Physiological control of intracellular ions. Use of stress ethylene and lichen substances. Differential expression of heat shock proteins, glutathione S-transferase, and ABC transporter. Antioxidant resistance mechanisms

more of the available water than annual rainfall and on some trees, and 50 in of rain has been recorded dripping down the trunk in a year (Oberlander 1956). By contrast, tropical rainforest occurs in low-altitude regions near the equator and comprises the Amazon and Orinoco basin in South America; the Central American isthmus; the Congo, Niger, and Zambezi basin of Central/West Africa and Madagascar; and the Indo-Malay-Borneo-New Guinea region (Odum 1971). These regions are characterized by greater than 80–90 in of rain a year, which is well distributed across seasons, with usually one or more shorter dry seasons. Temperature variation is small across the seasons as a whole and often less than the diurnal variation.

Wet forest is a surprising environment to be considered as extreme for lichens. Nevertheless, a combination of high thallus water contents over long periods, which can depress net photosynthesis (nPS), and intense competition from rapidly growing bryophytes and other plants can be challenging (Fig. 1.3). As a consequence, despite the large number of individual species present, 50% of all known lichen species being recorded from tropical areas (Mattick 1954), individual lichen species are rarely dominant in these forests. The lichens of wet forests exhibit a range of morphological and physiological adaptations (Kappen 1988), light being a strong limiting factor (Awasthi and Agrawal 1970). Hence, lichen morphology often varies with light regime, thalli in shade being less pigmented, having a thinner cortex, fewer apothecia, and more abundant soredia (Riehmer 1932). Such species also appear to be able to tolerate higher temperatures while wet (Kappen 1988). Some species possess a 'tomentum' on the surface (Kappen 1988), which may be water-holding adaptation, reducing the chance that the thallus will become saturated for long periods. In some species, e.g. *Peltigera praetextata* (Flörke ex Summerf.)

Fig. 1.3 The wet forest of the Olympic Peninsula, Washington State, USA. Lichens in these environments are subjected to extremely wet conditions and intense competition from bryophytes and higher plants

Zopf, the upper thallus surface is impermeable to CO_2 and covered by a thick water film after rain (Matthes 1980), whereas in species of *Sticta* and *Pseudocyphellaria*, only the upper surface allows exposure to rain or runoff, the lower surface being sealed at high water contents to enable CO_2 to diffuse through pores (Green et al. 1981). The extent of the depression of nPS at high thallus water contents varies among species in the wet forest, a 15% reduction being observed in species of *Sticta* while 45% is more typical of *Pseudocyphellaria*. In addition, lichens were soaked, sprayed, and shaken and then allowed to dry slowly so that their response could be compared with samples of thalli continuously exposed to a wet environment (Lange et al. 1993). Several types of photosynthetic response were observed at high water contents including species exhibiting no depression, little depression, or large degrees of depression suggesting considerable variations in the ability of different species to adapt to very wet conditions.

A major problem faced by lichens in the wet forest is competition from bryophytes and higher plants (Fig. 1.3), and as stress-tolerant organisms, they are unlikely to be successful in communities in which competition is intense (Armstrong and Welch 2007). Nevertheless, a large number of lichen species have been recorded especially in tropical forests (Mattick 1954). One possibility is that lichens use chemical substances as a defensive strategy. Over 500 secondary metabolites have been reported in lichens of which 350 are unique to the symbiosis (Lawrey 1995). Most of these metabolites are weak phenolic compounds produced by the fungal partner and which accumulate in the outer walls of the hyphae. For many years, there have been claims that such compounds, when leached from the thallus, may suppress adjacent lichens and mosses (allelopathy) (Lawrey 1995). A prediction of Grime's hypothesis (Grime 1979) is that the production of secondary compounds in stress-tolerant organisms is essentially defensive. In addition, Rogers (1990) concluded that competitors produce offensive allelochemicals while stress-tolerant organisms produce defensive (antimicrobial) allelochemicals. Hence, Hilmo (1994) and Glenn et al. (1995) reported that the presence of usnic acid and other secondary compounds may reduce grazing by microarthropods and generalist herbivores, while John (1989) suggested that crustose lichens use allelochemicals as a contact avoidance strategy. Hence, defensive allelochemicals could be part of an adaptive strategy to survive the intense competition of wet forests.

1.3.2 Desert

Either desert regions are characterized as having less than 10 in of rainfall a year, or if more than 10 in is present, there is a greater uneven distribution across the seasons (Odum 1971). The scarcity of rainfall can be attributable to three main causes: (1) high subtropical pressure, e.g. in the Saharan and Australian deserts; (2) location in rain shadows, e.g. in western North American deserts; or (3) high altitude, e.g. the Tibetan, Bolivian, and Gobi deserts. Most deserts receive some rain and therefore have a sparse cover of vegetation with the possible exceptions of the extremely arid regions of central Sahara and north Chile (Odum 1971).

Availability of water is clearly the main problem facing desert lichens, and although the dry conditions may favour some species, lichens may be absent in some sandy deserts (Kappen 1973, 1988). There is a balance in lichen thalli between exposing the maximum water absorbing surface and being able to restrict water loss (Rundel 1988), and there are a number of adaptations reflecting this dilemma. First, there is an increase in the frequency of thin, crustose species as conditions become increasingly dry, which exposes the maximum degree of horizontal surface accompanied by internal anatomical adaptations to increase the density and thickness of hyphal walls (Rundel 1988). Second, there are changes in pigmentation, many species being either light coloured (Galun 1963) or possessing light-coloured powdery 'pruina' on a darker thallus (Weber 1962). In extremely dry conditions, dark-coloured lichens containing blue-green algae as the algal partner ('cyanolichens') have been reported (Marton and Galun 1981), while on the gravel plains of Namaqualand, orange/yellow lichens are frequent, a pigmentation which is protective against light (Kappen 1983). Third, dry thalli regardless of pigmentation can resist high temperatures of up to 70 °C without heat damage (Lange 1953). Fourth, increasing cortical thickness relative to the other layers has been reported in some desert lichens reaching a maximum where thalli are frequently exposed to sand blast (Kappen 1988). Fifth, some taxa, e.g. species of *Buellia*, may exhibit an 'inversion' of thallus structure in which the sensitive algal layer is located towards the base of the lichen rather than near its surface, an adaptive response to high light intensity (Broady 1986).

A number of physiological adaptations have been observed in desert lichens. Hence, in *Chondropsis semiviridis* (F. Muell. Ex. Nyl.) Nyl., growing in South Australia, the optimum temperature for nPS is approximately 30 °C, one of the highest recorded in lichens, indicating a specific adaptation of the algal partner (Rogers 1971). In the Sonoran Desert in North America, temperature optima for nPS are significantly lower, and there are adaptations to the cooler dimmer conditions characteristic of winter rain periods in this region (Nash et al. 1982a, b). Cyanolichens in the Judean Desert and Arava valley occur exclusively in rain tracks as they cannot utilize water vapour for photosynthesis (Llimona 1982). By contrast, *Ramalina maciformis* (Delise) Bory can use water vapour in the Negev desert very efficiently, nPS being recorded at very low water potentials (Lang and Bertsch 1965). Many lichens can use atmospheric water vapour (Lange and Kilian 1985), and hence, some desert lichens take advantage of occasional periods of fog (Kappen 1988). Such lichens often have a large surface area to trap moisture and hence have a fruticose, pendulous, or tufted growth form. Such adaptation is also seen in some essentially crustose species, e.g. *Caloplaca coralloides* (Tuck.) Hulting, which adapt a pulvinate, branched, or coralloid form.

Some desert lichens exhibit changes in ecological behaviour in response to the extreme conditions. Hence, in the Negev desert, thalli frequently occur on northerly exposed rocks or at more shaded sites, while in the Namib, south-westerly exposed sites are favoured or sites located under translucent rocks (Vogel 1955). In addition, in the Negev, a differential distribution with reference to small pebbles has been observed with different taxa present either at the top or sides of the stones (Kidron 2002). The top of the pebbles receive twice the average daily dew compared with their margins, but there is greater moisture at the margin after rain as a result of capillary action in soil adjacent to the pebble. In the driest parts of the Atacama Desert in Chile, weakly welded rhyolitic ignimbrite is frequently colonized by lichens that grow within the rock ('endolithic lichens'), the porous interior of the rock which, as well as encouraging hydration, also protects the thalli from damaging UV radiation and excessive levels of visible light (Wierzchos et al. 2013). In addition, the endolithic species *Verrucaria rubrocincta* Breuss inhabits the 'caliche plates', resulting from subsurface precipitates dominated by calcite, exposed on the surface of the Sonoran Desert where temperatures may exceed 60 °C in summer and fall to 0 °C in winter (Garvie et al. 2008). An upper micrite layer within the rock is highly reflective, thereby reducing light intensity levels experienced by the alga.

1.3.3 The Arctic

The Arctic is characterized by low average heat input, a long severe winter with temperatures usually less than −40 °C, and a short growing season of less than 60 days (Odum 1971). The soil remains frozen except for the upper few inches in summer months, the predominantly frozen deeper layers known as 'permafrost'. Over most of the region, the vegetation is essentially a wet grassland composed of lichens, grasses, sedges, and dwarf shrubs. In addition, there is a gradient from tundra through Arctic 'desert' to regions with permanent ice. The vegetation is often divided into two zones: (1) low Arctic composed of a thick spongy mat of living, non-decomposed vegetation often saturated with water and associated with scattered often frozen ponds and (2) high Arctic which may be completely bare apart from rare patches of lichens and grasses (Odum 1971). A relatively new stress factor for Arctic lichens is climate change which can cause acute but short-lasting winter warming periods (Bjerke et al. 2011).

Arctic lichens face many stress factors and morphological adaptation may be in response to high radiation, low temperature, short periods when metabolic activity is possible, drought or wetness, long-lasting snow cover, and the erosive action of wind (Kappen 1988). Most species are crustose or fruticose, and there is considerable morphological variation. Hence, in *Cladonia uncialis* (L.) Web., variations in morphology occur along a gradient from Southern Finland into the Arctic with the ratio of length to thickness of internodes and area of algal cell layer being at their lowest but relative percentage of pycnidia and perforate axils highest in the Arctic (Kärenlampi and Pelkonen 1971). Microtopography can also have a significant effect with degree of slope as an important factor (Link and Nash 1984), and many species are finding shelter against the cooling of the wind in small depressions. Some of these species are darkly pigmented, an adaptation to absorb heat and therefore to increase availability of water by melting snow more rapidly (Nordhagen 1928). By contrast, lichens in deeper depressions are light coloured often 'greenish', which may be an adaptation to lower light conditions under snow. The Arctic can be strongly heated in summer but subject to slower rates of evaporation. Hence, widely distributed taxa such as *R. geographicum* (Armstrong 2013) may change their ecological behaviour, being characteristic of well-lit southern-facing surfaces in temperate regions (Armstrong 2002) but north-facing rocks in the Arctic (Pitman 1973). In addition, fruticose lichens often occur in dense light-coloured mats which dissipate solar radiation and cause cooler soil surface temperatures, thus conserving moisture (Kershaw 1978).

Arctic lichens exhibit a variety of physiological adaptations to take advantage of short, favourable environmental events after being dormant for long periods (Kappen 1988). Hence, the growth of the fruticose species *Cetrariella delisei* (Bory ex Schaer.) Kärnefelt and A. Thell. was measured in the high Arctic (Uchida et al. 2006). Positive nPS was recorded when thallus water contents were high, photosynthetic rates being lower on clearer days due to lower thallus water content. Most Arctic species are tolerant of freezing with many taxa exhibiting CO_2 uptake (Kallio and Heinonen 1971) and recovery of photosynthesis and respiration after storage at low temperatures (Lange 1966). Soluble carbohydrates in cryptograms are often highest in spring and summer especially trehalose, a non-reducing disaccharide which accumulates as a result of heat, cold, or osmotic stress (Avonce et al. 2006) and which may be involved in acclimation to low temperature and partial dehydration (Monteil 2000; Armstrong and Smith 2009). Hence, trehalose is present at higher concentration in the more vulnerable marginal prothallus of *R. geographicum* compared with the central areolae (Armstrong and Smith 2009).

Arctic species subjected to higher light conditions often exhibit the same strategies as more temperate species, i.e. they show high efficiency of nPS in winter. Nevertheless, availability of nitrogen may be an important limiting factor for Arctic lichens and in other cold temperature regions (Bliss 1962; Russel 1940; Haag 1974). Lichens capable of fixing nitrogen are particularly common in the polar regions either being cyanophilous or possessing cephalodia as in species of *Peltigera*, *Nephroma*, *Solorina*, and *Stereocaulon*.

1.3.4 Alpine Regions

The challenges posed by an alpine environment are generally similar to those of the Arctic (Billings and Mooney 1968; Billings 1973; Bliss 1962). However, the growing season is longer, i.e. up to 7 months below 3000 m at favourable sites and up to 2 months at higher latitudes. On clear days, alpine environments are characterized by alternating high temperatures and frost (Kershaw 1983). In addition, there is a significant difference in day length with greater angles of the sun and of both long-wave and global radiation (Thomson 1982). Diurnal temperature changes are generally more extreme in alpine environments, and differences with the Arctic increase significantly with decreasing latitude (Bliss 1956; Vareschi 1956).

An advantage of mountainous environments is the ability to investigate the influence of environmental gradients with increasing altitude. Nevertheless, here have been relatively few such studies involving lichens. The lichen flora of Nepal between altitudes of 200 and 7400 m was studied by Baniya et al. (2010) who found the highest richness of species at 3100–3400 and 4000–4100 m. Crustose lichens reached a peak of species richness at higher altitudes, viz. 4100–4200. By contrast, Pinokiyo et al. (2008) found that in Northeast India, medium altitudes had the greatest diversity of taxa. More specifically, Fahselt (1998) found that *Rhizocarpon superficiale* (Schaerer) Vainio and *Lecidea tessellata* Flörke occurred at different elevations in Canada, with apothecia production being significantly lower at higher altitudes where growth conditions were more limiting. Altitude could limit the variety of algal partners available to the fungal partner and therefore could influence a wide variety of different lichens (Piercey-Normore and Dediche 2011).

Lichens may exhibit several adaptations to the increasingly severe conditions experienced at high altitude. Hence, *Cetraria nivalis* (L.) Ach. accumulates the UV-absorbing phenolic usnic acid in the upper cortex at higher altitudes (Bjerke et al. 2002). In addition, lichens at high altitudes experience alternating periods of desiccation and hydration, and pools of polyols are likely to contribute to the protection of cellular constituents and the preservation of intracellular structures during desiccation (Aubert et al. 2007). Hence, high rates of nPS when wet at low temperatures help alpine lichens to take advantage of brief periods of hydration such as melting ice. In addition, in Chile, high altitude may change the haplotype of the algal partner compared with lower altitudes where there is higher water availability in the form of fog, condensation, and precipitation. In *Pseudevernia furfuracea* (L.) Zopf., changes in protein composition occur with altitude, with certain sizes of proteins increasing and others decreasing (Strobl et al. 1994). In addition, changes in fatty acid composition have been reported with decreasing levels of unsaturation with altitude possibly reflecting changes in cellular membrane composition (Piervittori et al. 1994).

A number of morphological changes in the crustose lichen *Rhizocarpon geographicum* (L.) DC. have been observed at sites on Mount Pilchuck, Washington State, USA (Fig. 1.4). At the highest sites, including the summit, there were fewer marginal areolae, a wider peripheral prothallus, smaller central areolae, less spacing between areolae, and more apothecia per central areole (RA Armstrong, unpublished data). Marginal areolae were either randomly distributed or clustered on the

Fig. 1.4 Morphological changes in thalli of the crustose lichen *Rhizocarpon geographicum* (L.) DC. in response to altitude on Mount Pilchuck, Washington State: (**a**) lower altitude, (**b**) higher altitude

Fig. 1.5 The SO_2 pollution-tolerant lichen *Lecanora conizaeoides* Nyl. Ex Cromb., growing in Central Birmingham in 1976. In the 1960s and 1970s, this species was often the only lichen to be recorded consistently on trees in the centre of large cities in the UK. The surface of the thalli is 'super hydrophobic' but with breathable pores

marginal prothallus but with no clear relationship with altitude. Similarly, there was a more marked decline in the frequency of marginal areolae with size at the highest altitudes but with no simple relationship between the shape of the size distribution and altitude being log-normal at all sites. Hence, the exposure of a greater area of dark prothallus at the highest altitudes may be an adaptation to absorb heat and melt snow similar to that seen in some darkly pigmented Arctic species (Nordhagen 1928). The SO_2 pollution-tolerant lichen *Lecanora conizaeoides* Nyl. Ex Cromb. grows in Central Birmingham in 1976. In the 1960s and 1970s, this species was often the only lichen to be recorded consistently on trees in the centre of large cities in the UK. The surface of the thalli is 'super hydrophobic' but with breathable pores (Fig. 1.5).

1.3.5 Antarctica

Lichens are the dominant feature of the Antarctic continent and its adjacent islands (Llano 1965; Ahmadjian 1970; Longton 1979; Lindsay 1978; Smith 1984, 1995). Annual rainfall is not a limiting factor in the maritime Antarctic, but there is a marked moisture gradient from the coast to inland regions with degree of salinity, nutrient availability, oxygenation, and ice cover as additional variables (Convey et al. 2014). Drifting snow can also be important in part of the region, but snowfall is rare in the south of the Antarctic circle resulting in a 'polar desert' (Smiley and Zumberge 1971). Consequently, there is a decline in species diversity, coverage, and growth rates from the maritime Antarctic to the continental dry valleys (Raggio et al. 2016) with water availability and length of the growing period as the most likely determining factors.

Adaptations of Antarctic lichens are often similar to those seen in arctic and alpine species (Ahmadjian 1970). Hence, Schroeter and Scheidegger (1995) examined the effects of extreme cooling in water-saturated thalli of *Umbilicaria aprina* Nyl. from the continental Antarctic. When thalli were slowly cooled to subzero temperatures, ice nucleation occurred at −5.4 °C followed by extracellular freezing leading to cytorrhysis in algal cells and vacuolation in fungi hyphae, changes which were reversible on warming. Reproduction in Antarctic lichens is often via vegetative diaspores such as soredia rather than lichenization, and early development is very slow (Ott 2004). Cortical thickening occurs commonly with the addition of 'dead' layers of hyphae and structures to ensure strong fixation to the ground to withstand wind (Dodge 1973). There is a dominance of crustose species but fruticose lichens also occur, being stunted in form compared with their temperate relatives (Dodge 1973). Some crustose species also develop a 'stipitate' or 'pulvinate' type of growth, which may be an adaptation to the higher humidity as a result of fog in maritime regions (Kappen 1988). In addition, Antarctic species of *Usnea* usually grow vertically on rocks, but there are also relatively unattached prostrate forms which may result from competition with rapidly growing bryophytes (Kappen 1985). As in the Arctic, many of the continental species are darkly pigmented, while yellow to green thalli often predominate in cracks and fissures (Hertel 1984).

The endolithic growth form found in deserts also occurs in Antarctica, with *Lecidea phillipsiana* Filson being a common species in the east of the region (Friedmann 1977). Three types of endolithic organisms have been described from Antarctica: (1) 'chasmoendoliths' which occupy fissures and cracks in rocks but the organism may be partially exposed on the surface, (2) 'cryptoendoliths' which occupy pores and pre-existing structural cavities, and (3) 'euendoliths' that bore into relatively soluble rock substrates such as those rich in carbonate (Lawrey 1984). The dominant flora of the dry valleys is often chasmoendolithic and cryptoendolithic lichens which occupy a narrow zone of the subsurface of the rock 10 mm thick and form colonies from a few cm to a metre in diameter (Friedmann 1982). The lichens exhibit a similar structure to those that live on the surface, but a true fungal zone is absent, with instead the fungal hyphae filling the available pore space. In cross section of a typical rock, there is a black zone just below the surface containing the alga *Trebouxia* and below that a white zone of fungal tissue, a green layer of non-lichenised green algae, and finally, in some species, a layer of cyanobacteria. The subsurface layers are often solubilized by fungal hyphae resulting in the upper surface peeling away to expose the lichen tissue. Further penetration of the rock results in more rock layers being lost, and the consequence of this 'biogenic weathering' is a characteristic pockmarked surface.

The physiological adaptations of Antarctic lichens are variable depending on location (Kappen 1988). Contrary to the maritime Antarctic, snow-covered regions often retain the severe cold of winter for long periods preventing any early warming (Pannewitz et al. 2003). Hence, the lichens are at subzero temperatures for a prolonged period, and major activity occurs only when snow finally disappears. However, CO_2 gas exchange experiments suggest that nPS and dark respiration can occur at subzero temperatures in continental Antarctica (Schroeter and Scheidegger 1995). Nevertheless, there is a substantial decline in growth rates from the warmer, wetter peninsula to the colder dry valleys (Sancho et al. 2007) with higher photosynthetic rates and a wider temperature range of response than their counterparts on the continent. An extremely low nPS rate is particularly characteristic of endolithic species in the dry valleys. In studies of species of *Umbilicaria* and *Caloplaca*, there is a considerable difference in duration of activity in relation to microclimate, as measured by chlorophyll 'a' fluorescence, between the maritime regions and dry valleys, with *Caloplaca* species in particular exhibiting strategies to improve thallus hydration (Raggio et al. 2016). Successful adaptation of lichens to continental Antarctic conditions also involves the algal partner as a result of the strong adverse effects of radiation on nPS (Cao et al. 2015). Hence, no photoinhibition was observed in entire thalli of *Umbilicaria decussata* (Vill.) Zahlbr., when the photoprotective pigment melanin was present (Sadowsky and Ott 2016). In addition, the effect of high light exposure was studied on *Umbilicaria antarctica* E. Frey and M. Lamb, causing increased levels of oxidized glutathione and the conversion of violaxanthin to zeaxanthin, both of which are involved in antioxidant resistance (Bartak et al. 2004).

1.3.6 Chemically Rich Environments

Lichens are characteristic of several chemically rich environments including highly eutrophicated sites and those influenced by salinity, pollution, and heavy metals. As many species are unable to grow successfully in these conditions, they can also be regarded as extreme environments for lichens.

Several studies suggest that the degree of nutrient enrichment of a substratum, especially by birds, has a significant influence on lichens (Hale 1967). Bird droppings can influence lichens by smothering the thalli, altering the pH, or adding inhibitory and stimulatory compounds (Armstrong 1984). Hence, experiments in which droppings from a variety of birds were applied as a thick paste or as a suspension in deionized water had differential effects on species of foliose lichens (Armstrong 1984), e.g. increasing the radial growth of *X. conspersa*, a species common on well-lit nutrient-enriched rocks, but inhibiting the growth of *Parmelia saxatilis* (L.) Ach., a species relatively rare on bird perching stones. Uric acid, the most abundant nitrogenous component of bird droppings, did not influence growth when applied as a suspension to either species (Armstrong 1984). Hence, tolerance of a lichen to these conditions may depend on the ability to adapt to either the increased pH or excess inorganic chemical ions in the bird droppings.

Chemical factors may also be important in influencing lichens on maritime rocks where a distinct zonation is often present (Fletcher 1976). Ramkaer (1978) found that the response of four different lichen fungi to salinity correlated well with their zonation on maritime rocks. In addition, some coastal lichens can use hygroscopic salt films to condense liquid water out of humid air when dew-point conditions are not reached (Folmann 1967; Rundel 1978). Calcium has also been shown to be an important ion in these environments (Fletcher 1976). Hence, thalli of *X. parietina*, a common species of the submesic zone of the supralittoral (Fletcher 1976) and many nutrient-enriched sites inland (Brodo 1973), degenerate when transplanted to nutrient-poor inland sites (Armstrong 1990b). Transplanted thalli, however, grow successfully inland when supplied with calcium carbonate added as a paste to the thalli at intervals over a year. Adaptation to a relatively constant supply of calcium may therefore be necessary for this species. This hypothesis is also supported by experiments showing that *X. parietina* thalli lose potassium ions when treated with distilled water or a saline solution (Matos et al. 2011), and an application of a 0.250 mM solution of calcium to the medium prevents this loss in the light (Fletcher 1976; Beckett 1996). The maritime environment may also influence lichens some distance from the sea. Hence, in *Ramalina canariensis* J Steiner, there is a logarithmic decrease in extracellular ions including chloride, sodium, and magnesium with distance from the coast (Figueira et al. 1999). The intracellular fraction of these ions was relatively independent of the surface and wall-bound fractions reflecting physiological control by the organism and a likely adaptive mechanism in coastal environments (Figueira et al. 1999). In addition, many coastal lichens can maintain nPS at very low water potentials when moistened by salt spray (Rundel 1988).

Despite the reinvasion of many urban sites by lichens in the last 20 years, sites subject to SO_2 air pollution have significantly lower lichen diversity than rural sites, a testament to the powerful effect of air pollution on growth. In addition, SO_2-tolerant species such as *Lecanora conizaeoides* Nyl. Ex Cromb. have declined in the UK since the 1960s as a result of the reduction in pollution over this period (Massara et al. 2009). Lichens exhibited several adaptations to these conditions. First, surface hydrophobicity is a feature of lichens tolerant to SO_2 (Hauk et al. 2008), and *L. conizaeoides* appears to be 'superhydrophobic' as a result of the presence of hydrophobic compounds and a rougher thallus surface which sheds water (Shirtcliffe et al. 2006). Gas channels are also present to allow photosynthesis resulting in essentially a 'breathable' surface (Shirtcliffe et al. 2006). Second, fluctuations in stress ethylene concentration have been observed in some lichens including *Hypogymnia physodes* (l.) Nyl., when exposed to sulphur at low pH, a response which correlates with their degree of tolerance (Garty et al. 1995). Third, the presence of secondary lichen substances such as fumarprotocetraric acid may be associated with tolerance to acidic air pollution (Hauck et al. 2009).

Lichens may accumulate heavy metals under various conditions including from metal-enriched rocks, industrially polluted sites, and near metallurgic plants and waste dumps. There are three heavy metal uptake mechanisms in lichens: (1) extracellular via ion exchange, (2) intracellular accumulation, or (3) trapping metal-rich particles (Richardson 1995). Metal accumulation is often selective with lichens preferentially accumulating zinc at various sites (Glenn et al. 1991, 1995). Uptake via ion exchange is assumed to occur at anionic sites in association with the surface of cell walls (Puckett et al. 1973). Hence, potentially toxic metals taken up by this mechanism may not influence lichen metabolism as they may not enter fungal or algal cells. Nevertheless, metals may be capable of penetrating cells and accumulating to toxic levels in the cytoplasm under some conditions (Lawrey 1984). Adaptations to the presence of high concentrations of heavy metals in lichens may include the development of metal tolerance attributable to inherent cytoplasmic tolerance, cytoplasmic immobilization, or transport of ions to regions external to the plasmalemma and cell wall. Rustichelli et al. (2008) found that in *Physcia adscendens* (Th. Fr.) Oliv. Em.Bitt., there was increased expression of heat shock proteins and glutathione S-transferase (GST), while ATP-binding casette transporters (ABC transporters) were induced significantly only after longer exposure to increasing cadmium concentration. Hence, differential expression of a variety of protein systems may be important in heavy metal tolerance in this species. In addition, accumulation of heavy metals in *Diploschistes muscorum* (Scop.) R. Sant. induced oxidative stress, membrane damage, and elevation of superoxide dismutase (SOD) and glutathione (GSH) suggesting that antioxidant resistance may be involved (Cuny et al. 2004).

1.3.7 Extraterrestrial Environments

Given the ability of lichens to survive and adapt to a range of extreme conditions, there has been considerable speculation regarding whether they could survive in extraterrestrial environments such as outer space and Mars (Hale 1967; Armstrong 2004). This aspect of lichen adaptation has now been investigated in a series of experiments carried out at the European BIOPAN Facility (EBF) (Raggio et al. 2011) and the International Space Station (ISS) (Brandt et al. 2015, 2016). Lichens in space are exposed to the combined effects of insolation, UV irradiation, cosmic radiation, extremely low temperatures, and a vacuum (Brandt et al. 2016). Hence, after long-term exposure of 559 days to these conditions, *Xanthoria elegans* (Link) Th. Fr. showed considerable resistance with post-exposure activity 50–80% that of previous values in the alga and 60–90% in the fungus (Brandt et al. 2015). Growth and proliferation of isolated algal cells were also demonstrated after exposure. Nevertheless, desiccation-induced breakdown of cell integrity was also observed which was more severe under space conditions than in a 'simulated' Martian environment. In addition, when photosynthesis was specifically studied (Meesen et al. 2014), activity was demonstrated post-exposure with high viability of algal cells and low rates of impairment. However, a strong impairment of both photosynthesis and photoprotective mechanisms was observed when isolated alga were irradiated emphasizing the importance of the protection afforded within the lichen symbiosis resulting from morphological adaptation, the presence of lichen compounds, and the ability of lichens to survive desiccation (Meesen et al. 2014). By contrast, in *Aspicilia fruticulosa* (Eversm.) Flagey, during a 10-day spaceflight, solar electromagnetic radiation exposure of between 100 and 200 nm caused a reduction in chlorophyll 'a' yield which recovered after 72 h reactivation of the lichen, indicative of the lichens ability to repair space damage (Raggio et al. 2011).

By contrast, Mars is an arid, barren, and rocky planet unprotected from UV light and subjected to freezing temperatures and frequent sandstorms. Measurements of temperature on Mars have confirmed that the daytime surface temperature may vary from 26.6 °C during rare sunny days to −93 °C at the poles in winter, with air temperature rarely rising above zero and decreasing markedly with altitude above the surface. In addition, the atmosphere of Mars is composed mainly of CO_2 with trace quantities of nitrogen, argon, oxygen, and carbon monoxide (Nier et al. 2003). Very low levels of moisture are present in the atmosphere spectroscopic observations suggesting one thousandth of that in the atmosphere above the Sahara desert on Earth. Subsequent studies using theoretical climate models and simulated Martian environments (Kuznetz and Gan 2002), however, demonstrated that liquid water may be stable for extended periods of time on the surface under present-day conditions. These studies culminated in the discovery of water ice near the edge of the southern polar cap by Mars Odyssey using the Thermal Emission Imaging System (THEMIS) (Titus et al. 2003). Hence, surface water ice may be widespread around and under the CO_2 polar cap and possibly in other regions.

The endolithic lichens of deserts and the dry valleys of Antarctica are often regarded as possible analogues of lichen survival on Mars. Hence, dehydrated Antarctic cryptoendolithic communities and colonies of rock-inhabiting black fungi were exposed to simulated Martian conditions for 18 months on the ISS (Onofri et al. 2015), less than 10% of the colonies proliferated while 60% of cells and rock communities remained intact. Hence, endolithic lichens on Mars may be able to tolerate the low temperatures of the surface, and as in Antarctica, the rock subsurface is likely to be warmer and subjected to smaller fluctuations than the surface of the rocks. Nevertheless, the lack of a ready supply of surface water would be a significant problem for the lichens. In certain areas, however, water ice on the surface may melt and penetrate the boulders, with the lichens remaining in a dehydrated condition during the long intervening periods. In Antarctica, carbon dioxide exchange takes place very slowly through a relatively thick surface crust (Kappen et al. 1981; Kappen and Friedmann 1983), and this could presumably also take place in Martian rocks. In addition, in regions of high light intensity, approximately 1% of the light reaches the lichen zone inside Antarctic rocks, with the harmful UV being screened out by the dark-pigmented fungal layer, and this process would be even more important on Mars. The main source of nitrogen for endolithic lichens is abiotically fixed nitrogen by atmospheric electric discharge, with the fixed nitrogen then being conveyed to the rock by atmospheric precipitation. However, there are only trace amounts of nitrogen gas in the Martian atmosphere (Nier et al. 2003), and hence, it is unclear how Martian endolithic lichens would obtain their nitrogen supply. One possibility is that cyanobacteria in the rocks could fix sufficient nitrogen from the trace levels available to supply small populations of endolithic lichens.

1.4 Conclusions

Lichens are often described as 'pioneer organisms' or 'extremists', a testament to their ability to adapt and thrive in some of the most extreme environments on Earth. This tolerance has even resulted in speculation that lichens may have existed on Mars in the past and that endolithic lichens might even survive today within the rocks, as in some extremely arid deserts (Wierzchos et al. 2013) or the dry valleys of Antarctica (Friedman 1977, 1982). Even in the most extreme environments on Earth, however, lichens require at least some short favourable periods for growth and reproduction. Hence, the tolerance exhibited by lichens under extraterrestrial conditions for relatively short periods does not indicate that they could survive permanently in environments in which favourable growing periods may be rare or absent.

Lichens in deserts and the polar regions exhibit many of the characteristics of Grime's stress-tolerant organisms, i.e. slow growth rates, significant longevity, low demand for nutrients, and adaptations to stressful conditions (Grime 1979). Nevertheless, lichen species also occur in considerable frequency in the tropics where competition is intense (Mattick 1954). Hence, despite being stress-tolerant organisms, lichens may need additional mechanisms to enable them to survive in the highly competitive wet forests. Various substances unique to the symbiosis could be used as defensive allelochemicals, enabling a large number of species to survive in relative low abundance in the wet forests.

A number of studies suggest that the lichen symbiosis is significantly more tolerant of extreme conditions than its constituent algae and fungi. This resistance is likely to be attributable to the various morphological and physiological adaptations specific to the symbiosis such as a layered thallus structure, the presence of lichen compounds, and a physiology enabling survival in extremely dry conditions (Meessen et al. 2014). In addition, the alga and fungus appear to be able to induce increased regulation of protective systems in the other (Kranner et al. 2005).

Adaptations of lichens to extreme conditions often involve changes in morphology and physiology, and there is a close interrelationship between the two. Hence, the tomentum in some lichens of the wet forest (Kappen 1988) may be water-holding adaptation, preventing the thallus becoming saturated for long periods, thus enabling more optimal nPS for longer periods. In addition, the lower cortex may be sealed to prevent water uptake but permeable to CO_2 (Green et al. 1981), a similar adaptation to that of the SO_2-tolerant lichen *L. conizaeoides* (Shirtcliffe et al. 2006). Some desert species exhibit an 'inversion' of thallus structure in which the algal layer is located ventrally rather than dorsally, thus increasing protection of the alga and enabling photosynthesis at higher light intensities (Broady 1986). Moreover, dark pigmentation or increased exposure of a black prothallus is found in most extremely cold environments and may be an adaptation to enable the thallus to absorb heat and melt covering snow, thus enabling photosynthesis to begin more rapidly (Nordhagen 1928).

There are similar or convergent ecological adaptations in very different types of extreme environment. Hence, crustose lichens are frequent in both deserts and in the drier polar regions, a growth form which exposes the maximum degree of horizontal surface. By contrast, fruticose lichens are frequent in areas where humidity is high, e.g. continental Antarctica or where fog occurs, a growth form which maximizes S/V ratio. The endolithic growth form appears to be a response to extreme aridity, and hence, such species are a feature of very dry deserts (Wierzchos et al. 2013) and the dry valleys of Antarctica (Friedman 1977). Change to a more darkly pigmented thallus occurs in deserts, arctic/alpine regions, and in Antarctica (Nordhagen 1928;

Hertel 1984). In less arid deserts, more lightly pigmented thalli occur, with such forms often occurring in depressions or cracks in colder regions (Hertel 1984). The presence of pools of polyols for stress resistance is likely to be a common feature of most stressful environments (Farrar 1973, 1976) although the relative importance of individual polyols may vary. Photoprotection by various mechanisms may also be common especially in deserts and at high altitude.

References

Ahmadjian V (1970) Adaptation of Antarctic terrestrial plants. In: Holdgate MW (ed) Antarctic ecology, vol 2. Academic, New York, pp 801–811

Armstrong RA (1974) The descriptive ecology of saxicolous lichens in area of South Merionethshire, Wales. J Ecol 62:33–45. https://doi.org/10.2307/2258878

Armstrong RA (1976) The influence of the frequency of wetting and drying on the radial growth of three saxicolous lichens in the field. New Phytol 77:719–724. https://doi.org/10.1111/j.1469-8137.1976.tb-04666.x

Armstrong RA (1981) Field experiments on the dispersal, establishment and colonization of lichens on a slate rock surface. Environ Exp Bot 21:116–120

Armstrong RA (1984) The influence of bird droppings and uric acid on the growth of five species of saxicolous lichens. Environ Exp Bot 4:95–99. https://doi.org/10.1016/0098-8472(84)90065-0

Armstrong RA (1990a) Dispersal, establishment, and survival of soredia and fragments of the lichen *Hypogymnia physodes* (L.) New Phytol 114:239–245

Armstrong RA (1990b) The influence of calcium and magnesium on the growth of the lichens *Xanthoria parietina* and *Parmelia saxatilis*. Environ Exp Bot 30:51–57. https://doi.org/10.1016/0098-8472(90)90008-R

Armstrong RA (1991) The influence of climate on the dispersal of lichen soredia. Environ Exp Bot 31:239–245

Armstrong RA (2002) The effect of rock surface aspect on growth, size structure, and competition in the lichen *Rhizocarpon geographicum*. Environ Exp Bot 48:187–194. https://doi.org/10.1016/S00098-8472(02)00040-0

Armstrong RA (2004) Could lichens live on Mars? Microbiologist 4:30–33

Armstrong RA (2005) Radial growth of *Rhizocarpon* section *Rhizocarpon* lichen thalli over six years at Snoqualmie pass in the Cascade range, Washington state. Arct Antarct Alp Res 37:411–415. https://doi.org/10.1657/1523-0430(2005)037[0411:RGORSR]:RGORSRJ2.0.CO;2

Armstrong RA (2013) Development of areolae and growth of the peripheral prothallus in the crustose lichen *Rhizocarpon geographicum*: an image analysis study. Symbiosis 60:7–15. https://doi.org/10.1007/s13199-013-0234-2

Armstrong RA (2015) The influence of environmental factors on the growth of lichens in the field. In: Upreti DK, Divakar PK, Shukla V, Bajpal R (eds) Recent advances in lichenology. Springer International Publishing. AG, New Delhi, pp 1–18

Armstrong RA, Bradwell T (2010) Growth of crustose lichens: a review. Geografiska Ann Series A Phys Geogr 92A:3–17

Armstrong RA, Bradwell T (2011) Growth of foliose lichens: a review. Symbiosis 53:1–16. https://doi.org/10.1007/s13199-011-0108-4

Armstrong RA, Smith SN (1987) Development and growth of the lichen *Rhizocarpon geographicum*. Symbiosis 3:287–300

Armstrong RA, Smith SN (2009) Carbohydrates in the hypothallus and areolae of the crustose lichen *Rhizocarpon geographicum* (L.) DC. Symbiosis 49:95–100

Armstrong RA, Welch AR (2007) Competition in lichen communities. Symbiosis 43:1–12

Aubert S, Juge C, Boisson AM, Gout E, Bligny R (2007) Metabolic processes sustaining the revi-viscence of lichen *Xanthoria elegans* (Link) in high mountain environments. Planta 226:1287–1297. https://doi.org/10.1007/s00425-007-0563-6

Avonce N, Mendoza-Vargas A, Morett E, Iturriaga G (2006) Insights on the evolution of trehalose biosynthesis. BMC Evol Biol 6:109–124

Awasthi DD, Agrawal MR (1970) An enumeration of lichens from the tropical and subtropical regions of Darjeeling District, India. J Indian Bot Soc 69:122

Bailey RH (1976) Ecological aspects of dispersal and establishment in lichens. In: Brown DH, Bailey RH, Hawksworth DL (eds) Progress and problems in lichenology. Academic, London, pp 215–247

Baniya CB, Solhoy T, Gauslaa Y, Palmer MW (2010) The elevation gradient of lichen species richness in Nepal. Lichenologist 42:83–96. https://doi.org/10.1017/S0024282909008627

Bartak M, Hajek J, Vrabilkova H, Dubova J (2004) High-light stress and photoprotection in *Umbilicaria antarctica* monitored by chlorophyll fluorescence imaging and changes in zeaxanthin and glutathione. Plant Biol 6:333–341. https://doi.org/10.1055/s-2004-820877

Beckett RP (1996) Some aspects of the water relations of the coastal lichen *Xanthoria parietina* (L.) Th. Fr. Acta Physiol Plant 18:229–234

Beschel RE (1961) Dating rock surfaces by lichen growth and its application to the glaciology and physiography (lichenometry). In: Raasch GO (ed) Geology of the Arctic. University of Toronto Press, Toronto, pp 1044–1062

Billings WD (1973) Arctic and alpine vegetation: similarities and susceptibility to disturbance. Bioscience 23:697–704. https://doi.org/10.2307/1296827

Billings WD, Mooney HA (1968) The ecology of arctic and alpine plants. Biol Rev 43:481–529. https://doi.org/10.1111/j.1469-185X.1968.tb00968x

Bjerke JW, Lerfall K, Elvebakk A (2002) Effects of ultraviolet radiation and PAR on the content of usnic and divaricatic acids in two arctic-alpine lichens. Photochem Photobiol Sci 1:678–685. https://doi.org/10.1039/b203399b

Bjerke JW, Bokhorst S, Zielke M, Callaghan TV, Bowles FW, Phoenix GK (2011) Contrasting sensitivity to extreme winter warming events of dominant sub-Arctic heathland bryophyte and lichen species. J Ecol 99:1481–1488. https://doi.org/10.1111/j.1365-2745.2011.01859.x

Bliss LC (1956) A comparison of plane development in microenvironments of Arctic and Alpine tundras. Ecol Monogr 26:303–337

Bliss LC (1962) Adaptations of arctic and alpine plants to environmental conditions. Arctic 15:117–144

Brandt A, de Vera JP, Onofri S, Ott S (2015) Viability of the lichen *Xanthoria elegans* and its symbionts after 18 months of space exposure and simulated Mars conditions on the ISS. Int J Astrobiol 14:411–425. https://doi.org/10.1017/S1473550414000214

Brandt A, Posthoff E, de Vera JP, Onofri S, Ott S (2016) Characterisation of growth and ultrastructural effects of the *Xanthoria elegans* photobiont after 1.5 years of space exposure on the International Space Station. Orig Life Evol Biosph 46:311–321. https://doi.org/10.1007/s11084-015-9470-1

Broady PA (1986) In: Pickard J (ed) Ecology and taxonomy of the terrestrial algae of the Vestfold Hills. Antarctic oases. Academic, Sydney, pp 165–202

Brodo IM (1973) Substrate ecology. In: Ahmadjian V, Hale ME (eds) The lichens. Academic, London

Cao SN, Zhang J, Zheng HY, Liu CP, Zhou QM (2015) Photosynthetic performance in Antarctic lichens with different growth forms reflect the diversity of lichenized algal adaptations to microhabitats. Polish Polar Res 36:175–188. https://doi.org/10.1515/popore-2015-0012

Clayden SR (1998) Thallus initiation and development in the lichen *Rhizocarpon lecanorinum*. New Phytol 139:685–695

Convey P, Chown SL, Clarke A, Barnes, Bokhurst S, Cummings V, Ducklow HW, Frati F, Green TGA, Gordon S, Griffiths HJ, Howard-Williams C, Huiskes AHL, Laybourn-Parry J, Lyons WB, McMinn A, Morley SA, Peck LS, Quesada A, Robinson SA, Schiaparelli S, Wall DH

(2014) The spatial structure of Antarctic biodiversity. Ecol Monogr 84:203–244. https://doi.org/10.1890/12-2216.1

Cuny D, Van Haluwyn C, Shiralli P, Zerimech F, Jerome L, Haguenoer JM (2004) Cellular impact of metal trace elements in terricolous lichen *Diploschistes muscorum* (Scop.) R Sant., – identification of oxidative stress biomarkers. Water Air Soil Pollut 152:55–69

Dodge CW (1973) Lichen Flora of the Antarctic continent and Adjacent Islands. Phoenix, Canaan

Fahselt D (1998) Production of ascocarps by high Arctic lichens in relation to altitude. In: Glenn MG, Harris RC, Dirig R, Cole MS (eds) Lichenographia Thomsoniana: North American lichenology in honour of JW Thomson. Mycotaxon Ltd, Ithaca, pp 389–397

Farrar JF (1973) Lichen physiology. In: Ferry BW, Baddeley MS, Hawksworth DL (eds) Progress and problems in lichenology. Athlone Press, University of London, London, pp 238–282

Farrar JF (1976) The lichen as an ecosystem: observation and experiment. In: Brown DH, Bailey RH, Hawksworth DL (eds) Progress and problems in lichenology. Academic, London, pp 385–406

Figueira R, Sousa AJ, Brown DH, Catarino F, Pacheco AMG (1999) Natural levels of saline elements in lichens: determination of cellular fractions and their importance as saline tracers. Lichenologist 31:183–196. https://doi.org/10.1006/lich.1998.0179

Fletcher A (1976) Nutritional aspects of marine and maritime lichen ecology. In: Brown DH, Bailey RH, Hawksworth DL (eds) Progress and problems in lichenology. Academic, London, pp 359–384

Follman G (1967) Zur Bedentung der Salzbestänbung für die Krustenflecten. Ber Dtsch Bot Ges 80:206–208

Friedmann EI (1977) Microorganisms in Antarctic desert rocks from dry valleys and Dufek Massif. Antarc J US 12:26–30

Friedmann EI (1982) Endolithic microorganisms in the Antarctic cold desert. Science 215:1045–1053

Galun M (1963) Autecological and synecological observations on lichens in the Negev, Israel. Isreal J Bot 12:179–187

Garty J, Kauppi M, Kauppi A (1995) Differential responses of certain lichen species to sulphur-containing solutions under acidic conditions as expressed by the production of stress-ethylene. Environ Res 69:132. https://doi.org/10.1006/enrs.1995.1034

Garvie LAJ, Knauth LP, Bungartz F, Kionowski S, Nash TH (2008) Life in extreme environments: survival strategy of the endolithic desert lichen *Verrucaria rubrocincta*. Naturwissenschaften 95:705–712. https://doi.org/10.1007/s00114-008-0373-0

Glenn MG, Orsi EV, Hemsley ME (1991) Lichen metal contents as correlates of air filter measurements. Grana 30:44–47

Glenn MG, Gomez-Bolea A, Lobello R (1995) Metal content and community structure of cryptogram bioindicators in relation to vehicular traffic in Montseny Biosphere Reserve (Catalonia, Spain). Lichenologist 27:291–304

Green TGA, Snelgar WP, Brown DH (1981) Carbon dioxide exchange in lichens. Carbon dioxide exchange through cyphellate lower cortex of *Sticta latifrons* rich. New Phytol 88:421–426. https://doi.org/10.1111/j.1469-8137.1981.tb04090.x

Grime JP (1979) Plant strategies and vegetation processes. Wiley, London

Haag RW (1974) Nutrient limitations to plant production in two tundra communities. Can J Bot 52:103–116

Hale ME (1967) The biology of lichens. Edward Arnold, London

Hauk M, Jurgens SR, Brinkman M, Heminghaus S (2008) Surface hydrophobicity causes SO_2 tolerance in lichens. Anns Bot 101:531–539. https://doi.org/10.1093/aob/mcm306

Hauk M, Jurgens SR, Huneck S, Leuschner C (2009) High acidity tolerance in lichens with fumarprotocetraric, periatolic or thamnolic acids correlated with low pK(a1) values of these lichen substances. Environ Pollut 157:2776–2780. https://doi.org/10.1016/j.envpol.2009.04.022

Hertel H (1984) Über Saxicole, Lecideoide Flechten der Subantarckis. In: Hertel H, Oberwinkler F, Cramer J (eds) Beiträge zur lichenologie, Festschrift J Poelt. Vaduz, pp 399–499

Hilmo O (1994) Distribution and succession of epiphytic lichens on *Picea abies* branches in a boreal forest, Central Norway. Lichenologist 26:149–169
Honegger R (1978) Ascarpontogenie, Axusstrukter und funktion bei Vertretern der Gattung *Rhizocarpon*. Beriche Dtsch Botanischen Ges 91:579–594
James PW, Hawksworth DL, Rose F (1977) In: Seaward MRD (ed) Lichen communities in the British isles: a preliminary conspectus. Academic, London, pp 295–419
John EA (1989) An assessment of the role of biotic interactions and dynamic processes in the organisation of a species in a saxicolous lichen community. Can J Bot 67:2025–2037
Kallio P, Heinonen S (1971) Influence of short-term low temperature on net photosynthesis in some subarctic lichens. Rep Kevo Subarct Res Stn 8:63–72
Kappen L (1973) Response to extreme environments. In: Ahmadjian V, Hale ME (eds) The lichens. Academic, New York, pp 311–380
Kappen L (1983) Ecology and physiology of the Antarctic fruticose lichen *Usnea sulphurea* (Koenig) Th. Fries. Polar Biol 1:249–255. https://doi.org/10.1007/BF00443196
Kappen L (1985) Water relations and net photosynthesis of *Usnea*. A comparison between *Usnea faciata* (maritime Antarctic) and *Usnea sulphurea* (continental Antarctic). In: Brown DH (ed) Lichen physiology and cell biology. Plenum Press, New York, pp 41–56
Kappen L (1988) Chapter II.B.2: Ecophysiological relationships in different climatic regions. In: Galun M (ed) Handbook of lichenology, vol 2. CRC Press, Boca Rato
Kappen L, Friedmann EI (1983) Ecophysiology of lichens in the dry valleys of southern Victoria-land, Antarctic. 2. CO_2 gas exchange in cryptoendolithic lichens. Polar Biol 1:227–232. https://doi.org/10.1007/BF00443193
Kappen L, Friedmann EI, Garty J (1981) Ecophysiology of lichens in the dry valleys of southern Victoria-land, Antarctic. 1. Microclimate of the cryptoendolithic-lichen habitat. Flora 171:216–235
Kärenlampi L, Pelkonen M (1971) Studies on the morphological variation of the lichen *Cladonia uncialis*. Rep Kevo Res Stn 7:47–53
Kershaw KA (1978) The role of lichens in boreal tundra transition areas. Bryologist 81:294–306. https://doi.org/10.2307/3242190
Kershaw KA (1983) The thermal operating-environment of a lichen. Lichenologist 15:191–207. https://doi.org/10.1017/S002428298300286
Kidron GJ (2002) Causes of the two patterns of lichen zonation on cobbles in the Negev Desert Israel. Lichenologist 34:71–80. https://doi.org/10.1006/lich.2001.0366
Kranner I, Cram WJ, Zorn M, Wornik S, Yoshimura I, Stabentheiner E, Pfeifhofer HW (2005) Antioxidants and photoprotection in a lichen as compared with its isolated symbiotic partners. Proc Natl Acad Sci U S A 102:3141–3146. https://doi.org/10.1073/pnas.0407716102
Kuznetz LH, Gan DC (2002) On the existence and stability of liquid water on the surface of Mars today. Astrobiology 2:183–195. https://doi.org/10.1089/15311070260192255
Lange OL (1953) Hitze- und Trokenresistenz der Flechten in Beziehung zu ihrer Verbreitung. Flora. (Jena) 140:39–97
Lange OL (1966) CO_2-Gaswechsel der Fechte *Cladonia alcicornis* nach Langfristigem Aufenthalt bei tiefen Temperaturen. Flora 156(B):500–502
Lange OL, Bertsch A (1965) Photosynthese der Wüstenflechte *Ramalina maciformis* nach Wasserdampfaufnahme aus dem Luftraum. Naturwissenschaften 52:215
Lange OL, Killian E (1985) Reaktvierung der Photosynthese trockener Flechten durch Wasserdampfanfriahire aus dem Luftraum. Flora 176:7–23
Lange OL, Budel B, Heber U, Meyer A, Zellner H, Green TGA (1993) Temperate rain-forest lichens in New Zealand, high thallus water content can severely limit photosynthetic CO2 exchange. Oecologia 95:303–313
Larsen DW (1984) Thallus size as a complicating factor in the physiological ecology of lichens. New Phytol 97:87–97
Lawrey JD (1984) Biology of lichenized fungi. Praeger Publishers, New York
Lawrey JD (1995) Lichen allelopathy-a review. ACS Symp Ser 582:26–38

Li H, Wei JC (2016) Functional analysis of thioredoxin from the desert lichen-forming fungus, *Endocarpon pusillum* Hedwig, reveals its role in stress tolerance. Sci Rep 6:27184. https://doi.org/10.1038/srep27184

Lindsay DC (1978) The role of lichens in Antarctic ecosystems. Bryologist 81:268–276. https://doi.org/10.2307/3242188

Link SO, Nash TH (1984) An analysis of an arctic lichen community with respect to slope on siliceous rocks at Anaktuvuk Pass, Alaska. Bryologist 87:162–166. https://doi.org/10.2307/3243126

Llano GA (1965) The flora of Antarctica. In: Hatherton T (ed) Antarctica. Methuen & Co, London, pp 331–350

Llimona X (1982) Lichens of the arid Mediterranean area and North Africa. J Hattori Bot Lab 33:345–359

Longton RE (1979) Vegetation ecology and classification in the Antarctic zone. Can J Bot 57:2264–2278

Marton K, Galun M (1981) The cyanophilous lichen population of the Arava valley and the Judean Desert (Israel). Isreal J Bot 30:125–155

Massara AC, Bates JW, Bell JNB (2009) Exploring causes of the decline of the lichen *Lecanora conizaeoides* in Britain: effects of experimental N and S applications. Lichenologist 41:673–681

Matos P, Cardosa-Vihena J, Figueira R, Sousa A (2011) Effects of salinity stress on cellular location of elements and photosynthesis in *Ramalina canariensis* Steiner. Lichenologist 43:155–164. https://doi.org/10.1017/S0024282910000757

Matthes U (1980) Photosynthetischer CO_2-Gaswechsel Verschiedener Flechtenarten in Abhängigkeit vom Wassergehalt des Thallus. Diplomarbeit, Würzburg

Mattick F (1954) Die Flechten der Tropen. Congrès Int de Botanique, Botanisches Museum Berlin Dahlem, Rapport, p 21

Meessen J, Backhaus T, Sadowsky A, Mrkalj M, Sanchez FJ, de la Torre R, Ott S (2014) Effects of UVC254 nm on the photosynthetic activity of photobionts from the astrobiologically relevant lichens *Buellia frigida* and *Circinaria gyrosa*. Int J Astrobiol 13:340–352. https://doi.org/10.1017/S1473550414000275

Monteil PO (2000) Soluble carbohydrates (trehalose in particular) and cryoprotection in polar biota. Cryo-Lett 21:83–90

Nash TH, Moser TJ, Bertke CC, Link SO, Sigal LL, White SL, Fox CA (1982a) Photosynthetic patterns of Sonoran desert lichens. I. Environmental considerations and preliminary field measurements. Flora 172:335–345

Nash TH, Lange OL, Kappen L (1982b) Photosynthetic patterns of Sonoran desert lichens. II A multivariate laboratory analysis. Flora 172:419–426

Nier AO, Hanson WB, Seiff WB, McElroy MB, Spencer NW, Duckett RJ et al (2003) Composition and structure of the Martian atmosphere: preliminary results from Viking 1. J Mass Spect 38:3–5. (reprinted from Science 193:786–788, 1976)

Nordhagen R (1928) Die Vegetation und Flora des Sylenegebietes. I. die Vegetation. Skr Nor Vidensk Akad Oslo 1:1–612

Oberlander GT (1956) Summer fog precipitation on the San Francisco Peninsula. Ecology 37:851–852

Odum EP (1971) Fundamentals of ecology. WB Saunders Co, Philadelphia

Onofri S, de Vera JP, Zucconi L, Selbmann L, Scalzi G, Venkateswaran KJ, Rabbow E, de la Torre R, Homeck G (2015) Survival of Antarctic cryptoendolithic fungi in simulated Martian conditions on board the International Space Station. Astrobiology 15:1052–1324. https://doi.org/10.1089/ast.2015.1324

Ott S (2004) Early stages of development in *Usnea antarctica* Du Rietz in the South Shetland Islands, northern maritime Antarctica. Lichenologist 36:413–423. https://doi.org/10.1017/S0024282904014380

Pannewitz S, Schlensog M, Green TGA, Sancho LG, Schroeter B (2003) Are lichens active under snow in continental Antarctica? Oecologia 135:30–38

Piercey-Normore MD, Deduke C (2011) Fungal farmers or algal escorts: lichen adaptation from the algal perspective. Mol Ecol 20:3708–3710. https://doi.org/10.1111/j.1365-294X.2011.05191.x

Piervittori R, Alessio F, Maffei M (1994) Fatty-acid variations in the lichen *Xanthoria parietina*. Phytochemistry 36:853–856
Pinokiyo A, Singh KP, Singh JS (2008) Diversity and distribution of lichens in relation to altitude within a protected biodiversity hot spot, north-east India. Lichenologist 40:47–62. https://doi.org/10.1017/S0024282908007214
Pitman GTK (1973) A lichenometrical study of snowpatch variation in the Frederikshåb district, southwest Greenland and its implication for the study of climatic and glacial fluctuations. Bull Grønl Undersøkelse 104:1–31
Puckett KJ, Nieboer E, Gorzinski MJ, Richardson DHS (1973) The uptake of metal ions by lichens: a modified ion-exchange process. New Phytol 72:329–342
Raggio J, Pintado A, Ascaso C, de la Torre R, de los Rios A, Wierzchos J, Horneck G, Sancho LG (2011) Whole lichen thalli survive exposure to space conditions: results of Lithopanspermia experiment with *Aspicilia fruticulosa*. Astrobiology 11:281–292. https://doi.org/10.1089/ast.2010.0588
Raggio J, Green TGA, Sancho LG (2016) In situ monitoring of microclimate and metabolic activity in lichens from Antarctic extremes: a comparison between South Shetland Islands and the McMurdo Dry Valleys. Polar Biol 39:113–122. https://doi.org/10.1007/s00300-015-1676-1
Ramkaer K (1978) The influence of salinity on the establishment phase of rocky shore lichens. Bot Tidsskr 72:119–123
Richardson DHS (1995) Metal uptake by lichens. Symbiosis 18:119–127
Richardson DHS, Hill DJ, Smith DC (1968) Lichen physiology XI. The role of the alga in determining the pattern of carbohydrate movement between lichen symbionts. New Phytol 67:469–486
Riehmer R (1932) Eine Okologie Afrikanischer Rindenflechten. (Quoted in L Kappen 1988)
Rogers RW (1971) Distribution of the lichen *Chondropsis semiviridis* in relation to its heat and drought resistance. New Phytol 70:1069
Rogers RW (1990) Ecological strategies of lichens. Lichenologist 22:149–289
Rundel PW (1978) Ecological relationships of desert fog lichens. Bryologist 81:277–293
Rundel PW (1988) Chapter II.B.2: Water relations. In: Galun M (ed) Handbook of lichenology, vol 2. CRC Press, Boca Raton
Russel RS (1940) Physiological and ecological studies on an arctic vegetation. III. Observations on carbon assimilation, carbohydrate storage and stomatal movement in relation to the growth of plants on Jan Mayen Island. J Ecol 28:289–309
Rustichelli C, Visioli G, Kostecka D, Vurro E, Sanita di Troppi L, Marmiroli N (2008) Proteomic analysis of the lichen *Physcia adscendens* exposed to cadmium stress. Environ Pollut 156:1121–1127. https://doi.org/10.1016/j.envpol.2008.04.010
Sadowsky A, Ott S (2016) Symbiosis as a successful strategy in continental Antarctica: performance and protection of *Trebouxia* photosystem II in relation to lichen pigmentation. Polar Biol 39:139–151. https://doi.org/10.1007/s00300-015-1677-0
Sancho LG, Green TGA, Pintado A (2007) Slowest to fastest: extreme range in lichen growth rates supports their use as an indicator of climate change in Antarctica. Flora 202:667–673
Schroeter B, Scheidegger C (1995) Water relations in lichens at subzero temperatures – structural changes and carbon-dioxide exchange in the lichen *Umbilicaria aprina* from continental Antarctica. New Phytol 131:273–285. https://doi.org/10.1111/j.1469-8137.1995.tb05729.x
Shirtcliffe NJ, Pyatt FB, Newton MI, McHale G (2006) A lichen protected by a super-hydrophobic and breathable structure. J Plant Physiol 163:1193–1197. https://doi.org/10.1016/jplph.2005.11.007
Smiley TL, Zumbergh JH (1971) Polar deserts. Science 174:79–80
Smith RIL (1984) Terrestrial plant biology of the sun-Arctic and Antarctica. In: Laws RM (ed) Antarctic ecology, vol 1. Academic, New York, pp 61–162
Smith RIL (1995) Colonization by lichens and the development of lichen-dominated communities in the maritime Antarctic. Lichenologist 27:473–483. https://doi.org/10.1016/S0024-2829(95)80007-7
Smith DC, Douglas E (1987) The biology of symbiosis. Edward Arnold, Victoria

Smith DC, Molesworth S (1973) Lichen physiology XIII. Effects of rewetting dry lichens. New Phytol 72:525–533. https://doi.org/10.1111/j.1469-8137.1973.tb04403.x

Strobl A, Turk R, Thalhamer J (1994) Investigations on the protein composition of the lichen *Pseudevernia furfuracea* (L.) Zopt. Var Ceratea (Ach.) Hawksw. From different altitudes. Phyton-Anns Bot 34:67–83

Thomson JW (1982) Lichen vegetation and ecological patterns in the high Arctic. J Hattori Bot Lab 53:361–364

Titus TN, Kieffer HH, Christensen PR (2003) Exposed water ice discovered near the south pole of Mars. Science 290:1048–1051. https://doi.org/10.1126/science.1080497

Uchida M, Nakatsubo T, Kanda H, Koizumi H (2006) Estimation of annual primary production of the lichen *Cetrariella delisei* in a glacier foreland in the high Arctic, Nylalesund, Svalbard. Polar Res 25:39–49. https://doi.org/10.1111/j.1751-8369.2006.tb00149.x

Vareschi V (1956) Algunos aspectos de la ecologia vegetal de la zona mas alto de la Sierra Nevada de Merida. Biol Fac Cienc For 3:9–16. (Quoted in L Kappen 1988)

Vogel S (1955) Niedere 'Fensterpfanzen' in der Südafrikanisch Wüste, Eine Ökologische Schilderung. Beitr Biol Pfanz 32:45–135

Weber WA (1962) Environmental modification and the taxonomy of the crustose lichens. Sven Bot Tidskr 56:293–333

Wierzchos J, Davila AF, Artieda O, Camara-Gallego B, Rios AD, Nealson KH, Valea S, Garcia-Gonzalez MT, Ascaso C (2013) Ignimbrite as a substrate for endolithic life in the hyper-arid Atacama desert: implications for the search for life on Mars. Icarus 224:334–346. https://doi.org/10.1016/j.icarus.2012.06.009

Adaptive Mechanisms of Desiccation Tolerance in Resurrection Plants

2

Farah Deeba and Vivek Pandey

Abstract

The present changing climate scenario suggests an increased aridity in many areas of the globe in coming years, so the research into plant responses to water stress has become increasingly important. On a global basis, drought, in addition to high temperature and radiation, poses the most important environmental constraint to plant survival and crop productivity.

Desiccation tolerance is not synonymous with drought tolerance. Desiccation tolerance is defined as the ability of a living structure to survive drying to equilibrium with low (<5%) RH and maintain low intracellular water concentrations (WCs), while drought tolerance is defined as survival at low environmental water availability while maintaining high internal water contents (WCs). Desiccation tolerance, which is one mechanism of drought tolerance, involves an integrated mechanism where morphological adaptations are complemented with physiological, biochemical and genetic tolerance. With the advent of genomic and proteomic tools, the knowledge of the adaptive strategies involved has been greatly improved.

Keywords

Water stress · Osmoregulation · Genomics · Proteomics · Stable isotope · Thermal imaging

---

F. Deeba
Biotechnology Department, CSIR-Central Institute of Medicinal and Aromatic Plants, Khurram Nagar, Lucknow, India

V. Pandey (✉)
Plant Ecology & Environmental Sciences Division, CSIR-National Botanical Research Institute, Rana Pratap Marg, Lucknow, India
e-mail: v.pandey@nbri.res.in

V. Shukla et al. (eds.), *Plant Adaptation Strategies in Changing Environment*,
https://doi.org/10.1007/978-981-10-6744-0_2

2.1 Introduction

The present changing climate scenario depicts an increased aridity in many areas of the globe (Petit et al. 1999), so the research into plant responses to water stress has become increasingly important. Globally, drought, in addition to high temperature and radiation, poses the most important environmental constraint to plant survival and crop productivity (Boyer 1982).

Desiccation tolerance is not synonymous with drought tolerance (Alpert 2005). Desiccation tolerance is defined as the ability of a living structure to survive drying to equilibrium with low (<5%) RH and maintain low intracellular water concentrations. Drought tolerance is defined as survival at low environmental water availability while maintaining high internal water contents. A drought-tolerant organism that is not desiccation tolerant will die if it loses much of its internal water, whereas a desiccation-tolerant organism will survive under similar circumstances. Thus, desiccation tolerance is one mechanism of drought tolerance.

Desiccation tolerance is a phenomenon of drying to equilibrium of an air dry state or less than 5% of cellular water content. This condition is lethal for all the living being including animals and plants. However, there are some species of animals and plants that can withstand the complete water loss. For example, among animals, desiccation tolerance is commonly found in three phyla, namely, nematodes (Wharton 2003), rotifers (Ricci 1998; Ricci and Carprioli 2005) and tardigrades (Wright et al. 1992; Wright 2001). Juvenile tolerance is also known in two more phyla: one is the encysted embryos of one crustacean genus (Clegg 2005) and the other is the larvae of the fly (Kikawada et al. 2005). However, among plants, bryophytes are best examples for desiccation tolerance (Proctor and Tuba 2002; Oliver et al. 2005), but in adult pteridophytes and angiosperms, this phenomenon is found rarely (Porembski and Barthlott 2000) but common in their spores, seeds and pollens (Dickie and Prichard 2002; Tweddle et al. 2003; Farmsworth 2004; Illing et al. 2005). Prokaryots and fungi are lesser known for desiccation tolerance, but many bacteria (Potts 1994; Guerrero et al. 1999; Billi and Potts 2002), terrestrial microalgae (Ong et al. 1992; Agrawal and Pal 2003) and lichens (Palmqvist 2000; Beckett et al. 2003; dela Kranner et al. 2003) and some yeast (Sales et al. 2000) tolerate desiccation as does at least one intertidal microalgae (Abe et al. 2001).

Vegetative organs can tolerate dryness to ca 4–13% of relative water content in desiccation-tolerant plants, while it's almost impossible for desiccation-sensitive plants to survive below 20–50% of water content. The reason might be the extra energy required for the protection and repair mechanisms, metabolic processes, biomass production and competitive abilities in desiccation-sensitive plants as compared to desiccation-tolerant plants.

Resurrection plants are known mostly for their poikilohydrous nature which means plants adjust their water content with the relative humidity in the environment. These plants remain alive in dehydrated state until water becomes available and resume all their physiological activities on rehydration. The leaves of resurrection plants shrink their volume by curling up inside or outside along the midrib or main axis. They can remain alive in a quiescent phase up to the air-dried state for months, and this state can be compared with dormancy in seeds in several aspects. Resurrection plants revive their growth on rainfall and reproduce before other species can do so (Scott 2000). For example, seeds of sacred lotus (*Nelumbo nucifera*) germinated around 75% after 1300 years of storage (Shen-Miller et al. 2002). However, very few plants exhibit this ability of desiccation tolerance in some organs such as leaves, but desiccation-tolerant plants are capable of surviving under these circumstances.

Mosses and liverworts remain in quiescent state at air dryness and revived after 20–25 years, while mature angiosperms and pteridophytes revived after 5 years (Alpert and Oliver 2002). There were no significant DNA damage in cyanobacterium *Nostoc commune* even after 13 years of being dry, and its growth could be resumed after 55 years under herbarium storage (Shirkey et al. 2003).

There are around 330 species of resurrection plants, which include nine pteridophyte and ten angiosperm families (Proctor and Pence 2002). Resurrection plants have been found both in monocotyledonous and dicotyledonous families, but there is no report of desiccation-tolerant gymnosperms yet. The Scrophulariaceae and Myrothamnaceae families majorly contribute for dicot species, whereas the monocot plants are more represented by the families including the plants *Xerophyta viscosa*, *X. humilis*, *Sporobolus stapfianus* and *Eragrostis nindensis* (Ingram and Bartels 1996; Alpert and Oliver 2002; Moore et al. 2009; Cushman and Oliver 2011; Oliver et al. 2011a, b). Desiccation tolerance is supposed to be linked with the limitation of size, as available references of desiccation-tolerant flowering plants are limited to certain height (Bewley and Krochko 1982). The largest resurrection plant, which is known till date, is a small woody shrub *Myrothamnus flabellifolia* (Sherwin et al. 1998). Most often the known resurrection plants are herbaceous in nature.

The mechanisms of desiccation tolerance differ between the extant lower orders and the angiosperms. In the lower plants, the process of desiccation occurs very rapidly, and protection prior to drying seems to be minimal and constitutive. The strategy of survival is considered to be based mainly on rehydration-induced repair processes (Oliver et al. 1998; Alpert and Oliver 2002).

Although some damages are probably irreversible in angiosperm vegetative tissues, there appears to be some complex and considerable protection mechanisms during drying which are thought to minimise the need for extensive repair (Gaff 1989; Farrant 2000; Scott 2000; Alpert and Oliver 2002; Vicre et al. 2003, 2004; Bartels 2005; Illing et al. 2005; Farrant 2007). These mechanisms include accumulation of sucrose and other oligosaccharides (reviewed in Pammenter and Berjak 1999; Scott 2000; Farrant 2007), the production of late embryogenesis abundant (LEA) proteins (Russouw et al. 1995; Illing et al. 2005), the upregulation of "housekeeping" antioxidants and the appearance of novel antioxidants that are apparently unique to desiccation-tolerant organisms (Illing et al. 2005; Farrant 2007). All of these contribute to protecting the subcellular *milieu* (reviewed by Berjak 2006). The phytohormone abscisic acid (ABA) plays an important role in mediating the responses to environmental stresses such as desiccation, salt and low temperature (Leung and Giraudat 1998). Increased levels of ABA in response to desiccation have been reported in many physiological studies. The plant hormone ABA is supposed to accumulate on trigger by drought which ultimately recruits various stress-associated genes that are thought to induce protective mechanisms. The characterisation of ABA-deficient and ABA-insensitive mutants from non-tolerant plants has contributed towards the understanding of the gene expression regulation in desiccation tolerance (Leung and Giraudat 1998; Bray 1997). These mutants have led to the isolation of genes involved in ABA synthesis and various components involved in desiccation signalling during stress conditions.

With the increasing problems of drought and growing population, water shortage is going to be a big issue for the upcoming future. A better understanding of plant's responses to drought will not only provide a better chance for improved management practices and breeding efforts in agriculture and but also be useful for assessing the immediate fate of natural vegetation (also crops) under climate change. In this chapter we have discussed various aspects involved in desiccation tolerance that enable resurrection plants to survive in the extreme environmental conditions.

2.2 Defintion of Water Stress and Desiccation Tolerance

Berjak (2006) emphasised that desiccation tolerance involves not only potential of extreme water loss from an organism but also to remain in this state for prolonged period. Drought tolerance is the survival under low water availability while maintaining a higher internal water content but not for prolonged period. Desiccation tolerance is the survival of organism under dehydration to an overall water content, equal to, or less than, 0.1 g (water) per gram dry mass ($g\ g^{-1}$). Anhydrobiosis is synonymous with desiccation tolerance in literature, but Berjak (2006) strictly used this term for complete water loss which is not the case of water content of around 0.1 $g\ g^{-1}$.

Desiccation tolerance is the result of a dynamic process and supposed to be mediated by several protective measures that ensure to prevent the plant system from lethal damage.

As discussed earlier, orthodox seeds are desiccation tolerant because this is the prerequisite of seeds to overcome the process of desiccation tolerance for completion of their life cycle. It is generally considered as an approach for seed survival and dissemination of a particular species.

Plants have developed two major strategies to cope with desiccation stress: stress avoidance and tolerance. Avoidance strategy involves the formation of seeds before drought conditions prevail; it is also achieved by specialised adaptations at morphological level that may be, for example, the development of specialised leaf surfaces to decrease the rate of transpiration, reduce leaf area, sunken stomata or increase root length and density to use water more efficiently. On the other hand, stress tolerance appears to be the result of synchronised execution of physiological and biochemical alterations at the cellular and molecular levels: i.e. the accumulation of various osmolytes and late embryogenesis abundant (LEA) proteins coupled with an efficient antioxidant system. Many of these mechanisms have been characterised. They have been found to exist both in desiccation-tolerant and non-tolerant plants.

2.3 Geographical Distribution and Nature of Desiccation Tolerance

Desiccation-tolerant vascular plants are geographically widely but unevenly distributed. Nonetheless there is an inclined richness of species from temperate regions towards the tropics, where most of the species occur. They are found abundantly in seasonally wet regions than constantly wet or arid regions in the tropics. Desiccation-tolerant angiosperms are mostly located in three areas of the southern hemisphere: Southern Africa, eastern South America and Western Australia (Gaff 1977, 1987; Porembski 2000). The plants from the families, Scrophulariaceae and Velloziaceae, mostly occupy the desiccation-tolerant vascular flora here, but Poaceae and Cyperaceae also have many species. The desiccation-tolerant Myrothamnaceae (comprising two species) is an endemic family present in Africa and Madagascar. There is characteristic vegetation on outcrops in this region: ferns and the genus *Streptocarpus* (Gesneriaceae) but less rich in desiccation-tolerant species. Remarkably, the latter family is represented by a few desiccation-tolerant lithophytes in Southern Europe (e.g. *Ramonda pyrenaica*). Among the ferns and fern allies, the genera *Selaginella*, *Asplenium* and *Pellaea* are dominant. On Australian inselbergs, the genus *Borya* (Boryaceae) is represented by a number of desiccation-tolerant species. Most occur in the western part of the continent. The Poaceae are also well-represented among Australian resurrection plants; *Micraira* is the most species-rich genus. It has been suggested that the Australian resurrection plants are less desiccation tolerant than their Southern African counterparts (Gaff 1981). This might indicate a shorter time period for adaptation within the genera of desiccation-tolerant species in Australia. India and Sri Lanka are well-known for the widespread occurrence of inselbergs, but there are no detailed accounts of their vegetation. Gaff and Bole (1986) reported desiccation-tolerant grasses (e.g. *Tripogon*) from shallow soils in rocky areas in India.There are no desiccation-tolerant angiosperms reported from North America, but ferns clearly dominate among the desiccation-tolerant vascular plants on North American inselbergs. There are several lichens and bryophytes which are desiccation tolerant, probably common on all continents, including Antarctica (Davey 1997).

Mostly the desiccation-tolerant species are rosette plants and woody shrubs and perennial herbs measuring up to 2 m in height, and probably succulents (Gaff 1986; Barthlott and Porembski 1996; Porembski 2000). Morphological characteristics may be similar with xerophytes (e.g. fibrous, needle-like leaves in *Borya nitida*; Gaff and Churchill 1976), mesophytes (thin and broad leaves in *Boea hygroscopica*; Gaff 1981) or even hydrophytes (aerenchyma in floating leaves of *Chamaegigas intrepidus*; Gaff and Geiss 1986). There are no desiccation-tolerant trees known till date. Sherwin et al. (1998) suggested that inability to reverse the cavitation of xylem may interdict desiccation-tolerant plants from growing more than a few metres tall.

Desiccation-tolerant lichens, mosses and bryophytes differ in the timing of resuming their all physiological activities. In general highly desiccation-tolerant lichens and bryophytes take less than 15 min to rehydrate, and within 24 h, their full photosynthetic functions recover (Alpert and Oechel 1987; Csintalan et al. 1998, 1999).

However, it takes 12 h to several days rehydration time for desiccation-tolerant ferns and angiosperms (Gaff 1997). There is one exception of aquatic species of *Craterostigma*, which can rehydrate in 1.5 h (Gaff and Giess 1986). Lichens and bryophytes readily absorb moisture and can rehydrate with dew also (Lange et al. 1994). There is no desiccation-tolerant vascular plant known to have this ability. Lichens can even absorb enough water from water vapour to regain positive CO_2 uptake (Hahn et al. 1993). That's the reason that lichens are prevalently distributed in coastal deserts where fog is a major source of moisture (Thompson and Iltis 1968; Rundel 1978).

2.4 Types of Desiccation-Tolerant Plants

2.4.1 Fully Desiccation-Tolerant vs. Modified Desiccation-Tolerant Plants

Vascular plants have shown desiccation tolerance abilities in vegetative tissues with some 350 species, which contributes less than 0.2% of the total flora (Porembski and Barthlott 2000), but this figure keeps on expanding. As all the vascular plants supposedly have desiccation-tolerant spores or seeds, it seems desiccation tolerance is probably universal. Majorly, vegetative desiccation-tolerant plants constitute two categories based on their rate of drying, namely, (1) fully desiccation-tolerant plants including the less complex clades that constitute the algae, lichens and mosses, while the other is (2) modified desiccation-tolerant plants including most of the higher angiospermic resurrection plants. Fully desiccation-tolerant plants are called so, because they can tolerate the total loss of free protoplasmic water (Oliver et al. 1998) very rapidly. These plants rapidly equilibrate their internal water content with that of the environment because they possess very few morphological and physiological adaptations for water retention. The ability of rapid drying in this group of plants also strongly suggests that a constitutive protection mechanism is necessary for survival in this group of lower desiccation-tolerant plants (Oliver et al. 2000). This sets the basis for the hypothesis that the primitive tolerance mechanism involves a constitutively protection mechanism coupled with active cellular repair (Oliver et al. 2000). Modified desiccation-tolerant plants are more complex and relatively larger group of vegetative desiccation-tolerant plants. These plants usually dry slowly followed by a series of morphological and physiological mechanisms that reduce the rate of water loss to the extent required to establish tolerance. There are several evidences available for modified desiccation-tolerant plants which strongly suggest that they utilise preventive mechanisms that majorly rely on inducible cellular protection systems (Gaff 1989).

Fig. 2.1 Hydrated (**a**, **c**, and **e**) and dry (≤5% RWC (**b**, **d**, and **f**)) monocotyledonous resurrection plants *X. viscosa* (**a** and **b**) and *X. humilis* (**c** and **d**) and the grass *E. nindensis* (**e** and **f**). Scale bars: A, B, D, E, F = 10 cm; C = 1 cm (Image Source: Farrant et al. 2007)

2.4.2 Alternatives in the Modified Desiccation Tolerance: Homoiochlorophylly and Poikilochlorophylly

Based on retention of photosynthetic apparatus, desiccation-tolerant plants may be divided into two major categories, namely, (1) homoiochlorophyllous desiccation tolerance (HDT) and (2) poikilochlorophyllous desiccation-tolerant (PDT) plants. The HDT plants retain their photosynthetic apparatus intact along with chlorophyll while drying, but on the contrary PDT plants disintegrate the chloroplast structure, thus leading to loss of chlorophyll during drying which can be recovered following rehydration (Figs. 2.1 and 2.2).

Fig. 2.2 Hydrated (**a** and **c**) and dry (≤5% RWC (**b** and **d**)) dicotylendous resurrection plants C. wilmsii (**a** and **b**) and *M. flabellifolia* (**c** and **d**). Inset to D: cross section of dry leaves *of M. flabellifolia* showing leaf curling and retention of chlorophyll in the shaded adaxial surfaces and waxy anthocyanin in the outer abaxial surfaces. Scale bars = 1 cm (Image source: Farrant et al. 2007)

PDT plants follow an ordered dismantling of internal chloroplast structure during drying after which an ordered reconstruvtive assembly takes place upon rehydration. These processes can thus be thought of as not only being superimposed on an existing cellular protection mechanism of vegetative desiccation tolerance (Oliver et al. 2000) but as a distinct new DT strategy (Tuba et al. 1994). Mostly monocot plants belong to category of PDT (Gaff 1977, 1989; Bewley and Krochko 1982). Poikilochlorophylly is currently constituted by eight genera of five families (Cyperaceae, Liliaceae, Anthericaceae, Poaceae and Velloziaceae). The best physiologically studied PDT plants are the African *Xerophyta scabrida*, *X. viscosa*, *X. humilis* and the Australian *Borya nitida* (Gaff and Churchill 1976; Hetherington and Smillie 1982; Tuba et al. 1993a, 1996; Dace et al. 1998; Farrant 2000; Cooper 2001).

Much or all of the chlorophyll are retained by most of the HDT plants during the desiccation –rehydration cycle. The chlorophyll loss is variable in dicotyledons species which depends on species and mostly influenced by several environmental factors. There were no significant changes observed in chlorophyll content on drying and rewetting cycle in two HDT plants, namely, *Haberlea rhodopensis* and *Ramonda serbica* from Southeast Europe (Markovska et al. 1994), but *Ramonda nathaliae* exhibited loss of 20% and 70% of its chlorophyll when desiccated in the glasshouse and in natural habitats, respectively (Drazic et al. 1999). *Myrothamnus flabellifolia* retains its thylakoids, but half of its chlorophyll is lost on drying (Farrant et al. 1999).

In PDT plants the breakdown of photosynthetic apparatus due to desiccation is different from the process with that of leaf senescence. The systematic disintegration of photosyntheitic apparatus seems to beam a protective mechanism rather than damage to be repaired after rehydration. If the *Borya nitida* plant, which is a PDT plant, dried with a fast speed, it didn't get the sufficient time to break down chlorophyll and loses its viability (Gaff and Churchill 1976), and the same is true for *Xerophyta humilis* (Cooper 2001).

As we have discussed earliar in PDT plants, disintegration of the thylakoid membranes, which is a strictly organised process, leads to the formation of nearly similar-sized globuli called "desiccoplasts", which, unlike chromoplasts, are capable of regreening its tissues and start the process of photosynthesise after rehydration (Tuba et al. 1993b). There occupies a granular stroma inside desiccoplasts which contains lipid-filled translucent vesicles called as plastoglobuli, also containing lipoquinones and neutral lipids. The thylakoid material is represented by osmiophilic lipid material which is found stretched in desiccoplast (Tuba et al. 1993b). Most of the mitochondrial cristae also disappear during the process of desiccation, while the remaining ones seem to decompose within 30 min after rewetting in *X. villosa*. This is exhibited by a loss of almost 50% of insoluble or structural proteins which is found to be very less observed (generally *c.* 10%) in HDT plants (Gaff and Hallam 1974).

2.5 Morphological Adaptations

2.5.1 Leaf Curling and Cell Wall Folding

There are several changes in leaf architecture during the periods of water deficit. These modulations are generally slower responses but provide great potential for survival during desiccation. For example, many resurrection plants undergo leaf rolling towards inside in such a way that the epidermal hairs on abaxial surface now face the upper surface, which turn into a grey-green or purplish colouration. If this process doesn't occur, it can cause photoinhibition and production of reactive oxygen species (ROS). Leaf curling is followed by the accumulation of anthyocyanins and other phenolic compounds that in turn protect against solar radiation (Rakic et al. 2015; Mitra et al. 2013; Figs. 2.3, 2.4 and 2.5). The process of leaf folding is reversible, and it resumes its shape after rehydration and is considered to be related to cell wall folding. Homoiochlorophyllous angiosperm resurrection plants employ

Fig. 2.3 *Ramonda serbica* plants. (**a**) Well hydrated, (**b**) desiccated (Image source: Rakíc et al. 2015)

Fig. 2.4 *Selaginella bryopteris*. (**a**) In natural habitat, (**b**) fully hydrated, (**c**) desiccated and (**d**) rehydrated (Image source: Pandey et al. 2010; Deeba et al. 2009)

this strategy of chlorophyll shading and masking (Sherwin and Farrant 1998; Rakic et al. 2015; Mitra et al. 2013), and frond curling has also been noted in the DT fern *Polypodium polypodioides* (Helseth and Fischer 2005) and fern ally *Selaginella lepidophylla (*Lebkuecher and Eickmeier 1993*)*, *S. bryopteris* (Deeba et al. 2016), from the family Gesneriaceae *Ramonda serbica* (Rakic et al. 2015) and *Boea hygrometrica* (Mitra et al. 2013).

The loss of water from the cell causes conformational changes in the plant cell wall polysaccharides and proteins (Moore et al. 2008). The cell wall becomes highly flexible and gets folded during desiccation, which helps in reducing the loss due to plasmolysis (Jones and McQueen-Mason 2004). By doing so, plasma membrane is saved from damage which in turn maintains the coherence in the cell structures and the cell-to-cell communication through plasmodesmata (Neale et al. 2000; Jones and McQueen-Mason 2004). The process of cell wall folding causes shrinkage of

Fig. 2.5 *Boea hygrometrica*. (**a**) Desiccated state, (**b**) fully hydtated state (Image source: Mitra et al. 2013)

cells which in turn enables leaf curling and reversible rolling (Moore et al. 2008; Farrant et al. 2007). Plants also minimise the mechanical stress by increasing the number of vacuoles wherein the water in vacuoles is replaced by nonaqueous substances (Oliver et al. 2011b). Angiospermic resurrection plants undergo considerable shrinkage in their leaves and roots (Farrant et al. 2007), and the extent of shrinkage is found greater in dicots, because wall folding plays an important role in mechanical stabilisation in these plants.

The leaves of *Craterostigma plantagineum* show increased abundance of α-expansin transcript during drying and rehydration cycle which might be correlated with the modulations in wall extensibility (Jones and McQueen-Mason 2004). The wall loosening can be corrrelated well with the expansins via disruption of non-covalent bonds between polysaccharides (McQueen- Mason et al. 1995), and this provides an additional mechanism of wall folding in the *Craterostigma* species.

2.5.2 Root Adaptations

The roots are the first organ of plant to detect a drop in water availability. In these tissues the drought response is more rapid than in the leaves. In addition, shrinkage in roots is found to be very less as compared to leaves during dehydration as they are embedded in a solid soil. Major root network damage occurs if shrinkage of roots go beyond that of the soil. In *Craterostigma plantagineum*, it has been observed that secondary roots contract more than the soil, but this contraction was found to be very minor. Most of the secondary/tertiary roots survive desiccation but die rapidly after rehydration (24 h after rehydration). The primary roots, which constitute about 70% of the total root biomass in terms of mass, were found to be alive and functional because they contained functionally active enzymes (Norwood et al. 2003). Although drought stress did not affect the growth of the primary root in higher plants, due to lateral root meristem inhibition, its growth is significantly reduced (Deak and Malamy 2005). Crop plants employ another adaptive strategy of hydrotropism to counter the stress, and it has been shown in a study that amyloplast degradation due to drought stress in the columella cells of roots tends to increase the habit of hydrotropism in plants. Architecture of the root system is being modulated by chemical signals produced by hormonal crosstalk mediated by auxin, CK, GA and ABA in response to water stress (Basu et al. 2016). Response of xylem vessels also plays a major role in desiccation tolerance in resurrection plants (Sherwin et al. 1998). Xylem blockage occurs due to the formation of small air bubbles during the scarcity of water (Sperry and Tyree 1988). In *M. fabellifolia*, it has been suggested that narrow reticulate xylem vessels, which cavitate on desiccation, refilled on rehydration through the capillarity and root pressure (Sherwin et al. 1998).

In response to drought, ABA is synthesised in roots and is released to different plant parts. The release of ABA in turn activates the various genes which are required for metabolic processes, e.g. the accumulation of sucrose from either stored carbohydrates or through altered photosynthetic carbon partitioning. Generally it is known that ABA, which is released in roots, is transported to shoots where it modulates various metabolic functions (Schulze 1986).

Although there are few studies pertaining to desiccation tolerance responses in roots of resurrection plants, some studies have been undertaken to observe the changes in carbohydrate metabolism and mobilisation of reserve food (Norwood et al. 1999, 2003). It has been observed that oligosaccharides make up the majority of carbohydrate contents of the root. Gas chromatography (GC) analysis revealed that 97% of raffinose series oligosaccharides were in the form of stachyose. On drying, there is a small decrease in the raffinose series of oligosaccharides and dramatic rise in the amount of sucrose in the roots. Little changes occurred in other carbohydrate. Neglegible quantities of 2-octulose were detected in root sample (Norwood et al. 1999). Enzymes responsible for the carbohydrate metabolism and gluconeogenesis (aldolase and GAPDH) have been found to be increased in roots during desiccation. However, glycolysis and Krebs cycle's enzymes did not show any significant increase in activity (PFP, PFK and malate dehydrogenase) suggesting a minor role of respiration during the preparation of desiccation (Norwood et al. 1999). In higher plants upon mild drought stress, enzymes related to root morphology (e.g. xyloglucan endotransglucosylase) were found to be increased, which strongly correlates with the enhanced surface area for increased water uptake. The modulations in the expression of these proteins lead to more lateral development which in turn affects the process of photosynthesis (Sengupta and Reddy 2011). This enhanced lateral root development and root hair formations were found in lines possessing a QTL, qDTY12.1, only under drought condition. Drought-responsive specific traits of such roots in these plants contributed towards higher capacity for increased grain yield under drought. Moreover, various tissues in the roots of different cultivar or species undergo varied anatomical adaptations under drought stress conditions. For example, in order to increase the retention of water in sclerenchyma layer cells, there were found to be decreased suberisation and compaction in rice under drought stress condition (Basu et al. 2016).

It has been found in a study on *S. bryopteris* undergoing dehydration followed by two rehydrations (RI and RII) that almost all the identified proteins (barring one) showed higher abundance during dehydrations and on rehydration most of them came to their normal control values (Deeba et al. 2016). Roots exhibited higher abundance of the proteins involved in the process of signalling, stress and defence, protein and nucleotide metabolism, carbohydrate and energy metabolism, storage and epigenetic control. Most of these proteins remained upregulated on first rehydration, suggesting their role in recovery phase also (Deeba et al. 2016). It has been suggested that disturbance of the root system during the drying period leads to disruption of acquisition of the desiccation-tolerant state (Gaff 1997) in *Sporobolus stapfianus* suggesting that root signal(s) may be important for desiccation tolerance. In contrast to this, some resurrection plants possess the same capacity for desiccation tolerance as that of the whole plant (Deeba et al. 2009; Mitra et al. 2013).

2.6 Physiological Adaptations

2.6.1 Stomatal Control of Gas Exchange and Water Status of Leaves

In order to avoid extensive water loss, plants prefer to close their stomatal aperture as an initial response under water-deficit conditions, thus suggesting a preventive measure from cell dehydration, xylem cavitation and death. Plants develop these responses quickly or slowly, depending on the rate of water loss, and may result from shoot and/or root dehydration (Schulze 1986; Chaves 1991). The changes in turgidity of guard cell in relation to adjacent epidermal cell influence the opening and closing of stomata which in turn is dependent on metabolic energy coming from mesophyll photosynthesis, and changes in membrane permeability as well. Stomata respond to a number of stresses at any given moment, so it is very difficult to interpret its response under any specific condition of drought stress ranging from light intensity to CO_2 concentration.

Reduction in leaf surface area via leaf curling in response to dehydration is considered as a common response in resurrection plants which ultimately leads to the reduced transpiration, thus regulating temperature and limiting the damage caused by radiation incidence (Dinakar et al. 2012). The decreased transpiration due to reduced stomatal conductance was observed in *Barbacenia purpurea* (*Pleurostima purpurea*) upon water deficit (Suguiyama et al. 2014). It has been observed that a metabolite, fumarate, acts as stomatal controller which avoids water loss and thus controls gas exchange. Therefore, stomatal closure can be correlated well with increased fumarate content during winter season thus suggesting diminished transpiration and photosynthetic rate in *B. purpurea* (Aidar et al. 2010).

A net CO_2 assimilation fall has been observed in *X. scabrida* due to stomatal closure and loss of chlorophyll as observed by fluorescence imaging; this ultimately leads to a halt in photosynthesis (Tuba 2008).

When the leaves of *X. viscosa* had lost about 60% of their water content through the first 4 days of drying, no photosynthetic activity were found by that time (Tuba et al. 1996; Tuba 2008). The recovery of CO_2 assimilation during rehydration process in *Haberlea rhodopensis* and *Ramonda serbica* can be correlated well with an increased stomatal conductance and Rubisco activity. The photosynthetic activity was significantly reduced with the desiccation in *C. plantagineum*, while its rate is further resumed after rehydration (Dinakar et al. 2012).

Conductance rate of water through the xylem vessels directly influences the stomatal responses in plants under water stress conditions (Salleo et al. 2000). Alterations in leaf turgidity in terms of water would translate into a signal that might consequently lead to stimulate the changes in osmotic potential of guard cells during scarcity in water supply (Buckley and Mott 2002). Stomata regulate the xylem pressure and prevent them going beyond its cavitation threshold (Jackson et al. 2000; Buckley and Mott 2002).

Stomatal aperture is also regulated by other hormones besides ABA, either in isolation or acting in association with ABA. Stomatal sensitivity is decreased towards ABA with the increase in cytokinin concentration in the xylem that promotes stomatal opening directly (reviewed by Wilkinson and Davies 2002). On the contrary, root cytokinin (zeatin and zeatin riboside) was found to be decreased due to partial root zone drying, thus an increase in xylem ABA brings about a reduction in stomatal conductance (Stoll et al. 2000). Stomatal regulation through ABA involves both its long-distance transport and modulations at the guard cells to a given dose of hormones (Wilkinson and Davies 2002). Xylem sap and leaf tissue pH exhibit an increase due to high evaporative demands caused by a high deficit in water vapour pressure of the air, high light intensity and high leaf temperature.

2.6.2 Photosynthesis

In resurrection plants, photosynthesis is shut down during desiccation. The downregulation of photosynthesis-related gene expression (Bockel et al. 1998) along with the degradation of photosynthetic structures contributes a significant reduction in this process. As already explained earlier, despite changes observed in photosynthetic pigment distribution, homoiochlorophyllous species retain its chlorophyll and thylakoid membranes structures (Alamillo and Bartels 2001). However, dismantling of chlorophyll and thylakoid structures favours the poikilochlorophyllous species to remain safe by preventing the accumulation of reactive oxygen species (Tuba et al. 1998).

Reactive oxygen species (ROS) production and accumulation during water stress condition are contributed majorly by chloroplasts (Moran et al. 1994; Kranner and Lutzoni 1999). Closed stomata limit the intake of CO_2 to chloroplasts, thus inhibiting carbon fixation and causing overexcitation of chlorophylls which ultimately transfers excess energy to oxygen, giving rise to ROS (Inźe and Van Montagu 1995; Kranner and Lutzoni 1999; Navari-Izzo and Rascio 1999).

It is well known that photosynthesis is affected by a number of genes during desiccation. The decreased rubisco activity is largely dependent on the activity of

Rubisco activase and stromal ATP/ADP ratio than to the changes at protein level. Under normal conditions, rubisco activase transcripts were shown to be abundantly expressed in leaves suggesting the plant's preparation before dehydration in *C. plantagineum* (Dinakar et al. 2012). Different species vary in their responses to shutting down the photosynthesis during dehydration in resurrection plants. For this it is important to maintain the membrane intactness or their repairing for carrying on the electron transport reaction during dehydration. In desiccation-tolerant plants, thylakoid membranes and their functions were found to be recovered completely upon rehydration (Dinakar et al. 2012).The thylakoid membrane intactness and the ratio of pigment–protein complexes were maintained in dehydrated *Boea hygrometrica* and *H. rhodopensis* despite a decrease in photosynthetic carbon fixation and the PS II functions (Georgieva et al. 2007; Deng et al. 2003). Another protective mechanism was also observed in *R. serbica* leaves upon desiccation where the PS II activity was significantly decreased to maintain membrane integrity (Deng et al. 2008). The low levels of psbA transcripts, required for the initial assembly of PS II and also important component of reaction centre protein D1, are stably maintained during desiccation in *X. humilis*. The necessary transcript for the recovery of PS II is already stored in order to immediately translate within first few hours of rehydration (Collett et al. 2003). Light-harvesting complex (LHC) protein, which is a primary chlorophyll-binding protein and responsible for the maintenance of chloroplast stability, is considered to be induced under desiccation (Schneider et al. 1993; Neale et al. 2000). Other proteins which were found to be involved in chloplast stability in leaves of *Craterostigma plantagineum* (Bartels et al. 1992; Alamillo and Bartels 2001) and *Sporobolus stapfianus* (Neale et al. 2000) exhibited higher similarities with ELIPs (early light-inducible proteins). It has been suggested that the enzymes involved in Calvin cycle are more affected by dehydration with that of the membrane-bound electron transport reactions (Georgieva et al. 2007).

Homoiochlorophyllous plants rapidly resume photosynthesis upon rehydration (Drazic et al. 1999) because their photosynthetic structure is almost maintained in recoverable form during dehydration. At the same time, the event of photoexcitation of chlorophyll persists which is further responsible for ROS production. In order to limit the rise in ROS and to safeguard the cells from their consequences, various structural and biochemical modifications occur in dehydrating cells. Antioxidant system of chloroplast provides first level of defence in dehydrating cells. Various enzymes and antioxidants scavenge the free radicals produced in cells during the process of desiccation (Pandey et al. 2010).

Nonradiative dissipation of excessive excitation energy is another approach to limit the levels of ROS. The carotenoids of the xanthophyll cycle contribute in this process (Ruban and Horton 1999) through the de-epoxidation of violaxanthin to zeaxanthin that either directly or indirectly dissipates solar radiation as heat (Alamillo and Bartels 2001; Kranner et al. 2002).

The strategy which allows poikilochlorophyllous plants to degrade their photosynthetic structures during desiccation also offers an advantage of reducing the cellular sources of ROS in the dry state (Proctor and Tuba 2002). However, it takes time to resume their photosynthesis after rehydration because the entire photosynthetic apparatus needs to be rebuilt after rehydration (Sherwin and Farrant 1996; Tuba et al. 1994, 1998). A programmed event that transforms the chloroplast into a "desiccoplast" which is devoid of any pigment and thylakoid membrane follows well-defined metabolic pathways (Sherwin and Farrant 1998; Tuba et al. 1998). While this process happens, "desiccation respiration" of mitochondria provides energy for this organelle transformation. Unlike chloroplast, mitochondria do not undergo structural or functional alterations until a severe degree of cell dehydration is reached (Tuba et al. 1996). In order to meet the energy demands for rebuilding the chloroplast structure, respiration follows a very quick resumption in poikilochlorophyllous plants on rehydration (Vander Willigen et al. 2001). Despite all these preventive measures, poikilochlorophyllous plants also face the risk of ROS production which arises from respiration maintained in conditions of stress. Thus, to combat these deleterious effects, a preventive and protective system activates against free oxygen radicals during dehydration in poikilochlorophyllous resurrection plants. Large quantities of anthocyanins were produced in drying leaves of *Eragrostis nindensis* (Vander Willigen et al. 2001). Similar behaviour of leaf folding and increased anthocyanin production was also observed in *Xerophyta humilis* (Farrant 2000). There was an increased level of the antioxidative enzymes, e.g. ascorbate peroxidase (APX), glutathione reductase (GR) and superoxide dismutase (SOD), observed in *Xerophyta viscosa* during desiccation (Sherwin and Farrant 1998). A new desiccation-inducible antioxidant enzyme which corresponds to a form of 1-cys peroxiredoxin was found in leaves of *Xerophyta viscosa* (Ndima et al. 2001; Mowla et al. 2002). A group of highly conserved class of thiol-specific antioxidant enzymes is represented by peroxiredoxins (Prxs), which is present in all organisms ranging from archaebacteria to plants and mammals (Chae et al. 1994; Kozak 1999). Two conserved residues of cysteines are present at the N-terminus of this protein, which are responsible for peroxide reduction. In plants, the 2-cys Prxs which acts in conjugation with photosynthetic machinery in chloroplast is also responsible for scavenging free radicals (Baier and Dietz 1997). Because of the fact that 1-cys Prxs are exclusively expressed in seeds during storage and their complete disappearance after germination and their absence in vegetative tissues even after dehydration stress, this suggests that the main function of these enzymes is to protect the seed tissues that survive desiccation from oxidative stress (Haslekas et al. 1998). Now the interesting fact is the identification of the Prx in leaves of *Xerophyta viscosa* (XvPer1) undergoing dehydration which exhibits approx 70% sequence identity with the seed-specific 1-cys peroxiredoxins (Aalen et al. 1994; Haslekas et al. 1998; Lewis et al. 2000), and this was the only instance which shows 1-cys Prx noticeable in vegetative tissues. Its transcripts were observed only under dehydration and upon other stresses including heat and high light responsible for producing increased levels of ROS. This protein (XvPer1) also ensured its nuclear location in *X. viscosa*, thus indicating its protecting role (Mowla et al. 2002).

2.7 Biochemical Adaptations

2.7.1 Antioxidative Defences

Reactive oxygen species (ROS) possess readily available unpaired electron and thus considered as highly reactive molecule. Oxygen, being highly oxidising, gets readily converted into highly reactive oxygen and nitrogen species, namely, singlet oxygen ($^1O_2$), superoxide ($O_2^{\bullet-}$), the hydroxyl radical ($^{\bullet}OH$) and nitric oxide ($NO^{\bullet}$). The process of electron transfer to oxygen in ground state ($^3O_2$) during electron transport chains of respiration and photosynthesis is responsible for production of ROS (Ananyev et al. 1994; Navari-Izzo and Rascio 1999). At cellular level, ROS are also produced under non-stress conditions, but its level does not exceed beyond the safe values through the defence system of the cell including superoxide dismutase (SOD), catalase (CAT) and peroxidise (POX) and free radical scavengers, such as carotenoids, tocopherols, ascorbate and glutathione (Navari-Izzo et al. 1997a, b; Loggini et al. 1997; Alscher et al. 1997). If level of ROS exceeds its normal values, it can cause lipid peroxidation and protein and nucleic acid denaturation that finally lead to destruction of subcellular structures (Mundree et al. 2002; Walters et al. 2002; Vicre et al. 2004; Berjak 2006). The occurence of ROS, being an unavoidable consequence to desiccation stress, is frequent in both seeds (Hendry 1993; Kranner et al. 2006) and resurrection plants (Kranner et al. 2006). The antioxidative activity was found to be increased in response to dehydration or rehydration in both desiccation-sensitive and desiccation-tolerant plants, but generally it is important to mention that a well-defined control of oxidative damage is seen in the desiccation-tolerant species as compared to the sensitive ones which exhibit clear signs of oxidative damage (Dhindsa and Matowe 1981).

The activity of defence enzymes, namely, superoxide dismutase (SOD) and catalase, exhibited fourfold increase in the desiccation-tolerant moss *Tortula ruralis* than in the sensitive *Cratoneuron filicinum* (Dhindsa and Matowe 1981). Drought-tolerant *Tortula ruraliformis* exhibited induced response of superoxide dismutase (SOD) and catalase activities as compared to relatively sensitive *Dicranella palustris*, but peroxidise and ascorbate peroxidase (AP) activities were somewhat greater in the sensitive ones than in the DT species suggesting importance of antioxidants (tocopherols and glutathione) than that of active oxygen-processing enzymes in determining desiccation tolerance. During the process of rehydration after 24 h, the leaves of the DT grass *Sporobolus stapfianus* showed twofold induced activities of glutathione reductase (GR) and dehydroxy ascorbate reductase (DHAR), while AP decreased by half (Sgherri et al. 1994). However, the content of total ascorbate doubled, and glutathione increased 50-folds on drying in the dicotyledon *Boea hygroscopica*. In *S. bryopteris*, various defence enzymes like CAT, SOD and APX, along with invertase and proline, were found to be significantly increased during dehydration (Pandey et al. 2010).

Although the activities of antioxidative enzymes, namely, APX, GR and SOD, were found to be increased in *Craterostigma wilmsii*, *Myrothamnus flabellifolia* and *Xerophyta viscosa*, during drying or rehydration phase at some point, their detailed pattern of response varied greatly among these species (Sherwin and Farrant 1998; Farrant 2000). Despite having several evidences, it is slightly difficult to explain any stable universal pattern in responses of antioxidants and antioxidant enzyme systems to desiccation and rehydration in resurrection plants. Although dry *S. lepidophylla* and *P. polypodioides* employ the strategy of leaf curling to limit photodamage during rehydration, this was not found to be different from non-DT plants (Lebkuecher and Eickmeier 1993; Muslin and Homann 1992). Based on all evidences, it has been suggested that DT plants generally encounter the problem at source rather than by initiating high activity of defence enzymes after the event.

The chloroplast is also taken care by the glutathione system for protection from photo-oxidative damage. By forming glutathione–protein complex (GSSP), glutathione (GSH) not only provides protection to thiol groups from irreversible formation of intramolecular disulphide bonds, it also scavenges free radicals in response to desiccation (Kranner and Grill 1996; Navari-Izzo et al. 1997a; Kranner et al. 2002).

2.7.2 Proteomic Alterations

The "omics" technologies provide a platform for measuring quantitative abundance of biological molecules in such a way, thus making it possible to easily visualise and compare the differential responses among desiccation-tolerant and desiccation-sensitive species with that of their respective controls. Various approaches including transcriptomics, proteomics and metabolomics have enabled us to visualise and compare the complete digital information of the modulations taking place in transcripts, proteins and metabolites, respectively, during the process of desiccation followed by rehydration, thus providing a flow chart of metabolic events. Various "omics" approaches have been applied to study the molecular changes occuring during desiccation tolerance in resurrection plants (Table 2.1).

Some proteins which are specifically induced in response to desiccation provide cellular protection and help in recovery after desiccation in vascular plants. The late embryogenesis abundant proteins (LEAs), which are responsive to ABA and constitute a group of dehydrins, have been isolated as cDNA clones from the desiccation-tolerant *Craterostigma* tissues (Bartels et al. 1993; Ingram and Bartels 1996). These proteins provide cellular protection to seeds during desiccation and water deficits (Bray 1993). Similarly, *Sporobolus stapfianus* also exhibited several drying-induced cDNA coding for proteins related to LEAs (Blomstedt et al. 1998).

LEA proteins, including dehydrins (Close 1996), are hydrophilic in nature and help in preserving the membrane structure and also provide protein protection in conjugation with disaccharides. These proteins help sugars in maintaining cell vitrification and thus forming a glasslike situation with a very slow/negligible movement inside cell suggesting their primarily important role in determining the physical properties of intracellular glasses rather than their sugar component (Buitink and Leprince 2004).

Two groups of dehydrins, namely, LEA II or LEA D-11, and six different groups of LEA proteins were identified on the basis of expression pattern and sequence

Table 2.1 Various omics studies carried out in resurrection plants and sister group comparisons between desiccation-tolerant and desiccation-sensitive plants

Approach	Desiccation-tolerant species	Desiccation-sensitive species	References
Resurrection plants			
Transcriptome analysis	*Craterostigma plantagineum*		Rodriguez et al. (2010)
Transcriptome and metabolomic analysis	*Haberlea rhodopensis*		Gechev et al. (2013)
Proteomic analysis	*Selaginella tamariscina*		Wang et al. (2010)
Proteomic analysis	*Selaginella bryopteris*		Deeba et al. (2016)
Proteomic analysis	*Xerophyta viscosa*		Ingle et al. (2007)
Proteomic analysis	*Boea hygrometrica*		Jiang et al. (2007)
Proteomic analysis	*Sporobolus stapfianus*		Oliver et al. (2011a)
Metabolomic analysis	*Selaginella lepidophylla*		Yobi et al. (2013)
Sister group comparison			
Metabolomic comparison	*Sporobolus stapfianus*	*Sporobolus pyramidalis*	Oliver et al. (2011a)
EST sequencing and comparison	*Selaginella lepidophylla*	*Selaginella moellendorffii*	Iturriaga et al. (2006)
Metabolic comparison	*Selaginella lepidophylla*	*Selaginella moellendorffii*	Yobi et al. (2012)

similarity (Campbell and Close 1997; Cuming 1999). These proteins were found to be associated with different cellular compartments and form a complex including nucleoprotein complexes to endomembranes either in cytosol, nucleoplasm and chloroplasts at the subcellular level (Puhakainen et al. 2004). Dehydrins were found to act as surfactant on the macromolecules, thus providing a hydration buffer on them and protecting them from coagulation by preserving their structural integrity, and this could be possible because of their unique molecular features which include the abundance of glycine, the absence of cysteine and tryptophan and the presence of amphipathic α-helices (Close 1997). These proteins were also found to protect the macromolecules by sequestering ions and renaturing unfolded proteins and act as free radical scavengers (Bray 1993; Cuming 1999; Hara et al. 2004).

The LEA proteins have also found to be constitutively expressed or induced by several other stresses including cold, osmotic stress or exogenous abscisic acid or can even be expressed constitutively (Wise and Tunnacliffe 2004). Some LEA genes are constitutively recruited for ameliorating the deleterious effects of stress condition and thus also found to be induced abundantly in response to desiccation in *Haberlea rhodopensis* (Gechev et al. 2013). But unlike *Haberlea*, LEA genes were exclusively induced only during desiccation in leaves of *C. plantagineum* and not in fully hydrated tissues (Rodriguez et al. 2010). ELIPs (early light-inducible proteins) play an important role in chloroplast stability in homoiochlorophyllous drought-tolerant plants like *Craterostigma plantagineum* (Bartels et al. 1992) and *Sporobolus stapfianus* (Neale et al. 2000).

The ELIPs which are thylakoidal chlorophyll-binding proteins (discussed under photosynthesis section) exhibit similarity with the HLIPs (high light-induced proteins) and the LHC (light-harvesting complex) proteins of photosystems (Montańe and Kloppstech 2000). These proteins have a protective function against photo-oxidative damage of the photosynthetic apparatus (Hutin et al. 2003). These proteins are thought to bind chlorophylls, thus keeping free pigments under strict control during the high light stress (Hutin et al. 2003). These proteins might also be participating in the non-photochemical quenching of light energy by binding with zeaxanthin (Krol and co-authors 1999). A plastid-targeted protein (CpPTP) was also found to express in leaves of *Craterostigma plantagineum* during dehydration (Phillips et al. 2002). These proteins thought to have the ability to modulate the chloroplast genes expression by interacting with plastid DNA (Phillips et al. 2002).

For plant's ability to control the exchange of water between cell and environment, it is important to adjust the permeability of cell membrane during the drying periods. The channel proteins of cell membrane, called aquaporins, facilitate the movement of water across the membrane (Maurel and Chrispeels 2001). These proteins constitute a family of hydrophobic membrane proteins, which are also called as major intrinsic proteins (MIPs). These occupy different subcellular locations, such as the tonoplast intrinsic proteins (TIPs) and the plasma membrane intrinsic proteins (PIPs). (Tyerman et al. 2002). The zones with high water flow in rapid cell division and enlargement areas are enriched with these aquaporins (Suga et al. 2001). The expression and activity of aquaporins under water stress is under strict control (Yamada et al. 1997; Martinez-Ballestra et al. 2000). This directly indicates that aquaporin expression and activity plays an important role in desiccation tolerance mechanisms. One TIP and several PIPs have been found to be differentially expressed during dehydration in the leaves of *Craterostigma plantagineum* (Mariaux et al. 1998). The *Cp-PIPa* gene transcript exhibits the varied response during different stages of dehydration, one being expressed more in the early phase, while another predominates in the late stage of desiccation. This suggests its role for the preparation for rehydration. A TIP has also been reported to be expressed during desiccation in the leaves of *Sporobolus stapfianus* (Neale et al. 2000). A tonoplast aquaporin, namely, TIP 3;1, has also been reported to be specifically expressed in dried leaves of *Eragrostis nindensis* (Vander Willigen et al. 2004). In fact, TIP 3;1 proteins (formerly called α-TIPs), being unique to orthodox seeds (Maurel et al. 1995), had never been found in vegetative tissues. The TIP 3;1 is considered to be permeable to glycerol and small proteins also in addition to water and participate in the mobilisation of solutes during the process of germination (Maurel and Chrispeels 2001). The TIP3;1 helps in mobilising sugars, proteins and other solutes accumulated inside the small vacuoles of bundle sheath during rehydration in dried leaves of *Eragrostis nindensis* (Vander Willigen et al. 2004).

2.7.3 Osmolyte

Osmolytes constitute a group of organic solutes, including a variety of low molecular weight compounds, e.g. amino acids, sugars, polyhydric alcohols, tertiary sulphonium and quaternary ammonium compounds. Osmolytes help in maintaining functional physiological state of cell, and their biosynthesis is affected by environmental factors at any developmental stage of the plant. Despite maintaining osmoregulation, osmolytes play a major role in scavenging reactive oxygen species, thus protecting subcellular structures. Osmolytes in the cytoplasm provide protection to cell membranes, proteins and metabolic machinery, thus preserving subcellular structure from damage(s) of dehydration (Rhodes and Samaras 1994; Rathinasabapathi 2000; Slama et al. 2015).

A coordinated regulation of biosynthesis and catabolic pathways leads to osmolyte accumulation that is considered as a determining factor for their mode of action. Osmolytes are thought to function for rapid and short-term fluctuations (accumulation/degradation, e.g. proline), as well as for longer periods also (e.g. betaines) (Gagneul et al. 2007). Osmolyte production and consumption follow a metabolic dynamics in such a way that its terminal/end products are also as much important as of osmolyte. For example, proline biosynthesis and catabolism lead to rapid and efficient consumption or release of reducing power (Szabados and Savouré 2010).

Despite showing varied biochemical/metabolic nature, osmolyte accumulation also exhibits species-specific varied responses to seasonal patterns. Various studies have been undertaken pertaining to osmolyte production in different resurrection plants (Table 2.2). In addition, level of osmolytes is also influenced by changes in growth period, developmental stage and organ and environmental factors (Hare et al. 1998; Murakeözy et al. 2003). Growing conditions and nutritional and environmental challenges also influence the distribution of osmolytes among various cell compartments (Gagneul et al. 2007).

Osmolytes include a number of osmoprotective compounds, ranging from mono-, di-, oligo- and polysaccharides (glucose, fructose, sucrose, trehalose, raffinose); sugar alcohols (sorbitol, mannitol, glycerol, inositol and methylated inositols); amino acids (proline, pipecolic acid); other betaines (glycine betaine, β-alanine betaine, choline *O*-sulphate) and tertiary sulphonium compounds (dimethylsulphoniopropionate, DMSP) (Rhodes et al. 2002; Ashraf and Foolad 2007). With the influence of various abiotic stresses, these different osmoprotective compounds were found to accumulate in higher plants (Ahmad et al. 1981).

These compounds are metabolically inactive (safe) molecules produced from the primary metabolic pathways (Rhodes et al. 2002). Proline, synthesised mainly from glutamate and also ornithine, was found to hyperaccumulate in plants experiencing with salt and drought stress (Kishor et al. 2005).

Proline is responsible for reducing cytoplasmic acidosis and maintaining $NADP^+/NADPH$ ratios at the required level for the respiration and photosynthetic process (Hare and Cress 1997). For the adequate regeneration of NADPH, the pentose phosphate pathway requires higher concentrations of $NADP^+$, thus facilitating the supply of ribose-5-phosphate for photosynthesis and for purines synthesis as well (Kishor et al. 2005).

Table 2.2 Various osmolytes in vegetative tissues of resurrection plants under hydrated and desiccated conditions

Species	Hydrated condition	Dehydrated/desiccated condition	References
Craterostigma plantagineum, *Craterostigma wilmsii*	2-Octulose	Sucrose	Bianchi et al. (1991), and Cooper and Farrant (2002)
Haberlea rhodopensis	Sucrose raffinose	Sucrose, raffinose, maltose, verbascose, stachyose	Djilianov et al. (2011), and Gechev et al. (2013)
Myrothamnus flabellifolia	Fructose, glucose	Sucrose, arbutin, glucosylglycerol	Bianchi et al. (1993)
Boea hygroscopica	Glucose, fructose, sucrose, inositol	Sucrose	Bianchi et al. (1991)
Sporobolus stapfianus	Glucose, fructose, sucrose	Sucrose, raffinose, stachyose	Oliver et al. (2011a)
Xerophyta viscosa	Galactinol, Myoinositol	Sucrose, Raffinose	Peter et al. (2007)
Selaginella lepidophylla	Trehalose, sucrose and glucose	Trehalose, sucrose and glucose	Yobi et al. (2012)

Glycine betaine, one of the quaternary ammonium compounds, is considered as the highly abundant osmoprotective compound known to accumulate in plants under salinity, drought, heat and cold stress (Ashraf and Harris 2004; Hanson et al. 1991; Lokhande and Suprasanna 2012). Choline, the precursor molecule for glycine betaine synthesis, gets converted into glycine betaine through the sequential enzymatic action of choline monooxygenase (CMO) and betaine aldehyde dehydrogenase (BADH), respectively. There are other pathways known for glycine betaine synthesis, such as the direct N-methylation of glycine, but the choline pathway is considered to be found in all glycine betaine-accumulating plant species (Ashraf and Foolad 2007; Fitzgerald et al. 2009).

The production of organic osmolytes such as sucrose, polyols, glycine betaine or proline follows one common strategy which is found in both desiccation-tolerant and desiccation-sensitive species (Bray 1997; Geigenberger et al. 1997; Pelah et al. 1997; Clifford et al. 1998; Hare et al. 1998; Mundree et al. 2000; Gao et al. 2004). The preferential exclusion of these organic solutes on the surface of membrane and protein provides protection to them against moderate water loss (Hare et al. 1998; Hoekstra et al. 2001).

2.7.4 Sugars

Sugars are considered to be a major component of desiccation tolerance (Scott 2000; Oliver 1996: Alpert and Oliver 2002). Carbohydrate metabolism undergoes major changes upon drying that can be directly correlated with the desiccation tolerance.

Cellular protection in fully desiccation-tolerant mosses (*Tortula ruraliformis* and *T. ruralis*) is majorly employed by the free sugar, sucrose (Toldi et al. 2009), which accounts for approximately 10% of the dry weight for *T. ruralis* gametophytic cells that is sufficient to provide membrane protection during drying, at least in vitro.

Moreover, the content of sucrose does not change on a dry weight basis regardless of the fact that plant is subjected to drying or rehydration in the dark or light, suggesting that it is important for cells to maintain sufficient amounts of sucrose (Bewley et al. 1978). The fully desiccation-tolerant mosses that maintain high sucrose content constitutively do not require increasing its content on dehydration (Smirnoff 1992). However, dehydration-specific induction in the level of sucrose was found in modified desiccation-tolerant plants (Vicre et al. 2004; Ingram et al. 1997; Buitink and Leprince 2004). A comparative metabolic analysis of desiccation-tolerant and desiccation-sensitive species *Eragrostis nindensis* demonstrated that sucrose, being the major carbohydrate found upon desiccation in angiospermic resurrection plants (Cooper and Farrant 2002), exhibited enhanced accumulation in dried leaves of tolerant species, while the sensitive species did not accumulate sucrose (Illing et al. 2005).

Sucrose helps in maintaining the drying cells stabilised either by direct interactions with macromolecules and membranes or by reversibly immobilising the cytoplasm to the extent of slow-flowing glasslike vitrified liquid (Buitink and Leprince 2004).

A significant rise in sucrose metabolising enzyme, namely, class-I sucrose synthase genes, was found in *C. plantagineum* during the dehydration or abscisic acid (ABA) treatment, suggesting an increased glycolytic demand (Kleines et al. 1999). This increase was found to be paralleled by the increase in the abundance of cytosolic glyceraldehyde-3-phosphate dehydrogenase (GAPDH) during desiccation or ABA treatment (Velasco et al. 1994). This increase in mRNA levels of class-I sucrose synthase and GAPDH ultimately leads to the higher protein and enzyme contents. The eight-carbon carbohydrate, 2-octulose, is converted into the sucrose with the functional enzyme complex of transketolase–transaldolase in *C. plantagineum* in response to dehydration (Bernacchia et al. 1995), and this enzyme complex might alternatively enhance pentose-to-hexose flux, thus providing more hexose for glycolysis. This process might be a preparation for the cell to meet the demand of ATP and NADH during recovery (Kleines et al. 1999). Extremely high concentrations of the unusual C8 sugar, 2-octulose, are found under normal conditions in *C. plantagineum* and *C. Wilmsii*, acting as a temporary storage carbohydrate. This sugar is converted to sucrose during dehydration, providing much needed energy to plants. The conversion of octulose to sucrose shares up to 80% of the carbohydrates in desiccated leaves of *Craterostigma*. Although *C. plantagineum* plants synthesise starch during photosynthesis, the percent share of starch as storage material is very less as compared with that of 2-octulose which is small (Norwood et al. 2000). Similarly in *Sporobolus stapfianus*, with the concomitant decrease of glucose and fructose levels, sucrose was found to significantly increase upon desiccation (Ghasempour et al. 1998). Despite the induced accumulation of sucrose and raffinose, other carbohydrates including stachyose, maltotetraose and myoinositol have also been found to accumulate upon desiccation. Glucose-6-phosphate that stores phosphate during dehydration is responsible for the synthesis of maltotetraose and myoinositol (Oliver et al. 2011a). Maltose, verbascose and stachyose are among the other unique sugars apart from sucrose which were found to increase upon desiccation in *H. rhodopensis* (Gechev et al. 2013). The dehydration-specific accumulation of the arbutin, glucosylglycerol and trehalose was reported from *M. flabellifolia.* However, trehalose, sucrose and glucose are among the metabolites found to increase in *Selaginella lepidophylla* during desiccation which account for 50% of the total metabolites.

The oligosaccharides from raffinose family, being widely distributed nonstructural carbohydrates in the plant kingdom, have been considered to act as antistress agent throughout the whole life cycle of plants either in vegetative or reproductive tissues (Peterbauer et al. 2002; Peters et al. 2007; Taji et al. 2002; Pennycooke et al. 2003). These compounds have the ability to accumulate in large quantities without affecting primary metabolic processes (Peters et al. 2007).

The non-reducing sugar trehalose significantly contributes to vegetative desiccation tolerance in plants by providing protection against a variety of stresses in different organisms (Goddijn et al. 1997; Gaff 1989; Drennan et al. 1993; Bianchi et al. 1993). Being an unreactive sugar, trehalose is known for its strong stability which results from very low energy (1 kcal mol^1) of the glycoside oxygen bond joining the two hexose rings in it (Schiraldi et al. 2002). Thus, the primary metabolism and the energy homeostasis of the cell are not considered to be affected by trehalose.

Reports from the transgenic plants grown for the enhanced trehalose expression exhibited varied responses. In some cases plants had shown to increase tolerance against abiotic stresses in terms of sustained growth (Garg et al. 2002), but it cannot be generalised, because the expression of the *S. cerevisiae* trehalose-6-phosphate synthase gene in potato resulted in drought tolerance, but with slow growing, retarded plants (Stiller et al. 2008). Several studies on plants with modified trehalose expression had revealed that trehalose-inducible genes are also associated with jasmonic acid and ethylene-dependent stress signalling pathways (Bae et al. 2005a, b). The expression modulation of *Arabidopsis* trehalose- 6-phosphate synthase gene simultaneously influences the sugar sensing and photosynthate partitioning (Gomez et al. 2005; Kolbe et al. 2005).

2.8 Molecular Mechanism of Desiccation Tolerance

2.8.1 Transcription Factors

Generally proteins, functional byproduct of dehydration-induced genes, are responsible for preventing cellular damages and thus participate in antioxidant defence of cell machinery upon desiccation. There are various techniques and algorithms assembled in order to sequencing the entire transcriptome by deep RNA sequencing (RNA-Seq) even without a reference genome; therefore this is also applicable to resurrection plants. Various gene expression and EST sequence studies have been performed in variety of resurrection plants such as the moss Tortula ruralis (Zeng et al. 2002; Oliver et al. 2004); the club moss Selaginella lepidophylla and

Selaginella tamariscina (Iturriaga et al. 2006); the monocot species Sporobolus stapfianus (Neale et al. 2000), X. viscosa (Mundree et al. 2000; Mowla et al. 2002), X. humilis (Collett et al. 2003, 2004; Illing et al. 2005) and X. villosa (Collett et al. 2004); and the dicot species C. plantagineum (Bockel et al. 1998).

The identified transcripts of *C. plantagineum* and *H. rhodopensis* at diffrent physiological stages revealed highest similarities to the genes of *Vitis vinifera*, *Ricinus communis* and *Populus trichocarpa* (Dinakar and Bartels 2013).

The enzymes related to carbohydrate metabolism, proteins with protective properties, regulatory proteins such as transcription factors and kinases and signalling molecules are encoded by transcription factor induced upon desiccation stress (Dinakar and Bartels 2013).

The abundance of the transcripts generally varies between desiccation-tolerant and desiccation-sensitive plants. As an example, the LEA-like CDeT11-24 transcript exhibits varied response between desiccation-tolerant *C. plantagineum* (higher abundance) and desiccation-sensitive species of *Lindernia subracemosa* plants (lower abundance) suggesting unstable or lower rate of accumulation of this transcript in *L. subracemosa* during dehydration. The transcript expression level is influenced by promoter cis elements which was found to be absent in *L. brevidens*; that's why the accumulation of transcript CDeT11-24 exhibited lower abundance upon dehydration (van den Dries et al. 2011). This strategy strengthens these results which follow the hypothesis that high level of dehydration-specific gene expression is governed by selection of promoter cis elements. In *Sporobolus stapfianus* as well as in *X. humilis*, a large number of genes related to the antioxidant pathway were found to be significantly enhanced during desiccation (Neale et al. 2000; Collett et al. 2004). In addition to induced expression of antioxidant genes, expression of genes from other pathways also provided protection to some resurrection plants. The desiccation-induced antioxidant gene encoding 1-Cys peroxiredoxin (XvPer1) in *X. viscosa* exhibited sequence identity (70%) with the seed-specific dormancy-related 1-Cys peroxiredoxin in *Arabidopsis* (AtPer1; Mowla et al. 2002). The transcripts of thiamine biosynthesis and aldehyde dehydrogenase (CpALDH) were found to be upregulated upon desiccation and thus contributed to antioxidant defence in *C. plantagineum* (Rodriguez et al. 2010).

Several genes, for galactinol and stachyose synthase, are found constitutively present in *C. plantagineum* and *H. rhodopensis* leaves under normal conditions, but on dehydration these get additionally induced (Rodriguez et al. 2010; Gechev et al. 2013) suggesting the important role of oligosaccharides in providing protection to the cells upon dehydration. Various sugar metabolism-related transcripts encoding sucrose synthases (SuS) and sucrose–phosphate synthases (SPS) have also been found to be increased in *C. plantagineum* during dehydration (Ingram and Bartels 1996; Kleines et al. 1999). Various isoforms of transketolases, considered as key enzymes of the reductive and oxidative pentose phosphate pathways, are responsible for the synthesis of intermediates for sugar phosphate which ultimately help in different metabolic pathways of cell on desiccation. There were three isoforms of transketolase in *C. plantagineum* including *tkt3*, *tkt7* and *tkt10*; out of which one is constitutively expressed and also involved in octulose synthesis (Rodriguez et al. 2010; Willige et al.

2009), while other one, e.g. *tkt7*, was abundantly found in rehydrating tissues, and lastly, *tkt10* is exclusively expressed in fully hydrated tissues. Two isoforms of tkt were also observed in *H. rhodopensis*, but they exhibited differential expression in its carbohydrate metabolism with that of *C. plantagineum* (Gechev et al. 2013).

Expansins are the proteins responsible for the enhanced extensibility of cell wall, and the transcripts of this protein were found to be induced upon dehydration in *C. plantagineum* (Jones and McQueen-Mason 2004). Although some genes encoding for xyloglucan endotransglucosylases, pectin esterases and pectate lyases are constitutively expressed in *H. rhodopensis* in hydrated conditions, on approaching towards the desiccation phase, the expression of these genes were switched off, but simultaneously transcripts for laccase genes, which is involved in lignin biosynthesis, were only expressed in desiccated tissues suggesting a cell wall remodelling approach during desiccation (Gechev et al. 2013).

The above studies suggested that a catalogue of the regulatory genes can be prepared by transcriptome analysis which provides a profile of upregulated genes during desiccation, but there is another problem to understand at present how these genes interact with other pathways and which target genes they regulate.

2.9 Signalling Cascade

The environmental signals are perceived and followed by the secondary messenger production that in turn modulates the levels of intracellular Ca^{2+}, thus making a cascade of protein phosphorylation which ultimately assigns the proteins involved in cellular protection or transcription factors controlling the expression of stress-related genes. These signalling events ultimately lead to the production of the genes which may contribute in the generation of regulatory molecules like the phytohormone, abscisic acid (ABA). The phytohormone ABA accumulation under various kind of abiotic stress is one of the earliest responses observed in plants (Dinakar et al. 2012).

2.9.1 ABA Dependent

Various abiotic stresses including drought, low temperature and hypoxia induce the biosynthesis of abcisic acid (ABA) in the shoot and root of the plant. The epoxidation of zeaxanthin is the precursor event for ABA biosynthesis in roots which leads to the formation of epoxy xanthophyll precursors with the help of zeaxanthin epoxidase (ZEP) enzyme (Audran et al. 1998). Dehydration-induced expression of ZEP mRNA has been observed to increase by sevenfold in roots after 8 h of desiccation, while it does not report to increase in leaves (Audran et al. 1998; Bray 2002). There is sufficient amount of violaxanthin present in the leaves, so the gene induction for ABA biosynthesis is not required in leaves as compared to roots (Bray 2002).

The ABA hormone, which is considered to coordinate various metabolic processes during stress condition, also modulates the expression of most of the genes involved in acquisition of desiccation tolerance (Wilkinson and Davies 2002).

Dehydration-specific induction of ABA molecules serves as signalling molecule in most of the resurrection plants (Hartung et al. 1998). The genes which are thought to be induced by ABA contain a specific region of ABA-responsive elements (ABRE) in their promoter region (Mundree et al. 2002). Dehydration or the treatment of exogenous ABA stimulates the accumulation of *CDeT27-45*, a *LEA*-like gene in *C. plantagineum*, thus suggesting a model to study ABA-dependent signalling pathways (Bartels et al. 1990; Piatkowski et al. 1990). Dehydration-specific induction in the level of ABA varies with the different species of resurrection plants; as an example, generally it leaps by three- to sevenfold in the leaves upon desiccation (Bernaccia and Furini 2004), but it may increase by 20 times as compared with the control values in drying leaves of *Chamaegigas intrepidus* (Schiller et al. 1997). Moreover, exogenously applied ABA can also stimulate the adaptive responses in plants which was supposed to be induced by dehydration (Kleines et al. 1999; Frank et al. 2000; Kirch et al. 2001; Mowla et al. 2002), and dedifferentiated callus tissue of *Craterostigma plantagineum* also exhibited desiccation tolerance by exogenously applied ABA (Bartels and Salamini 2001; Bernacchia and Furini 2004). Some workers have identified the ABA-responsive region of the *CDeT27-45* promoter and also proposed that *CDeT27-45* expression depends on DNA-binding activity by nuclear proteins (Michel et al. 1994; Nelson et al. 1994). It has also been found that a novel SAP-domain transcription factor (CpR18) binds with the *CDeT27-45* promoter region (Hilbricht et al. 2002). Abcisic acid is found to regulate the expression of a gene called *CDT-1* which is supposed to activate a pathway inducing desiccation tolerance in *C. plantagineum* (Furini et al. 1997). The transcription factors, namely, homeodomain–leucine zipper (HD-ZIP) proteins, were shown to be induced by ABA; however, some genes of HD-ZIP family (CPHB-1 and CPHB-2) exhibited differential responses to ABA, one being nonresponsive to ABA (CPHB-1) and other one increased their accumulation in response to ABA in *C. plantagineum* (Frank et al. 1998). ABA not only modulates the regulatory proteins but also influences the expression of functional proteins such as LEA proteins that collectively lead to desiccation tolerance (Piatkowski et al. 1990; Ingram and Bartels 1996). The application of exogenous ABA or dehydration induces a novel gene family which was found to be involved in maintenance of chloroplast integrity, and its expression is very rapid in response to these treatments. Moreover, the expressions of several aquaporins are found to be upregulated by dehydration and ABA as well in *C. plantagineum* (reviewed in Ramanjulu and Bartels 2002).

Late embryogenesis abundant (LEA) proteins are also known to be ABA responsive and accumulate in vegetative organs during desiccation stress conditions. The vegetative tissues of resurrection plants were found to accumulate high levels of dehydrins under both drought and ABA induction. As dehydrins provide protection during the early phase of dehydration, that's why they accumulate in the cytoplasm and plastids during the first phases of cell drying (Rascio and Rocca 2005).

The process of leaf development is influenced by expression of expansin which can also manipulate the leaf shapes (Pien et al. 2001). However, the process of expansin accumulation in response to desiccation is not evident. The tomato and barley mutants with ABA deficiency exhibited reduced leaf expansion with an increase in xylem sap pH which might act as drought signal governed by ABA-mediated mechanism (Bacon et al. 1998).

2.9.2 ABA Independent

ABA plays various roles in induction of desiccation tolerance, but there are several other signalling pathways and metabolic processes which are not under the control of ABA; they are considered as ABA independent. There are some dehydration-induced genes which are not regulated by ABA. As an example, the *CPHB-1* gene of a transcription factor and phospholipase D genes both are induced in response to drying but are shown to be insensitive to ABA in *Craterostigma plantagineum* (Frank et al. 1998; Frank et al. 2000). Similarly, the expression of the *XVSAP1* gene is only influenced by dehydration but did not show any effects of ABA in *Xerophyta viscosa* (Garwe et al. 2003). Furthermore, it has also been shown that the mechanism of desiccation tolerance follows an ABA-independent pathway in the leaves of *Sporobolus stapfianus* which were dried down to 60% RWC (Neale et al. 2000).

These observations have indicated that, in addition to ABA-mediated processes, other signalling pathways must also exist that serve to protect the deleterious effects of dehydration in order to provide the drought tolerance in resurrection plants but that involve yet unidentified molecular signals (Bray 2002).

2.10 Genetic Transformation of Desiccation-Tolerant Plants

There are several studies available on genetic engineering of drought tolerance pertaining to various traits including osmolytes metabolism, protective proteins, ROS scavenging proteins etc. (Umezawa et al. 2006). However, gene bank only provides information based on putative functions of gene on the basis of sequence similarity. It is mandatory to obtain in vivo proof after scruitinising transgenic plants with either higher or lower abundances of genes associated with desiccation tolerance (Toldi et al. 2006). Although genetic transformation procedures are having some limitation with vegetative desiccation-tolerant plants because of the extreme sensitivity to tissue culture level manipulations during the establishment of their plant regeneration and genetic transformation systems (Table 2.3), there are some resurrection plants for which plant regeneration protocol has been established, e.g. *C. plantagineum* (Toldi et al. 2002: Furini et al. 1994), *R. myconi* (Toth et al. 2004), *H. rhodopensis* (Djilianov et al. 2005) and *Lindernia brevidence* (Smith-Espinoza et al. 2007). The problems frequent necrosis, hyperhydrated leaves and polyphenol secretion into the culture media under suboptimal conditions are some of the limitations during the plant regeneration and genetic transformation process. Therefore, in order to develop non-lethal selection

Table 2.3 Advantages and disadvantages of the most abundant protective agents associated with vegetative desiccation tolerance in terms of practical application

Protection agent	Metabolism	Advantage	Disadvantage
Sucrose	Carbohydrate	Carbon and energy source, osmoprotectant	Possible interference with growth and osmoregulation
Trehalose	Carbohydrate	Osmoprotectant, signal metabolite	Possible interference with sugar sensing and photosynthetic performance
RFOs	Carbohydrate	Osmoprotectant, antistress agent, metabolically inactive, synergistic effects	None
LEAs	Protein	Hydration buffer, protein and membrane protectant	The most specific and most determinative LEAs are still missing
ABA	Mevalonate	Growth regulator, signal metabolite	Possible interference with plant development and growth, contrasting effects

Source: Toldi et al. (2009)

strategies and modified tissue culture media with relatively lesser amount of macro- and micro-elements and different antioxidant agents, it is mandatory to establish low stress transformation technologies for nearly all resurrection plants (Toldi et al. 2006). The biochemical preinduction of *vir* genes enhances the colonisation of wounded surface by the transforming *Agrobacterium* cells which is shown to be an additional strategy for increasing the transforming frequency. This could be possible because of the application of low-pH liquid medium with acetosyringone, aldose-type sugar source and organic nitrogen during the infection phase. Therefore, for a reliable genetic transformation method for resurrection plants, it is important to establish these modifications along with the appropriate tissue culture system. These knowledges can be further used to induce resistance in recalcitrant species (Toldi et al. 2009).

2.11 Evaluation of Plant Responses to Drought: Methodological Tools

There are several areas covered for evaluating plant's responses to drought including leaf gas exchange process, growth analysis and in vitro determination of enzymes, metabolites and gene expression patterns. However, evaluation of plant's responses to stress is revolutionised over the years by the methodological advances in various aspects.

Advances in genomics and proteomics provided a platform for understanding the molecular mechanisms underlying plant response to stress, but there are other tools, for example, stable isotope and thermal or fluorescence imaging, which can be used to study the plant responses within their natural habitat under any stress condition.

2.11.1 Genome-Wide Tools

The study of global transcriptome response of plants to various stresses including abiotic and biotic stress could be possible because of the advancements in high-throughput DNA sequencing methods which possibly made the sequencing complete for the genome projects including *Arabidopsis* (the first genome) and draft sequence of rice genome for both indica (Yu et al. 2002), japonica (Goff et al. 2002) and *Selaginella moellendorffii* (one of smallest plant genomes known to date: ~110 Mb; Wang et al. 2005; Banks et al. 2011). The DNA chips (DNA microarray technology) is considered as very promising tool in the current scenario of stress genomics studies in which cDNA gene probes (generally produced by expressed sequence tags (ESTs)) are being spotted onto a glass slide with the help of printing robots. Complementary DNAs (cDNAs) taken from control samples as well as from stressed samples are fluorescently tagged and mixed in equal amounts and agonistically hybridised to the probes in the microarray slide. The amount of hybridised probes in control is quantified relatively with the stressed one with a high resolution scanner (Brown and Botstein 1999). By using this approach, it was studied that how many of key regulatory genes have changed expression wise for the induction of CAM metabolism in *Mesembryanthemum crystallinum* and there were 60 functionally unknown ESTs found in *M. crystallinum* leaves (Maroco et al. 2000).

2.11.2 Proteomic Tools

Genome-wide tools provide a high-throughput data for the differential changes in gene expression for evaluating the plant response to drought. There is a drawback of this technique by which gene expression is prevented to translate it into functional proteins through posttranscriptional and posttranslational events. Therefore, this is a prerequisite of researchers to obtain a platform based on differential abundance of proteins. Traditionally, a technique of two-dimensional polyacrylamide gel electrophoresis (2D PAGE) is used to separate proteins according to their isoelectric points in their first dimension and at molecular weight basis in the second dimension. Since this technique includes several extraction and separation steps during its processing and proteins have large degree of diversity, thus this is considered as having a drawback of low reproducibility. This is also a labour-intensive process and involves the issues with loading capacity and limitation in protein detection on 2D gel. These flaws in 2D GE lead the development of new alternative due to the requirement of high-throughput proteome analysis in the present post-genomic era. There is another technique named difference gel electrophoresis (2D-DIGE) which is known to overcome many of these problems and provide relatively more precise and sensitive quantitative proteomics studies. CyDyes are normally cyanine dyes with a reactive group of *N*-hydroxysuccinimidyl ester which binds covalently with the ε-amino groups of lysine residues in proteins. Dye concentrations are optimised in such a way that only one dye molecule attaches per protein. Due to the limitations of the current in-gel digestion as well as the detection limits for various mass spectrometers, it is almost difficult to identify the lesser abundant proteins on 2D-DIGE. There is yet another advanced combination of multidimensional chromatography which involves the protein separation coupled with mass spectrometry techniques for the functional characterisation of proteins (Washburn and Yates 2000. A variety of protein ligands are also used for microarrays (protein chips) which allow for the investigation of a high-throughput functional determination of proteins (MacBeath and Schreiber 2000). Fluorescent tags are used to label the proteins in a mixture and identified then by their positional coding or by further mass spectrometry, and this technique is useful where ligands are unknown for the proteins (Zhou et al. 2001). There are other studies pertaining to the quantitative proteomic changes during drought (Koh et al. 2015) in *Brassica napus* and combined effects of drought with heat stress (Zhao et al. 2016) in maize leaf. Such type of studies can be conducted in resurrection plants as well.

2.11.3 Stable Isotope

Stable isotope has become a powerful tool to study plant responses to the environment changes, from cellular to ecosystem level. Isotopic effects originate by analysing the differences in physical and chemical properties between heavy (e.g. ^{13}C or ^{18}O) and lighter isotope (e.g. the more common ^{12}C or ^{16}C). After the confirmation that plant material depletion occurs in heavier isotope of carbon (^{13}C) in terms of CO_2 in the air, it's the first time when the natural abundances of stable isotopes get into the notice in plant science in the middle 1900s. Lately, it was found that discrimination against the heavier carbon isotope was more in C_3 plants as compared to C_4 and CAM plants. The modelling of ^{13}C fractionation during photosynthesis in C_3 plants has been studied in details during the early 1980s (Farquhar et al. 1982). There were other heavy element isotope studies including oxygen ($^{18}O/^{16}O$) and hydrogen (D/H), which also contributed to the studies of plant water relations. Theoretical and practical knowledge of the use of stable isotope has exponentially increased with the advancements in mass spectrometry techniques (Griffiths 1998). The discrimination of the carbon isotopic composition occurs during photosynthesis as well as in respiration, so this is considered a complex process to study.

Utilisation of glycine and serine were studied as nitrogen sources in the roots of *Zea mays* and a resurrection plant *Chamaegigas intrepidus* by using NMR spectroscopy and stable isotope labelling. The various pathways of these amino acid utilisation were studied by using *C. intrepidus* and maize (*Zea mays*) root tips which were incubated with [^{15}N]glycine, [^{15}N]serine and [$^{2\text{-}13}C$]glycine (Hartung and Ratcliffe 2002).

2.11.4 Thermal and Fluorescence Imaging

There are various setbacks of destructive sampling or in situ measurements faced during the past for stress monitoring in plants. Therefore, it is a need of present time to develop almost accurate and non-obtrusive methods. Improved capacity for data processing and analysis of several imaging techniques in the present time has allowed us to monitor the changes in physiological state of plants in a nonintrusive way (Buschmann and Lichtenthaler 1998). With these techniques it is possible to study the spatio-temporal localisation stress responses of plants, thus providing a better understanding of the dynamics of plant performance under field conditions. There are various advanced imaging technologies, including thermal imaging, fluorescence imaging and multi- or hyperspectral reflectance imaging. In order to study plant–water relations, thermal imaging is used to measure leaf surface temperature specifically for stomatal conductance, because the rate of transpiration is a determining factor of leaf temperature. Fluorescence signals which are good indicator of blue or actinic light absorption from plants are measured by fluorescence imaging. Moreover, the physiological parameters of plants including photoassimilation and non-photochemical quenching are estimated by combining an actinic light source with brief saturating blue pulses.

Advances in these recent technologies extended their usage to field conditions (Jones et al. 2009). Screening of a large number of phenotypes at a time is made possible because of the high working efficiency of these techniques, suggesting their importance in crop breeding programme, in crop management techniques (e.g. field irrigation scheduling) or in supporting models that simulate the impact of global climate change on natural and agroecosystems.

The process of thermography enables the measurements of the leaf temperature and their canopies which is directly dependent on rate of transpiration and thus can be used as an indicator of stomatal conductance and stress intensity (Chaerle and Van Der Straeten 2001). This technique is ahead over direct methods of stomatal conductance measurements, in its ability to focus the spatial heterogeneity of stress response over a leaf or canopy. The investigation of any kind of "hypersensitive response" to any infection before it became apparent is quantified with the help of this technique (Chaerle and Van Der Straeten 2001). It has also been used to screen for differential distribution and responses in stomatal apertures to identify mutants with altered stomatal function (Merlot et al. 2002).

The drawback of the thermographic approach is in the acquisition of temperature data that are truly representative of the crop (Jones 1999). Using image analysis software along with thermal imaging can solve this problem under controlled conditions. However, stomatal behaviour study or any kind of stress responses in the field is very difficult to interpret because of the variation in environmental conditions (Jones 1999). The photosynthetic activity and pigment contents have been studied by chlorophyll fluorescence imaging in the resurrection plants *Ramonda serbica* and *Ramonda nathaliae* following dehydration and rehydration (Gashi et al. 2013). In the resurrection plant *Haberlea rhodopensis*, it has been shown that lumenal substance protects the thylakoid against the damaging combined effects of light and drought Georgieva et al. (2012).

In order to powerfully probe the photosynthetic functioning in plants, the modulated fluorometers have been used for the measurements of chlorophyll fluorescence which provide greater potential when combined with reflectance or thermal imagery (Schreiber et al. 1986). For getting a detailed spatial analysis of stress-induced inhibition of photosynthesis, fluroscence imaging is used for preparing quantitative maps of photosynthetic electron transport activity over the leaf (Meyer and Genty 1999). This technique has been proven very helpful to screen and identify mutants in altered PSII photochemistry and xanthophyll cycle (Niyogi et al. 1998).

2.12 Conclusion and Future Perspective

In the present chapter, it has been shown that the resurrection plants possess an extraordinary mechanism of desiccation tolerance for surviving in a very extreme water conditions in such a way that after almost complete water loss in vegetative organs, they resume their biological activities after rehydration. This outstanding capability of resurrection plants could be possible because of simultaneous interaction of various mechanisms and pathways required to cope with the numerous problems as a result of cell drying. There are various studies on morphological, physiological and molecular standpoints in resurrection plants which contribute greatly for improving the knowledge of their adaptive strategies. However, there are still a number of desiccation-related genes and proteins providing cell protection against water loss damage which need to be elucidated. It has been shown that environmentally induced adaptive strategies of desiccation-tolerant resurrection plants share its traits with the developmentally induced desiccation tolerance of seeds, suggesting that almost similar line of tolerance mechanisms is followed by both. This is also in agreement with the fact that inductive desiccation tolerance in seeds as well as in resurrection plants evolved directly or indirectly, from a primitive kind of constitutive desiccation tolerance found in bryophytes.

Transcriptome and proteome analysis revealed various genes and proteins with potential role in triggering desiccation tolerance; similarly metabolic profiling provided detailed descriptions of metabolic processes involved in the functioning of stress responses. Performing systematic analyses of resurrection plants is still an intricate task technically as well as intellectually. However, with the help of transcriptome and proteome, partial profiling of candidate genes could be possible in elucidating their functions for molecular breeding programmes. Although sugar is very important for most of the metabolic process, manipulating its expression (upregulation or downregulation) could lead to serious threat for balancing the osmotic adjustment, energy metabolism and growth and development. On the contrary, transgenic plants with enhanced trehalose synthesis do not seem to obstruct with the primary metabolism unlike sucrose, but with the other pathways including jasmonic acid/ethylene signalling. This can mislead the signals which ultimately disturb the plant responses in general.

As far as the specific LEA genes are concerned, although, known to provide tolerance against various abiotic stressors, they have already been quite extensively utilised in molecular breeding programmes, no field trials have still been conducted under real field situations. This aspect needs further studies for identifying the most effective types of LEA genes.

Several signalling and metabolic genes need to be modified in order to achieve desiccation tolerance, but the constraints for the entire high-throughput technology are the limited capacity of conventional genetic transformation techniques. Therefore, the development of alternative in planta transformation appears promising, at least in the case of some model plants (Toldi et al. 2009) and is currently available for both monocotyledonous and dicotyledonous plants. Moreover epigenetic mechanisms should be studied in resurrection plants to explain their role in modifying response of these plants to desiccation stress.

References

Aalen RB, Opsahl-Ferstad HG, Linnestad C, Olsen OA (1994) Transcripts encoding an oleosin and a dormancy-related protein are present in both the aleurone layer and the embryo of developing barley (*Hordeum vulgare* L.) seeds. Plant J 5:385–396

Abe S, Kurashima A, Yokohama Y, Tanaka J (2001) The cellular ability of desiccation tolerance in Japanese intertidal seaweeds. Bot Mar 44(2):125–131

Adamska I (1997) ELIPs. Light-induced stress proteins. Physiol Plant 100:794–805

Agrawal SC, Pal U (2003) Viability of dried vegetative cells or filaments, survivability and/or reproduction under water and light stress, and following heat and UV exposure in some blue-green and green algae. Folia Microbiol 48(4):501–509

Ahmad I, Larher F, Stewart G (1981) The accumulation of Δ′-acetylornithine and other solutes in the salt marsh grass *Puccinellia maritima*. Phytochemistry 20(7):1501–1504

Aidar ST, Meirelles ST, Pocius O, Delitti WBC, Souza GM, Gonçalvez AN (2010) Desiccation tolerance in *Pleurostima purpurea* (Velloziaceae). Plant Growth Regul 62:193–202

Alamillo JM, Bartels D (1996) Light and stage of development influence the expressionof desiccation-induced genes in the resurrection plant Craterostigma plantagineum. Plant Cell Environ 19:300–310

Alamillo JM, Bartels D (2001) Effects of desiccation on photosynthesis pigments and the ELIP-like dsp 22 protein complexes in the resurrection plant *Craterostigma plantagineum*. Plant Sci 160:1161–1170

Alpert P (2005) The limits and frontiers of desiccation-tolerant life1. Integr Comp Biol 45(5):685–695

Alpert P, Oechel WC (1987) Comparative patterns of net photosynthesis in an assemblage of mosses with contrasting microdistributions. Am J Bot 74:1787

Alpert P, Oliver MJ (2002) Drying without dying. In: Black M, Prichard HW (eds) Desiccation and survival in plants: drying without dying. CAB International, Wallingford

Alscher RG, Donahue JL, Cramer CL (1997) Reactive oxygen species ant antioxidant: relationship in green cells. Physiol Plant 100:224–233

Ananyev G, Renger G, Wacker U, Klimov V (1994) The photoproduction of superoxide radicals and the superoxide dismutase activity of photosystemII. The possible involvement of cytochrome b559. Photosynth Res 41:327–338

Armengaud P, Thiery L, Buhot N, Grenier-de March G, Savouré A (2004) Transcriptional regulation of proline biosynthesis in *Medicago truncatula* reveals developmental and environmental specific features. Physiol Plant 120(3):442–450

Ashraf M, Foolad M (2007) Roles of glycine betaine and proline in improving plant abiotic stress resistance. Environ Exp Bot 59(2):206–216

Ashraf MPJC, Harris PJC (2004) Potential biochemical indicators of salinity tolerance in plants. Plant Sci 166(1):3–16

Aubert S, Hennion F, Bouchereau A, Gout E, Bligny R, DORNE AJ (1999) Subcellular compartmentation of proline in the leaves of the subantarctic Kerguelen cabbage Pringlea antiscorbutica R. Br. In vivo13C-NMR study. Plant Cell Environ 22(3):255–259

Audran C, Borel C, Frey A, Sotta B, Meyer C, Simonneau T, Marion-Poll A (1998) Expression studies of the zeaxanthin epoxidase gene inNicotiana plumbaginifolia. Plant Physiol 118(3):1021–1028

Bacon MA, Wilkinson S, Davies WJ (1998) pH-regulated leaf cell expansion in droughted plant is absicic acid dependent. Plant Physiol 118:1507–15015

Bae H, Herman E, Bailey B, Bae HJ, Sicher R (2005a) Exogenous trehalose alters Arabidopsis transcripts involved in cell wall modification, abiotic stress, nitrogen metabolism, and plant defense. Physiol Plant 125(1):114–126

Bae H, Herman E, Sicher R (2005b) Exogenous trehalose promotes non-structural carbohydrate accumulation and induces chemical detoxification and stress response proteins in *Arabidopsis thaliana* grown in liquid culture. Plant Sci 168(5):1293–1301

Baier M, Dietz K-J (1997) The plant 2-Cys peroxiredoxin BAS1 is a nuclear encoded chloroplast protein. Its expressional regulation, phylogenetic origin and implications for its specific physiological function in plants. Plant J 12:179–190

Banks J, Nishiyama T, Hasebe M et al (2011) The selaginella genome identifies genetic changes associated with the evolution of vascular plants. Science 332:960–963

Bartels D (2005) Desiccation tolerance studied in the resurrection plant *Craterostigma plantigineum*. Integr Comp Biol 45:696–701

Bartels D, Salamini F (2001) Desiccation tolerance in the resurrection *Craterostigma plantagineum*. A contribution to the study of drought tolerance at the molecular level. Plant Physiol 127(4):1346–1353

Bartels D, Schneider K, Terstappen G, Piatkowski D, Salamini F (1990) Molecular cloning of abscisic acid-modulated genes which are induced during desiccation of the resurrection plant Craterostigma plantagineum. Planta 181(1):27–34

Bartels D, Hanke C, Schneider K, Michel D, Salamini F (1992) A desiccation-related Elip-like gene from the resurrection plant Craterostigma plantagineum is regulated by light and ABA. EMBO J 11:2771–2778

Bartels D, Alexander K, Schneider R, Elster R, Velasco J, Alamillo G, Nelson D, Salamini F (1993) Desiccation-related gene products analyzed in a resurrection plant and in barley embryos. In: Close TJ, Bray EA (eds) Plant responses to cellular dehydration during environmental stress. Curr Top Plant Physiol A: Soc Plant Physiol Ser 10:119–127

Barthlott W, Porembski S (1996) Ecology and morphology of Blossfeldia liliputana (Cactaceae): a poikilohydric and almost astomate succulent. Bot Acta 109:161–166

Basu S, Ramegowda V, Kumar A, Pereira A (2016) Plant adaptation to drought stress. F1000Research:5

Beckett RP, Minibayeva FV, Vylegzhanina NN, Tolpysheva T (2003) High rates of extracellular superoxide production by lichens in the suborder Peltigerineae correlate with indices of high metabolic activity. Plant Cell Environ 11:1827–1837

Berjak P (2006) Unifying perspectives of some mechanism basic to desiccation tolerance across life forms. Seed Sci Res 16:1–15

Bernacchia G, Furini A (2004) Biochemical and molecular responses to water stress in resurrection plants. Physiol Plant 121(2):175–181

Bernacchia G, Schwall G, Lottspeich F, Salamini F, Bartels D (1995) The transketolase gene family of the resurrection plant Craterostigma plantagineum: differential expression during the rehydration phase. EMBO J 14(3):610

Bewley JD, Krochko JE (1982) Desiccation-tolerance. In: Physiological plant ecology II. Springer, Berlin

Bewley JD, Halmer P, Krochko JE, Winner WE (1978) Metabolism of a drought-tolerant and a drought-sensitive moss. Respiration, ATP synthesis and carbohydrate status. Dry biological systems 185–203

Bianchi G, Murelli C, Vazzana C (1991) Changes of low-molecular weight substances in *Boea hygroscopica* in response to desiccation and rehydration. Phytochemistry 30(2):461–466

Bianchi G, Gamba A, Limiroli R, Pozzi N, Elster R, Salamini F, Bartels D (1993) The unusual sugar composition in leaves of the resurrection plant *Myrothamnus flabellifolia*. Physiol Plant 87(2):223–226
Billi D, Potts M (2002) Life and death of dried prokaryotes. Res Microbiol 153(1):7–12. Biol Med 40:2155–2165
Blomstedt CK, Gianello RD, Hamill JD, Neale AD, Gaff DF (1998) Isolation and characterization of cDNAs associated with the onset of desiccation tolerance in the resurrection grass, Sporobolus stapfianus. Plant Growth Regul 24:153–161
Bockel C, Salamini F, Bartels D (1998) Isolation and characterization of genes expressed during early events of the dehydration process in the resurrection plant *Craterostigma plantagineum*. J Plant Physiol 152:158–166
Boyer JS (1982) Plant productivity and environment. Science 218:443–448
Bray EA (1993) Molecular responses to water deficit. Plant Physiol 103:1035–1040
Bray EA (1997) Plant responses to water deficit. Trends Plant Sci 2:48–54
Bray EA (2002) Abscisic acid regulation of gene expression during water-deficit stress in the era of the Arabidopsis genome. Plant Cell Environ 25(2):153–161
Brown PO, Botstein D (1999) Exploring the new world of the genome with DNA microarrays. Nat Genet 21:33–37
Buckley TN, Mott KA (2002) Dynamics of stomatal water relations during the humidity response: implications of two hypothetical mechanisms. Plant Cell Environ 25:407–419
Buitink J, Leprince O (2004) Glass formation in plant anhydrobiotes: survival in the dry state. Cryobiology 48(3):215–228
Buschmann C, Lichtenthlaer HK (1998) Principals and characteristics of multi colour fluorescence imaging of plants. J Plant Physiol 152:297–314
Campbell SA, Close TJ (1997) Dehydrins: genes, proteins, and associations with phenotypic traits. New Phytol 137:61–74
Chae HZ, Robison K, Poole LB, Church G, Storz G, Rhee SG (1994) Cloning and sequencing of thiol-specific antioxidant from mammalian brain: alkyl hydroperoxide reductase and thiol-specific antioxidant define a large family of antioxidant enzymes. Proc Natl Acad Sci U S A 91(15):7017–7021
Chaerle L, van der Straeten D (2001) Seeing is believing: imaging techniques to monitor plant health. Biochim Biophys Acta Gene Struct Expr 1519:153–166
Chaves MM (1991) Effects of water deficits on carbon assimilation. J Exp Bot 42:1–16
Clegg JS (2005) Desiccation tolerance in encysted embryos of the animal extremophile, Artemia. Integr Comp Biol 45(5):715–724
Clifford SC, Arndt SK, Corlett JE, Joshi S, Sankhla N, Popp M, Jones HG (1998) The role of solute accumulation, osmotic adjustment and changes in cell wall elasticity in drought tolerance in *Ziziphus mauritiana* (Lamk.) J Exp Bot 49(323):967–977
Close TJ (1996) Dehydrins: emergence of a biochemical role of a family of plant dehydration proteins. Physiol Plant 97:795–803
Close TJ (1997) Dehydrins: a commonalty in the response of plants to dehydration and low temperature. Physiol Plant 100:291–296
Collett H, ButowtR SJ, Farrant JM, Illing N (2003) Photosynthetic genes are differentially transcribed during the dehydration–rehydration cycle in the resurrection plant, Xerophyta humilis. J Exp Biol 54:2593–2595
Collett H, Shen A, Gardner M, Farrant JM, Denby KJ, Illing N (2004) Towards transcript profiling of desiccation tolerance in Xerophyta humilis: construction of a normalized 11 k X. humiliscDNA set and microarray expression analysis of 424 cDNAs in response to dehydration. Physiol Plant 122:39–53. https://doi.org/10.1111/j.1399-3054.2004.00381.x
Cooper K (2001) The effect of drying rate of the resurrection species Craterostigma wilmsii (homoiochlorophyllous) and Xerophyta humilis (poikilochlorophyllous). MSc thesis, University of Cape Town, South Africa
Cooper K, Farrant JM (2002) Recovery of the resurrection plant *Craterostigma wilmsii* from desiccation. Protection versus repair. J Exp Bot 53(375):1805–1813

Csintalan Z, Takács A, Proctor MCF, Lichtenthaler HK, Tuba Z (1998) Desiccation and rehydration responses of desiccation-tolerant moss and lichen species from a temperate semidesert grassland. J Hattori Bot Lab 84:71–80

Csintalan Z, Proctor MCF, Tuba Z (1999) Chlorophyll fluorescence during drying and rehydration in the mosses Rhytidiadelphus loreus (Hedw.)Warnst., Anomodon viticulosus (Hedw.) Hook. & Tayl. and Grimmia pulvinata (Hedw.) Sm. Ann Bot 84:235–244

Cuming AC (1999) LEA proteins. In: Casey R, Shewry PR (eds) Seed proteins. Kluwer Academic Publishers, Dordrecht, pp 753–780

Cushman JC, Oliver MJ(2011) Understanding vegetative desiccation tolerance using integrated functional genomics approaches within a comparative evolutionary framework. In: Plant desiccation tolerance. Springer, Berlin/Heidelberg

Dace H, Sherwin HW, Illing N, Farrant JM (1998) Use of metabolic inhibitors to elucidate mechanisms of recovery from desiccation stress in the resurrection plant Xerophyta humilis. Plant Growth Regul 24:171–177

Davey MC (1997) Effects of physical factors on photosynthesis by the Antarctic liverwort Marchantia berteroana. Polar Biol 17(3):219–227

Deak KI, Malamy J (2005) Osmotic regulation of root system architecture. Plant J 43(1):17–28

Deeba F, Pandey V, Pathre U, Kanojiya S (2009) Proteome analysis of detached fronds from a resurrection plant Selaginella bryopteris–response to dehydration and rehydration. J Proteome Bioinforma 2:108–116

Deeba F, Pandey AK, Pandey V (2016) Organ specific proteomic dissection of Selaginella bryopteris undergoing dehydration and rehydration. Front Plant Sci 7:425. http://doi.org/10.3389/fpls.2016.00425

Deng X, Hu ZA, Wang HX, Wen XG, Kuang TY (2003) A comparison of photosynthetic apparatus of the detached leaves of the resurrection plant Boea hygrometrica with its nontolerant relative Chirita heterotrichia in response to dehydration and rehydration. Plant Sci 165:851–861. Desiccation and survival in plants: drying without dying. CAB International, Wallingford

Devaux F, Marc P, Bouchoux C, Delaveau T, Hikkel I, Potier MC, Jacq C (2001) An artificial transcription activator mimics the genome-wide properties of the yeast Pdr1 transcription factor. EMBO Rep 2(6):493–498

Dhindsa RS, Matowe W (1981) Drought tolerance in two mosses: correlated with enzymatic defense against lipid peroxidation. J Exp Bot 32:79–91

Dickie JB, Pritchard HW (2002) Systematic and evolutionary aspects of desiccation tolerance in seeds. Desiccation and survival in plants: drying without dying, vol 1. CAB International, Wallingford, pp 239–259

Dinakar C, Bartels D (2013) Desiccation tolerance in resurrection plants: new insights from transcriptome, proteome and metabolome analysis. Front Plant Sci 4:482. http://doi.org/10.3389/fpls.2013.00482

Dinakar C, Djilianov D, Bartels D (2012) Photosynthesis in desiccation-tolerant plants: energy metabolism and antioxidative stress defense. Plant Sci 182:29–41. https://doi.org/10.1016/j.plantsci.2011.01.018

Djilianov D, Genova G, Parvanova D, Zapryanova N, Konstantinova T, Atanasov A (2005) In vitro culture of the resurrection plant Haberlea rhodopensis. Plant Cell Tissue Organ Cult 80:115–118

Djilianov D, Ivanov S, Moyankova D, Miteva L, Kirova E, Alexieva V, Joudi M, Peshev D, Van den Ende W (2011) Sugar ratios, glutathione redox status and phenols in the resurrection species *Haberlea rhodopensis* and the closely related non-resurrection species *Chirita eberhardtii*. Plant Biol 13:7676–7776

Drazic G, Mihailovic N, Stevanovic B (1999) Chlorophyll metabolism in leaves of higher poikilohydric plants Ramonda service Panc. and Ramonda nathaliae Panc. et Petrov. during dehydration and rehydration. J Plant Physiol 154:379–384

Drennan PM, Smith MT, Goldsworthy D, Van Staden J (1993) The occurrence of trehalose in the leaves of the desiccation-tolerant angiosperm *Myrothamnus flabellifolius* welw. J Plant Physiol 142(4):493–496

Farnsworth E (2004) Hormones and shifting ecology throughout plant development. Ecology 85(1):5–15

Farquhar GD, O'Leary MH, Berry JA (1982) On the relationship between carbon isotope discrimination and intercellular carbon dioxide concentration in leaves. Aus J Plant Physiol 9:121–137

Farrant JM (2000) A comparison of mechanisms of desiccation tolerance among three angiosperm resurrection plant species. Plant Ecol 151:29–39

Farrant JM (2007) Mechanisms of desiccation tolerance in angiosperm resurrection plants. In: Jenks MA, Wood AJ (eds) Plant desiccation tolerance. CABI Press, Wallingford

Farrant JM, Cooper K, Kruger LA, Sherwin HW (1999) The effect of drying rate on the survival of three desiccation-tolerant angiosperm species. Ann Bot 84:371–379

Farrant JM, Brandt WB, Lindsey GG (2007) An overview of mechanisms of desiccation tolerance in selected angiosperm resurrection plants. Plant Stress

Fitzgerald TL, Waters DL, Henry RJ (2009) Betaine aldehyde dehydrogenase in plants. Plant Biol 11(2):119–130

Frank W, Phillips J, Salamini F, Bartels D (1998) Two dehydration-inducible transcripts from the resurrection plantCraterostigma plantagineumencode interacting homeodomain-leucine zipper proteins. Plant J 15(3):413–421

Frank W, Munnik T, Kerkmann K, Salamini F, Bartels D (2000) Water deficit triggers phospholipase D activity in the resurrection plant Craterostigma plantagineum. Plant Cell 12(1):111–123

Fujita Y, Fujita M, Satoh R, Maruyama K, Parvez MM, Seki M, Yamaguchi-Shinozaki K (2005) AREB1 is a transcription activator of novel ABRE-dependent ABA signaling that enhances drought stress tolerance in Arabidopsis. Plant Cell 17(12):3470–3488

Furini A, Koncz C, Salamini F, Bartels D (1994) Agrobacterium-mediated transformation of the desiccation-tolerant plant *Craterostigma plantagineum*. Plant Cell Rep 14:102–106

Furini A, Koncz C, Salamini F, Bartels D (1997) High level transcription of a member of a repeated gene family confers dehydration tolerance to callus tissue of Craterostigma plantagineum. EMBO J 16(12):3599–3608

Gaff DF (1977) Desiccation tolerant vascular plants of Southern Africa. Oecologia 31:95–109

Gaff DF (1981) The biology of resurrection plants. In: Pate JS, McComb AJ (eds) The biology of Australian plants. University of Western Australia Press, Perth, pp 114–146

Gaff DF (1987) Desiccation tolerant plants in South America. Oecologia 74(1):133–136

Gaff DF (1989) Responses of desiccation tolerant 'resurrection' plants to water stress. In: Kreeb KH, Richter H, Hinckley TM (eds) Structural and functional responses to environmental stress. SPB Academic Publishing, The Hague, p 308

Gaff DF (1997) Mechanisms of desiccation tolerance in resurrection vascular plants. In: Basra AS, Basra RK (eds) Mechanisms of environmental stress resistance in plants. Harwood Academic Publishers, London

Gaff DF, Bole PV (1986) Resurrection grasses of India. Oecologia (Berlin) 71:159–160

Gaff DF, Churchill DM (1976) Borya nitida Labill. – an Australian species in the Liliaceae with desiccation-tolerant leaves. Aust J Bot 24:209–224

Gaff DF, Giess W (1986) Drought resistance in water plants in rock pools of Southern Africa. Dinteria 18:17–37

Gaff DF, Hallam ND (1974) Resurrecting desiccated plants. Roy Soc NZ Bull 12:389–393

Gagneul D, Aïnouche A, Duhazé C, Lugan R, Larher FR, Bouchereau A (2007) A reassessment of the function of the so-called compatible solutes in the halophytic Plumbaginaceae Limonium latifolium. Plant Physiol 144(3):1598–1611

Gao XP, Wang XF, Lu YF, Zhang LY, Shen YY, Liang Z, Zhang DP (2004) Jasmonic acid is involved in the water-stress-induced betaine accumulation in pear leaves. Plant Cell Environ 27(4):497–507

Garg AK, Kim JK, Owens TG, Ranwala AP, Do Choi Y, Kochian LV, Wu RJ (2002) Trehalose accumulation in rice plants confers high tolerance levels to different abiotic stresses. Proc Natl Acad Sci 99(25):15898–15903

Garwe D, Thomson JA, Mundree SG (2003) Molecular characterization of XVSAP1, a stress-responsive gene from the resurrection plant *Xerophyta viscose* Baker. J Exp Bot 54:191–201

Gashi B, Babani F, Kongjika E (2013) Chlorophyll fluorescence imaging of photosynthetic activity and pigment contents of the resurrection plants Ramonda serbica and Ramonda nathaliae during dehydration and rehydration. Physiol Mol Biol Plants 19(3):333–341. https://doi.org/10.1007/s12298-013-0175-5

Gechev T, Dinakar C, Benina M, Toneva V, Bartels D (2012) Molecular mechanisms of desiccation tolerance in resurrection plants. Cell Mol Life Sci 69:3175–3186. https://doi.org/10.1007/s00018-012-1088-0

Gechev TS, Benina M, Obata T, Tohge T, Sujeeth N, Minkov I et al (2013) Molecular mechanisms of desiccation tolerance in the resurrection glacial relic Haberlea rhodopensis. Cell Mol Life Sci 70:689–709. https://doi.org/10.1007/s00018-012-1155-6

Geigenberger P, Reimholz R, Geiger M, Merlo L, Canale V, Stitt M (1997) Regulation of sucrose and starch metabolism in potato tubers in response to short-term water deficit. Planta 201(4):502–518

Georgieva T, Christov NK, Djilianov D (2012) Identification of desiccation-regulated genes by cDNA-AFLP in Haberlea rhodopensis: a resurrection plant. Acta Physiol Plant 34:1055–1066

Georgieva K, Szigeti Z, Sarvari E et al (2007) Photosynthetic activity of homoiochlorophyllous desiccation tolerant plant Haberlea rhodopensis during dehydration and rehydration. Planta 225:955–964

Georgieva K, Sárvári É, Keresztes Á (2010) Protection of thylakoids against combined light and drought by a lumenal substance in the resurrection plant Haberlea rhodopensis. Ann Bot 105(1):117–126. http://doi.org/10.1093/aob/mcp274

Ghasempour HR, Gaff DF, Williams RPW, Gianello RD (1998) Contents of sugars in leaves of drying desiccation tolerant flowering plants, particularly grasses. Plant Growth Regul 24(3):185–191

Goddijn OJ, Verwoerd TC, Voogd E, Krutwagen RW, De Graff PTHM, Poels J, Pen J (1997) Inhibition of trehalase activity enhances trehalose accumulation in transgenic plants. Plant Physiol 113(1):81–190

Goff SA, Ricke D, Lan TH, Presting G, Wang R, Dunn M, Glazebrook J, Sessions A et al (2002) A draft sequence of the rice genome (*Oryza sativa* L. ssp. japonica). Science 296:92–100

Gomez LD, Baud S, Graham IA (2005) The role of trehalose-6-phosphate synthase in Arabidopsis embryo development. Biochem Soc Trans 33(Pt 1):280–282

Goyal K, Walton LJ, Tunnacliffe A (2005) LEA proteins prevent protein aggregation due to water stress. Biochem J 388(1):151–157

Guerrero R, Haselton A, Solé M, Wier A, Margulis L (1999) Titanospirillum velox A huge, speedy, sulfur-storing spirillum from Ebro Delta microbial mats. Proc Natl Acad Sci USA 96(20):11584–11588

Guo P, Baum M, Grando S, Ceccarelli S, Bai G, Li R, Valkoun J (2009) Differentially expressed genes between drought-tolerant and drought-sensitive barley genotypes in response to drought stress during the reproductive stage. J Exp Bot 60(12):3531–3544

Hahn SC, Tenhunen JD, Popp PW, Meyer A, Lange OL (1993) Upland tundra in the foothills of the brooks range, Alaska: diurnal CO_2 exchange patterns of characteristic lichen species. Flora 188:125–143

Hanson AD, Rathinasabapathi B, Chamberlin B, Gage DA (1991) Comparative physiological evidence that β-alanine betaine and choline-O-sulfate act as compatible osmolytes in halophytic Limonium species. Plant Physiol 97(3):1199–1205

Hara M, Fujinaga M, Kuboi T (2004) Radical scavenging activity and oxidative modification of citru dehydrin. Plant Physiol Biochem 42:657–662

Hare PD, Cress WA, Van Staden J (1998) Dissecting the roles of osmolyte accumulation during stress. Plant Cell Environ 21(6):535–553

Hartung Ratcliffe RG (2002) Utilization of glycine and serine as nitrogen sources in the roots of Zea mays and Chamaegigas intrepidus. J Exp Bot 53(379):2305–2314

Hartung W, Schiller P, Dietz KJ (1998) Physiology of poikilohydric plants. In: Progress in botany. Springer, Berlin

Haslekas C, Stacy RAP, Nygaard V, Culianez-Macia FA, Aalen RB (1998) The expression of a peroxiredoxin antioxidant gene, AtPer1, in Arabidopsis thaliana is seed-specific and related to dormancy. Plant Mol Biol 36:833–845

Helseth LE, Fischer TM (2005) Physical mechanisms of rehydration in Polypoidium polypodioides, a resurrection plant. Phys Rev 71(6):061903

Hendry GAF (1993) Oxygen and free radical processes in seed longevity. Seed Sci Res 3:141–153

Hetherington SE, Smillie RM (1982) Humidity-sensitive degreening and regreening of leaves of Borya nitida Labill. as followed by changes in chlorophyll fluorescence. Aust J Plant Physiol 9:587–599

Hilbricht T, Salamini F, Bartels D (2002) CpR18, a novel SAP-domain plant transcription factor, binds to a promoter region necessary for ABA mediated expression of the CDeT27-45 gene from the resurrection plant *Craterostigma plantagineum* Hochst. Plant J 31(3):293–303

Hoekstra FA, Golovina EA, Buitink J (2001) Mechanisms of plant desiccation tolerance. Trends Plant Sci 6(9):431–438

Hutin C, Nussaume L, Moise N, Moya I, Kloppstech K, Havaux M (2003) Early light-induced proteins protect *Arabidopsis* from photooxidative stress. Proc Natl Acad Sci U S A 100:4921–4926

Illing N, Denby KJ, Collett H, Shen A, Farrant JM (2005) The signature of seeds in resurrection plants: a molecular and physiological comparison of desiccation tolerance in seeds and vegetative tissues. Integr Comp Biol 45(5):771–787

Ingle RA, Schmidt UG, Farrant JM, Thomson JA, Mundree SG (2007) Proteomic analysis of leaf proteins during dehydration of the resurrection plant *Xerophyta viscosa*. Plant Cell Environ 30:435–446

Ingram J, Bartels D (1996) The molecular basis of dehydration tolerance in plants. Annu Rev Plant Biol 47(1):377–403

Ingram J, Chandler JW, Gallagher L, Salamini F, Bartels D (1997) Analysis of cDNA clones encoding sucrose-phosphate synthase in relation to sugar interconversions associated with dehydration in the resurrection plant *Craterostigma plantagineum* Hochst. Plant Physiol 115(1):113–121

Inźe D, Van Montagu M (1995) Oxidative stress in plants. Curr Opin Biotechnol 6:153–158

Iturriaga G, Cushman M, Cushman J (2006) An EST catalogue from the resurrection plant *Selaginella lepidophylla* reveals abiotic stress-adaptive genes. Plant Biol 170:1173–1184

Jackson RB, Sperry JS, Dawson TE (2000) Root water uptake and transport: using physiological processes in global predictions. Trends Plant Sci 5:482–488

Jiang GQ, Wang Z, Shang HH, Yang WL, ZA H, Phillips J, Deng X (2007) Proteome analysis of leaves from the resurrection plant Boea hygrometrica in response to dehydration and rehydration. Planta 225(6):1405–1420

Jones HG (1999) Use of thermography for quantitative studies of spatial and temporal variation of stomatal conductance over leaf surfaces. Plant Cell Environ 22:1043–1055

Jones L, McQueen-Mason S (2004) A role for expansins in dehydration and rehydration of the resurrection plant *Craterostigma plantagineum*. FEBS Lett 559:61–65

Jones HG, Serraj R, Loveys BR, Xiong L, Wheaton A, Price AH (2009) Thermal infrared imaging of crop canopies for the remote diagnosis and quantification of plant responses to water stress in the field. Funct Plant Biol 36:978–989

Kikawada T, Minakawa N, Watanabe M, Okuda T (2005) Factors inducing successful anhydrobiosis in the African chironomid Polypedilum vanderplanki: significance of the larval tubular nest. Integr Comp Biol 5:710–714

Kirch HH, Vera-Estrella R, Golldack D, Quigley F, Michalowski CB, Barkla BJ, Bohnert HJ (2000) Expression of water channel proteins in Mesembryanthemum crystallinum. Plant Physiol 123:111–124

Kirch HH, Nair A, Bartels D (2001) Novel ABA-and dehydration-inducible aldehyde dehydrogenase genes isolated from the resurrection plant Craterostigma plantagineum and Arabidopsis thaliana. Plant J 28(5):555–567

Kishor PK, Sangam S, Amrutha RN, Laxmi PS, Naidu KR, Rao KRSS, Sreenivasulu N (2005) Regulation of proline biosynthesis, degradation, uptake and transport in higher plants: its implications in plant growth and abiotic stress tolerance. Curr Sci 88:424–438

Kleines M, Elster RC, Rodrigo MJ, Blervacq AS, Salamini F, Bartels D (1999) Isolation and expression analysis of two stress-responsive sucrose-synthase genes from the resurrection plant Craterostigma plantagineum (Hochst.) Planta 209(1):13–24

Koh J, Chen G, Yoo MJ, Zhu N, Dufresne D, Erickson JE, Chen S (2015) Comparative proteomic analysis of *Brassica napus* in response to drought stress. J Proteome Res 14(8):3068–3081

Kolbe A, Tiessen A, Schluepmann H, Paul M, Ulrich S, Geigenberger P (2005) Trehalose 6-phosphate regulates starch synthesis via posttranslational redox activation of ADP-glucose pyrophosphorylase. Proc Natl Acad Sci U S A 102(31):11118–11123

Kozak M (1999) Initiation of translation in prokaryotes and eukaryotes. Gene 234:187–208

Kranner I, Grill D (1996) Significance of thiol disulphide exchange in resting stages of plant development. Bot Acta 109:8–14

Kranner I, Lutzoni F (1999) Evolutionary consequences of transition to a lichen symbiotic state and physiological adaptation to oxidative damage associated with poikilohydry. In: Plant responses to environmental stress

Kranner I, Beckett RP, Wornik S, Zorn M, Pfeifhofer HW (2002) Revival of a resurrection plant correlates with its antioxidant status. Plant J 31:13–24

Kranner I, Zorn M, Turk B, Wornik S, Beckett RP, Batič F (2003) Biochemical traits of lichens differing in relative desiccation tolerance. New Phytol 160(1):167–176

Kranner I, Birtić S, Anderson KM, Pritchard HW (2006) Glutathione halfcell reduction potential: a universal stress marker and modulator of programmed cell death? Free Radic Biol Med 40:2155–2165

Kŕol M, Ivanov MG, Jansson S, Kloppstech K, Huner NPA (1999) Greening under high light or cold temperature affects the level of xanthophyllcycle pigments, early light-inducible proteins, and light-harvesting polypeptides in wild-type barley and the *chlorina f2* mutant. Plant Physiol 120:193–203

Lange OL, Meyer A, Zellner H, Heber U (1994) Photosynthesis and water relations of lichen soil crusts: field measurements in the coastal fog zone of the Namib Desert. Funct Ecol 1:253–264

Lebkuecher LG, Eickmeier WG (1993) Physiological benefits of stem curling for resurrection plants in the field. Ecology 74:1073–1080

Leung J, Giraudat J (1998) Abscisic acid signal transduction. Annu Rev Plant Physiol Plant Mol Biol 49:199–221

Lewis ML, Miki K, Ueda T (2000) *FePer* 1, a gene encoding an evolutionarily conserved 1-Cys peroxiredoxin in buckwheat (*Fagopyrum esculentum* Moench), is expressed in a seed-specific manner and induced during seed germination. Gene 246:81–91

Loggini B, Navari-Izzo F, Izzo R (1997) Has glutathione a key role in the resistance to oxidative stress in durum wheat? Phyton 37:151–156

Lokhande VH, Suprasanna P (2012) Prospects of halophytes in understanding and managing abiotic stress tolerance. Environmental adaptations and stress tolerance of plants in the era of climate change. Springer, New York, pp 29–56

MacBeath G, Schreiber SL (2000) Printing proteins as microarrays for high-throughput function determination. Science 289:1760–1763

Mariaux JB, Bockel C, Salamini F, Bartels D (1998) Desiccation and abscisic acid-responsive genes encoding major intrinsic proteins (MIP) from the resurrection plant Craterostigma plantagineum. Plant Mol Biol 38:1089–1099

Markovska YK, Tsonev TD, Kimenov GP, Tutekova AA (1994) Physiological changes in higher poikilohydric plants Haberlea rhodopensis Friv. and Ramonda serbica Panè. During drought and rewatering at different light regimes. J Plant Physiol 144:100–108

Maroco JP, Pereira JS, Chaves MM (2000) Growth, photosynthesis and water use efficiency of two C4 Sahelian grasses subjected to water deficits. J Arid Environ 45:119–137

Martinez-Ballestra MDC, Martinez V, Carvajal M (2000) Regulation of water channel activity in whole roots and in protoplasts from roots of melon plants grown under saline conditions. Aust J Plant Physiol 27:685–691
Maurel C, Chrispeels MJ (2001) Aquaporins a molecular entry into plant water relations. Plant Physiol 125:135–138
Maurel C, Kado RT, Guern J, Chrispeels MJ (1995) Phosphorylation regulates the water channel activity of the seed-specific aquaporin alpha-TIP. EMBO J 14:3028–3035
McQueen-Mason SJ, Cosgrove DJ (1995) Expansin mode of action on cell walls: analysis of wall hydrolysis, stress relaxation, and binding. Plant Physiol 107:87–100
Merlot S, Mustilli AC, Genty B, North H, Lefebvre V, Sotta B, Vavasseur A, Giraudat J (2002) Use of infrared thermal imaging to isolate arabidopsis mutants defective in stomatal regulation. Plant J 30:601–609
Meyer S, Genty B (1999) Heterogenous inhibition of photosynthesis over the leaf surface of Rosa rubiginosa L. during water stress and abscisic acid treatment: induction of a metabolic component by limitation of CO2 diffusion. Planta 210:126–131
Michel D, Furini A, Salamini F, Bartels D (1994) Structure and regulation of an ABA-and desiccation-responsive gene from the resurrection plant Craterostigma plantagineum. Plant Mol Biol 24:549–560. https://doi.org/10.1007/BF00023553
Mitra J, Xu G, Wang B, Li M, Deng X (2013) Understanding desiccation tolerance using the resurrection plant Boea hygrometrica as a model system. Front Plant Sci 4(446):2013
Montané MH, Kloppstech K (2000) The family of light-harvesting-related proteins (LHCs, ELIPs, HLIPs): was the harvesting of light their primary function? Gene 258:1–8
Moore RB, Obornik M, Janouskovec J, ChrudimskyT VM, Green DH, Wright SW et al (2008) A photosynthetic alveolate closely related to apicomplexan parasites. Nature (Lond) 451:959–963
Moore JP, Le NT, Brandt WF, Driouich A, Farrant JM (2009) Towards a systems-based understanding of plant desiccation tolerance. Trends Plant Sci 14(2):110–117
Moran JF, Becana M, Iturbe-Ormaetxe I, Frechilla S, Klucas RV, Aparicio-Tejo P (1994) Drought induces oxidative stress in pea plants. Planta 194:346–352
Mowla SB, Thomson JA, Farrant JM, Mundree SG (2002) A novel stress-inducible antioxidant enzyme identified from the resurrection plant *Xerophyta viscosa* Baker. Planta 215:716–726
Mundree SG, Whittaker A, Thomson JA, Farrant JM (2000) An aldose reductase homolog from the resurrection plant *Xerophyta viscosa* Baker. Planta 211(5):693–700
Mundree SG, Baker B, Mowla S, Peters S, Marais S, Vander Willigen C, Govender K, Maredza A, Muyanga S, Farrant JM, Thomson JA (2002) Physiological and molecular insights into drought tolerance. Afr J Biotechnol 1(2):28–38
Murakeözy ÉP, Nagy Z, Duhazé C, Bouchereau A, Tuba Z (2003) Seasonal changes in the levels of compatible osmolytes in three halophytic species of inland saline vegetation in Hungary. J Plant Physiol 160(4):395–401
Muslin EH, Homann PH (1992) Light as a hazard for the desiccation-resistant "resurrection" fern *Polypodium polypodioides* L. Plant Cell Environ 15:81–89
Navari-Izzo F, Rascio N (1999) Plant response to water-deficit conditions. Handbook of plant and crop stress. In: Pessarakli M (ed) Handbook of plant and crop stress. Marcel Dekker, New York
Navari-Izzo F, Meneguzzo S, Loggini B, Vazzana C, Sgherri CLM (1997a) The role of the glutathione system during dehydration of Boea hygroscopica. Physiol Plant 99:23–30
Navari-Izzo F, Quartacci MF, CLM S (1997b) Desiccation tolerance in higher plants related to free radical defences. Phyton 37:203–214
Ndima T, Farrant JM, Thomson JA, Mundree SG (2001) Molecular characterisation of XVT8, a stress-responsive gene from the resurrection plant, *Xerophyta viscosa* Baker. Plant Growth Regul 35:137–145
Neale AD, Blomstedt CK, Bronson P, Le TN, Guthridge K, Evans J, Gaff DF, Hamill JD (2000) The isolation of genes from the resurrection grass Sporobolus stapfianus which are induced during severe drought stress. Plant Cell Environ 23:265–277
Nelson D, Salamini F, Bartels D (1994) Abscisic acid promotes novel DNA-binding activity to a desiccation-related promoter of Craterostigma plantagineum. Plant J 5(4):451–458

Nielsen TH, Rung JH, Villadsen D (2004) Fructose-2, 6-bisphosphate: a traffic signal in plant metabolism. Trends Plant Sci 9(11):556–563
Niyogi KK, Grossman AR, Björkman O (1998) Arabidopsis mutants define a central role for the xanthophyll cycle in the regulation of photosynthetic energy conversion. Plant Cell 10:1121–1134
Norwood M, Truesdale MR, Richter A, Scott P (2000) Photosynthetic carbohydrate metabolism in the resurrection plant *Craterostigma plantagineum*. J Exp Bot 51(343):159–165
Norwood M, ToldiO RA, Scott P (2003) Investigation into the ability of roots of the poikilohydric plant Craterostigma plantagineum to survive dehydration stress. J Exp Bot 54(391):2313–2321
Oliver MJ (1996) Desiccation tolerance in vegetative plant cells. Physiol Plant 97:779–787
Oliver MJ, O'mahony P, Wood AJ (1998) "To dryness and beyond"–preparation for the dried state and rehydration in vegetative desiccation-tolerant plants. Plant Growth Regul 24(3):193–201
Oliver MJ, Tuba Z, Mishler BD (2000) The evolution of vegetative desiccation tolerance in land plants. Plant Ecol 151:85–100
Oliver MJ, Dowd SE, Zaragoza J, Mauget SA, Payton PR (2004) The rehydration transcriptome of the desiccation-tolerant bryophyte Tortula ruralis: transcript classification and analysis. BMC Genomics 5:89. https://doi.org/10.1186/1471-2164-5-89
Oliver MJ, Velten J, Mishler BD (2005) Desiccation tolerance in bryophytes: a reflection of the primitive strategy for plant survival in dehydrating habitats? Integr Comp Biol 45(5):788–799
Oliver MJ, Jain R, Balbuena TS, Agrawal G, Gasulla F, Thelen JJ (2011a) Proteome analysis of leaves of the desiccation-tolerant grass, Sporobolus stapfianus, in response to dehydration. Phytochemistry 72(10):1273–1284
Oliver MJ, Guo L, Alexander DC, Ryals JA, Wone BW, Cushman JC (2011b) A sister group contrast using untargeted global metabolomic analysis delineates the biochemical regulation underlying desiccation tolerance in Sporobolus stapfianus. Plant Cell 23(4):1231–1248
Ong BL, Lim M, Wee YC (1992) Effects of desiccation and illumination on photosynthesis and pigmentation of an edaphic population of Trentepohlia odorata (Chlorophyta). J Phycol 28(6):768–772
Palmqvist K (2000) Carbon economy in lichens. New Phytol 148:11–36
Palmqvist K, Tansley (2000) Carbon economy in lichens. New Phytol 1:11–36
Pammenter NW, Berjak P (1999) A review of recalcitrant seed physiology in relation to desiccation-tolerance mechanisms. Seed Sci Res 9:13–37
Pandey V, Ranjan S, Deeba F, Pandey AK, Singh R, Shirke PA, Pathre UV (2010) Desiccation-induced physiological and biochemical changes in resurrection plant, Selaginella bryopteris. J Plant Physiol 16:1351–1359
Pelah D, Wang W, Altman A, Shoseyov O, Bartels D (1997) Differential accumulation of water stress-related proteins, sucrose synthase and soluble sugars in Populus species that differ in their water stress response. Physiol Plant 99(1):153–159
Pennycooke JC, Jones ML, Stushnoff C (2003) Down-regulating α-galactosidase enhances freezing tolerance in transgenic petunia. Plant Physiol 133(2):901–909
Peterbauer T, Mucha J, Mach L, Richter A (2002) Chain elongation of Raffinose in pea seeds-isolation, characterization, and molecular cloning of a multifunctional enzyme catalyzing the synthesis of stachyose and verbascose. J Biol Chem 277:194–200
Peters S, Mundree SG, Thomson JA, Farrant JM, Keller F (2007) Protection mechanisms in the resurrection plant *Xerophyta viscosa* (Baker). Both sucrose and raffinose family oligosaccharides (RFOs) accumulate in leaves in response to water deficit. J Exp Bot 58:1947–1956
Petit JR, Jouzel J, Raynaud D, Barkov NI, Barnola JM et al (1999) Climate and atmospheric history of the past 420,000 years from the Vostok ice core, Antarctica. Nature 399:429–436
Phillips JR, Hilbricht T, Salamini F, Bartels D (2002) A novel abscisic acid and dehydration-responsive gene family from the resurrection plant *Craterostigma Plantagineum* encodes a plastid- targeted protein with DNA-binding acivity. Planta 215:258–266
Piatkowski D, Schneider K, Salamini F, Bartels D (1990) Characterization of five abscisic acid-responsive cDNA clones isolated from the desiccation-tolerant plant Craterostigma plantagineum and their relationship to other water-stress genes. Plant Physiol 94(4):1682–1688

Pien S, Wyrzykowska J, McQueen-Mason S, Smart C, Fleming A (2001) Local expression of expansin induces the entire process of leaf development and modifies leaf shape. Proc Natl Acad Sci U S A 98:11812–11817

Porembski S (2000) Granitic and gneissic outcrops (inselbergs) as center of diversity for desiccation tolerant vascular plants. Pl Ecol 151:19–28

Porembski S, Barthlott W (2000) Granitic and gneissic outcrops (inselbergs) as centers of diversity for desiccation-tolerant vascular plants. Plant Ecol 151:19–28

Potts M (1994) Desiccation tolerance of prokaryotes. Microbiol Storage Soil Biol Biochem 23:313–322

Proctor MCF, Pence VC (2002) Vegetative tissues: bryophytes, vascular resurrection plants, and vegetative propogules. In: Black M, Pritchard HW (eds) Desiccation and survival in plants: drying without dying. CABI Publishing, Wallingford

Proctor MC, Tuba Z (2002) Poikilohydry and homoihydry: antithesis or spectrum of possibilities? New Phytol 3:327–349

Provart NJ, Gil P, Chen W, Han B, Chang HS, Wang X, Zhu T (2003) Gene expression phenotypes of *Arabidopsis* associated with sensitivity to low temperatures. Plant Physiol 132:893–906

Puhakainen T, Hess MW, Makela P, Svensson J, Heino P, Palva ET (2004) Overexpression of multiple dehydrin genes enhances tolerance to freezing stress in Arabidopsis. Plant Mol Biol 54:743–753. https://doi.org/10.1023/B:PLAN.0000040903.66496.a4

Rakić T, Gajić G, Lazarević M, Stevanović B (2015) Effects of different light intensities, CO2 concentrations, temperatures and drought stress on photosynthetic activity in two paleoendemic resurrection plant species *Ramonda serbica* and *R. nathaliae*. Environ Exp Bot 109:63–72

Ramanjulu S, Bartels D (2002) Drought- and desiccation induced modulation of gene expression in plants. Plant Cell Environ 25:141–151

Rascio N, La Rocca N (2005) Resurrection plants: the puzzle of surviving extreme vegetative desiccation. Crit Rev Plant Sci 24(3):209–225. https://doi.org/10.1080/07352680591008583

Rathinasabapathi B (2000) Metabolic engineering for stress tolerance. Installing osmoprotectant synthesis pathways. Ann Bot 86(4):709–716

Rhodes D, Samaras Y (1994) Genetic control of osmoregulation in plants. In: Strange K (ed) Cellular and molecular physiology of cell volume regulation. CRC Press, Boca Raton

Rhodes NM, Truesdale MR, Richter A, Scott P (1999) Metabolic changes in leaves and roots during dehydration of the resurrection plant *Craterostigma plantagineum*. (Hochst). S Afr J Bot 65:421–442

Rhodes D, Nadolska-Orczyk A, Rich P J (2002) Salinity, osmolytes and compatible solutes. In: Salinity. Environment-plants-molecules. Springer, Dordrecht, pp 181–204

Ricci C (1998) Anhydrobiotic capabilities of bdelloid rotifers. Hydrobiologia 87:321–326

Ricci C, Caprioli M (2005) Anhydrobiosis in bdelloid species, populations and individuals. Integr Comp Biol 45(5):759–763

Rodriguez MCS, Edsgärd D, Hussain SS, Alquezar D, Rasmussen M, Gilbert T et al (2010) Transcriptomes of the desiccation-tolerant resurrection plant Craterostigma plantagineum. Plant J63:212–228. https://doi.org/10.1111/j.1365-313X.2010.04243.x

Ruban AV, Horton P (1999) The xanthophyll cycle modulates the kinetics of nonphotochemical energy dissipation in isolated lightharvesting complexes, intact chloroplasts and leaves. Plant Physiol 119:531–542

Rundel PW (1978) Ecological relationships of desert fog zone lichens. Bryologist 81:277–293

Russouw PS, Farrant J, Brandt W, Maeder D, Lindsey GG (1995) Isolation and characterization of a heat-stable protein from pea (*Pisum sativum*) embryos. Seed Sci Res 5:137–144

Sales K, Brandt W, Rumbak E, Lindsey G (2000) The LEA-like protein HSP 12 in Saccharomyces cerevisiae has a plasma membrane location and protects membranes against desiccation and ethanol-induced stress. Biochim Biophys Acta (BBA) Biomembranes 2:267–278

Salleo S, Nardini A, Pitt F, Lo Gullo MA (2000) Xylem cavitation and hydraulic control of stomatal conductance in Laurel (Laurus nobilis L.). plant. Cell Environ 23:71–79

Savenstrand H, Olofsson M, Samuelsson M, Strid A (2004) Induction of early-inducible protein gene expression in *Pisum sativum* after exposure to low levels of UV-B radiation and other environmental stresses. Plant Cell Rep 22:532–536

Schiller P, Heilmeier H, Hartung W (1997) Abscisic acid (ABA) relations in the aquatic resurrection plant Chamaegigas intrepidus under naturally fluctuating environmental conditions. New Phytol 136(4):603–611

Schiraldi C, Di Lernia I, De Rosa M (2002) Trehalose production. Exploiting novel approaches. Trends Biotechnol 20(10):420–425

Schneider K, Wells B, Schmelzer E, Salamini F, Bartels D (1993) Desiccation leads to the rapid accumulation of both cytosolic and chloroplastic proteins in the resurrection plant Craterostigma plantagineum. Hochst Planta 189:120–131

Schreiber U, Schliwa U, Bilger W (1986) Continuous recording of photochemical and non-photochemical chlorophyll fluorescence quenching with a new type of modulation fluorometer. Photosynth Res 10:51–62

Schulze E-D (1986) Whole-plant responses to drought. Aust J Plant Physiol 13:127–141

Scott P (2000) Resurrection plants and the secrets of eternal leaf. Ann Bot 85:159–166

Sengupta D, Reddy AR (2011) Water deficit as a regulatory switch for legume root responses. Plant Signal Behav 6(6):914–917

Sgherri CLM, Loggini B, Puliga S, Navari-Izzo F (1994) Antioxidant system in Sporobolus stapfianus: changes in response to desiccation and rehydration. Phytochemistry 35:561–565

Shen-Miller J, Schopf JW, Harbottle G, Cao RJ, Ouyang S, Zhou KS, Southon JR, Liu GH (2002) Long-living lotus: germination and soil γ-irradiation of centuries-old fruits, and cultivation, growth, and phenotypic abnormalities of offspring. Am J Bot 2:236–247

Sherwin HW, Farrant JM (1996) Differences in rehydration of three desiccation-tolerant angiosperm species. Ann Bot 78:703–710. https://doi.org/10.1006/anbo.1996.0180

Sherwin HW, Farrant JM (1998) Protection mechanism against excess light in the resurrection plants Craterostigma wilmsii and Xerophyta viscosa. Plant Growth Regul 24(203):210

Sherwin HW, Farrant JM (2008) Protection mechanisms against excess light in the resurrection plants Craterostigma wilmsii and Xerophyta viscosa. Plant Growth Regul 24(3):203–210

Sherwin HW, Pammenter NW, February ED, Vander Willigen C, Farrant JM (1998) Xylem hydraulic characteristics, water relations and wood anatomy of the resurrection plant Myrothamnus fabellifolia Welw. Ann Bot 81:567–575

Shirkey B, McMaster NJ, Smith SC, Wright DJ, Rodriguez H, Jaruga P, Birincioglu M, Helm RF, Potts M (2003) Genomic DNA of Nostoc commune (Cyanobacteria) becomes covalently modified during long-term (decades) desiccation but is protected from oxidative damage and degradation. Nucleic Acids Res 31:2995–3005

Slama I, Abdelly C, Bouchereau A, Flowers T, Savouré A (2015) Diversity, distribution and roles of osmoprotective compounds accumulated in halophytes under abiotic stress. Ann Bot 115(3):433–447

Smirnoff N (1992) The carbohydrates of bryophytes in relation to desiccation tolerance. J Bryol 17(2):185–191

Smirnoff N (1993) The role of active oxygen in the response of plants to water deficit and desiccation. New Phytol 125:214–237

Smith-Espinoza C, Bartels D, Phillips J (2007) Analysis of a LEA gene promoter via *Agrobacterium-mediated* transformation of the desiccation tolerant plant Lindernia brevidens. Plant Cell Rep 26:1681–1688

Sperry JS, Tyree MT (1988) Mechanism of water stress-induced xylem embolism. Plant Physiol 88(3):581–587

Stiller I, Dulai S, Kondrák M, Tarnai R, Szabó L, Toldi O, Bánfalvi Z (2008) Effects of drought on water content and photosynthetic parameters in potato plants expressing the trehalose-6-phosphate synthase gene of Saccharomyces cerevisiae. Planta 227(2):299

Stoll M, Loveys B, Dry P (2000) Hormonal changes induced by partial rootzone drying of irrigated grapevine. J Exp Bot 51:1627–1634

Suga S, Imagawa S, Maeshima M (2001) Specificity of the accumulation of mRNAs and protein of the plasma membrane and tonoplast aquaporins in radish organs. Planta 212:294–304

Suguiyama VF, Silva EA, Meirelles ST, Centeno DC, Braga MR (2014) Leaf metabolite profile of the Brazilian resurrection plant Barbacenia purpurea Hook. (Velloziaceae) shows two time-dependent responses during desiccation and recovering. Front Plant Sci 5. 10.3389/fpls.2014.00096 FPLS

Szabados L, Savouré A (2010) Proline: a multifunctional amino acid. Trends Plant Sci 15(2):89–97

Taji T, Ohsumi C, Iuchi S, Seki M, Kasuga M, Kobayashi M, Yamaguchi-Shinozaki K, Shinozaki K (2002) Important roles of drought-and cold-inducible genes for galactinol synthase in stress tolerance in Arabidopsis thaliana. Plant J 29(4):417–426

Thomson JW, Iltis HH (1968) A fog-induced lichen community in the coastal desert of southern Peru. Bryologist 1:31–34

Toldi O, Tóth S, Ponyi T, Scott P (2002) An effective and reproducible transformation protocol for the model resurrection plant *Craterostigma plantagineum* Hochst. Plant Cell Rep 21:63–69

Toldi O, Djilianov D, Scott P (2006) Transgenic approach for functional analysis of genes associated with desiccation tolerance in resurrection plants. In: Teixeira da Silva JA (ed) Floriculture ornamental and plant biotechnology, vol 3. Global Science Book, Falmer, pp 209–217

Toldi O, Tuba Z, Scott P (2009) Vegetative desiccation tolerance: is it a goldmine for bioengineering crops? Plant Sci 176:187–199. https://doi.org/10.1016/j.plantsci.2008.10.002

Toth S, Scott P, Sorvari S, Toldi O (2004) Effective and reproducible protocols for in vitro culturing and plant regeneration of the physiological model plant Ramonda myconi (L.) Rchb. Plant Sci 166:1027–1034

Tuba Z (2008) Notes on the poikilochlorophyllous desiccation-tolerant plants. Acta Biol Szeged 52(1):111–113

Tuba Z, Lichtenthaler HK, Maroti I, Csintalan ZS (1993a) Resynthesis of thylakoids and functional chloroplasts in the desiccated leaves of the poikilochlorophyllous plant Xerophyta scabrida upon rehydration. J Plant Physiol 142:742–748

Tuba Z, Lichtenthaler HK, Csintalan ZS, Pócs T (1993b) Regreening of desiccated leaves of the poikilochlorophyllous *Xerophyta scabrida* upon rehydration. J Plant Physiol 142:103–108

Tuba Z, Lichtenthaler HK, Zs C, Nagy Z, Szente K (1994) Reconstitution of chlorophylls and photosynthetic CO2 assimilation upon rehydration in the desiccated poikilochlorophyllous plant Xerophyta scabrida. Planta 192:414–420

Tuba Z, Lichtenthaler HK, Csintalan Z, Nagy Z, Szente K (1996) Loss of chlorophylls, cessation of photosynthetic CO2 assimilation and respiration in the poikilochlorophyllous plant Xerophyta scabrida. Physiol Plant 96:383–388

Tuba Z, Proctor MCF, Csintalan Z (1998) Ecophysiological responses of homoichlorophyllous and poikilochlorophyllous desiccation-tolerant plant *Xerophyta scabrica* at present day CO_2 concentration. Plant Physiol Biochem 35:381–386

Tweddle JC, Dickie JB, Baskin CC, Baskin JM (2003) Ecological aspects of seed desiccation sensitivity. J Ecol 91(2):294–304

Tyerman SD, Niemietz CM, Bramley H (2002) Plant aquaporins: multifunctional water and solute channels with expanding roles. Plant Cell Environ 25:173–194

Umezawa T, Fujita M, Fujita Y, Yamaguchi-Shinozaki K, Shinozaki K (2006) Engineering drought tolerance in plants discovering and tailoring genes to unlock the future. Curr Opin Biotechnol 17:113–122

van den Dries N, Facchinelli F, Giarola V, Phillips JR, Bartels D (2011) Comparative analysis of LEA-like 11-24 gene expression and regulation in related plant species within the Linderniaceae that differ in desiccation tolerance. New Phytol 190:75–88. https://doi.org/10.1111/j.1469-8137.2010.03595.x

Vander Willigen C, Pammenter NW, Mundree S, Farrant J (2001) Some physiological comparisons between the resurrection grass, Eragrostis nindensis, and the related desiccation-sensitive species, E curvula. Plant Growth Regul 35:121–129

VanderWilligen C, Pammenter NW, Mundree SG, Farrant JM (2004) Mechanical stabilization of desiccated vegetative tissues of the resurrection grass Eragrostis nindensis: does a TIP 3;

1 and/or compartmentalization of subcellular components and metabolites play a role? J Exp Bot 397:651–661

Velasco R, Salamini F, Bartels D (1994) Dehydration and ABA increase mRNA levels and enzyme activity of cytosolic GAPDH in the resurrection plant Craterostigma plantagineum. Plant Mol Biol 26(1):541–546

Vicre M, Farrant JM, Gibouin D, Driouich A (2003) Resurrection plants: how to cope with desiccation? Recent Res Dev Plant Biol 3:69–93

Vicre M, Farrant JM, Driouich A (2004) Insights into the mechanisms of desiccation tolerance among resurrection plants. Plant Cell Environ 27:1329–1340

Walters C, Farrant JM, Pammenter NW, Berjak P (2002) Desiccation and damage. Black M, Pritchard HW Desiccation and survival in plants – drying without drying, CABI Publishing Wallingford

Wang W, Tanurdzic M, Luo M et al (2005) Construction of a bacterial artificial chromosome library from the spikemoss *Selaginella moellendorffii*: a new resource for plant comparative genomics. BMC Plant Biol 5:10

Wang X, Chen S, Zhang H, Shi L, Cao F, Guo L et al (2010) Desiccation tolerance mechanism in resurrection fern-ally *Selaginella tamariscina*revealed by physiological and proteomic analysis. J Proteome Res 9:6561–6577

Washburn MP, Yates JR 3rd (2000) Analysis of microbial proteome. Curr Opin Microbiol 3:292–297

Wharton DA (2003) The environmental physiology of Antarctic terrestrial nematodes. J Comp Physiol B 173(8):621–628

Wilkinson S, Davies WJ (2002) ABA-based chemical signalling: the co-ordination of responses to stress in plants. Plant Cell Environ 25(2):195–210

Willige BC, Kutzer M, Tebartz F, Bartels D (2009) Subcellular localization and enzymatic properties of differentially expressed Transketolase genes isolated from the desiccation tolerant resurrection plant *Craterostigma plantagineum.* Planta 229:659–666

Wise MJ, Tunnacliffe A (2004) POPP the question: what do LEA proteins do? Trends Plant Sci 9:13–17. http://dx.doi.org/10.1016/j.tplants.2003.10.012

Wright JC (2001) Cryptobiosis 300 years on from van Leuwenhoek: what have we learned about tardigrades? Zool Anz A J Comp Zool 240(3):563–582

Wright JC, Westh P, Ramløv H (1992) Cryptobiosis in tardigrada. Biol Rev 67(1):1–29

Yamada S, Komori T, Myers PN, Kuwata S, Kubo T, Imaseki H (1997) Expression of plasma membrane water channel genes underwater stress in Nicotiana excelsior. Plant Cell Physiol 38:1226–1231

Yobi A, Wone BWM, Xu W, Alexander DC, Guo L, Ryals JA et al (2012) Comparative metabolic profiling between desiccation-sensitive and desiccation tolerant species of *Selaginella* reveals insights into the resurrection trait. Plant J 72:983–999. https://doi.org/10.1111/tpj.12008

Yobi A, Wone BWM, Xu W, Alexander DC, Guo L, Ryals JA et al (2013) Metabolic profiling in *Selaginella lepidophylla* at various hydration states provides new insights into the mechanistic basis of desiccation tolerance. Mol Plant 6:369–385. https://doi.org/10.1093/mp/sss155

Yu J1, Hu S, Wang J, Wong GK, Li S, Liu B, Deng Y, Dai L, Zhou Y, Zhang X, Cao M, Liu J, Sun J, Tang J, Chen Y, Huang X, Lin W, Ye C, Tong W, Cong L, Geng J, Han Y, Li L, Li W, Hu G, Huang X, Li W, Li J et al (2002) A draft sequence of the rice genome (Oryza sativa L. ssp). Science 296(5565):79–92. indica

Zeng Q, Chen X, Wood AJ (2002) Two early light-inducible protein (ELIP) cDNAs from the resurrection plant Tortula ruralis are differentially expressed in response to desiccation, rehydration, salinity, and high light. J Exp Bot 53:1197–1205

Zhao F, Zhang D, Zhao Y, Wang W, Yang H, Tai F, Hu X (2016) The difference of physiological and proteomic changes in maize leaves adaptation to drought, heat, and combined both stresses. Front Plant Sci.:7 58(8):1947–1956

Zhou H, Watts JD, Aebersold R (2001) A systematic approach to the analysis of protein phosphorylation. Nat Biotechnol 19:375–378

Dehydration and Freezing Resistance of Lichenized Fungi

3

H. Harańczyk, A. Casanova-Katny, M. Olech, and K. Strzałka

Abstract

The poikilohydrous nature of lichens provides them the ability to resist low temperature, deep dehydration and deficit in light irradiance. The process of water uptake can be correlated with the resistance for dehydration below water percolation threshold (fractal exponent characteristic for approximately two-dimensional lattice). Gaseous phase hydration kinetics presents tightly bound water and mobile loosely bound water fractions, differentiated in hydration/dehydration rate and in proximity to thallus surfaces.

Some lichens have the ability to hydrate from snow to the level sufficient for activation of photosynthesis. Freezing tolerance of lichens combines resistance to low temperature and duration of freezing. Antarctic lichens reveal the lowest temperature of net photosynthesis. In number of species, water supercools;

H. Harańczyk (✉)
Institute of Physics, Jagiellonian University, Cracow, Poland
e-mail: hubert.haranczyk@uj.edu.pl

A. Casanova-Katny
Veterinary school, Faculty of Natural Resources, Catholic University of Temuco, Temuco, Chile
e-mail: angecasanova@gmail.com

M. Olech
Institute of Botany, Jagiellonian University, Cracow, Poland

Institute of Biochemistry and Biophysics, Polish Academy of Sciences, Warsaw, Poland
e-mail: maria.olech@uj.edu.pl

K. Strzałka
Malopolska Centre of Biotechnology, Jagiellonian University, Cracow, Poland

Faculty of Biochemistry, Biophysics and Biotechnology, Jagiellonian University, Cracow, Poland
e-mail: kazimierzstrzalka@gmail.com

V. Shukla et al. (eds.), *Plant Adaptation Strategies in Changing Environment*,
https://doi.org/10.1007/978-981-10-6744-0_3

however, some lichen taxa tolerate ice nucleation activity (INA) for temperatures at which the active photosynthesis process still occurs.

A number of factors contribute the lichen freezing resistance. The decrease in hydration rate of freezing loosely bound water pool; active transfer of freezing, loosely bound water pool to non-freezing one; presence of tightly bound non-freezing water fraction; and non-cooperative immobilization of supercooled bound water are important factors to the lichen freezing resistance together in the, at higher hydration levels INA for supercooled bound water, diffusion-induced migration of supercooled water molecules to ice microcrystallites (which is not phase growth in thermodynamic system but is implied by the structure).

Freezing resistance and dehydration resistance direct lichens for testing Panspermia theory. Several endolithic lichens may survive the space conditions; however, their survival during re-entry from the space was not yet demonstrated.

Keywords
Lichenized fungi · Extremotolerant organisms · Lithopanspermia · Bound water · NMR · DSC · Fluorescence spectroscopy

3.1 Introduction

Lichens or lichenized fungi are symbiotic organisms constituing of a fungus being aheterotropic partner, belonging to *Ascomycota* or Basidiomycota and are often called *mycobiont*, and the autotrophic partner, either a microalga, a cyanobacterium or in some cases both, is called *mycobiont* (Honegger 1991). Different photobionts occupying thallus formed of mycobiont may show different responses to the same factor. According to Schroeter (1994), photobionts belonging to *Cyanobacteria* and green alga present in the thallus of *Placopsis contortuplicata* show different responses in PSII photochemical efficiency as an effect of hydration. It is generally accepted that the names given to lichens refer to *mycobiont* species, while photobionts are described by their own scientific names. The fungal partner forms *thallus* which is a structure giving shape to the lichen organism and supplies water and minerals to the photobiont. The photobionts produce carbohydrates through the process of photosynthesis and share it with the fungal cells.

Lichenization is a common phenomenon among the fungi; as out of the existing fungal species, about 20% are lichenized (Hawksworth 1988). Phycobiont being an important source of carbohydrate production through photosynthesis benefits the fungus by providing essential nutrient. The factors such as water relation in lichen thallus, response to strong light and exchange of carbon dioxide clearly indicate that the benefits among the partners are not equal and far from mutualism but rather parasitic in nature. Mycobiont has an influence on the photobionts and may reduce the number of their cells resulting in lower biomass production. In addition, the mycobiont suppresses the sexuality of photobionts which has significant impact on photobiont evolution (Kappen 1994). However, symbiotic interactions between mycobiont and photobiont lichens enable an organism to be resistant to harsh environmental conditions, which may colonize areas exposed to very low temperatures (polar regions, high mountains), be resistant to acute water deficiency, and survive-long periods in low light irradiance (growing in eroded rocks) or darkness (polar nights). Even lichen can survive if frequently these stressors act simultaneously.

Lichens are poikilohydric in nature and are unable to regulate the water loss and prevent desiccation different from the high plants which have roots, cuticle and epidermis. Lichens uptake water from the environment through their whole surface, in balance with the air humidity. All metabolic processes occurring in lichen cells, including photosynthesis, respiration and CO_2 exchange, depend totally on the moisture present in the environment. The primitive water uptake system in lichens, together with the ability to adapt acute water stress, makes them extremely efficient and resistant to unfavourable conditions and enables to colonize areas where other higher groups of plants could not grow and survive (Kappen 1988). Elevated level of carotenoids, lutein and neoxanthin, found in several Antarctic lichens helps them to protect from antioxidative stress, frequently accompanying other types of stresses (Strzalka et al. 2011). The unique symbiosis of mycobiont and photobionts makes lichens advantageous over higher plants in harsh climatic conditions, but it results in slow and retarded growth, as species of *Usnea* and *Umbilicaria* grow only up to 20 cm in Antarctica and Antarctic maritime, respectively (Kappen et al. 1996b). However, in spite of limitations of retarded and slow growth rate, it is evident that lichens achieved evolutionary success populating very hostile areas of the planet.

3.2 Thallus Porosity, Water Percolation and Water-Accessible Surface

During deep dehydration process, the lichen thallus does not collapse, maintaining its native shape and most of its volume, but the mycobiont morphology changes. Dry lichen thallus resembles porous system. The porosity of lichen thalli of *Umbilicariacae* family, as estimated using mercury intrusion porosimetry, decreased from 30% for *Lasallia hispanica* down to 10% for *Umbilicaria cinereorufescens*, with the increase of thallus volume occupied by cell walls and gelatinous substances (for 12 *Umbilicaria* species $r^2 = 0.68$) (Valladares et al. 1998).

To detect whether a thallus may dehydrate even below a water percolation threshold (Stauffer and Aharony 1991), the static electric conductivity, evaluated from imaginary part (dielectric loss factor) of complex dielectric spectrum, *Cladonia mitis* and *Himantormia lugubris* thallus cells, reacting positively 50–60% and 40–50%, respectively, in tests using the fluorescein bi-acetate and ethidium bromide were used. The detected conductivity, σ, was a superposition of the dry thallus conductivity, σ_0, and proton (or residual ion) conductance contained in water shell covering thallus surfaces, $\sigma - \sigma_0$. With the thallus dehydration, the thallus conductance near percolation threshold decreases with a power function $(\sigma - \sigma_0)/\sigma_0 = (\Delta m - \Delta M/m_0)^\alpha$, where α is a fractal exponent, characteristic for the dimensionality of percolating lattice (Harańczyk et al. unpublished data; Harańczyk 2003), and m_0 dry mass

Table 3.1 Percolation thresholds and fractal exponents for some lichen thalli (Harańczyk 2003)

	Site	$\Delta M/m_0$	α	p/p_0
Himantormia lugubris	King George Island	0.099	1.46	0.35
Cladonia mitis	Northern Sweden	0.0926	1.18	0.34

of the sample after heating at 70 °C for 24 h (Gaff 1977), which was a very mild dehydration protocol, as TG analyses show a water loss extending up to 400 K (Jänchen et al. 2015).

Fractal exponent equal $\alpha = 1.18$ is characteristic for percolation of two-dimensional lattice, while exceeding two dimensional value (1.46) signalizes the spatial curvature of inner thallus surfaces (Table 3.1). The hydration level for percolation threshold, $\Delta M/m_0$, often happens at storage of lichen in herbarium. From the state of anabiosis during storage in herbarium, lichens easily return to active life. The local tetragonal arrangement of bonds is roughly saved as in liquid water many of hydrogen bonds may be distorted. Thus, the majority of water molecules in monomolecular layer of water present on dry biological surface are bound to the surface by one hydrogen bond, and so there is not many water molecules bonded with more than four hydrogen bonds to their neighbours. There are not so many two-dimensional lattices with the number of bonds matching to this range, namely, Bethe lattices and honeycomb lattice, which may approximate the lattice of water molecules on the surface of ice. For liquid water, the average number of hydrogen bonds per molecule equals 2.3 (molecular dynamics simulations at $t = +10$ °C) (Stillinger 1980), for others, 3.1. All of them (including Bethe lattices) have close values of site percolation threshold. The percolation density (dense packing spheres model) does not apply for water molecules on the surface (Scher and Zallen 1970), so the approximated percolation threshold, p_c, equals 0.6185 for *H. lugubris* thallus ($n_{cal} = 2.6$) and 0.6215 (for $n_{cal} = 2.6$) or 0.6859 (2.5) for *C. mitis* thallus (Harańczyk 2003).

The expressed in units of dry mass, mass of water monolayer bound on the thallus surfaces, $\Delta m/m_0$, for H. lugubris equals 0.160 ± 0.010, for *C. mitis* 0.149 ± 0.037 or 0.135 ± 0.017 for different samples. The estimated water-accessible surface area of lichen thallus per dry mass unit for *H. lugubris* thallus equals $S_{Lh} = 447.2$ m^2 g^{-1} and for *C. mitis* thallus between 416.4 m^2 g^{-1} and 377.3 m^2 g^{-1} (Harańczyk unpublished data; Harańczyk 2003).

The relative water-accessible surfaces of thalli are close by high (450 m^2 g^{-1}); it is close (taking into account the difference in density) to the total water-accessible surface of commercially available porous model system, controlled pore glass CPG74 (with pore diameter of 74 Å and porosity ca. 54%) (Harańczyk et al. 1991).

3.3 Lichen Resistance on Dehydration

Lichen thalli have the ability to hydrate both gaseous and aqueous phase. In both cases, the degree of hydration reaches relatively high level. It has been shown that *Usnea sphacelata* (= *U. sulphurea*) hydrates to a maximum level of $\Delta m/m_0 = 1.60$

(Kappen 1983); rainwater-soaked *Usnea aurantiaco-atra* (=*U. fasciata*) takes water up to $\Delta m/m_0 = 1.64$, whereas in *Ramalina terebrata*, the $\Delta m/m_0$ reaches the value of 2.63 (Kappen 1985). For lichens populating rocky sites in Southern Norway (Hvaler island), in situ hydration level of *Lasallia pustulata* ranged between $\Delta m/m_0 = 0.5$ and 4.0, whereas for *Umbilicaria spadochroa*, the respective value varied between 1.0 and 5.0 (Kappen et al. 1996a).

Single species of lichen exhibit varied maximal hydration level at different latitudes. The circumpolar Arctic lichen, *Flavocetraria nivalis* (=*Cetraria nivalis*), showed maximal hydration level of thallus as well as its drying rate which changed with increasing latitude of their location. The hydration level increases from ≈2.9 to ≈4.0, and the drying rate decreases significantly (Schipperges et al. 1995). *Turgidosculum complicatulum* (=*Mastodia tessellata*) growing in the supralittoral maritime Antarctic on the exposed rock surfaces takes water to significantly higher level when sprayed with seawater comparing to freshwater (Huiskes et al. 1996).

Hydration level has a pronounced effect on photosynthesis, as net photosynthesis almost linearly depends on $\Delta m/m_0$ at low hydration levels. On the other hand, this process shows a depression in highly hydrated thalli.

The rate of CO_2 fixation depends on water potential φ, as *Ramalina menziesii* and *Dendrographa minor* showed a gradual decline in CO_2 fixation rates parallel to decreasing water potential. The dehydration to the gaseous phase under controlled relative humidity, p/p_0, or under osmotic stress generated by incubation of the thalli in sea salt or sorbitol solutions showed the CO_2 fixation between ranges of 77–85% (−35 to −20 Mpa). In *Pseudocyphellaria anthrapsis*, a species with a cyanobacterial photobiont, the thallus hydration with liquid water is inevitable for net photosynthesis. Water potential equal to −3.5 Mpa generated by sorbitol or salt solutions was necessary for net photosynthesis; on the other hand, no CO_2 uptake was detected in moist air (Nash et al. 1990). Lichens containing photobionts of cyanobacterial and green algal nature differ in their hydration characteristics as the former required liquid water for activation of photosynthetic processes (Lange et al. 1986, 1988).

The lichen thalli achieved maximum net photosynthesis depending on the CO_2 diffusion within the thallus, and the rate of this process depends, in turn, on hydration level. Generally, a high hydration level causes increased resistance for CO_2 diffusion in thallus (Green et al. 1994). As shown in field measurements performed for *Ramalina terebrata*, thallus hydrated to $\Delta m/m_0 = 1.12$ exhibits net photosynthesis lower than the thallus hydrated to the value of 0.92, optimal hydration level being 0.87 (Kappen et al. 1986). The maximum of CO_2 exchange and maximum net photosynthesis differ among lichen species. For *Usnea sphacelata* (= *U. sulphurea*) and for *Usnea aurantiaco-atra* (= *U. fasciata*), the maximum of CO_2 exchange occurs at $\Delta m/m_0 = 0.70$ (Kappen 1985), whereas optimal net photosynthesis in *Umbilicaria aprina* occurs at $\Delta m/m_0 = 1.2$, for *Umbilicaria decussata* at $\Delta m/m_0 = 1.0$, and at $\Delta m/m_0 = 0.85$ for *Usnea antarctica* and *Usnea sphacelata* (Kappen and Breuer 1991). From the measurements performed in laboratory conditions at the temperature of +5 °C, the maximum CO_2 exchange for *Lasallia pustulata* was recorded at $\Delta m/m_0 = 1.50$, whereas for *Umbilicaria spadochroa* at 0.90

(Kappen et al. 1996a). Green alga *Prasiola crispa ssp. antarctica* a photobiont of *Turgidosculum complicatulum* (=*Mastodia tessellata*) exhibits the optimal net photosynthesis at the hydration level of $\Delta m/m_0 = 5.20$, as compared with $\Delta m/m_0 = 3.00$ for the optimal photosynthesis in the thallus of this species (Huiskes et al. 1996). *Ramalina bourgeana* and *Teloschistes lacunosus*, two fruticose lichen in semiarid regions, show a maximal photosynthesis at $\Delta m/m_0$ = *1.15 and 1.05, respectively*. It has been also demonstrated that at the higher hydration levels, net photosynthesis drops rapidly to the value of 20% of the maximal level at $\Delta m/m_0 \approx 2.3$ for *T. lacunosus* and at $\Delta m/m_0 \approx 2.6$ for *R. bourgeana* (Del-Prado and Sancho 2000).

In natural conditions, mild desiccation to gaseous phase may result in total removal of loosely bound water fraction from the thallus and resulted into significant reduction of the tightly bound water fraction. The metabolic processes of lichen inhibited at very low hydration level, and their recovery occurs at the certain level of rehydration which differs among lichen species. The recovery of metabolic processes upon rehydration in *Teloschistes lacunosus* occurs at $\Delta m/m_0 \approx 0.2$; for *Ramalina bourgeana* (Del-Prado and Sancho 2000), this happens at $\Delta m/m_0 \approx 0.25$ and for *Usnea aurantiaco-atra* (= *Usnea fasciata*) (Kappen 1985) at $\Delta m/m_0 = 0.2$–0.3. The hydration value $\Delta m/m_0 = 0.3$ was found for *Flavocetraria nivalis* (=*Cetraria nivalis*) (Kappen et al. 1995), 0.3–0.4 for *Ramalina terebrata* (Kappen 1985) and 0.4 for *Usnea sulphurea* (Kappen 1983, 1985). The photosynthesis or dark respiration processes become activated at relative humidity of $p/p_0 = 70\%$ in *Dendrographa minor*, $p/p_0 = 80\%$ in *Ramalina menziesii* and $p/p_0 = 85\%$ in *Pseudocyphellaria antraspis* (Nash et al. 1990). According to Bertsch (1966a, b), Lange (1969) and Nash et al. (1990), aerophilic green algae and lichens start photosynthetic activity at the relative humidity of $p/p_0 = 77$–85%.

The recovery of life activity after rehydration is relatively a fast process as it has been reported that air-dry *Usnea sphacelata* (= *U. sulphurea*) after 60 h of exposition to high air humidity ($p/p_0 > 96\%$) restores net photosynthesis up to 50% and dark respiration up to 60%. There is a dependence between light intensity and the level of rehydration. At 6 °C *U. sphacelata* (= *U. sulphurea*) hydrates to $\Delta m/m_0 = 0.70$, in darkness, but when exposed to light (360 $\mu Em^{-2}\ s^{-1}$), the $\Delta m/m_0 = 0.55$ (Kappen 1983).

It is generally accepted that sugars and sugar alcohols play an adaptive role in certain type of stresses. It has been reported by Hamada et al. (1994) that mycobionts of the lichen species, *Evernia esorediosa*, *Ramalina subbreviuscula* and *Ramalina sublitoralis*, growing in dry habitat accumulate large quantities of glucose, fructose, mannitol and arabitol which may have an adaptive significance.

On the other hand, Melick and Seppelt (1994) reported that thalli of *Umbilicaria decussata* growing on wet habitats of Clark Peninsula, Windmill Islands and East Antarctica (level of thallus hydration $\Delta m/m_0 \approx 2.0$) did not show any difference in the content of low-molecular-weight carbohydrates and pigment from the thalli populating dry habitats ($\Delta m/m_0 \approx 0.17$). The two populations differed however in the proportion of algal products (ribitol) relative to fungal products (arabitol and mannitol) which was greater in the population from wet habitats. Lichens from wet habitats also exhibit slightly elevated tissue freezing point (−13 °C, as determined by DTA), compared to −16 °C for the thalli populating dry habitats.

^{1}H-NMR is a powerful and widely used technique for studying hydration/dehydration processes in living organisms, tissues and artificial systems. This technique revealed anomalous (non-linear) form of the proton signal/relaxation dependency for dehydrated lichen thalli at initial stages of rehydration. Such a form is characteristic for the presence of water-soluble solid fraction, which dissolves with the increased level of hydration.

For a rehydrated *Cetraria aculeata* thallus in the initial hydration range of $\Delta m/m_0 \leq 0.80$), the averaged saturation concentration for water-soluble solid fraction is equal to $c_s = (57 \pm 12)\%$, a value for averaged carbohydrate and polyol saturation concentration $(65.7 \pm 8.3)\%$. The total mono-, disaccharide and polyol content in *C. aculeata* thallus is low as compared to the polysaccharide fraction in lichen thallus which consists mainly of lichenin (about 50% of total) and of iso-lichenin (about 10–20%). During the rehydration process, lichenized fungus shifted from the cryptobiotic to the metabolically active state, and the necessary amount of lichenin and/or iso-lichenin was converted to mono- and disaccharides through the enzymatic action of lichanase hydrolises (Harańczyk et al. 2016) according to

$$C_6H_{10}O_5 + x \cdot H_2O \underset{\text{lichenase}}{\rightarrow} x \cdot C_6H_{12}O_6$$

Usnea antarctica thallus rehydrated up to the level of $\Delta m/m_0 \leq 0.45$ with the saturation concentration of water-soluble solid fraction $c_s = (81.0 \pm 4)\%$. According to Harańczyk et al. (2016), *Turgidosculum complicatulum* (= *Mastodia tessellata*, so-called borderline lichen) (Pérez-Ortega et al. 2010) thallus rehydrated in the range $\Delta m/m_0 \leq 0.6$; the saturation concentration of water-soluble solid fraction calculated from ^{1}H-NMR data was equal to $c_s = 0.67$ (for relaxation data) and $c_s = 0.60 \pm 0.13$ (for NMR spectra), which resembles the values for some sugars (e.g. sucrose, galactose or xylose). In contrast to *T. complicatulum* thallus, for its free-living photobiont *Prasiola crispa* rehydrated in the range $\Delta m/m_0 \leq 0.5$, the hydration dependency of ^{1}H-NMR data is a linear function, showing that the water-soluble solid fraction is not present in photobiont but in a mycobiont thallus only (Bacior et al. 2017).

3.4 Classification of Bound Water Fractions in Dry Thallus, Effects at Gaseous Phase Hydration/Dehydration and Sorption Isotherm

From the gaseous phase (in moist chamber), lichen thalli rehydrate to the relatively high hydration level, e.g. *Usnea aurantiaco-atra* rehydrates to the saturation value $\Delta m/m_0 = 0.85 \pm 0.03$, *Ramalina terebrata* to $\Delta m/m_0 = 1.2 \pm 0.05$ and *Himantormia lugubris* to $\Delta m/m_0 = 0.85 \pm 0.05$ (Kappen and Redon 1987). The maximal hydration level from the gaseous phase (for $p/p_0 = 100\%$) for *U. antarctica* thallus equals $\Delta m/m_0 = 1.10$ (Harańczyk et al. 2006), for *L. puberulum* thallus $\Delta m/m_0 = 0.64$ (Harańczyk et al. 2009), for *R. terebrata* $\Delta m/m_0 = 1.6$ (Harańczyk et al. 2012a), for *U. aprina* $\Delta m/m_0 = 0.82$ (Harańczyk et al. 2008), for *C. aculeata* $\Delta m/m_0 = 1.0$

Table 3.2 The minimal hydration level at $p/p_0 = 0\%$, A_0^h, and the dehydration time t_d [h]; air-dry hydration, $\Delta m/m_0$, for lichen thalli and for free-living photobiont, *Prasiola crispa*

	Site	A_0^h	t_d [h]	$\Delta m/m_0$	Ref
Usnea antarctica	King George Island	0.035		0.10	Haranczyk et al. (2006)
Leptogium puberulum	King George Island	0.012 ± 0.00	12.1 ± 2.6	0.072 ± 0.011	Haranczyk et al. (2009)
Ramalina terebrata	King George Island	0.048 ± 0.004		0.096	Haranczyk et al. (2012a)
Umbilicaria aprina	Schirmacher Oasis, Queen Maud Land, continental Antarctica	0.048 ± 0.004		0.096 ± 0.010	Haranczyk et al. (2008)
Cetraria aculeata	Penguin Island	0.031 ± 0.003	14.11 ± 0.65	0.088 ± 0.011	Haranczyk et al. (2016)
Turgidosculum complicatulum	King George Island	0.028 ± 0.003		0.080 ± 0.013	Bacior et al. (2017)
Prasiola crispa	King George Island	0.028 ± 0.003		0.066 ± 0.012	Bacior et al. (2017)

(Harańczyk et al. 2016), for *T. complicatulum* $\Delta m/m_0 = 1.3$ and for its free-living photobiont *P. crispa* $\Delta m/m_0 = 0.95$ (Bacior et al. 2017).

The air-dry lichen thallus dehydrates to the gaseous phase with the one exponential kinetics (Table 3.2) to the minimal hydration level characteristic for dry air ($p/p_0 = 0\%$) (Harańczyk et al. 2006, 2009, 2012a, 2016; Bacior et al. 2017).

Gaseous phase hydration kinetic courses of fruticose lichens allow to differentiate several bound water fractions, namely, (1) very tightly bound water (hydrating with a very short time (less than 10 min) and the signal amplitude A_0^h which in most investigated species is close to the hydration level of primary water binding sites, recorded from the sorption isotherms, (2) tightly bound water with the amplitude of the signal A_1^h and the hydration time t_1^h finally appearing at the higher relative humidity of the environment (3) loosely bound water fraction, which appears for a higher relative air humidity, with the hydration time t_2^h and the signal amplitude saturating gradually with the increased air humidity (Table 3.3).

Bound water fractions present in rehydrated dry lichen thallus may be classified according to their proximity to inner surfaces (the intra- and extracellular water fractions are not yet differentiated for so low hydration levels). Hydration and dehydration kinetics is expressed as a superposition of at least two exponential functions, plus one component saturating for the time less than 10 min hydrating very fast. The hydration kinetics is sensitive to water-binding strength to the thallus surfaces or to previously bound water molecules, whereas NMR experiments report on bound water fractions concerning their mobility.

^{1}H-NMR experiment differentiates water fraction with very restricted mobility and more mobile, loosely bound or free water fractions, which appear at higher hydration levels of rehydrated lichen thallus. The mobile water (loosely bound water) component appears in *U. antarctica* thallus $\Delta m/m_0 > 0.10$ (Harańczyk et al. 2006), in *L. puberulum* thallus $\Delta m/m_0 > 0.232$ (Harańczyk et al. 2009), in *R.*

Table 3.3 Hydration kinetic parameters and primary water-binding site population (from sorption isotherms)

	Site	A_0^h	A_1^h	t_1^h [h]	t_2^h [h]	$\Delta M/m_0$	Ref
Usnea antarctica	King George Island	0.040 ± 0.011	0.087 ± 0.028	3.53 ± 0.98	69.7 ± 21.3	0.0678	Haranczyk et al. (2006)
Leptogium puberulum	King George Island	0.008 ± 0.004		1.6 ± 0.3	25.8 ± 3.2	0.043 ± 0.007	Haranczyk et al. (2009)
Ramalina terebrata	King George Island	0.007 ± 0.001	0.059 ± 0.007	1.24 ± 0.24	14.4 ± 2.1	0.046	Haranczyk et al. (2012a)
Umbilicaria aprina	Schirmacher Oasis, Queen Maud Land, continental Antarctica	0.054 ± 0.011	0.051 ± 0.038	4.7 ± 2.6	31.0 ± 1.9	0.054	Haranczyk et al. (2008)
Cetraria aculeata	Penguin Island	0.04 ± 0.02	0.037 ± 0.005	0.43 ± 0.10	14.2 ± 6.2	0.046 ± 0.003	Haranczyk et al. (2016)
Turgidosculum complicatulum	King George Island	0.028 ± 0.003	0.038 ± 0.010	1.45 ± 0.21	23.2 ± 5.0	0.055	Bacior et al. (2017)
Prasiola crispa	King George Island	0.022 ± 0.002	0.048 ± 0.012	0.37 ± 0.14	42.6 ± 3.2	0.131	Bacior et al. (2017)

terebrata at $\Delta m/m_0 > 0.12$ (which is the hydration level at $p/p_0 = 52\%$) (Harańczyk et al. 2012a), for *C. aculeata* $\Delta m/m_0 > 0.10$ (Harańczyk et al. 2016), for *T. complicatulum* $\Delta m/m_0 > 0.15$ and for its free-living photobiont *P. crispa* $\Delta m/m_0 > 0.2$ (Bacior et al. 2017).

3.5 Water Uptake from Snow via Gaseous Phase

Snow cover plays important protective role for higher plants (Tieszen et al. 1981). Snow, apart from reducing extremes of temperature, ensures moisture in plant environment allowing at the same time light transmission sufficient for moderate under-snow photosynthesis (Kappen and Breuer 1991; Marchand 1987). It has not been elucidated yet whether the developmental processes such as germination from the seed, growth and leaves formation, which are observed in higher plants under snow (Salisbury et al. 1973), proceed along with photosynthetic activity. On the basis of their studies on *Claytonia lanceolata* leaves, Kimball and Salisbury (Kimball and Salisbury 1974) postulate that photosynthetic activity may occur only after the majority of snow covering the plant vanished. This assumption is supported by a lack of full set of enzymes engaged in photosynthetic activity as it was in Arctic herbaceous plants before snowmelt (Tieszen 1974).

For lichens populating Antarctic continent, temperature rarely exceeds 8 °C during their photosynthetically active period when snow and moisture are almost the exclusive sources of water (Kappen 1985). Snow cover of the thickness of 1–4 cm does not inhibit photosynthetic activity of lichens.However, for *Usnea sphacelata* covered in snow during period of strong irradiance, above 600 $\mu Em^{-2}\ s^{-1}$, net photosynthesis can be reversibly decreased for several hours (Kappen et al. 1991).

Sensitivity to snow cover varies among lichen species, as in many of them, prolonged snow cover leads to strong injury (Benedict 1990, 1991), whereas other lichens (like chionophilous species, e.g. crustose lichen *Carbonea assenties* (= *Lecidea sciatrapha*) populating maritime Antarctic), are evolutionary and physiologically adapted to the presence of snow cover (Kappen et al. 1990). *Usnea antarctica* populates mainly windblown and snow-free sites in the maritime Antarctic, but it can also be found in snow-covered sites in the Antarctic continent (Hancock and Seppelt 1988).

Thallus hydration level as a function of snow cover thickness was investigated in several lichen species. It was found that the hydration level of *Cetraria nivalis* reaches the value $\Delta m/m_0 = 1.40–2.20$ when snow cover was 10–20 cm and 1.70–2.05 under snow cover of 1–5 cm, whereas for *Usnea sphacelata*, these values were $\Delta m/m_0 = 0.50–1.10$ and 0.50–1.95, respectively (Kappen et al. 1995). The m_0 value is the dry mass of thallus obtained after heating at 80 °C to constant mass. For *Usnea antarctica*, the recorded level of hydration from snow cover equals $\Delta m/m_0 = 0.75–1.65$ (Kappen and Breuer 1991).

There is a clear dependence between hydration from snow and temperature. *Umbilicaria aprina* rehydrated exclusively from snow performed net photosynthesis and dark respiration down to –4 °C (Schroeter and Scheidegger 1995). Both

Usnea sphacelata and *Umbilicaria aprina* influenced hydration associated with snowfall and may activate photosynthetic CO_2 fixation below 0 °C (down to −10 °C for the former and to −17 °C for the latter species, respectively) (Kappen 1989; Schroeter et al. 1994).

The dependence of hydration level due to snow cover on temperature has been demonstrated for *Umbilicaria aprina.* When air-dry thallus ($\Delta m/m_0 = 0.09$) of this lichen species was covered with snow and kept in the dark for 16 h or more, its hydration level rose to 0.56 at −4.5 °C, to 0.50 at –6 °C and only to 0.25 at −14 °C. No ice nucleation activity (INA) in *Umbilicaria aprina* thallus hydrated from snow was observed. Although the lichens may effectively hydrate from snow, the hydration of thalli from liquid water is four times higher reaching more than 200% of m_0. Respiration and photosynthetic activity in lichens are not directly proportional to degree of hydration. Schroeter and Scheidegger (Schroeter and Scheidegger 1995) monitored CO_2 release in *U. aprina* at –4 °C from snow up to $\Delta m/m_0 = 0.20$ which did not show significant rates of CO_2 exchange. At the hydration level of $\Delta m/m_0 \approx 0.60$, an equilibrium with a photosynthetic rate of +1.1 μmole CO_2 kg^{-1} was reached after 20 h (Schroeter and Scheidegger 1995).

The uptake of water by lichens from snow at freezing temperature may occur only from gaseous phase, from water vapour sublimating from snow, and this process to occur requires a gradient of water potential between dehydrated lichen thallus and surrounding snow which is necessary. In the absence of snow, at −18 °C, Antarctic endemic crustose lichen *Buellia frigida* hydrates from the gaseous phase up to $\Delta m/m_0$ = ca. 0.15, whereas *Circinaria gyrosa* from Central Spain hydrates up to ca. 0.1 (Jänchen et al. 2015).

3.6 Freezing Tolerance of Lichens

There is a great diversity among plants in respect to frost tolerance. Apart from cold-sensitive species, there are also plants that are extremely frost tolerant surviving temperatures of –40 °C or lower. Numerous vascular plants such as trees of boreal forest and evergreen coniferous species in north areas of Canada and Siberia have extreme tolerance of frost. On the other hand, there are only two taxa present in the polar regions of Southern Hemisphere, e.g. Antarctic Peninsula (Casanova-Katny and Cavieres 2012), where *Deschampsia antarctica* shows the highest cold tolerance with −27 °C (Bravo et al. 2001). As a rule, with increasing latitude, the proportion of non-vascular plants to vascular ones increases significantly. Lichens are distinct than vascular plants as they adapted better in short and cold vegetation seasons, thus reaching the highest latitude of 86°S and exhibiting their dominance (Green et al. 2011; Kappen 1993).

Freezing tolerance in lichens together with their resistance to low temperature enables them for adaptation to long duration of the freezing period. The experiments performed with air-dry *Alectoria ochroleuca* kept for 3.5 years at –60 °C and then hydrated for 12 h in distilled water at +6 °C revealed no difference in photosynthesis and respiratory activity as compared with freshly collected lichen (Larson 1978).

Table 3.4 Freezing tolerance of lichens

			No injury if frozen *t* [°C]	
Species	Region	Month	Rapidly	Slowly
Xanthoria elegans (=*Caloplaca elegans*)	Antarctica	Nov.	−196	−196
Rhizoplaca melanophthalma	Antarctica	Nov.	−196	−196
Xanthoria mawsoni	Antarctica	Nov.	−78	−196
Buellia frigida	Antarctica	Nov.	−78	−196
Umbilicaria decussate	Antarctica	Nov.	−78	−196
Usnea capillacea	New Zealand	Nov.	−78	−196
Sticta marginifera	New Zealand	Nov.	−30	−50
Lobaria pulmonaria	Central Europe	Sept.	−78	−196
Cladonia rangiferina	Central Europe	Sept.	−78	−196
Umbilicaria vellea	Central Europe	Dec.	−50	−196
Ramalina maciformis	Negev Desert	Nov.	−78	−196
Cladonia convoluta	Mediterranean region	Mar.	−50	−50
Roccella fucoides	Mediterranean region	Apr.	−10	−20
Dendrographa minor	California	June	–	−12

Kappen (1993); After: Lange and Kappen (1972), and Nash et al. (1987a, b). After Harańczyk (2003), modified

Cladonia foliacea (= *C. alcicornis*) thallus returns relatively fast (20 h incubation at +10 °C) to its control level of CO_2 uptake after prolonged (96 and 110 weeks) freezing at −15 °C (Lange 1966).

The unusual resistance of lichens to extremely low temperatures depends on the rate of their freezing (Table 3.4). The viability of isolated photobionts of genus *Trebouxia*, a most frequent photobiont found in lichen species well adapted to extreme environments, was decreased PS II activity at shock freezing in liquid nitrogen, while not affected by slow cooling rate (0.5 $°C \cdot min^{-1}$) (Hájek et al. 2012). Although dependent on freezing rate, the freezing resistance cannot occur without mechanisms controlling freezing of water bound in tissues. In isolated *Trebouxia* photobionts, 1 h of freezing at −25 °C significantly reduces PS II activity for Antarctic *Pleopsidium chlorophanum*, for *Umbilicaria decussata* and for European *Fulgensia bracteata* (although further decrease is not detected for longer freezing), whereas for Antarctic *Buellia frigida*, *Usnea lambii* and *Umbilicaria antarctica*, the decrease in PSII activity is not observed (Sadowsky and Ott 2012).

The isolated photobionts are more stressed by the excess of light than by the decrease of the temperature, and the effect of cortex light-screening pigments of photobiont shows that photoprotection is significantly increased in the symbiotic state (Sadowsky and Ott 2016). It seems that freezing-tolerant plants avoid intracellular formation of hexagonal crystals of ice, lethal to the cells, by transferring water to extracellular spaces where it may freeze without damaging cell structures (Burke et al. 1976).

Although the molecular mechanisms of freezing resistance in lichen have not been fully elucidated yet, it is evident that sugars and polyols play a significant role in cryoprotection of the tissue (Kaurin et al. 1981). On the other hand, it has been found that in Antarctic bryophytes, the repeated freeze-thaw cycles (up to $n = 16$) decrease the level of carbohydrates, and this decrease is not significantly correlated with freezing temperature of tissue as evidenced by DSC measurements (Melick and Seppelt 1992).

3.7 Photosynthetic CO_2 Uptake Below 0 °C

Net photosynthesis maximum and its level depend on habitat of lichens. *Umbilicaria nylanderiana* population from Mediterranean areas exhibits increased level of net photosynthesis with its maximum at +15 °C as compared to Antarctic population of the same species in which net photosynthesis maximum occurs at +3 °C (Sancho et al. 2000).

Photosynthetic activity at the temperatures below 0 °C occurs in many lichen species. Lichen *Usnea sphacelata* (= *U. sulphurea*) exhibits positive net photosynthesis at −5 °C when it grows on high-light sites and at −3 °C when it grows on shade sites (Kappen 1983). Significant fixing of ^{14}C-labelled CO_2 at −10 °C has been reported for *Cladonia foliacea* (= *C. alcicornis*), *Peltigera subcanina* and *Stereocaulon alpinum* (Lange and Metzner 1965), whereas Lange and Kappen demonstrated the positive net photosynthesis in lichens even at −18 °C.

The lowest temperature at which net photosynthesis occurs in field conditions (−17 °C) was reported by Schroeter et al. (1994), for *Umbilicaria aprina* growing in its natural habitat (Cape Geology, Antarctica). The dark respiration significantly diminished already at −7 °C and completely ceased at −11 °C. Decreasing temperature showed correlation with decrease in light saturation of maximum photosynthetic rate, from 400 $\mu Em^{-2}\ s^{-1}$ at above −6 °C down to 150 $\mu Em^{-2}\ s^{-1}$ at −17 °C. Similar dependence was found for the light compensation point of net photosynthesis which decreased with decreasing temperature (from 35 $\mu Em^{-2}\ s^{-1}$ at +1 °C to 20 $\mu Em^{-2}\ s^{-1}$ at −5 °C), vanishing completely with the cessation of dark respiration. Two phases of maximum net photosynthesis decline with temperature were observed: a fast phase taking place between −1 °C and −9 °C with decline to the level of about 10% of the rate occurring at +1 °C and a second phase during which photosynthesis declined slowly but steadily, reaching very low rates at −17 °C. Factors such as increased resistance of CO_2 diffusion and ice nucleation occurring in water-saturated thalli of *Umbilicaria aprina* at −5.4 °C could contribute to these results (Schroeter and Scheidegger 1995). Table 3.5 sets up the lowest recorded temperatures of photosynthetic CO_2 uptake during laboratory measurements and during field measurements.

3.8 PS II Activity Below 0 °C

Umbilicaria antarctica and *Xanthoria elegans* from the maritime Antarctic (Galindez Island) reveal the primary photochemical processes (photosynthetic activity of PS II) for the temperatures down to at least −15 °C; however, below

Table 3.5 Lowest recorded temperatures for photosynthetic activity of lichens

Species	t [°C]	Method	Region	References
(A) Laboratory measurements				
Leptogium puberulum	−2	IRGA	South Shetland Islands	Schlensog et al. (1997)
Usnea sphacelata (= *U. sulphurea*)	<−6	IRGA open cycle	Antarctic continent	Kappen (1983)
Light form	<−3		(Southern Victoria Land)	
Shade form				
Cryptoendolithic lichen[a] sample 790/47	<−4.2	IRGA open cycle	Antarctic continent	Kappen and Friedmann (1983)
Sample 790/50	Est.–6, −8		(Southern Victoria Land)	
Peltigera subcanina	<−10.6	^{14}C fixation	Southern Germany	Lange and Metzner (1965)
Cladonia foliacea (= *C. alcicornis*)	<−10.7	^{14}C fixation	Southern Germany	Lange and Metzner (1965)
Stereocaulon alpinum	<−10.9	^{14}C fixation	Southern Germany	Lange and Metzner (1965)
Xanthoria mawsonii	−16.5	IRGA closed cycle	Antarctic continent	Lange and Kappen (1972)
Rhizoplaca melanophthalma (= *Lecanora melanophthalma*)	−16.5	IRGA closed cycle	Antarctic continent	Lange and Kappen (1972)
Xanthoria parietina	−18	IRGA closed cycle	Southern Finland	Kallio and Heinonen (1971)
Usnea acromelanea Stirt. (= *Neuropogon acromelanus)*	−18.5	IRGA closed cycle	Antarctic continent	Lange and Kappen (1972)
Flavocetraria nivalis (=*Cetraria nivalis*)	−20	IRGA closed cycle	Kevo, Finland	Kallio and Heinonen (1971)
Cladonia convoluta	−22	IRGA closed cycle	Southern France	Lange (1965)
Cladonia foliacea (= *C. alcicornis*)	−24	IRGA closed cycle	Southern Germany	Lange (1965)
(B) Field measurements				
Himantormia lugubris	−5	IRGA	King George Island, South Shetland Islands	Kappen et al. (1987)
Caloplaca regalis	−5	IRGA	King George Island, South Shetland Islands	Kappen et al. (1987)
Usnea antarctica	<−5	IRGA open system	King George Island, South Shetland Islands	Schroeter et al. (1995)
Usnea aurantiaco-atra (= *U. fasciata*)	−6	IRGA	King George Island, South Shetland Islands	Kappen et al. (1987)

(continued)

Table 3.5 (continued)

Species	t [°C]	Method	Region	References
Hypogymnia physodes	−6	IRGA open system	Solling, Germany	Schulze and Lange (1968)
Ramalina farinacea	−9	Ivanov method	Sinaia Mts., Romania	Atanasiu (1971)
Letharia vulpina	−9	Ivanov method	Sinaia Mts., Romania	Atanasiu (1971)
Usnea submollis	−10	Ivanov method CO_2 absorption and $Ba(OH)_2$ titration	Sinaia Mts., Romania	Atanasiu (1969)
Usnea sphacelata	<−10	IRGA open system	Antarctic continent	Kappen (1989)
Umbilicaria aprina	−17	IRGA, minicuvette	Antarctic continent	Schroeter et al. (1994)

After Harańczyk (2003), modified
IRGA infrared gas analysis
[a]Identified as *Buellia*

–10 °C, the substantial inhibition of chlorophyll fluorescence is detected. Both species differ in spatial localization of photochemical activity in the thallus, i.e. in *U. antarctica* the active areas are located between irregularly distributed patches, whereas in *X. elegans* in close to central parts of the thallus (Barták et al. 2007).

Ribitol is the main high-energy compound exported from producer (photobiont) to a consumer (mycobiont), and moreover it is an effective cryoprotectant. In foliose lichens *Lasallia pustulata* and *Umbilicaria hirsuta* from the sites in Central Europe, at subzero temperatures (−5 °C), external ribitol added to the thallus maintained primary photosynthetic processes in hydrated thalli, while untreated thalli exhibited temperature-dependent photosynthesis inhibition (however, the localization of external ribitol was not controlled in this experiment). The stimulation effect was not observed for higher temperatures (0 °C and +5 °C) or for dehydrated thalli (Hájek et al. 2009).

Thalli of *Usnea antarctica; Usnea aurantiaco-atra*, from the maritime Antarctic (James Ross Island and King George Island, respectively); and *Umbilicaria cylindrica* from central part of Spitsbergen island (Svalbard), cooled down from +20 to –50 °C (cooling rate 2 $°C \cdot min^{-1}$), show the decrease in chlorophyll fluorescence with the decreased temperature (between –10 and –30 °C for *U. cylindrica*, between –10 and –20 °C for *U. antarctica* and broad gradual change between room temperature and –30 °C for *U. aurantiaco-atra*) (Hájek et al. 2016).

3.9 Supercooled Water Behaviour, Ice Nucleation Activity and Ice Growth in Lichen Thallus

In majority of lichen species, water significantly supercools; however, some of them tolerate ice nucleation inside their thallus, at the relatively high temperatures (for the temperatures at which the active photosynthesis process still occurs).

Thallus extracts of numerous lichens display biological ice nucleation activity (INA) at much higher temperatures than the low temperature limit of their photosynthetic activity. In *Umbilicaria aprina* from continental Antarctica, INA resulting in extracellular freezing of water is detected at –5.4 °C in water-saturated thalli (Schroeter and Scheidegger 1995). In contrast, *U. aprina* collected from temporary streams flowing from melting glaciers at Schirmacher Oasis, Droning Maud Land, does not reveal the mechanism of stimulated ice nucleation. Freezing point is characteristic for heterogeneous ice nucleation and increases from ca. –30 to ca. –20 °C when the thallus hydration level increased from $\Delta m/m_0$ = 0.2 to $\Delta m/m_0$ = 0.8, respectively. Melting temperature is ca. 8 °C higher than freezing temperature (Harańczyk et al. 2012b). The mechanism of seasonal switching between the freeze tolerance and freezing avoidance, resulting in different INA, was observed for arthropods (Block 1995). Even for two species of overwintering larvae of beetles *Dendroides canadensis* (Horwath and Duman 1984) and *Cucujus clavipes* (Duman 1984), the process of permanent change in survival strategy was detected; however, for lichenized fungi, potential durability of this change is still an open question.

Certain strains of epiphytic bacteria are other organisms which demonstrate active ice nucleation at relatively high temperature (Lindow 1982a, b; 1983). It is generally accepted that for INA high temperature means the temperature of –5 °C or higher (Lindow et al. 1978, 1982) and the nuclei active above this temperature are termed Type I nuclei (Phelps et al. 1986; Yankofsky et al. 1981). So far bacteria active in ice nucleation were not isolated from lichens; therefore, it is thought that ice nuclei formation in these organisms results from the activity of one or both lichen symbionts (Kieft 1988).

According to report of Kieft and Ahmadjian (1989), Type I nuclei are present rather in the mycobiont than in photobiont partner of lichen. However, some of the lichen symbionts show INA only in low temperature, e.g. for the mycobiont *Cladonia bellidiflora* INA has been detected at −8.3 °C while for *Cladonia rangiferina* at −5.0 °C.

Ice nuclei isolated from lichens growing at different altitudes differ in their activity. Accordingly ice nuclei from *Xanthoparmelia* growing at the sites of about 2300 m and higher (N. Mexico) were active at the temperatures above –3 °C, whereas for *Xanthoparmelia* collected from about 1600 m, the nuclei were active at lower temperature (Kieft 1988).

For *Rhizoplaca chrysoleuca*, the extracted biological ice nuclei (active at ≈ –4 °C) showed stability for 10 min heating at 60 °C, and they were active from pH 1.5 to 12 (Kieft and Ruscetti 1990). Their nature seems to be proteinaceous because of their sensitivity to treatment of proteases, guanidine hydrochloride and high concentration of urea (2 M and 5 M, respectively) which completely abolish INA above

−10 °C. On the other hand, no effect on INA was observed after chloroform treatment used for lipid removal. According to Kieft and Ruscetti (1990), protein monomers in thallus aggregate forming high-temperature ice nuclei.

Worland et al. (1996) described the effect of particle size on INA. They provided an evidence that in *Usnea aurantiaco-atra* particles of the diameter less than 120 μm were active at higher temperature with first freezing event occurring at −5.1 °C and 75,000 g^{-1} nuclei at −7 °C. For the larger particles, these values were the following: for the particles 160 μm–1400 g^{-1} nuclei at −7 °C and for the particles 240 μm – first freezing event at −8.3 °C. When comparing lichens, mosses and higher plants from sub-Antarctic South Georgia and the maritime Antarctic, it appeared that at −7 °C the mean number of nuclei per gram ranged from 257,000 g^{-1} to 16,220 g^{-1} in the order: lichens>mosses>higher plants.

The benefits for lichens resulting from the occurrence of the Type I ice nuclei may be twofold. Such nuclei may enhance uptake of water molecules from the gaseous phase and its deposition as ice either in or on the thallus of lichen. The other possibility, which may exist in parallel with the former one, is that induction of freezing at relatively elevated temperature decreases frost damage by inducing early formation of smaller ice crystals, possibly with extracellular location (Kieft 1988).

Even partially digested *Usnea fasciata* thallus may probably play supplementary role at INA for Antarctic perimylopid beetle, *Hydromedion sparsutum*. Its gut material shows the supercooling point at −(3.0 ± 1.0)°C, which is among the highest recorded for any terrestrial arthropods and whose guts are probable site of nucleation, as the supercooling point [−(3.9 ± 0.6)°C] of the excised gut is close to the whole insect's supercooling point. Narrow range of supercooling points suggests the presence of an efficient ice nucleator. Gut contents of *Hydromedion* consist mostly of partly digested plant material, bacteria and fungi. Two potential food plants (mosses *Polytrichum alpinum* and *Andreaea* sp.) show ice nucleating activity at little lower temperatures (−4.0 and −5.0 °C, respectively; and −5.0 °C is similar to lichen *Usnea fasciata*), suggesting that ingested plant material probably may not be directly responsible for ice nucleation. Heating to 75 °C significantly reduces the ice nucleation activity, suggesting the proteinaceous nature of ice nucleator (Worland et al. 1993).

Cladonia mitis thallus rehydrated to $\Delta m/m_0 > 0.157$ from the gaseous phase two fractions of bound water are detected in ^{1}H-NMR FID experiment. With the temperature decreased down to −20 °C, the freezing, loosely bound (more mobile) water pool changes into the nonfreezing tightly bound water fraction which is detected even at −60 °C. For the thallus hydrated to $\Delta m/m_0 = 0.388$ (and to 0.448 in T_1 measurement), the cooperative freezing of bound water is observed at a temperature ca. −8 °C. For the thallus hydrated to $\Delta m/m_0 = 0.195$, freezing is not yet observed (Harańczyk et al. 2000). The temperature dependencies of ^{1}H-NMR spectra (for which only mobile water signal was detected) for *C. mitis* thalli showed the gradual decrease in mobility of supercooled loosely bound water occurring with the temperature decreased between −20 °C and −40 °C and the absence of cooperative water freezing for the sample hydrated even to $\Delta m/m_0 = 0.193$ (Harańczyk et al. 2003a).

In the thalli of Antarctic lichens from Bunger Oasis, *Himantormia lugubris* and *Usnea aurantiaco-atra*, and *Cladonia mitis* from Sweden, the ^{1}H-NMR spectra allow to differentiate the nonfreezing (immobilized) tightly bound water fraction and freezing (mobile) loosely bound water. For the temperatures above –20 °C, the process of transfer of loosely bound water fraction to nonfreezing less mobile water pool is observed. For the temperatures below –20 °C, continuous broadening of ^{1}H-NMR monitors the non-cooperative immobilization of supercooled water which does not freeze inside the thallus (Harańczyk et al. 2003b).

In the gaseous phase hydration process over the D_2O surface, dry *Himantormia lugubris* thallus rehydrates up to $\Delta m/m_0 = 0.422 \pm 0.014$, which is the value almost four times less than the one needed to activate the photosynthesis process in H_2O. The ^{1}H-NMR spectra show the presence of two fractions of residual bound H_2O and the increase of non-freezable water fraction on the cost of freezing loosely bound water pool, occurring with the decreased temperature. No continuous decrease in ^{1}H-NMR signal was detected which signalizes that the residual H_2O remains supercooled beside nucleating D_2O crystallites (Harańczyk et al. 2001).

For *U. aprina* collected from temporary streams flowing from melting glaciers at Schirmacher Oasis, Droning Maud Land, the ^{1}H-NMR measurements show transfer of freezing loosely bound water fraction to nonfreezing tightly bound water fraction occurring at temperatures down to –40 °C. The overall signal from supercooled water decreases with the decreased temperature which may be caused by the diffusion of supercooled water molecules to ice microcrystallites already present in thallus (detected using DSC for the hydration levels above $\Delta m/m_0 = 0.202$). The extrapolated from DSC (zero transition enthalpy) minimal hydration level at which ice nucleation occurs in *U. aprina* thallus equals $\Delta m/m_0 = 0.127 \pm 0.027$, which is the number regardless the long time diffusion of water molecules in incubated thallus. Freeze-thawing cycles ($n \leq 22$) do not restore INA for these *U. aprina* thalli. The $t_m - t_f$, = ca. 10.5 °C for the first cycle, and for $n > 5$ remains constant $t_m - t_f = (8.0 \pm 0.2)$°C (Harańczyk et al. 2012b).

The thallus of Antarctic *Cetraria aculeata* observed by ^{1}H-NMR and rehydrated to $\Delta m/m_0 = 0.499$ shows the cooperative water freezing for the temperatures between −6.5 °C and −15.6 °C, which responds for a water mass equal to $\Delta m/m_0 = 0.12$, whereas the thallus hydrated to $\Delta m/m_0 = 0.964$ freezes for the temperatures even higher (between 0 °C and −6.5 °C), responding for freezing of $\Delta m/m_0 = 0.55$ mass of water. For lower hydration levels, the signal from bound water decreases smoothly with the decreased temperature, signalizing continuous decrease in supercooled water mobility or diffusion-induced binding to small ice crystallites beyond the spectrometer resolution.

DCS scans show that the cooperative water freezing for *C. aculeata* thalli occurs for the hydration level $\Delta m/m_0 \leq 0.56 \pm 0.25$ and for melting $\Delta m/m_0 \leq 0.346 \pm 0.075$. Beside main peak for melting the characteristic low temperature "shoulder" is detected, which was also observed in liposomes (Bronshteyn and Steponkus 1993), and which may respond for melting episodes in isolated compartments of thallus, e.g. photobiont cells. The transition enthalpy of both melting acts increases linearly with the thallus hydration level, which suggests that photobiont and mycobiont hydrate evenly with the increased hydration level.

Cooled down to –60 °C and subsequently incubated at –23 °C (still below the cooperative water temperature but at more intense thermal diffusion) for 120 min,

C. aculeata thallus reveals cooperative water freezing even if hydrated to $\Delta m/m_0 = 0.34$ or $\Delta m/m_0 = 0.23$ (when no freezing is observed). The estimated lowest hydration level of *C. aculeata* at which cooperative water freezing occurs equals $\Delta m/m_0 = 0.167$. This effect is not a typical phase growth for thermodynamic system, but the effect characteristic for microheterogeneous biological system, here lichenized fungus, in vivo, because the diffusion of separated supercooled water molecules to ice microcrystallites is needed (Nowak et al. in press).

In the lichen thalli experiencing temperatures below 0 °C, several molecular processes take place, which are passively or actively used by the organism to survive extremely low temperature. The presence of at least five these processes respond for the unusual resistance of lichens to decreased temperature: (1) the slowing down of the hydration rate of freezing loosely bound water pool, which allows to avoid the accidental ice nucleation in thallus, (2) active transfer of freezing loosely bound water pool to non-freezing tightly bound water fraction, (3) non-cooperative immobilization of water bound in thallus occurring continuously with the decreased temperature, (4) cooperative ice nucleation of supercooled bound water occurring at the higher hydration levels and (5) diffusion-induced migration of supercooled single water molecules to ice microcrystallites already present in thallus.

3.10 The Extremes of Lichen Freezing Tolerance, Lithopanspermia Hypothesis

That some lichenized fungi may reveal the unusual resistance for outer space conditions was suggested by the space-simulating experiments, e.g. about 50% ascospores of mycobionts of the lichens *Fulgensia bracteata*, *Xanthoria elegans* and *Xanthoria parietina* exposed to space-simulating conditions (up to 16 h at 10^{-3} Pa, UV radiation 160 nm$\leq \lambda \leq$400 nm) recover, with the ascospores of *X. elegans* showing the highest survival (79% germination rate) (de Vera et al. 2004). The crustose Antarctic endemite, *Buellia frigida*, populating rocky habitats up to 2015 m a.s.l. exposed for 7 days under simulated Mars atmosphere (CO_2 gas composition, at 10–5 Pa, temperature cycling between –10 and +45 °C (50 cycles of 8 h each), temperature extremes of –25 °C and +60 °C and monochromatic UV λ = 254 nm at varied fluences, survived such conditions. The mycobiont showed lower decrease in viability than the photobiont (more than 83% and 69%, respectively) (Meeβen et al. 2015).

The unusual resistance to extreme environmental conditions made lichenized fungi interesting objects for testing the Panspermia theory. Its modern variant, the lithopanspermia hypothesis, proposes that microorganisms populating the surface or pores of rock might have been transported in the Solar System between the planets of the Earth type. The process should be composed from three phases: (1) ejection from the planetary surface to outer space, e.g. by volcanism, (2) interplanetary travel and (3) re-entry into the atmosphere (Fajardo-Cavazos et al. 2005).

The survival capability of lichens exposed to outer space conditions (16 days in space) was tested on bipolar species *Rhizocarpon geographicum* and on *Xanthoria elegans* (collected above 2000 m, mountains of central Spain) during BIOPAN-5 experiment of ESS on Earth-orbiting FOTON-M2 Russian satellite. The photosynthetic activity after exposure was nearly the same as before. The lichen upper cortex seems to provide sufficient protection against solar irradiation. After extreme dehydration, the thallus recovers, in full, its metabolic activity within 24 h (Sancho et al. 2007, 2008).

The possibility of lithopanspermia process, i.e. the transport of living organisms by the meteorites in the Solar System between the planets of the Earth type (phase II and phase III), was tested. Rock-colonizing endolithic lichenized fungi *Rhizocarpon geographicum* (from 2020 m, Sierra de Gredos, Spain), epilithic crustose *Xanthoria elegans* (from 2400 m, Sierra Nevada, Spain; samples with fruiting bodies from Sanetsch glacier and the Gornergrat glacier, Zermatt (between 2000 and 3300 m) and the vagrant lichen *Circinaria fruticulosa* (=*Aspicilia fruticulosa*) (from clayey soils, 260 m, Guadalajara, Central Spain) and endolithic and endoevaporitic communities of cyanobacteria and bacteria (colonizing halite from Salar Grande, Atacama Desert, Chile) with their natural rock substrate were exposed to space for 10 days on board the *Biopan* device of the European Space Agency (ESA). *Biopan* was closed during launch and re-entry. All three tested species of lichenized fungi were resistant to cosmic space conditions, which included full spectrum of electromagnetic irradiation from the Sun, and selected ranges of electromagnetic irradiation, e.g. *Circinaria fruticulosa* (=*Aspicilia fruticulosa*) chlorophyll fluorescence recovered to the preflight level at 72 h of reactivation (Raggio et al. 2011). In contrast, halite-populating bacteria were severely damaged by these stress factors.

After short space exposure experiments, the *X. elegans* thallus was exposed to space conditions (combined: insolation, UV radiation, cosmic radiation, low temperature and vacuum) for prolonged period. The experiment was performed for 1.5 years on board the International Space Station (ISS). Additionally a subset of the samples was exposed to simulated Martian environmental conditions by applying Mars-analogue atmosphere (10^3 Pa, 95.3% CO_2, 2.7% N_2, 1.6% Ar, 0.15% O_2 and ca. 370 ppm water vapour) and solar radiation filters. The lichen photobiont shows an average viability rate of 71%, whereas its mycobiont was even more resistant with a viability equal to 84% (Brandt et al. 2014).

Several lichen species may be considered as candidates for the experiments in outer space as they reveal the sufficient resistance on harsh environmental conditions; among them eight species were tested in outer space or simulating outer space experiments. These are *Circinaria gyrosa*, *Rhizocarpon geographicum*, *Xanthoria elegans*, *Buellia frigida*, *Pleopsidium chlorophanum*, *Fulgensia bracteata*, *Xanthoria parietina* and *Peltigera aphthosa*. Some of them are resistant to space exposition, hypervelocity impact, UV irradiation and Martian conditions; however, some important aspects of full Panspermia process were not yet tested. It is very promising for these considerations that the resistance results from different adaptations for different lichen species, a combination of pigmented cortex, algal arrangement and mucilage (crustose *B. frigida*, crustose *R. geographicum*, crustose *X. elegans*), subcortex and algal clustering (fruticose *C. gyrosa*), or pigmented cortices and basal thallus protrusions (crustose *P. chlorophanum*) (Meeβen et al. 2013a, b).

In *lithopanspermia/stone* experiment, the possibilities of lichen survival during the re-entry from the space were tested on granite rock colonized with *R. geographicum*; however, in so performed experiment, the exposed stone was transformed into a glassy, nearly homogeneous material, and no cell survived (de la Torre et al. 2010). Although Fajardo-Cavazos et al. (2009) showed that *Bacillus subtilis* spores may survive the spallation effect at (3) phase of lithopanspermic life transfer, there is still an open question on the survival of prolonged heating of the life-transporting rock at re-entry from the outer space.

Acknowledgements The research was carried out with the equipment purchased, thanks to the financial support of the European Regional Development Fund in the framework of the Polish Innovation Economy Operational Program (contract no. POIG.02.01.00-12-023/08). Angélica Casanova-Katny thanks INACH grant RT-27-16.

References

Atanasiu L (1969) Photosynthesis and respiration of some lichens during winter. Rev Roum Biol Ser Bot 14:165–168

Atanasiu L (1971) Photosynthesis and respiration in some lichens in relation to winter low temperatures. Rev Roum Biol Ser Bot 16:105–110

Bacior M, Nowak P, Harańczyk H, Patryas S, Kijak P, Ligęzowska A, Olech M (2017) Extreme dehydration observed in Antarctic *Turgidosculum complicatulum* and in *Prasiola crispa*. Extremophiles 21:331–343

Barták M, Váczi P, Hájek J, Smykla J (2007) Low-temperature limitation of primary photosynthetic processes in Antarctic lichens *Umbilicaria antarctica* and *Xanthoria elegans*. Polar Biol 31:47–51

Benedict JB (1990) Lichen mortality due to late-lying snow. Results of a transplant study. Arct Alp Res 22:81–89

Benedict JB (1991) Experiments on lichen growth. II. Effects of a seasonal snow cover. Arct Alp Res 23:189–199

Bertsch A (1966a) CO_2 – Gaswechsel und Wasserhaushalt der aerophilen Grünalge *Apatococcus lobatus*. Planta (Berl) 70:46–72

Bertsch A (1966b) Uber den CO_2 – Gaswechsel einiger Flechten nach Wasserdampfaufnahme. Planta (Berl) 68:157–166

Block W (1995) Insects and freezing. Sci Prog 78:349–372

Brandt A, de Vera J-P, Onofri S, Ott S (2014) Viability of the lichen Xanthoria elegans and its symbionts after 18 months of space exposure and simulated Mars conditions on the ISS. Int J Astrobiol 1–15, doi:10:1017/S1473550414000214

Bravo LA, Ulloa N, Zúñiga GE, Casanova A, Corcuera LJ, Alberdi M (2001) Cold resistance in Antarctic angiosperms. Physiol Plant 111:55–65

Bronshteyn VL, Steponkus PL (1993) Calorimetric studies of freeze-induced dehydration of phospholipids. Biophys J 65:1853–1865

Burke MJ, Gusta LV, Quamme HA, Weiser CJ, Li PH (1976) Freezing and injury in plants. Ann Rev Plant Physiol 27:507–528

Casanova-Katny MA, Cavieres L (2012) Antarctic Moss carpets facilitate growth of *Deschampsia antarctica* but not its survival. Polar Biol 35:1869–1878

de la Torre R, Sancho LG, Horneck G, de los Ríos A, Wierzchos J, Olsson-Francis K, Cockell CS, Rettberg P, Berger T, de Vera J-P, Ott S, Martinez Frías J, Gonzalez Melendi P, Mercedes Lucas M, Reina M, Pintado A, Demets R (2010) Survival of lichens and bacteria exposed to outer space conditions – results of the *Lithopanspermia* experiments. Icarus 208:735–748 de Vera J-P, Horneck G, Rettberg P, Ott S (2004) The potential of the lichen symbiosis to cope with the extreme conditions of outer space II: germination capacity of lichen ascospores in response to simulated space conditions. Adv Space Res 33:1236–1243

Del-Prado R, Sancho LG (2000) Water relations and photosynthetic performance of fruticose lichens from the semiarid Southeast of Spain. Flora 195:51–60

Duman JG (1984) Change in overwintering mechanism of the cucujid beetle *Cucujus clavipes*. J Insect Physiol 30:235–239

Fajardo-Cavazos P, Link L, Melosh HJ, Nicholson WL (2005) *Bacillus subtilis* spores in simulated meteorite survive hypervelocity atmosphere entry, implications for lithopanspermia. Astrobiology 5:726–736

Fajardo-Cavazos P, Langenhorst F, Melosh HJ, Nicholson WL (2009) Bacterial spores in granite survive hypervelocity launch by spallation: implications for Lithopanspermia. Astrobiology 9(7):647–657

Gaff DF (1977) Desiccation tolerant vascular plants of Southern Africa. Oecologia (Berl) 31:95–109

Green TGA, Lange OL, Cowan IR (1994) Ecophysiology of lichen photosynthesis: the role of water status and thallus diffusion resistances. Crypt Bot 4:166–178

Green TGA, Sancho LG, Pintado A, Schroeter B (2011) Functional and spatial pressures on terrestrial vegetation in Antarctica forced by global warming. Polar Biol 34:1643–1656

Hájek J, Vaczi P, Barták M (2009) Photosynthetic electron transport at low temperatures in the green algal foliose lichens *Lasalia pustulata* and *Umbilicaria hirsuta* affected by manipulated levels of ribitol. Photosynthetica 47(2):199–205

Hájek J, Vaczi P, Barták M, Jahnova L (2012) Interspecific differences in cryoresistance of lichen symbiotic algae of genus Trebouxia assessed by cell viability and chlorophyll fluorescence. Cryobiology 64(3):215–222

Hájek J, Barták M, Hazdrová J, Forbelská M (2016) Sensitivity of photosynthetic processes to freezing temperature in extremophilic lichens evaluated by linear cooling and chlorophyll fluorescence. Cryobiology 73:329–334

Hamada N, Okazaki K, Shinozaki M (1994) Accumulation of monosaccharides in lichen mycobionts cultured under osmotic conditions. Bryologist 97:176–179

Hancock RJ, Seppelt RD (1988) Habitat specifity and morphological variation in two Antarctic *Usnea* species. Polarforschung 58:171–179

Harańczyk H (2003) On water in extremely dry biological systems. Wydawnictwo Uniwersytetu Jagiellońskiego, Seria Rozprawy Habilitacyjne Uniwersytetu Jagiellońskiego Nr 361, stron 276; ISSN 0239-782X; ISBN 83-233-1713-5

Harańczyk H, Soga KG, Rumm RJ, Pintar MM (1991) Can we see, by proton spin relaxation, a percolation transition upon drying controlled pore size glass? Mag Res Imag 9:723–726

Harańczyk H, Gaździński S, Olech M (2000) Freezing protection mechanism in *Cladonia mitis* as observed by proton magnetic relaxation, new aspects in cryptogamic research, contribution in honour of Ludger Kappen. Bibl Lichenologica 75:265–274

Harańczyk H, Grandjean J, Olech M (2001) Low temperature effect in D_2O-hydrated Antarctic lichen *Himantormia lugubris* as observed by ^{1}H NMR. Molec Phys Rep 33:220–224

Harańczyk H, Grandjean J, Olech M (2003a) Freezing of water bound in lichen thallus as observed by ^{1}H NMR. I. Freezing of loosely bound water in *Cladonia mitis* at different hydration levels. Colloids Surf B Biointerfaces 28(4):239–249

Harańczyk H, Grandjean J, Olech M, Michalik M (2003b) Freezing of water bound in lichen thallus as observed by ^{1}H NMR. II. Freezing protection mechanisms in a cosmopolitan lichen *Cladonia mitis* and in Antarctic lichen species at different hydration levels. Colloids Surf B Biointerfaces 28(4):251–260

Harańczyk H, Pietrzyk A, Leja A, Olech MA (2006) Bound water structure on the surfaces of *Usnea antarctica* as observed by NMR and sorption isotherm. Acta Phys Polonica A109:411–416

Harańczyk H, Bacior M, Olech MA (2008) Deep dehydration of *Umbilicaria aprina* thalli observed by proton NMR and sorption isotherm. Antarct Sci 20:527–535

Harańczyk H, Bacior M, Jastrzębska P, Olech MA (2009) Deep dehydration of Antarctic lichen *Leptogium puberulum* Hue observed by NMR and sorption isotherm. Acta Phys Polon A115:516–520

Harańczyk H, Nowak P, Bacior M, Lisowska M, Marzec M, Florek M, Olech MA (2012a) Bound water freezing in *Umbilicaria aprina* from continental Antarctica. Antarct Sci 24(4):342–352

Harańczyk H, Pater Ł, Nowak P, Bacior M, Olech MA (2012b) Initial phases of Antarctic *Ramalina terebrata* Hook f. & Taylor thalli rehydration observed by proton relaxometry. Acta Phys Polon A121:480–484

Haranczyk H, Nowak P, Lisowska M, Florek-Wojciechowska M, Lahuta LB, Olech MA (2016) A method of water-soluble solid fraction saturation concentration evaluation in dry thalli of Antarctic lichenized fungi, in vivo. Biochem Biophys Rep 6:226–235

Hawksworth DL (1988) Lichen components. IIa The fungal partner. In: Galun M (ed) CRC handbook of lichenology, vol 1. CRC Press, Boca Raton, pp 35–38. *taken from:* Hawksworth DL, Sutton BC, Ainsworth GC (1983), Ainsworth and Bisby's Dictionary of the Fungi, 7th ed., Commonwealth Mycological Institute, Kew, England

Honegger R (1991) Functional aspects of the lichen symbiosis. Ann Rev Plant Physiol 42:553–578

Horwath KL, Duman JG (1984) Yearly variations in the overwintering mechanisms of the cold hardy beetle *Dendroides canadensis*. Physiol Zool 57:40–45

Huiskes AHL, Gremmen NJM, Francke JW (1996) The benefits of being a lichen. Circumpolar J 1–2:77–80

Jänchen J, Meeßen J, Herzog TH, Feist M, de la Torre R, de Vera J-PP (2015) Humidity interaction of lichens under astrobiological aspects: the impact of UVC exposure on their water retention properties. Int J Astrobiol 14(3):445–456

Kallio P, Heinonen S (1971) Influence of short-term low temperature on net photosynthesis in some subarctic lichens. Rep Kevo Subarctic Res Station 8:63–72

Kappen L (1983) Ecology and physiology of the antarctic fruticose lichen Usnea sulphurea (Koenig) Th. Fries Polar Biol 1:249–255

Kappen L (1985) Water relations and net photosynthesis of *Usnea*. A comparison between *Usnea fasciata* (maritime Antarctic) and *Usnea sulphurea* (continental Antarctic). In: Brown DH (ed) Lichen physiology and cell biology. Plenum Press, New York, pp 41–56

Kappen L (1988) Ecophysiological relationships in different climatic regions. In: Galun M (ed) CRC handbook of lichenology, vol 2. CRC Press, Boca Raton, p 37

Kappen L (1989) Field measurements of carbon dioxide exchange of the Antarctic lichen *Usnea sphacelata* in the frozen state. Antarct Sci 1:31–34

Kappen L (1993) Plant activity under snow and ice, with particular reference to lichens. Arctic 46:297–302

Kappen L (1994) The lichen, a mutualistic system – some mainly ecophysiological aspects. Crypt Bot 4:193–202

Kappen L, Friedmann EI (1983) Ecophysiology of lichens in the dry valleys of Southern Victoria Land, Antarctica. II CO2 gas exchange in cryptoendolithic lichens. Polar Biol 1:227–232

Kappen L, Bölter M, Kühn A (1986) Field measurements of net photosynthesis of lichens in the Antarctic. Polar Biol 5:255–258

Kappen L, Bolter M, Kuhn A (1987) Photosynthetic activity of lichens in natural habitats in the maritime Antarctic. Progress and problems in lichenology in the eighties. Bibl Lichenol 25:297–312

Kappen L, Redon J (1987) Photosynthesis and water relations of three maritime Antarctic lichen species. Flora 179:215–229

Kappen L, Schroeter B, Sancho LG (1990) Carbon-dioxide exchange of Antarctic crustose lichens measured with a CO_2/H_2O porometer. Oecologia 82:311–316

Kappen L, Breuer M, Bolter M (1991) Ecological and physiological investigations in continental Antarctic cryptogams. III. Photosynthetic production of *Usnea sphacelata*: diurnal courses, models, and the effect of photoinhibition. Polar Biol 11:393–401

Kappen L, Sommerkorn M, Schroeter B (1995) Carbon acquisition and water relations of lichens in polar regions – potentials and limitations. Lichenologist 27:531–545

Kappen L, Schroeter B, Hestmark G, Winkler JB (1996a) Field measurements of photosynthesis of Umbilicarious lichens in winter. Bot Acta 109:292–298
Kappen L, Schroeter B, Scheidegger C, Sommerkorn M, Hestmark G (1996b) Cold resistance and metabolic activity of lichens below 0°C. Adv Space Res 18:119–128
Kaurin A, Juttila O, Hansen J (1981) Seasonal changes in frost hardiness in cloudberry Rubus chamaemorus in relation to carbohydrate content with special reference to sucrose. Physiol Plant 52:310–314
Kieft TL (1988) Ice nucleation activity in lichens. Appl Environ Microbiol 54:1678–1681
Kieft TL, Ahmadjian V (1989) Biological ice nucleation activity in lichen mycobionts and photobionts. Lichenologist 21:355–362
Kieft TL, Ruscetti T (1990) Characterization of biological ice nuclei from a lichen. J Bacteriol 172:3519–3523
Kimball SL, Salisbury FB (1974) Plant development under snow. Bot Gaz 135:147–149
Lange OL (1965) Der CO_2 Gaswechsel von Flechen bei tiefen Temperaturen. Planta 64:1–19
Lange OL (1966) CO_2 -Gaswechsel der Fleche *Cladonia alcicornis* nach langfristigem Aufenthalt bei tiefen Temperaturen. Flora Abt B 156:500–502
Lange OL (1969) Experimentell-okologische Untersuchungen an Flechten der Negev-Wüste. I. CO_2 -Gaswechsel von *Ramalina maciformis* (Del.) Bory unter kontrolierten Bedingungen in Laboratorium. Flora Abt B 158:324–359
Lange OL, Kappen L (1972) Photosynthesis of lichens from Antarctica. In: Llano GA (ed) Antarctic terrestrial biology, Antarctic Research Series, vol 20. American Geographical Union, Washington, DC, pp 83–95
Lange OL, Metzner H (1965) Lichtabhängiger Kohlenstoff-Einbau in Flechen bei tiefen Temperaturen. Naturwissenschaften 8: 191
Lange OL, Kilian E, Ziegler H (1986) Water vapour uptake and photosynthesis of lichens: performance differences in species with green and blue-green algae as phycobionts. Oecologia 71:104–110
Lange OL, Green TGA, Ziegler H (1988) Water related photosynthesis and carbon isotope discrimination in species of lichen genus *Pseudocyphellaria* with green or blue-green photobionts in photosymbiodemes. Oecologia 75:494–501
Larson DW (1978) Patterns of lichen photosynthesis and respiration following prolonged frozen storage. Can J Bot 56:2119–2123
Lindow SE (1982a) Epiphytic ice nucleation active bacteria as incitants of frost-injury and plant disease. In: Lacy G, Mount M (eds) Phytopathogenic prokaryotes. Academic, New York, pp 344–362
Lindow SE (1982b) Population dynamics of epiphytic ice nucleation active bacteria on frost sensitive plants and frost control by means of antagonistic bacteria. In: Li PH, Sakai A (eds) Plant cold hardiness and freezing stress, vol 2. Academic, New York, pp 395–416
Lindow SE (1983) The role of bacterial ice nucleation in frost injury to plants. Annu Rev Phytopathol 21:363–384
Lindow SE, Arny DC, Upper CD (1978) Distribution of ice nucleation – active bacteria on plants in nature. Appl Environ Microbiol 36:831–838
Marchand PJ (1987) Life in the cold, an introduction to winter ecology. University Press of New England, Hanover
Meeßen J, Sánchez FJ, Brandt A, Balzer E-M, de la Torre R, Sancho LG, de Vera J-P, Ott S (2013a) Extremotolerance and resistance of lichens: comparative studies on five species used in Astrobiological research I. Morphological and anatomical characteristics. Orig Life Evol Biosph 43:283–303
Meeßen J, Sánchez FJ, Sadowsky A, de la Torre R, Ott S, de Vera J-P (2013b) Extremotolerance and resistance of lichens: comparative studies on five species used in Astrobiological research II. Secondary lichen compounds. Orig Life Evol Biosph 43(6):501–526
Meeßen J, Wuthenow P, Schille P, Rabbow E, de Vera JP, Ott S (2015) Resistance of the lichen Buellia frigida to simulated space conditions during the preflight tests for BIOMEX-viability assay and morphological stability. Astrobiology 15(8):601–615

Melick DR, Seppelt RD (1992) Loss of soluble carbohydrates and changes in freezing point of Antarctic bryophytes after leaching and repeated freeze-thaw cycles. Antarct Sci 4:399–404

Melick DR, Seppelt RD (1994) The effect of hydration on carbohydrate levels, pigment content and freezing point of *Umbilicaria decussata* at a continental Antarctic locality. Crypt Bot 4:212–217

Nash TH III, Kappen L, Loesch R, Larson DW, Matthes-Sears U (1987a) Cold resistance of lichens with Trentepohlia- or Trebouxia- photobionts from the North American West Coast. Flora 179:241–251

Nash TH III, Kappen L, Loesch R, Matthes-Sears U, Larson DW (1987b) Cold resistance of lichens. Progress and problems in lichenology in the eighties. Bibl Lichenol 25:313–317

Nash TH III, Reiner A, Demmig-Adams B, Kilian E, Kaiser WM, Lange OL (1990) The effect of atmospheric desiccation and osmotic water stress on photosynthesis and dark respiration of lichens. New Phytol 116:269–276

Nowak P, Harańczyk H, Kijak P, Marzec M, Fitas J, Lisowska M, Baran E, Olech MA (in press) "Bound water behaviour in *Cetraria aculeata* thalli during freezing", submitted to Polar Biology, POBI-D-16-00226

Pérez-Ortega S, de los Ríos A, Crespo A, Sancho LG (2010) Symbiotic lifestyle and phylogenetic relationships of the bionts of *Mastodia tessellata* (Ascomycota, incertae sedis). Am J Bot 97(5):738–752

Phelps P, Giddings TH, Prochoda M, Fall R (1986) Release of cell-free ice nuclei by *Erwinia herbicola*. J Bacteriol 167:496–502

Raggio J, Pintado A, Ascaso C, De La Torre R, De Los RA, Wierzchos J, Horneck G, Sancho LG (2011) Whole lichen thalli survive exposure to space conditions: results of Lithopanspermia experiment with Aspicilia fruticulosa. Astrobiology 11(4); AST-2010-0588-Raggio_1P, https://doi.org/10.1089/ast.2010.0588

Sadowsky A, Ott S (2012) Photosynthetic symbiont in Antarctic terrestrial ecosystems: the physiological response of lichen photobionts to drought and cold. Symbiosis 58(1–3):81–90

Sadowsky A, Ott S (2016) Symbiosis as a successful strategy in continental Antarctica: performance and protection of Trebouxia photosystem II in relation to lichen pigmentation. Polar Biol 39(1:Special Issue: SI):139–151

Salisbury FB, Kimball SL, Bennett B, Rosen P, Weidner M (1973) Active plant growth at freezing temperatures. Space Life Sci 4:124–138

Sancho LG, Valladares F, Schroeter B, Kappen L (2000) Ecophysiology of Antarctic versus temperate populations of a bipolar lichen: the key role of the photosynthetic partner. In: Davison W, Howard-Williams C, Broady P (eds) Antarctic ecosystems: models for wider ecological understanding. The Caxton Press, Christchurch, pp 190–194

Sancho LG, de la Torre R, Horneck G, Ascaso C, de los Rios A, Pintado A, Wierzchos J, Schuster M (2007) Lichens survive in space: results from the 2005 LICHENS experiment. Astrobiology 7(3):443–454

Sancho LG, de la Torre R, Pintado A (2008) Lichens, new and promising material from experiments in astrobiology. Fungal Biol Rev 22(3):103–109

Scher H, Zallen R (1970) Critical density in percolation processes. J Chem Phys 53:3759–3761

Schipperges B, Kappen L, Senesson M (1995) Intraspecific variations of morphology and physiology of temperate to Arctic populations of *Cetraria nivalis*. Lichenologist 27:517–529

Schlensog M, Schroeter B, Sancho LG, Pintado A, Kappen L (1997) Effect of strong irradiance on photosynthetic performance of the melt-water dependent cyanobacterial lichen *Leptogium puberulum* (Collemataceae) Hue from the maritime Antarctic. In: Kappen L (ed) New species and novel aspects in ecology and physiology of lichens. In honour of O.L. Lange.: Bibl. Lichenol. vol 76, pp 235–246

Schroeter B (1994) In situ photosynthetic differentiation of the green algal and the cyanobacterial photobiont in the crustose lichen *Placopsis contortuplicata*. Oecologia 98:212–220

Schroeter B, Scheidegger C (1995) Water relations in lichens at subzero temperatures: structural changes and carbon dioxide exchange in the lichen *Umbilicaria aprina* from continental Antartctica. New Phytol 131:273–285

Schroeter B, Green TGA, Kappen L, Seppelt RD (1994) Carbon dioxide exchange at subzero temperatures. Field measurements on Umbilicaria aprina in Antarctica. CryptBot 4:233–241

Schroeter B, Olech M, Kappen L, Heitland W (1995) Ecophysiological investigations of *Usnea antarctica* in the maritime Antarctic. I. Annual microclimatic conditions and potential primary production. Antarct Sci 7:251–260

Schulze ED, Lange OL (1968) CO_2 -Gaswechsel der Fleche *Hypogymnia physodes* bei tiefen Temperaturen im Freiland. Flora Abt B 158:180–184

Stauffer D, Aharony A (1991) Introduction to percolation theory. Taylor & Francis, London

Stillinger FH (1980) Water revisited. Science 209:451–457

Strzalka K, Szymanska R, Suwalsky M (2011) Prenyllipids and pigments content in selected Antarctic lichens and mosses. J Chil Chem Soc 56:808–811

Tieszen LL (1974) Photosynthetic competence of the subnivean vegetation of an arctic tundra. Arct Alp Res 6:253–256

Tieszen LL, Lewis MC, Miller PC, Mayo J, Chapin III FS, Oechel W (1981) An analysis of processes of primary production in tundra growth forms. In: Bliss LC, Heal OW, Moore IJ (eds) Tundra ecology: a comparative analysis, pp 285–356

Valladares F, Sancho LG, Ascaso C (1998) Water storage in the lichen family *Umbilicariacae*. Bot Acta 111:99–107

Worland MR, Block W, Rothery P (1993) Ice nucleation studies of two beetles from sub-antarctic South Georgia. Polar Biol 13:105–112

Worland MR, Block W, Oldale H (1996) Ice nucleation activity in biological materials with examples from antarctic plants. Cryo-Letters 17:31–38

Yankofsky SA, Levin Z, Bertold T, Sandlerman N (1981) Some basic characteristics of bacterial freezing nuclei. J Appl Meteorol 20:1013–1019

Adaptation in Algae to Environmental Stress and Ecological Conditions

4

Sanjeeva Nayaka, Kiran Toppo, and Sushma Verma

Abstract

Algae (including *Cyanobacteria*) are aquatic organism and ubiquitous in distribution. They are found not only in fresh and marine waterbodies but also on terrestrial habitats such as soil, tree trunk and man-made substrates. As they heavily depend on water, it becomes a limiting factor for their survival. However, algae are found growing abundantly in extreme habitats indicating their adaptation to the harsh environment which we try to explore in this chapter. The response of algae to desiccation stress is widely studied. They produce specialized spores that would remain dormant during harsh period and revive once the favourable conditions return. Their thick cell walls would have further protective layers of chemical substances and also mucilage sheath which helps in the delay of desiccation. Algae produce and accumulate varieties of organic osmolytes that protect them from desiccation, high irradiation and UV light. Algae also have *de novo* biosynthesis mechanism to manage the damage occurred due to desiccation. Algae occurring in colder habitats have substances in their cells that would withstand sub-zero temperatures. Algae growing in saline habitats accumulate salt and maintain ionic balance with the cellular concentration. Although algae are also found in hot springs, not many studies are available to explain their adaptive strategies.

Keywords

Desiccation · Hot spring · Resting spores · Extreme environments · Evolution · Astaxanthin

S. Nayaka (✉) • K. Toppo • S. Verma
Algology Section, CSIR-National Botanical Research Institute, Lucknow, India
e-mail: nayaka.sanjeeva@gmail.com

V. Shukla et al. (eds.), *Plant Adaptation Strategies in Changing Environment*,
https://doi.org/10.1007/978-981-10-6744-0_4

4.1 Introduction

Algae are aquatic, oxygen-evolving, photosynthetic autotrophs that are unicellular, are colonial or are constructed of filaments or composed of simple tissues (Guiry 2012). As algae contain photosynthetic pigments, they are considered as plants, but unlike vascular plants, they lack true stems, roots, leaves, vascular tissue and flowers. Algae include organisms that are not closely related to each other and are mostly classified based on their pigments, reserve food material and mode of reproduction. While some of the authors consider only eukaryotes as algae (Allaby 1992; Nabors 2004), some also include prokaryotes *Cyanobacteria* (blue-green algae) (Fritsch 1935, 1945).

Algae are important component of biodiversity and an integral part of the ecosystem. Their contribution to the existence of life on earth is immense. Algae are the primary producers in the aquatic environment, and the oxygen that all organisms depend on was generated by numerous cyanobacteria during the Achaean and Proterozoic Eras. It is estimated that marine algae contribute to about 70–80% of the oxygen in the atmosphere. The chloroplast with which plants photosynthesize is actually a cyanobacterium living within the plant's cells. More than 100 species of algae serve as food to human beings. Use of seaweeds as food in China or Japan and high-protein-yielding *Chlorella* and *Spirulina* products available in the market are few examples. Some algae yield antibiotics, for example, chlorellin is obtained from green alga *Chlorella*, that inhibits the growth of certain bacteria. Because of high iodine contents, brown algae are used in the manufacture of various goitre medicines (McCledon 1993). Due to their ability to fix atmospheric nitrogen algae, especially *Cyanobacteria* are important source of biofertilizers. Many algae yield certain chemical products that have huge industrial applications such as agar-agar, alginates and carrageenin (Borowitzka and Hallegraeff 2007). Sometimes algae can also be harmful; algal bloom that increases the biological oxygen demand of water and obstructs light penetration results in death of aquatic animals. In the sea, red tides are best examples for mass destruction of fishes and contaminating seafood with toxins (Hallegraeff 1993).

4.2 Habitat Range for Algal Colonization

Algae are ubiquitous in distribution and represent a huge diversity in the world with an estimate of 72,500 species (Guiry 2012). They are free-living, prominent in waterbodies and also common in terrestrial environments such as soil, tree bark and rock. Sometimes algae are extremophiles, found growing in extreme environments such as hot springs, deserts, cold waterbodies, polar regions, permafrost, high or low pH and high levels of CO_2 (Seckbach 2007). Algae are also symbiotic, living in association with other organisms. For example, alga and fungus associate to form lichens wherein both the partners lose their original identity to behave as single entity (Nash 1996). Dinoflagellates are often endosymbionts in the cells of the corals, where they accelerate host-cell metabolism by generating sugar and oxygen through photosynthesis using incident light and the carbon dioxide produced by the host (Lesser et al. 2016). Some algae are parasites of plants such as *Cephaleuros* that causes red rust on tea and coffee leaves (Joubert and Rijkenberg 1971).

4.3 Adaptation to Desiccation Stress

Algae are poikilohydric organisms and don't have control over their water balance. They equilibrate with surrounding moisture, and when fully hydrated, intact cells will be able to function physiologically (Karsten and Holzinger 2012). Therefore, availability of water or moisture is the most important limiting factor for their survival. Water availability includes precipitation, condensation and water vapour. Algae growing in terrestrial habitats such as soil, rock, ephemeral ponds, ditches, seasonal streams, bark and man-made substrates experience severe desiccation, especially during summer. However, in places like alpine regions, water availability fluctuates from fluid droplets after rain or snow to extended periods of dryness or freezing (Holzinger and Karsten 2013). These algae adapt several survival strategies to exist in such stressed environment and are referred as desiccation-tolerant organism. The desiccation tolerance can be defined as the ability to survive drying to about 10% remaining water content (equivalent to 50% of relative humidity) at 20 °C (Oliver et al. 2010).

4.3.1 Stress Avoiders

Avoiding desiccation would be one of the strategies of the unicellular or filamentous algae. They aggregate into colonies and form multilayered mats and biofilms. While the outer layer is exposed and susceptible to damage, the cells underneath are protected, which can be referred as self-shading. The outer layer may be bleached or change its colour and acts as light screen in case of low-light-requiring communities. This strategy is more useful when more than one community is involved in grouping as in case of biological soil crusts. The whole community is protected by the contribution of each individual (Holzinger and Pichrtova 2016). Lüttge and Büdel (2010) observed that green algal biofilms on tree bark showed a pronounced desiccation tolerance up to 80 days and the recovery of photosynthetic activity was faster after shorter periods of dehydration. Similarly, green algae from biological soil crusts in the desert can survive at least 4 weeks (Gray et al. 2007). Desert algae survived desiccation for at least 4 weeks and recovered to high levels of photosynthetic quantum yield within 1 h of rehydration when experimented in laboratory condition (Gray et al. 2007).

The primary effect of dehydration in algae may be shrinkage process which may reduce the cells to less 60% of their original value (Karsten and Holzinger 2012). In case of *Klebsormidium* cells subjected to desiccation experiment, overall the cytoplasm appeared extremely dense and cell organelles including the nucleus and chloroplast were visible. The number of plastoglobules in chloroplasts was increased indicating capability of cell organelles to reorganize during desiccation (Karsten and Holzinger 2012). The plastoglobules are lipoprotein subcompartments of the chloroplast which contain biosynthetic metabolic enzymes that are responsible for desiccation tolerance (Fenández-Marín et al. 2013).

4.3.2 Resting Spores

Algae form specialized cells as strategy to survive adverse conditions. These stress-tolerant cells may be dormant zygotes, oospores and zygospores resultant of sexual reproduction and capable of continuing life cycle of the alga (Fig. 4.1a–c). The zygotes have thicker cell wall and inner wall of cortical cells, sometimes encrusted with lime and referred as gyrogonites (Leliaert et al. 2012). In *Coleochaete*, the inner zygote cell wall layers contain material similar to sporopollenin, which provides protection from both desiccation and UV radiation (Delwiche et al. 1989; Kroken et al. 1996). The dormant zygote is revived by the change in environmental conditions such as moisture, temperature and light. The oospores are overwintering cells that can survive anoxic condition on lake bottom in case of *Charophyceae* (Holzinger and Pichrtova 2016). The zygospores reported in *Coleochaetophyceae* (Delwiche et al. 1989) and *Zygnematophyceae* (Stancheva et al. 2012) are triple layered and contain crucial stress tolerance enzymes and acetolysis-resistant materials. Other specialized spores may be parthenospores, akinetes, hormogonia and aplanospores. The parthenospores are resultant of incomplete conjugation, and they are the gametes that did not find compatible sexual partner. The akinetes are developed directly from vegetative cells and have thick cell wall due to deposition of wall material (Fig. 4.1d, e). Hormogonia are motile filaments of cells produced by *Cyanobacteria* when exposed to environmental stress. Hormogonia are common in *Nostocales* and *Stigonematales* formed during asexual reproduction (Fig. 4.1f, g). They are capable of establishing symbiotic association with plant for nitrogen fixation. Aplanospores are formed within the vegetative cells while protoplast is shirking (Kadlubowaska 1984). *Haematococcus*, a green alga, is known to produce aplanospores (also called as haematocysts, hypnoblasts) during stress conditions. Suseela and Toppo (2006) observed that during summer, several seasonal ponds and lakes in Palampur, India, turned red due to the presence of *H. pluvialis* (Fig. 4.1h). As the water starts to dry, green algae lose their flagella and become spherical, and their cell wall becomes thick and turns red producing aplanospores rich in astaxanthin. The astaxanthin is a pink (or red)-coloured ketocarotenoid, which protects the alga from adverse environmental changes such as photo-oxidation and cell membranes and other sensitive structures against free radical attack (Suseela and Toppo 2006).

4.3.3 Dormant Vegetative Cells

In several cases, algae do not produce specialized cells to survive stress. They remain dormant in vegetative state with reduced physiological activity. They can be considered as true desiccation-tolerant species. The stress-tolerant vegetative cells

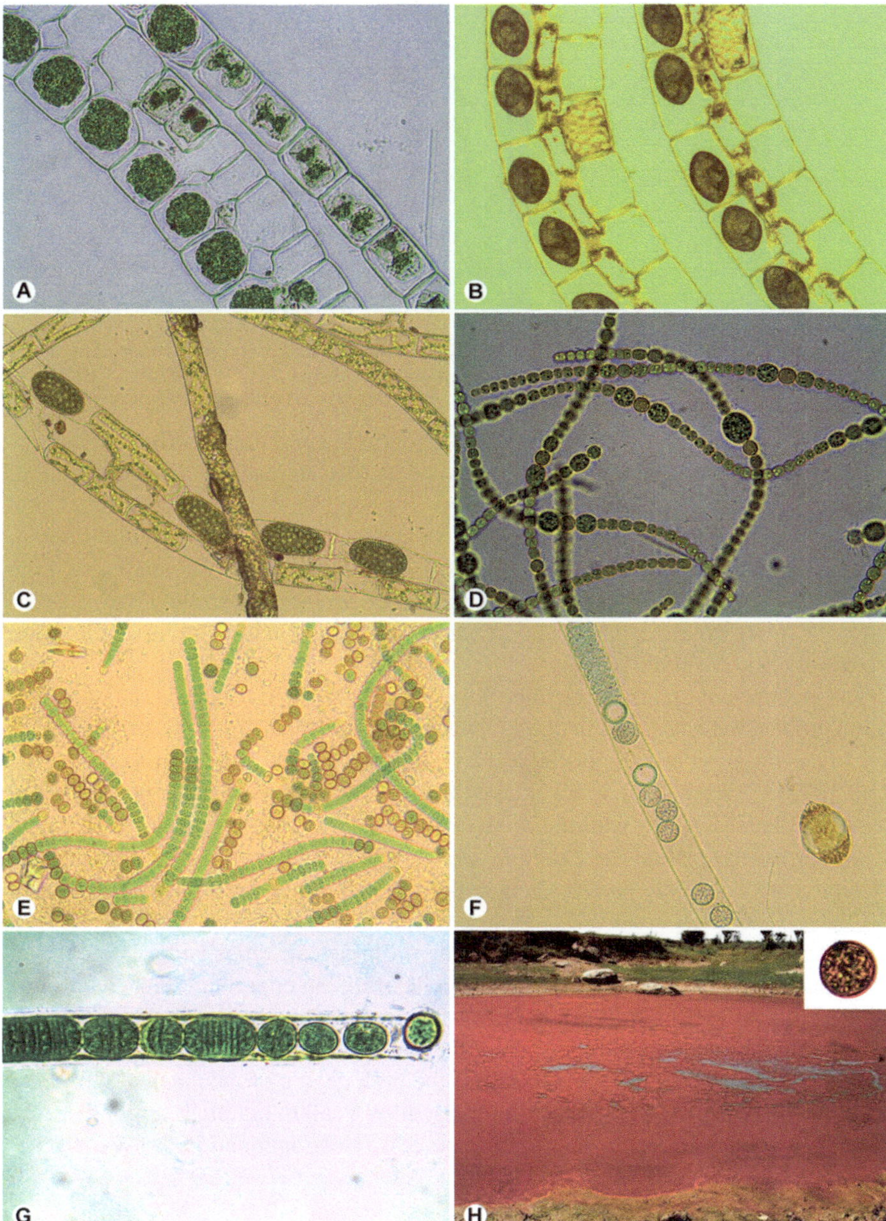

Fig. 4.1 Resting spores seen in algae. (**a**) Zygospore formation in *Zygnema* sp. (**b** and **c**) Different species of *Spirogyra*. (**d** and **e**) Akinetes in *Anabaena* spp. (**f** and **g**) Hormogonia (**f**) *Lyngbya* sp. (**g**) *Oscillatoria* sp. (**h**) Lake turned red in Palampur, Himachal Pradesh, due to the presence of *Haematococcus pluvialis*; inset aplanospore

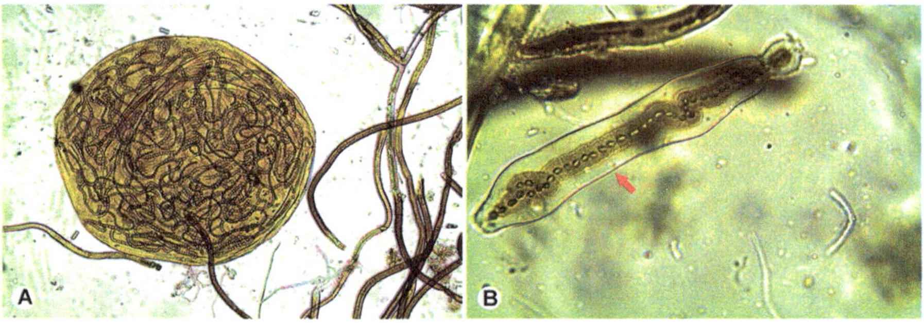

Fig. 4.2 Thick mucilage sheath protecting *Cyanobacteria*. (**a**) *Nostoc* colony enveloped by mucilage. (**b**) Single filament of *Nostoc* sp. with thick mucilage sheath

resume their physiological activity when favourable condition returns. Such hardened stress-tolerant vegetative cells are called as 'pre-akinete' or 'mature cell' and are proven strong for desiccation as well as freezing (Herburger et al. 2015). Upon desiccation, carbohydrate and lipid contents increased, and protein content decreased (Morison and Sheath 1985). However, recent proteomic analyses of the lichen green alga showed that dehydration caused an increase in the relative abundance of few proteins, while the proteins associated with Calvin cycle were decreased (Gasulla et al. 2013).

The presence of external mucilage cover on the algal cells not only delays desiccation process but also helps in quickly absorbing moisture (Fig. 4.2a, b). The mucilage layer has also been suggested to play a significant role in UV protection (Lütz et al. 1997). These cells may also have thick cell wall which further facilitates the stress tolerance. The cell wall of the algae is made of complex substances that may be similar to land plants. For example, charophyte cell wall contains polymers that are similar to cellulose, pectins, hemicellulose, arabinogalactan proteins, extensin and lignin (Sørensen et al. 2011; Domozych et al. 2012). Some of these cell wall components might facilitate prolonged water-holding capacities and tolerate desiccation stress. Callose found in cell wall is a flexibilizing compound that can protect algae from desiccation-induced damage when they follow the shrinkage of the protoplast. In some algae such as *Klebsormidium* in desiccated samples, undulating cross-walls were observed. This indicates the high degree of mechanical flexibility and structural integrity in the dried state that allows cell to maintain turgor pressure for prolonged period during dehydration process (Holzinger and Pichrtova 2016).

4.3.4 Physiological Changes

Physiologically, desiccation results in reduction of the cellular water potential and increases cellular ionic concentration (Holzinger and Karsten 2013). Desiccation also directly affects photosynthesis, the electron transport system in thylakoid membranes in algae, and hence may result in photoinhibition or even photodamage

(Wieners et al. 2012). There will be rapid reduction of photosynthesis during desiccation and quick recovery after rewetting. The loss of water in the presence of light increases the stress leading to generation of reactive oxygen species (ROS) (Fernandez-Marin et al. 2016). ROS are the most possible source of damage to nucleic acids, proteins and lipids. Particularly, the hydroxyl radical is extremely reactive and easily hydroxylates purine and pyrimidine bases of DNA, thus enhancing mutation rates (Halliwelli 1987). To fight the oxidative stress tolerant species upregulate photoprotection mechanism through photoinhibition (Roach and Krieger-Liszkay 2014). Some high-light-tolerating algae also contain increased concentration of zeaxanthin in PSII antenna (Stamenkovic et al. 2014) and phenolic compounds (Pichrtova et al. 2013). The CO_2 exchange is another important physiological parameter that is affected during desiccation. In case of alga *Apatococcus lobatus*, most favourable carbon assimilation was measured at 97–98% relative humidity (RH); however, lower limit of carbon assimilation was observed at 68% RH (Bertsch 1966).

4.3.5 Stress-Tolerant Substances

Algae produce and accumulate a variety of metabolites such as organic osmolytes that protect them from stress conditions and act as antioxidants, cryoprotectants and heat protectants. These organic osmolytes are chemically diverse, comprising sugars (sucrose, glucose, raffinose, xylose, galactose, mannose, inositol, fructose, glycerol, mannitol, sorbitol), amino acids and their derivatives (Welsh 2002; Yancey 2005; Kaplan et al. 2012). Among them, sucrose seems to be the major organic osmolyte found in several stress-tolerant algae. For example, the extraction of soluble sugars from Antarctic *Zygnema* yielded 95% of the sucrose (Hawes 1990). Concentration of composition of amino acid has been observed during desiccation.

In the algae growing in high UV radiation (cold desert, alpine areas), mycosporine-like amino acids are produced as UV screening compounds. Mycosporines are colourless, water-soluble substances with peak absorbance between 310 and 360 nm (Shick and Dunlap 2002). They can function also as antioxidants and are involved in osmotic regulation (Oren 2007). The phenolics are the other group of substances that are useful in screening UV. Some phenolic substances cause strong pigmentation of the vacuoles, and so they protect photosynthetic apparatus from excessive PAR irradiation (Holzinger and Pichrtova 2016). Purple pigmentation (purpurogallin derivatives) can be seen in some algae such as *Mesotaenium berggrenii* growing in alpine and Arctic glacier (Remias et al. 2012). Production of red astaxanthin by *H. pluvialis* is already discussed earlier, and recently, Shah et al. (2016) discussed in detail the biochemistry, pathway, function and use of this substance. During low nitrogen concentration and high solar irradiation, algae such as *Dunaliella* may produce more than 12% of its dry weight as β-carotene. β-Carotene is a provitamin A carotenoid and a natural antioxidant, which has been used as food and animal feed additive and cosmetic ingredient (Seckbach 2007).

4.3.6 *De novo* Mechanism

Algae have inbuilt repair and *de novo* biosynthesis mechanism to manage the damage occurred due to desiccation. DNA is found to be the only biomolecule that is steadily maintained and repaired, while other molecules such as proteins are degraded and *de novo* synthesized (Holzinger and Krasten 2013).

4.4 Adaptation for Cold

Winter season, temperate and alpine region of the terrestrial habitat impose cold stress to the algae. The cold deserts such as Leh, Ladakh, Antarctica and Arctic are other colder areas where algae are found growing. Algae are also seen in snow, ice, glaciers and permafrost. The occurrence of algae in these habitats indicates that they are psychrotrophs, well adapted to cold. Algae growing in cold desert suffer not only desiccation but also light stress with bright and high UV radiation. In *Zygnema* strains studied from Arctic and Antarctic habitats, a high amount of antheraxanthin and zeaxanthin was observed as light-tolerating mechanism (Pichrtova et al. 2013). Further, the highest de-epoxidation state was found in an Arctic strain, which went along with the highest concentration of phenolic compounds (Pichrtova et al. 2013). Glacier is one of the most extreme habitats on earth, and very few algae reported from here belong to the class *Zygnematophyceae* (Holzinger and Pichrtova 2016). These algae are found mostly in vegetative state, and spores or cysts are not produced even during overwintering. Alpine strains of *Klebsormidium* are markedly resistant to osmotic stress (Kaplan et al. 2012) and can survive freezing at −40 °C (Elster et al. 2008). We have observed that some of the algae, especially cyanobacteria (*Anabaena sp.*), growing in alpine areas are much darker in colour than in lower altitudes (Fig. 4.3a). The optimum temperature for growth of snow algae is generally below 10 °C. During summer months, the snow algae result into blooms giving red, orange, green or grey coloration to the snow depending upon the species. The colouration is due to the transformation of green vegetative cells to colourful resting spore stages. The pigments protect the algal cells from high light and UV radiation damage during the summer months. The spores also have thick walls and large amounts of reserve food in the form of lipid, polyols and sugars. Such spores are able to withstand sub-zero temperatures in winter and desiccation in summer. The cells of some species also secrete large amounts of mucilage which enable them to adhere to one another and to snow crystals and prevent them from being washed away by meltwater. As discussed earlier, the mucilage also forms a protective coat and delays desiccation. It may have an additional function as an UV shield. We have observed in the algal taxa, especially desmids collected from colder area had thick sheath of mucilage giving a feather-like appearance (Fig. 4.3b, c). The other strategies include algae incorporate polyunsaturated fatty acids with decreased chain-lengths into the membrane to maintain membrane fluidity at low temperatures. They also produce compatible solutes (e.g. trehalose) that help to reduce the freezing point of the intracellular fluid. This strategy also reduces cell desiccation as less water is needed to retain

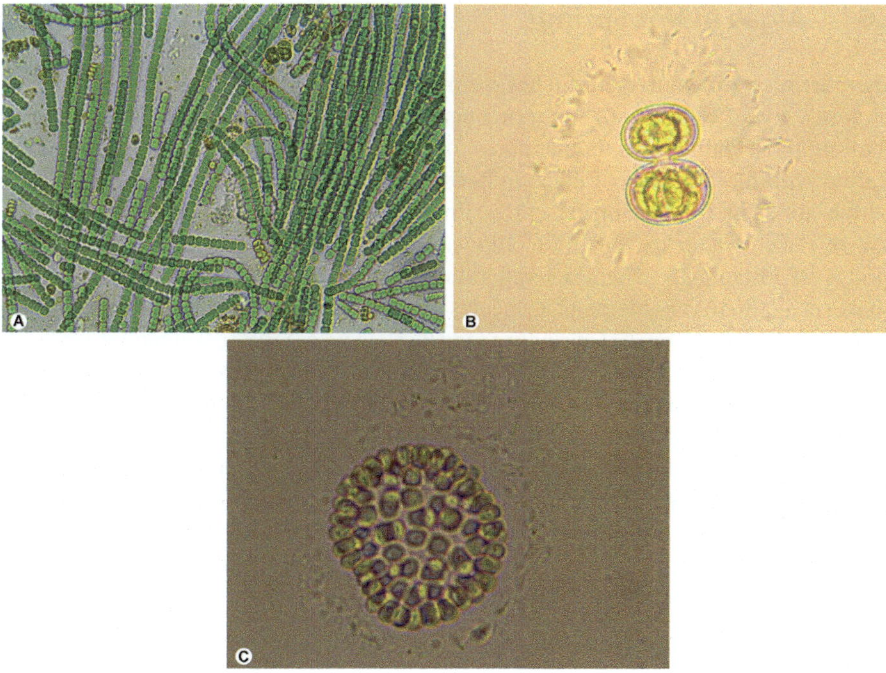

Fig. 4.3 Algae in alpine areas. (**a**) *Dark green* colouration of *Anabaena* sp. in higher-altitude areas, (**b**) feather-like mucilaginous structure around *Cosmarium* cells, and (**c**) *Sphaerocystis* sp.

the osmotic equilibrium (Welsh 2002). Further, extracellular compounds such as polymeric substances help to reduce ice nucleation around the cells (Vincent 2007).

4.5 Algae in Saline Habitats

According to evolutionary theories, algae originated from sea which has high-saline environment. However, during evolution, algae lost their degree of salt resistance (Fodorpataki and Bartha 2004). Apart from sea and ocean, algae also occur in saline water such as saline pond, lakes and brine channels in the sea. In these algae, the increase in the salt concentration in the environment is countered by accumulation of salt by algae to maintain osmotic balance. They uptake inorganic ions to balance the extracellular ion concentration and produce organic osmolytes as long-term survival strategies (Oren 2000). *Dunaliella salina*, an alga which grows in saline environment, is used as convenient model organism to study general mechanisms of salt adaptation in algae and plants. Katz et al. (2007) identified 55 novel membrane-associated proteins in *D. salina* and showed how changes in the structure and composition of the membrane may help this organism to adapt to high salt content. Algae such as *Scenedesmus opoliensis* excrete high amount of mucilage in which the individuals form extended aggregates (Fodorpataki and Bartha 2004).

4.6 Algae in Hot Springs

Algae are also known to occur in hot springs, and such thermophiles are considered as closest living relatives of microorganism present on early earth (Stetter 2006). They are most abundant at temperatures of 55 °C (131 °F) or below. Several studies dealing with the diversity of algae in hot springs are available; however, the information about their adaption is scarce. Prasad and Srivastava (1965) collected 24 taxa of *Cyanobacteria* from four (Tapovan, Badrinath, Manikaran, Bashisht) hot springs of Himalayas wherein temperature ranged from 45 to 97 °C. Recently, Bhakta et al. (2016) enumerated 50 species of algae (*Cyanobacteria*, Chlorophyceae, Bacillariophyceae) from hot springs in Odisha state where water temperature ranged from 35 to 60 °C. We have ourselves collected a total of seven genera of algae from Bashisht hot spring, Himachal Pradesh, and their further identification is in progress. The microbial diversity is lower in hot springs, but they exhibit remarkable genomic and metabolic flexibility. The diversity of microbes significantly depends on physicochemical parameters of water as well as geographical location. Badhai et al. (2015) observed that environmental physicochemical parameters of the hot spring not only control microbial composition and diversity but also play an important role in the dispersal of biological functions and adaptive responses of the communities.

4.7 Conclusion

Adaption is a dynamic evolutionary process wherein phenotypic or adaptive trait with functional role is passed on to the successive generation through natural selection. The evolution of land plants from aquatic algae was through gradual colonization of algae to moist habitats in proximity of water and from there to dry land. Later, the loss of vegetative desiccation tolerance led to the development of water-regulating structures such as xylem elements to transport water or stomata to regulate transpiration (Holzinger and Karsten 2013). Therefore, it can be concluded that terrestrial algae which have well adapted to harsh environmental conditions are in evolutionary process of achieving next cellular organization like plants. While algae produce several biological substances for their own protection, they also serve as important biomolecules for human welfare. Astaxanthin is one such highly beneficial substance having multiple usage such as antioxidant, treatment of Alzheimer's and Parkinson's disorders, etc. (Suseela and Toppo 2006).

Acknowledgements We are thankful to the director of the CSIR-NBRI, Lucknow, for providing infrastructural facilities, to Dr. Siljo Joseph for organizing the pictures, to members of Algology Laboratory for their support and to the Indo-US Science and Technology Forum, New Delhi, for financial support under i-Craft project.

References

Allaby M (ed) (1992) Algae. The concise dictionary of botany. Oxford University Press, Oxford

Badhai J, Ghosh TS, Das S (2015) Taxonomic and functional characteristics of microbial communities and their correlation with physicochemical properties of four geothermal springs in Odisha. India Front Microbiol 6:1–15

Bertsch A (1966) CO_2 Gaswechsel der Grunalge *Apatococcus lobatus*. Plant 70:46–72

Bhakta S, Das SK, Adhikary SP (2016) Algal diversity in hot springs of Odisha. Nelumbo 58:157–173

Borowitzka M, Hallegraeff GM (2007) Economic importance of algae. In: Algae of Australia: introduction. Algae of Australia. CSIRO Publishing/Australian Biological Resources, Canberra, pp 594–622

Delwiche CF, Graham LE, Thomson N (1989) Lignin-like compounds and sporopollenin Coleochaete, an algal model for land plant ancestry. Science 245:99–401

Domozych DS, Ciancia M, Fangel JU, Mikkelsen MD, Ulvskov P, Williats WGT (2012) The cell walls of green algae: a journey through the evolution and diversity. Front Plant Sci 3(82):1–7

Elster J, Degma P, Kovacik L, Valentova L, Sramkova K, Pareira B (2008) Freezing and desiccation injury resistance in the filamentous green alga *Klebsormidium* from Antarctic, Arctic Slovakia. Biology 63:839–847

Fernandez-Martin B, Holzinger A, Garcia-plazzaola JL (2016) Photosynthetic strategies of desiccation-tolerant organisms. In: Pessarakli (ed) Handbook of photosynthesis, 3rd edn. CRC Press, Boca Raton, pp 719–737

Fernández-Marín B, Kranner I, Sebastián MS, Artetxe U, Laza JM, Vilas JU, Pritchard HW, Nadajaran J, Míguez F, Becerril JM, García-Plazaola JI (2013) Evidence for the absence of enzymatic reactions in the glassy state. A case study of xanthophyll cycle pigments in the desiccation-tolerant moss *Syntrichia ruralis*. J Exp Bot 64(10):3033–3043

Fodorpataki L, Bartha C (2004) Salt stress tolerance of a freshwater green alga under different photon flux densities. Studia Universitatis Babes-Bolyali. Biology 49(2):85–93

Fritsch FE (1935) The structure and reproduction of the algae, vol 1. Cambridge University Press, London

Fritsch FE (1945) The structure and reproduction of the algae, vol 2. Cambridge University Press, London

Gasulla F, Jain R, Barreno E, Guera A, Balbuena TS, Thelen JJ (2013) The response of *Asterochloris erici* (Ahmadjian) Skaloud et Peksa to desiccation: a proteomic approach. Plant Cell Environ 36(7):1363–1378

Gray DW, Lewis LA, Cardon ZG (2007) Photosynthetic recovery following desiccation of desert green algae (Chlorophyta) and their aquatic relatives. Plant Cell Environ 30(10):1240–1255

Guiry MD (2012) How many species of algae are there? J Phycol 48:1057–1063

Hallegraeff GM (1993) A review of harmful algal blooms and their apparent global increase. Phycologia 32(2):79–99

Halliwell B (1987) Oxidative damage, lipid peroxidation and antioxidant protection in chloroplasts. Chem Phys Lipids 44:327–340

Hawes I (1990) Effects of freezing and thawing on a species of *Zygnema* (Chlorophyta) from the Antarctic. Phycologia 29:326–331

Herburger K, Lewis LA, Holzinger A (2015) Photosynthetic efficiency, desiccation tolerance and ultrastructure in two phylogenetically distinct strains of alpine *Zygnema* sp. (Zygnematophyceae, Streptophyta) role of pre akinete formation. Protoplasma 252:571–589

Holzinger A, Karsten U (2013) Desiccation stress and tolerance in green algae: consequences for ultrastructure, physiological and molecular mechanisms. Rev Front Plant Sci 4(327):1–18

Holzinger A, Pichrtova M (2016) Abiotic stress tolerance of Charophyte green algae: new challenges for omics techniques. Rev Front Plant Sci 7(678):1–17

Joubert JJ, Rijkenberg FHJ (1971) Parasitic green algae. Ann Rev Phytopathol 9:45–64

Kadlubowska JZ (1984) Conjugatophyceae I: chlorophyta VIII – zygnemales. In: Ettl H, Gerloff J, Heynig H, Mollenhauer D (eds) Susswasserflora von Mitteleuropa, Band 16 eds. Gustav Fisher, Jena, pp 1–532

Kaplan F, Lewis LA, Wastian J, Holzinger A (2012) Plasmolysis effects and osmotic potential of two phylogenetically distinct alpine strains of *Klebsormidium* (Streptophyta). Protoplasma 249:789–804

Karsten U, Holzinger A (2012) Light temperature and desiccation effects on photosynthetic activity and drought-induced ultra-structural changes in the green alga *Klebsormidium dissectum* (Streptophyta) from a high alpine soil crust. Microb Ecol 63:51–63

Katz A, Waridel P, Shevchenko A, Pick U (2007) Salt-induced changes in the plasma membrane proteome of the halotolerant alga *Dunaliella salina* as revealed by Blue Native Gel Electrophoresis and Nano-LC-MS/MS analysis. Mol Cellul Proteom 6(9):1459–1472

Kroken SB, Graham LE, Cook ME (1996) Occurrence and evolutionary significance of resistant cell walls in charophytes and bryophytes. Am J Bot 83:1241–1254

Leliaert F, Smith DR, Moreau H, Herron MD, Verbruggen H, Delchiche CF (2012) Phylogeny and molecular evolution of green algae. Crit Rev Plant Sci 31:1–46

Lesser MP, Stat M, Gates RD (2016) The endosymbiotic dinoflagellates (*Symbiodinium* sp.) of corals are parasites and mutualists. Coral Reefs 32(3):603–611

Lüttge U, Budel B (2010) Resurrection kinetics of photosynthesis in desiccation tolerant terrestrial green algae (Chlorophyta) on tree bark. Plant Biol 123:437–444

Lütz C, Seillitzz HK, Meindl U (1997) Physiological and structural changes in the chloroplast of the green alga *Micrasterias denticulata* induced by UV-B simulation. Plant Ecol 128:54–64

McClendon JF (1993) Iodine and goiter with especial reference to the far East. J Biol Chem 102:91–99

Morison M, Sheath R (1985) Response to desiccation by *Klebsormidium rivulare* (Ulotrichales, Chlorophyta) a Rhode Island stream. Phycologia 24:129–145

Nabors MW (2004) Introduction to botany. Pearson Education, San Francisco

Nash TH III (ed) (1996) Lichen biology. Cambridge University Press, Cambridge

Oliver MJ, Cushman JC, Koster KL (2010) Dehydration tolerance in plants. In: Sunkar R (ed) Plant stress tolerance, Methods in molecular biology, vol 639. Springer, New York, pp 3–24

Oren A (2000) Salts and brines. In: Whitton BA, Potts M (eds) Ecology of *Cyanobacteria*: their diversity in time and space. Kluwer Academic Publishers, Dordrecht, pp 281–306

Oren A (2007) Diversity of organic cosmetic compounds and osmotic adaptation in Cyanobacteria and algae. In: Seckbach J (ed) Algae and cyanobacteria in extreme environments. Springer, Berlin, pp 639–655

Pichrtova M, Remias D, Lewis L, Holzinger A (2013) Changes in phenolic compounds and cellular ultrastructure of Arctic and Antarctic strains of *Zygnema* (Zygnematophyceae, Streptophyta) after exposure to experimentally enhanced UV to PAR ratio. Microbiol Ecol 65:68–83

Prasad BN, Srivastava PN (1965) Thermal algae from Himalaya hot springs. Proc Nat Acad Sci India 31B(1&2):45–53

Remias D, Schwaiger S, Aigner S, Leya T, Stuppner H, Lutz C (2012) Characterization of an UV- and VIS-absorbing, purpurogallin-derived secondary pigment new to algae and highly abundant in *Mesotaenium berggrenii* (Zygnematophyceae, Chlorophyta) an extremophyte living on glaciers. FEMS Microb Ecol 79:638–648

Roach T, Krieger-Liszkay A (2014) Regulation of photosynthetic electron transport and photo inhibition. Curr Protein Pept Sci 15:351–362

Seckbach J (2007) Algae and cyanobacteria in extreme environments. Springer, Dordrecht

Shah MM, Liang Y, Cheng JJ, Daroch M (2016) Astaxanthin producing green microalga Haematococcus pluvialis: from single cell to high value commercial products. Front Plant Sci 7(531):1–28

Shick JM, Dunlap WC (2002) Mycosporine-like amino acids and related gadusols: biosynthesis accumulation and UV protective functions in aquatic organisms. Ann Rev Physiol 64:223–262

Sorensen L, Pettolino FA, Bacic A, Ralph J, Lu F, O'Neill MA (2011) The Charophycean green algae provide insights into the early origins of plant cell walls. Plant J 68:201–211

Stamenkovic M, Bischof K, Hanelt D (2014) Xanthophyll cycle pool size and composition in several *Cosmarium* strains (Zygnematophyceae, Streptophyta) are related to their geographic distribution patterns. Protist 165:14–30

Stancheva R, Sheath RG, Hall JD (2012) Systematics of the genus *Zygnema* (Zygnematophyceae, Charophyta) from Californian watersheds. J Phycol 48:409–422

Stetter KO (2006) Hyperthermophiles in the history of life. Philos Trans R Soc Lond Ser B Biol Sci 361(1474):1837–1843

Suseela MR, Toppo K (2006) *Haematococcus pluvialis* – a green alga, richest natural source of astaxanthin. Curr Sci 90(12):1602–1603

Vincent WF (2007) Cold tolerance in cyanobacteria and life in the cryosphere. In: Sechbach J (ed) Algae and cyanobacteria in extreme environments. Spinger, Dordrecht, pp 287–301

Welsh D (2002) Ecological significance of compatible solute accumulation by micro-organisms: from single cells to global climate. FEMS Microbiol Rev 24:263–290

Wieners PC, Mudimu O, Bilger W (2012) Desiccation induced non-radiative dissipation in isolated green lichen algae. Photosynth Res 113:239–247

Yancey P (2005) Organic osmolytes as compatible, metabolic and counteracting cytoprotectants in high osmolarity and other stresses. J Exp Biol 28:2819–2830

Biochemical and Molecular Responses in Higher Plants Under Salt Stress

5

Neerja Srivastava

Abstract

There are vast stretches of wasteland all over the world due to high concentration of salt in the soil. Salinity creates problem due to its effects on crop species which are predominantly salt sensitive. Although saline soils can be remediated through various means, these approaches are not very practical in vast areas and also unsustainable in the long run. Traditional means of ameliorating saline soil through excess irrigation water to leach salts below root zone have to be complemented with the genetic approaches like screening of germplasm and marker-assisted breeding for high salt tolerance. Therefore, genetic and physiological approaches should merge into the unifying more comprehensive approach to breeding for salt tolerance. Extending our knowledge on physiological mechanism of salt tolerance is of utmost importance in developing plants better adapted to saline soils. However, information on mechanism of salt tolerance is lacking.

Keywords

Salt stress · Salt tolerance · Salt stress-responsive genes · Marker-assisted breeding

5.1 Introduction

According to FAO (2009), world agriculture will face a lot of challenges like arrangement of more food for an additional 2.3 billion people by 2050, fight with poverty and hunger, efficient use of scarce natural resources, and adapting to climate change. However, with increase in food demand, the productivity of crops is

N. Srivastava (✉)
Department of Biochemistry, IBSBT, CSJMU, Kanpur, UP, India
e-mail: neerjas70@gmail.com

V. Shukla et al. (eds.), *Plant Adaptation Strategies in Changing Environment*,
https://doi.org/10.1007/978-981-10-6744-0_5

not increasing. In most of the cases, the loss of productivity is due to various abiotic stresses. An environmental factor that limits crop productivity or destroys mass is referred as a stress or disturbance (Grime 1979). A major area of concern with additional food requirements is crop losses due to these environmental stresses (Shanker and Venkateswarlu 2011). Various abiotic stresses like salt, drought, high or low temperature, flood, metal toxicity, ozone, UV radiations, herbicides, etc. are experienced by plants and are serious threat to crop production (Bhatnagar-Mathur et al. 2008; Ahmad and Prasad 2012a, b). These abiotic stresses directly reduce the crop yield up to 50% (Rodríguez et al. 2005; Acquaah 2007). The survival, biomass production, and reduction in yield of staple food crops up to 70% are greatly affected by major abiotic stresses like drought, high salinity, cold, and heat (Vorasoot et al. 2003; Kaur et al. 2008; Ahmad et al. 2010, 2012; Thakur et al. 2010; Mantri et al. 2012), which also threaten the food security worldwide. One of these major stresses, salinity in soil or water, is an important worldwide problem in crop production (Ashraf and Harris 2004). Earth's land has become salty due to accumulation of water-soluble salts especially sodium chloride (NaCl), sodium carbonate (Na_2CO_3), and partially calcium chloride ($CaCl_2$) (Nawaz et al. 2010). Development of saline soil takes place due to high amount of chloride or sulfate salts of sodium (Wyn Jones 1981). In fact, the most prominent potentially toxic ions of saline substrates are sodium and chloride (Hassanein et al. 2009). Salinity creates a problem due to its effects on crop species which are predominantly sensitive to the presence of high salt concentration in the soil. The most obvious effects of salt are observed in arid and semiarid regions, where limited rainfall, high evapotranspiration, and high temperature associated with poor water and soil management practices are the major contributing factors (Azevedo Neto et al. 2006; Nawaz et al. 2010). Approximately one-third of the land in the semiarid regions reflects some degree of salinity accumulation (Bresler et al. 1982). Although the general perception is that salinization only occurs in arid and semiarid regions, no climatic zone is free from this problem (Rengasamy 2006). More than 800 million hectare land worldwide is affected by either salinity (397 million hectares) or sodicity (434 million hectares), (FAO 2005; Munns 2005; Farooq et al. 2015). Nearly 20% of the world's cultivated land and nearly half of the world's irrigated lands are affected by salinity (Zhu 2001). According to Rengel (1992), increasing demands for quality water due to both population rise and industrial development are likely to increase usage of low-quality, brackish water (even sea water) in agriculture, thus aggravating the salinity problem. There will be more and more usage of marginal lands (including salt-affected soils) for crop production, because the best lands will already be farmed. Traditional means of ameliorating salt-affected soils through reclamations, drainage, and use of excess irrigation water to leach salts below the root zone have to be complemented with the genetic approaches by screening germplasm and breeding crops for higher salt tolerance. Genetic and physiological approaches should merge into the more comprehensive approach to breeding for salt tolerance. Extending our knowledge on physiological mechanism of salt tolerance is of utmost importance in developing plants better adapted to saline soils. However, information on mechanism of salt tolerance is lacking (Kwon et al. 1995).

Much of the physiological research into salinity has concentrated on three topics: water relations, photosynthesis, and the accumulation of a particular metabolite assuming that one or more of these processes would limit growth in saline conditions. Most of these studies have been descriptive and have established patterns of responses in many crops, but they have not elucidated mechanism at either the biochemical or whole plant level (Munns 1993).

5.2 Types and Causes of Salinity

Salinity is one of the most prominent abiotic stresses which limit the crop productivity. Salinity affects a very large area of land in the world which is increasing day by day. Salinity is more prominent problem in irrigated crop lands. Worldwide around 17% of the cultivated land is under irrigation, and irrigated agriculture contributes more than 30% of the total agriculture productivity (Hillel 2000). In the world, almost 20% of total irrigated land is salt affected (Pitman and Läuchli 2002). However, depending upon the sources, there is variation in this data. It is observed that 65% of the total land in the world is affected either salinity or sodicity (FAO 2008). This accounts for 831 M ha of land (Hasanuzzaman et al. 2013).

Soil salinity is caused by various factors. On the basis of these causes, it can be classified into two major forms:

1. Natural or primary salinity
2. Secondary or human-induced salinity

The primary salinity is caused by the long-term accumulation of salts in the soil or surface water. This is a natural process which is caused mainly by weathering of parent materials containing soluble salts through breakdown of rocks containing Cl^- of Na^+, Ca^{2+}, and Mg^{2+} and sometimes SO_4^{2-} and CO_3^{2-}. In addition deposition of sea salt carried by wind and rain is also a reason, which varies with the type of soil. Secondary salinity occurs due to anthropogenic activities that disrupt the hydrologic balance of the soil between water applied (irrigation or rainfall) and water used by crops (transpiration) (Munns 2005; Manchanda and Garg 2008). In many irrigated areas, the water table has risen due to excessive amounts of applied water together with insufficient drainage. Most of the irrigation systems of the world have caused secondary salinity, sodicity, or waterlogging (Manchanda and Garg 2008). In the irrigated land, after irrigation, the water applied to the soil is consumed either by the crop or evaporates directly from the moist soil. The excess salt is remained and accumulated in the soil which is called salinization. It is sometimes visible as a whitish layer of dry salt on the soil surface. In addition, salted groundwater may also contribute to salinization. Due to excessive irrigation and improper drainage, the water table rises which allow the salty groundwater to reach the upper soil layers and rhizosphere.

Szabolcs (1974) has classified the soil into different types based on the nature, characteristics, and plant growth relationship in salt-affected soils, which are:

(a) Saline soils – In this type of soil, the soluble salts are mainly NaCl and Na_2SO_4. Sometimes they also contain appreciable amount of Cl^- and SO_4^- of Ca^{2+} and Mg^{2+}. Sufficient amount of neutral soluble salts are also present in this type of soil which negatively affects growth of the most crop plants.
(b) Sodic soils – They are rich in Na + salts especially Na_2CO_3 which are capable of alkaline hydrolysis. These soils are also called as "alkali."

Further categories of salt-affected soils, which though less extensive are commonly found in the different parts of the world, are:

(c) Acid sulfate soils – This kind of soil is found within a 50 cm depth which is directly or indirectly caused by H_2SO_4 and has pH below 3.5–4.0. The H_2SO_4 is formed by the oxidation of pyrite (FeS_2) or other reduced sulfur compounds and is accelerated by brackish and saline mangrove swamp. This soil has not only high salinity but iron (Fe) and aluminum (Al) toxicities and deficiency of phosphorus (P) (Pons 1973; Abrol et al. 1988).
(d) Degraded sodic soils – These soils are formed by the washing out of salts and are an advanced stage of soil development. During this process, there is a formation of a dark and extremely compact layer because of the movement of dispersed clay and organic matter down the profile. This layer has a sharp distinct upper surface, and it merges gradually into the subsoil with increasing depth. These soils lost most of its original exchangeable Na^+ through leaching (Abrol et al. 1988).

The saline water is also classified into different types by Pitman and Lauchli (2002) on the basis of salt concentration.

1. Freshwater – having total dissolved salts below 500 mg/L and EC below 0.6 dSm^{-1}
2. Slightly brackish water – has total dissolved salts between 500 and 1000 mg/L and EC between 0.6 and 1.5 dSm^{-1}
3. Brackish water – having total dissolved salts between 1000 and 2000 mg/L and EC between 1.5 and 3.0 dSm^{-1}
4. Moderately saline water – has total dissolved salts between 2000 and 5000 mg/L and EC value between 3.0 and 8.0 dSm^{-1}
5. Saline water – has total dissolved salts between 5000 and 10,000 mg/L and EC between 8.0 and 15.0 dSm^{-1}
6. Highly saline water – has total dissolved salts between 10,000 and 35,000 mg/L and EC between 15 and 45 dSm^{-1}

5.3 Plant Responses to Salt Stress

The property of salinity tolerance is not a simple attribute, but it is an outcome of various features that depend on different physiological interactions, which are difficult to determine. The morphological appearance presented by the plant in response to salinity may not be enough to determine its effect, so it is important to recognize other physiological and biochemical factors, including toxic ions, osmotic potential, lack of elements, and other physiological and chemical disorders, as well as the interactions between these various stresses (Munns 1993, 2002a; Neumann 1997; Yeo 1998; Hasegewa et al. 2000; Qados 2011).

Various anatomical, ecological, physiological, and molecular responses of plants under salt stress have been studied by different workers (Tal 1984; Sachs and Ho 1986; Hurkman 1992; Wang et al. 1997). Plants reprogram their whole metabolism from growth and development to enhanced stress tolerance induction (Kosová et al. 2013). Various physiological and biochemical mechanisms have been developed by plants in order to survive in soils with high salt concentration.

It is clear therefore that there are various mechanisms of salinity tolerance and many of these can be present in a particular plant. Till date, there is no conclusive proof that these mechanisms are exclusive for one type of stress (i.e., ion exclusion prevents tolerance to the "osmotic phase" of salt toxicity) nor that a particular plant is committed to only one strategy (e.g., a plant may have ion exclusion as its primary tolerance mechanism at moderate salinity but then has tissue tolerance as its main tolerance mechanism when the exclusion processes are "swamped" at high salinity). It is quite possible that some of these tolerance mechanisms are more effective in particular circumstances. For example, Na^+ exclusion may be more effective in conditions of higher salinity (Munns et al. 2012), whereas "osmotic tolerance" may be more important in moderately saline conditions. Their interactions with other abiotic stresses, such as low water availability, may also be important (Roy et al. 2014).

5.4 Growth Responses

As far as growth is concerned, there is a plethora of data. It relates to every degree of tolerance, from the most sensitive glycophytes to the most resistant halophytes, and plants are normally categorized on the basis of their responses to salt (Greenway and Munns 1980). Arbitrarily, plants can be classified into two different classes on the basis of salt tolerance. Those plants that complete their life cycle in ≥200 mm concentration of salt, most commonly NaCl, are known as "halophytes" (Flowers and Colmer 2008), while all less tolerant species are termed as "glycophytes." However, there is a lot of variation in salt tolerance among plant species, from very sensitive species such as chickpea (*Cicer arietinum* L.) and rice (*Oryza sativa* L.) to the most tolerant halophytes (Greenway and Munns 1980). Halophytes have the ability to tolerate high concentration of Na^+ and Cl^- in their shoots, which would be damaging or even lethal in non-halophytes. Monocotyledonous halophytes contain lower Na^+ concentrations than dicotyledonous halophytes and have higher selectivity for K^+ uptake (Albert 1975; Flowers and Colmer 2008; Flowers et al. 2015).

In plants, one of the initial effects of salt stress is the reduction in growth rate. Plant growth can be affected by salinity stress in many ways. First, the water uptake capacity of the plant is reduced under the presence of salt in the soil which causes very fast reduction in the growth rate. The osmotic effect of the soil solution that contains salt causes first phase of the growth response and produces a package of effects similar to water stress (Munns 2002b). The time period over which a plant is exposed to salt decides mechanism by which salinity affects growth of the plant. Munns (2002b) summarized the sequential events of the plant grown in salinity. According to him, water is lost from cells in the first few seconds or minutes, and they become shrinked. The cells regain their original volume after a few hours, but the elongation rate is still reduced. This causes lower growth rates of leaf and root. Cell division rates are also affected over days and further contribute toward lower rates of leaf and root growth. Then, changes in vegetative development over weeks are observed, and changes in reproductive development can be seen over months. According to this, Munns (2005) developed the two-phase growth response to salinity (Hasanuzzaman et al. 2013) and categorized the effect of salt stress on growth impairment in all plants as well as in crops in two – short and long terms. In the category of short-term effects comes osmotic stress, lowering of external water potential, and reduction of water uptake by plants, all similar to the situation under drought (Wang et al. 2003). However, in the long-term effects, ion toxicity occurs when crops are not able to compartmentalize ions properly (Aghaei and Komatsu 2013).

Plant growth is inhibited by salt stress throughout the life cycle, but generally the most sensitive stage is seed germination. It seems that the main effect of salt is osmotic because salt stress mimics water stress in many ways. However, there must be some other specific toxic effects as some salts are more inhibitory than others. One of the most common salt sodium chloride is among the less toxic salts, but it is one of the most troublesome to agriculture (Bliss et al. 1984).

It can be observed from the result of the many studies, which looked at the effect of salt stress on the growth, that with the increase in the concentration of sodium chloride, there is decrease in plant length (Beltagi et al. 2006; Mustard and Renault 2006; Gama et al. 2007; Jamil et al. 2007a, b; Houimli et al. 2008; Rui et al. 2009; Memon et al. 2010). At different concentrations of NaCl, leaf area is negatively affected in many of these studies (Raul et al. 2003; Netondo et al. 2004; Mathur et al. 2006; Chen et al. 2007; Zhao et al. 2007; Yilmaz and Kina 2008; Rui et al. 2009; Qados 2011).

According to the different studies done by Raul et al. (2003), Jamil et al. (2005), Gama et al. (2007), Ha et al. (2008), Quados (2011), etc., as the concentration of salt increases, the harmful influence of salinity on leaf number also increases. The fresh weight and dry weight of the shoot system are affected either negatively or positively by changes in concentration and type of salt present as well as type of plant species (Bayuelo Jimenez et al. 2002; Jamil et al. 2005; Niaz et al. 2005; Saqib et al. 2006; Turan et al. 2007; Saffan 2008; Rui et al. 2009; Taffouo et al. 2009, 2010; Memon et al. 2010; Quados 2011).

There are also some reports about the anatomical changes of plant organs in the salinity stress. According to Kurth et al. (1986) and Lauchli and Schubert (1989), the cortical cells of the cotton roots were longer and narrower in salt-treated plants than those of cortical plants. They have also observed that cell production declined with the increase in salt concentration. Ewing (1981) reported that adding NaCl to the nutrient solution width of the cortex as well as the diameter of cortical cells increased the *Atriplex prostrata* stem.

Another significant factor which may contribute to the growth reduction in plants may be related to the inhibition by vascular tissue production under salinity stress (Shininger 1979; Ewing 1981). Lignifications of the stems of the halophyte *Suaeda maritima* have been shown to be negatively correlated with salinity (Hagege et al. 1988). Ewing (1981) also found that with increasing salinity, the width of anomalous secondary growth in the fourth internodes of *A. prostrata* decreases significantly (Wang et al. 1997).

5.4.1 Ion Accumulation

It is reported that sodium is an essential micronutrient element for higher plants having C4 pathway (Brownell 1965; Brownell and Crossland 1972). Survival of angiospermic halophytes at high salinity is invariably accompanied by high ion content. While glycophytes exclude ions in response to high salinity, the majority of these species in practice are leaf excluders which accumulate high levels of Na^+ in their roots and stems.

The concept of toxicity of various mineral elements in plants (including essential nutrients) is long standing in the study of plant mineral nutrition (e.g., Marschner 1995). Toxicity can be described, in an agronomic context, as the soil (i.e., external) concentration at which the additional supply of a particular element causes a decline in plant growth (Römheld 2012). According to him, for establishing a critical toxic concentration of a mineral element in soil, here is a need to define the criteria used to describe growth reduction like (90% of maximal growth). For establishing "tissue tests" as a diagnostic tool to determine the deficiency, sufficiency, or toxicity status of crops, internal (i.e., leaf tissue) concentrations to plant growth/yield responses are frequently related (Reuter and Robinson 1986) in the area of plant nutrition. These critical concentrations can then be used as a diagnostic guide for crops which might be based on economic yield, but it is difficult to do in a wild species especially in perennials, because survival is the functional criterion in these cases, especially where there may be competition from other plants and a variable environment (Flowers et al. 2015).

One of the deleterious effect of salt stress is nutrient imbalance and/or ion toxicity (Gorham et al. 1985), which may include low level of beneficial K^+ and Ca^{++} and higher level of toxic Na^+ and Cl^- and (Flowers et al. 1977; Greenway and Munns 1980). Plant cell can counter these deleterious effects of salt stress by two different processes. Firstly, they have to either exclude or sequester Na^+ and Cl^- to avoid their toxicity in the cytoplasm, and secondly, they have to maintain appropriate cellular level of K^+ and Ca^{2+} which is necessary for metabolic activities in the cell. Actually these two mechanisms are considered as reliable criteria for salinity tolerance (Greenway and Munns 1980; DuPont 1992; Munns 1993; Mansour et al. 2003).

Sodium is the chief toxic ion, as it not only interferes with the uptake of K^+ but also disturbs stomatal regulation and ultimately causes water loss and necrosis, while Cl^- results in impaired production of chlorophyll which is a chlorotic toxicity symptom. Many of the physiological disorders in plants are although induced through both Na^+ and Cl^-, but Cl^-, is more dangerous (Tavakkoli et al. 2010). In plant cell, for regulation of some enzyme activities in the cytoplasm, Cl^- is needed. It also regulates turgor and pH and is also a cofactor in photosynthesis. However, at higher concentration, it is toxic to plants. As for Cl-sensitive species, the critical levels for toxicity are reported to be 4–7 mg g^{-1}; for Cl-tolerant species, it is found to be 15–50 mg g^{-1} (Xu et al. 2000; White and Broadley 2001). There is significant reduction in growth and water use efficiency of plants by higher accumulation of Cl^- (Hasanuzzaman et al. 2013).

For ion toxicity in plants, when concentrations of Na^+ or Cl^- within a tissue inhibit the metabolism (and growth), it is defined as ion toxicity, while the response is decided by the concentrations within a specific compartment such as the cytoplasm. The concept of "tissue tolerance" was given by Yeo and Flowers (1983) on the basis of the differences in the severity of chlorosis that genotypes of rice showed and were not simply related to differences in the concentrations of Na^+ in these tissues. According to their concept, tissue tolerance is that concentration of Na^+ in the leaf at which chlorophyll concentration was reduced by 50% (Yeo and Flowers 1983). They assumed that the observed tissue damage and the dysfunction as reflected by the degree of chlorosis resulted from Na^+ toxicity. In addition, they also tried to establish a quantitative screening method that can determine quantitative trait loci (QTLs) for the selection of this trait in breeding more salt-resistant rice varieties. The halophytes do not have thousands of varieties which are available for rice, but varieties that differ at least to some degree in tolerance and have relatively similar leaf morphologies and physiologies are available. Therefore, it is very difficult to compare the changes in a metabolic function in relation to leaf ion concentration in diverse species of halophytes and is more complicated to interpret. Furthermore, it is not possible simply to relate cytoplasmic ion concentrations to whole-tissue concentrations because halophytes can tolerate vacuolar Na^+ and Cl^- concentrations $\geq$500 mm. Cytoplasmic ion concentrations could double when tissue concentrations only change by about 10% (a doubling of concentration in a tissue with 3% cytoplasm and 80% vacuole), because the cytoplasm is a very small proportion of the tissue. It is impossible to evaluate whether these ions reach toxic levels in halophytes or whether other factors are responsible for the reduction in growth and eventually death with rise in external salt concentration (Flowers et al. 2015).

Na^+ and Cl^- are accumulated by plants in their shoots at varying concentrations. Glycophytes generally will die at those concentrations in leaves at which halophytes survive like, Na^+ concentration in the leaves was 500–600 mM in the halophyte *S. maritima* based on a water content of 10 g g^{-1} dry weight (Flowers and Yeo 1986) at external salt concentration of 340 mM (Yeo and Flowers 1980). In some halophytes, growth continues at about 1.5 MNa^+ shoot tissue concentration on a tissue water basis (e.g., *Tecticornia* species; English and Colmer 2013), while rice (*Oryza sativa*) cannot tolerate Na^+ concentrations based on tissue water in the leaves much above 100 mm for long (Ul Haq et al. 2013). As vacuoles make up the major part of leaf volume (55% in mature sunflower, Fagerberg 1984; 72.5% in mesophyll cells of *S.*

maritima growing in 340 mm NaCl, Hajibagheri et al. 1984), the majority of Na^+ and Cl^- accumulated in them. This means tissue concentrations largely reflect vacuolar concentrations, and differences in tissue tolerance presumably reflect differences in abilities to accumulate Na^+ and Cl^- within vacuoles. Therefore, if these ions are not compartmentalized in the vacuoles, they will become toxic. However, according to Cheeseman (2013), this potential toxicity is associated with a sensitivity of many enzymes or cellular functions like signaling systems with higher concentrations of Na^+ and/or Cl^-, and the tolerable concentrations in cytoplasm are poorly defined (Flowers et al. 2015).

5.4.2 Water Relations

Water relations are integral part of any study of salt tolerance. There is a prevailing dogma that conductance and cell expansion are regulated by turgor, and hence plant growth in soils of low water potential is also regulated by it. Hundreds of studies on water relations and osmotic adjustments are based on this. In various studies changes in water relation of plants can be seen under salinity stress. It confirms that by increasing the negativity of the osmotic potential of the leaf sap, many plants undergo osmotic regulation under salt stress (Rodriguez et al. 1997; Gama et al. 2007, 2009; Kaymakanova and Stoeva 2008; Kaymakanova et al. 2008; Qados 2011).

Many plant processes are affected by salt increase in the root medium by decreasing leaf water potential (Romero–Aranda et al. 2001). Osmotic effects of salt on plants result from lowering of the soil water potential because of the increase in solute concentration in the root zone. This interferes with plant's ability to extract water from the soil and maintain turgor at very low soil water potentials. However, plants adjust osmotically (accumulate solutes) at low or moderate salt concentration (higher soil water potential) and maintain a potential gradient for the influx of water. A significant decrease in relative water content (RWC) in sugar beet varieties is caused by salt treatment (Ghoulam et al. 2002). A decrease in RWC indicates a loss of turgor that results in limited water availability for cell extension processes (Katerji et al. 1997). Steudle (2000) reported that in transpiring plants, water is thought to come from the soil to the root xylem through apoplastic pathway due to hydrostatic pressure gradient. However, this situation changes under salt-stressed condition, because of the restricted transpiration. In these situations, more of the water follows cell-to-cell path, flowing across membranes of living cells (Vysotskaya et al. 2010; Hasanuzzaman et al. 2013).

Plants need to adjust osmotically to maintain a positive turgor pressure for growing in saline soil. Therefore, total solute concentration must be greater than that of the external solution as turgor pressures have rarely been measured directly; for most halophytes, how much great it should be is not known. But generally turgor is lower (0.05–0.3 Mpa) than values in glycophytes with access to adequate water (around 0. MPa; Boyer 2009). In the cells of a plant growing in seawater of −2.3 MPa osmotic potential, maintaining a turgor pressure of 0.3 MPa requires solutes which can generate about −2.6 MPa cellular osmotic potential. This is equal to about

530 mM NaCl. On a tissue water basis, halophytes generally contain about >500 mM Na^+ and Cl^- concentrations (Flowers 1985) and some up to almost 2 M Na^+ (e.g., *Tecticornia* species at extreme salinity; English and Colmer 2013; Flowers et al. 2015).

The major components of the cellular osmotic potential in dicot halophytes are Na^+ and Cl^-. High concentrations of Na^+ and Cl^- in tissues are accommodated by intracellular compartmentation and the synthesis of compatible solutes. Vacuoles compartmentalized the bulk of these ions, and the osmotic potential of the cytoplasm is adjusted by organic solutes (Flowers and Colmer 2008). Vacuoles require ion exchangers and the H^+ pumps located on tonoplast that generate the electrochemical difference of H^+ across the tonoplast for successful Na^+ and Cl^- sequestration. H^+ pumps also contribute to the membrane potential that influences channel transport activity (Flowers and Colmer 2008; Munns and Tester 2008; Hasegawa 2013; Shabala 2013). Due to charge differences, sequestration of Na^+ is energetically more expensive than sequestration of Cl^-, because the potential inside the vacuole is positive relative to the cytoplasm (Rea and Poole 1993). The energetics also depend upon the concentration differences and type/mechanism of the transporters involved (Shabala 2013). If significant amounts of ATP are not to be utilized, then efficient retention of ions in the vacuoles is also required with these transport mechanisms, like very low membrane permeability to leakage of Na^+ and Cl^- back to the cytoplasm (Yeo 1981, 1983; Greenway and Munns 1983) to reload Na^+ (and Cl^-) at the tonoplast into the vacuoles. This kind of suggestion was also given for plasma membrane by Britto and Kronzucker (2006). This low permeability of the tonoplast may be achieved through control of Na^+-permeable SV (slow vacuolar) and FV (fast vacuolar) channels (Shabala 2013). There are many SV channels in the leaves of the halophyte *Suaeda maritima* which under physiological conditions largely remain closed because of the low gating frequency, which prevents the ions from leaking to the cytoplasm (Maathuis et al. 1992). In old leaves of *Chenopodium quinoa*, a similar situation occurs where there is a decrease in SV channel activity compared with young leaves (Bonales-Alatorre et al. 2013). At least in one halophyte *S. maritima* an additional feature of the tonoplast, the lipid composition is likely to contribute to low leakage of Na^+. It has a high phospholipid/protein ratio (1.1:1.0) with highly saturated fatty acids and a significant (30%) amount of cholesterol in the membrane sterols (Leach et al. 1990a, b; Flowers et al. 2015).

5.4.3 Effects on Photosynthesis

There are lots of studies on the effects of salt stress on photosynthesis. Mostly workers assumed that rate of photosynthesis limits growth in saline environment. This is based upon the frequent observations that photosynthesis is reduced in salt-stressed plants.

It has been observed by many workers that salinity reduces photosynthetic capacity in various species (Ashraf 2004; Dubey 2005). In a number of plant species, e.g., in cotton (Pettigrew and Meredith 1994), *Zea mays* (Crosbie and Pearce 1982), *Brassica* species (Nazir et al. 2001) and wheat (Raza et al. 2007; Nawaz et al. 2010), plant growth increased under normal stressed conditions because of higher photosynthetic capacity.

There is reduction in photosynthetic rates in plants under salt stress because of reduction in water potential. High concentrations of Na^+ and/or Cl^- accumulate in the chloroplasts which inhibit photosynthesis. Due to salt stress, either carbon metabolism or photophosphorylation may be affected because photosynthetic electron transport is relatively insensitive to salts (Sudhir and Murthy 2004). There is a positive growth inhibition with marked inhibition of photosynthesis under salt stress (Fisarakis et al. 2001). But many other reports do not show any or little relationship between growth and photosynthetic capacity (Rogers and Noble 1992; Hawkins and Lewis 1993). The photosynthetic rate depends on salt concentration as well as plant species or genotypes under salt stress. Sometimes at low salt concentration, photosynthesis is stimulated by salinity as in *Bruguiera parviflora*. The rate of photosynthesis increased at low salinity while decreased at high salinity, in comparison to stomatal conductance which remained unchanged at low salinity and decreased at high salinity (Parida et al. 2004). The reduction in photosynthetic rates under salt stress is due to some other factors like enhanced senescence, changes in enzyme activity induced by alterations in cytoplasmic structure, and negative feedback by reduced sink activity (Iyengar and Reddy 1996). The reduction in photosynthesis under stress is caused by stomatal conductance which restricts the availability of CO_2 for carboxylation reactions (Brugnoli and Björkman 1992). Loss of water through transpiration is minimized by stomatal closure which affects light harvesting and energy conversion systems thus altering chloroplast activity (Iyengar and Reddy 1996). Higher photosynthetic rates in plants are favored by increased CO_2 diffusion into leaves due to higher stomatal conductance. Altered photosynthetic pigment biosynthesis under salinity is one of the most notable effects (Maxwell and Johnson 2000). Chlorophyll content is decreased under salt stress and is a commonly reported phenomenon. The chlorophyll concentration was used as a sensitive indicator of cellular metabolic state in various studies (Chutipaijit et al. 2011; Hasanuzzaman et al. 2013). Seeman and Sharkey (1986) reported that salinity reduces photosynthetic capacity, and this reduction is independent of stomatal closure in *Phaseolus vulgaris*. This reduction was shown to be a consequence of a reduction in the efficiency of RuBPcase rather than a reduction in the leaf content of photosynthetic machinery. They have shown that salinity reduces the photosynthetic capacity of leaves by (a) reducing the pool of RuBP by an effect on RuBP regeneration capacity and (b) reducing the activity of RuBPcase by an unknown mechanism when RuBP is in limiting supply. By applying NaCl to intact plants of *Mesembryanthemum crystallinum*, a shift in the mode of carbon assimilation from that typical of C3 plants to that of CAM plants is observed which exhibit net carbon gain at night. During the induction period, activity of PEPcase, an enzyme associated with CAM, increased dramatically (Lauchli and Schubert 1989; Treichel et al. 1988; Von Willert et al. 1976). Bethke and Drew (1992) reported that leaves of young bell pepper exposed to 50 mol/m^3 NaCl after 10 or 14 days of salinization showed little change in photosynthetic ability, whereas those treated with 100 or 150 mol/m^3 NaCl had up to 85% inhibition with increase in CO_2 compensation point. Partial stomatal closure occurred with salinization, but reduction in photosynthesis was non-stomatal in origin. It is now frequently noted that the reduction of growth is greater than the decrease in potential photosynthesis and the reduction of shoot growth is much greater than shoot growth reductions. So such single-point comparisons should be interpreted with caution. Reduction in photosynthesis could be due to stomatal and non-stomatal limitations or combinations of both under salt stress. Photosynthetic capacity and Na^+ content in the leaves of rice (Yeo 1998) and wheat (James et al. 2002) are reduced by high accumulation of Na^+ and Cl^- in the leaves, while high Cl^- content in the citrus (Walker et al. 1981) and in the chloroplast of *Phaseolus vulgaris* (Seemann and Critchley 1985) was detrimental to photosynthesis. One of the main factors in plant growth and yield is photosynthesis. Effective exclusion or compartmentalization of the toxic ions decides tolerance of photosynthetic system to salinity. However, the type of species, level of stress, and duration of stress decide the extent of negative effects of salinity on photosynthesizing tissue or on growth (Nawaz et al. 2010).

5.4.4 Accumulation of Metabolites

The synthesis and accumulation of compatible solutes, a class of osmoprotective compounds, are the cellular response of organisms to both long- and short-term salinity stresses. They can build up to high cellular concentrations without inhibiting metabolic processes. They are different types of metabolites including quaternary amines (glycine betaine), amino acids (proline), soluble sugars such as trehalose, and sugar alcohols such as mannitol, sorbitol, and pinitol (Bose et al. 2014; Chinnusamy et al. 2006; Greenway and Munns 1980; Yeo 1998) which operate at various levels of the stress response. These compatible solutes known as osmolytes or osmoprotectants osmotically compensate the cell in the soil solution containing high amounts of NaCl by preventing the loss of water from the cell (Bonhert and Jensen 1996; Chen and Murata 2002). They also stabilize the quaternary structure of proteins and highly ordered state of membranes, thus preventing protein denaturation and membrane destabilization (Yancey et al. 1982). Some of them like mannitol also serve as ROS scavengers, those compounds which are too reactive to be detoxified, such as hydroxyl radical (Bose et al. 2014; Lodeyro and Carrillo 2015).

Halophytes accumulate organic solutes like sucrose, sugar alcohols, proline, and glycinebetaine (Flowers and Colmer 2008; Gil et al. 2013) and most probably osmotically adjust in the cytoplasmic compartments of vacuolated cells (Flowers et al. 1977; Wyn Jones et al. 1977; Greenway and Munns 1980) rather than the whole cell. By the use of ions than organic solutes, cellular osmotic adjustment is more efficiently achieved (Greenway and Munns 1983; Yeo 1983). Their synthesis would divert C and N from the supply of assimilates for growth processes (e.g., for N in *Spartina alterniflora*; Colmer et al. 1996). Evolutionary pressure has resulted in plants using Na^+ and Cl^- as osmotica in vacuoles while preventing these ions rising to toxic concentrations in the cytoplasm because large amounts of energy would be needed to synthesize these organic compounds in sufficient quantities to adjust the whole volume of cells (Flowers et al. 2010). To keep concentrations relatively low, entry of these ions into some critical cells and tissues, such as meristems and very young growing cells, might also be restricted (Greenway and Munns 1980; Flowers et al. 2010).

5.4.5 Enzymes Responses

Much of our understanding of biochemistry is dependent on the study of the activities of enzymes in vitro (Dixon and Webb 1964). During the late 1960s and early 1970s, a number of enzymes found in healthy plant tissues shown to be inhibited by the NaCl concentrations, (Johnson et al. 1968; Flowers 1972a, b; Greenway and Osmond 1972; Osmond and Greenway 1972) as tissue concentrations of Na^+ and Cl^- could not be uniform across the cell and higher concentrations of ions must be compartmentalized within vacuoles than in the cytoplasm (Flowers et al. 1977). This was supported by the presence of organic solutes (mentioned above) in tissues that are neutral with respect to enzyme activity (Wyn Jones et al. 1977; Gibson et al. 1984) or even protective (Pollard and Wyn Jones 1979; Manetas 1990). When ions accumulate in vacuoles, these organic solutes can balance osmotic imbalances across the tonoplast (Flowers et al. 2015). But it was clear even by the mid-1970s that salt-sensitive enzymes might not be as sensitive to Na^+ and/or Cl^- as originally reported. The response to NaCl added in vitro could be altered by the substrate concentration (Osmond and Greenway 1972; Flowers et al. 1976a, b), the preparative procedures used to extract enzymes from tissues (Blackwood and Miflin 1976; Flowers et al. 1976a, b), and the protein concentration (Manetas 1990). Some salt-sensitive enzymes like malate dehydrogenase (Flowers 1972a; Greenway and Osmond 1972) can tolerate, in vitro, NaCl concentrations of around 200 mm without significant loss of activity (Flowers et al. 1976b; Kalir and Flowers 1982). When protein concentration was increased tenfold to 400 $\mu g\ ml^{-1}$, inhibition of the activity of phosphoenolpyruvate carboxylase by NaCl (a salt-sensitive enzyme that has been extracted from *Salsola soda*, concentrated by ammonium sulfate precipitation and then desalted) decreased by 60% (Manetas 1990). Therefore, the functionality of enzyme pathways in the cytosol depends on stabilizing forces requiring ionic strengths of >100 mm (Cheeseman 2013; Flowers et al. 2015).

As there is a clear difference between the protein concentrations in vivo and that used in assays carried out in vitro, it is difficult to know the extent to which the response of enzymes in vitro mirrors that in vivo (Greenway and Osmond 1972; Manetas 1990). According to Cheeseman (2013), enzymes are likely to exist as complexes, and their behavior in the cytoplasm may be very different from that of enzymes in a dilute solution in vitro. Although the effects of NaCl on the activity of such complexes have not been examined, when a complex of enzymes such as the ribosomal system that catalyzes protein synthesis has been investigated, the stability of polysomes in both halophytes and glycophytes decreased as the concentration of KCl increased above 125 mm. Polysomes were less stable in Na^+ than in K^+ and also less stable in Cl^- than in acetate (Brady et al. 1984). Translation of mRNA from

Triticum aestivum, *Pisum sativum*, *Beta vulgaris*, *Chenopodium album*, and *Hordeum vulgare* and the halophyte *Atriplex nummularia* by wheat germ ribosomes (Gibson et al. 1984) or those of *S. maritima* (Flowers and Dalmond 1992) was inhibited by Cl^- at concentrations >80 mm. Tolerance to substitution of K^+ by Na^+ during elongation of message in halophytes *Atriplex isatidea*, *Inula crithmoides*, and *S. maritima* was more than glycophytes *P. sativum*, *O. sativa*, or *T. aestivum* by incorporation of [^{35}S]methionine into protein (Flowers and Dalmond 1992), which suggests about some adaptation of halophytes to a low cytosolic K^+/Na^+ ratio. Na^+ concentration of 100 mm would be tolerable (in the presence of 100 mm K^+) for ribosomes, and Cl^- is more inhibitory than Na^+ at least to this aspect of metabolism. It would be timely to look again at the effects of monovalent ions on the activities of particularly salt-sensitive proteins and some multiprotein complexes (Cheeseman 2013). It would be timely to look again at the effects of monovalent ions on the activities of particularly salt-sensitive proteins and some multiprotein complexes (Cheeseman 2013), although recent work on the enzymes of halophytes has concentrated on changes in expression patterns (Koyro et al. 2013). The response of organelles like mitochondria and chloroplasts to different salt concentrations in vitro could also give some insight on the response of enzyme complexes to changes in salt concentration as these organelles presumably contain proteins at a similar concentration in vivo and in vitro.

Salt stress is countered by expression of biosynthetic enzymes by increasing compatible solute production (Sakamoto and Murata 2000) and enhancing of Na^+ export from cells which establishes homeostasis, another approach against salinity. According to Zhu (2001) in *Arabidopsis*, overexpression of a plasma membrane Na^+/H^+ antiporter provides salt tolerance. Selective ion uptake, ion exclusion, and compartmentalization of Na^+ in vacuoles control the maintenance of low cytosolic salt concentrations in crops (Liu et al. 2012). Ion homeostasis in plant cells is disrupted through excess toxic ions like Na^+, in the cytosol and by the deficiency of essential ions such as K^+ which is the result of ionic stress (Zhu 2002). In this way, various ion transporters, pumps, and channels play important roles in these processes (Zhu 2003; Reddy and Reddy 2004; Aghaei and Komatsu 2013).

In the cytoplasm of potato (Aghaei et al. 2009) and cotton, antioxidant enzymes and soluble proteins (Gossett et al. 1994) protect cells from salt-induced oxidative stress. In tomato, plant growth regulators such as abscisic acid (ABA) and ethylene also play an important role in abiotic stress tolerance (Ouyang et al. 2007; Aghaei and Komatsu 2013).

5.4.6 Changes in Total Proteins Under Salt Stress

Many aspects of plant cell structure and metabolism are affected by salt stress. Cellular adjustment associated with reorganization of cell structure and metabolism osmotic and ionic aspects of salinity. Since proteins are involved in a wide array of cellular processes associated with salt acclimation like signaling, regulation of gene expression and protein metabolism, energy metabolism, redox metabolism, osmolyte metabolism, defense-related proteins, mechanical stress-related proteins, phytohormone, and lipid and secondary metabolism, they play a major role in salt stress acclimation and plant cellular adjustments. Differentially abundant proteins in genetically related plant materials revealing differential stress tolerance can be identified through comparative proteomic analysis (Wang et al. 2008; Vítámvás et al. 2012). Changes in protein posttranslational modifications (PTM) pattern as well as protein activity besides changes in protein relative abundance have been observed under salinity (Kosová et al. 2013).

In order to tolerate salt stress, plant cells may alter their gene expression which results in an increase, decrease, induction, or total suppression of some stress-responsive proteins such as malate dehydrogenase (NADP+) and pyruvate phosphate dikinase (Ngara et al. 2012). In response to salt stress, different families of proteins are newly synthesized, accumulated, or decreased in the crops and may be involved in signaling, translation, host-defense mechanisms, carbohydrate metabolism, and amino acid metabolism. Antioxidant enzymes and chaperonins may encounter salt stressors directly, while other groups of proteins like key enzyme in osmolyte synthesis encounter salt stress indirectly (Wang et al. 2004). Therefore, to know about the various mechanisms of plant response to salt stress and their roles in acquired stress tolerance in crops is of great importance (Aghaei and Komatsu 2013).

In carrot cells overexpression of *ornithine decarboxylase* gene indicated that these cells were significantly more tolerant to salt stress (Bohnert et al. 1995). Salt-tolerant varieties of rice showed higher levels of ABA-responsive proteins in their roots (Moons et al. 1995). Analysis of proteins under salt stress has a great value as proteins have important role in salt stress tolerance in crops. The responses of crops to salt stress as well as many other biotic and abiotic stresses were investigated through proteome analysis which is an important approach (Aghaei and Komatsu 2013).

Wang et al. (2007) identified 23 variety-specific salt-responsive proteins in a comparative study of salt-resistant and salt-sensitive wheat genotypes; ROS-scavenging enzymes like ascorbate peroxidase, dehydroascorbate reductase, peroxiredoxin, and superoxide dismutase were the major group of proteins which increased under salinity in rice which suggests that salt stress results in oxidative stress in rice (Salekdeh et al. 2002; Abbasi and Komatsu 2004). In pea cultivars exposed to NaCl, leaf mitochondrial Mn-SOD and chloroplast Cu/Zn-SOD activities increased under salt stress (Hernandez et al. 1995; Aghaei and Komatsu 2013).

Nucleoside diphosphate kinase and guanine nucleotide-binding protein involved in nucleotide metabolism and enoyl-ACP reductase involved in fatty acid metabolism in rice have been increased under salt stress (Dooki et al. 2006). GTP-binding protein, lectin-like protein, ribulose-1,5-bisphosphate carboxylase/oxygenase (RuBisCO) activase, ferritin, fructose bisphosphate aldolase, photosystem II oxygen-evolving complex protein, oxygen-evolving enhancer (OEE) protein 2, and SOD were also increased in rice (Abbasi and Komatsu 2004). In responses to salt stress in pea, PRPs, phytochelatins, chaperonins, and metallothioneins were also involved (Qureshi et al. 2007). There was significant increase in glutamine synthase, a key enzyme for proline biosynthesis and glycine dehydrogenase crucial for glycine betaine biosynthesis under salinity stress in wheat (*Triticum durum*) (Caruso et al. 2008; Aghaei and Komatsu 2013).

Chitinases, which are PRPs, were accumulated in leaf apoplast of tobacco plants under saline conditions (Dani et al. 2005). In transgenic variety of rice, protein sti1 which is one of the heat shock proteins appeared to be increased in response to salt stress (Kurek et al. 2002; Qureshi et al. 2007). These stress-responsive proteins are expressed in response to abiotic stresses such as heat, cold, drought, salinity, and oxidative stress (Wang et al. 2004). Heat shock proteins prevent aggregation of stress-denatured proteins and facilitate the refolding of proteins in order to restore their native biological functions (Wang et al. 2004; Aghaei and Komatsu 2013).

In *Suaeda aegyptiaca* leaves, oxidative stress tolerance-related proteins, glycine betaine synthesis, photosynthesis, adenosine triphosphate (ATP) production, protein degradation, cyanide detoxification, and chaperone activities were increased under salt stress(Askari et al. 2006). According to Aghaei et al. (2008b), when exposed to salt stress, beta-conglycinin and late embryogenesis-abundant protein were increased in soybean root and hypocotyl (Aghaei and Komatsu 2013).

Vacuolar Na^+/H^+ antiporter could confer salt tolerance in transgenic *Arabidopsis* and tomato plants (Ward et al. 2003). Salt tolerance is improved in maize by the overexpression of NHX1 (Neubert et al. 2005). Huang et al. (2004) identified a novel ethylene-responsive factor called TERF1 in tomato, functioning as a link between the ethylene and osmotic stress pathways. In response to salinity, enoyl-CoA hydratase and phosphoglycerate mutase-like protein were increased in tomato (Amini et al. 2007). In sorghum leaf, majority of the identified salt stress-responsive proteins (28 spots) were energy-related proteins (50.9%). There is an increased abundance of RuBisCO subunits, sedoheptulose bisphosphatase, phosphoribulokinase, ribulose 5-phosphate isomerase, OEE1, malate dehydrogenase, and ascorbate peroxidase under salt stress (Ngara et al. 2012). Osmotin is accumulated in salt adaptation (Qureshi et al. 2007), and its gene is responsible for osmoprotectant synthesis, the most important of which are proline and glycine betaine (Holmstrom et al. 2000). Hajheidari et al. (2005) in sugar beet and Aghaei et al. (2008a) in potato by proteomic approach found osmotin-like protein which increased in drought and salt conditions and can confer salt tolerance to these crops in response to osmotic stress that resulted from salt stress. NADP reductase, aminomethyltransferase, and decarboxylase subunit T in sugar beet can be considered as salt-responsive proteins in this crop which increased in response to salt stress (Wakeel et al. 2011). RuBisCO small subunits, phosphoglycerate kinase, which catalyzes a reduction step in Calvin cycle, and phosphoribulokinase, which catalyzes regeneration of primary CO_2 acceptor RuBP, were all decreased in salt-treated crops such as soybean (Sobhanian et al. 2010) and *Triticum durum* (Caruso et al. 2008; Aghaei and Komatsu 2013) in addition to those proteins which have increased in salt-stressed crops.

These salt-responsive proteins are involved in a variety of metabolic processes, like scavenging of ROS, signal transduction, transcription and translation, transporting, chaperones, cell wall biosynthesis, photosynthesis, processing and degradation, metabolism of energy, amino acids, and hormones (Aghaei and Komatsu 2013).

5.4.7 Changes in Cell Wall Properties

The role of the plant cell wall in plant life emerged with the compartment to compartment proteomics. It appeared that it is not only a physical barrier between the plant cell and the environment but also a very flexible and responsive part of the cell, functionally involved in growth and differentiation, signaling and response to pathogenic attack, and different stresses (Hématy et al. 2009). It is composed of about 95% carbohydrates with cellulose, pectin, and hemicelluloses being the major ones. Also there is a significant amount of proteins both structural and functional of which role in plant cell stress response is arising in recent years. Very few thorough studies on the cell wall proteome of several monocots such as maize (Zhu et al. 2006) and rice (Cho et al. 2009) were conducted in recent years, but still most of the data is based on *Arabidopsis* (Albenne et al. 2013). However, a significant part of these proteins remain with elusive function. About 11% of the cell wall proteins from *Arabidopsis thaliana* are referred to as with "unknown function" in a summary of a number of proteomics studies (Jamet et al. 2008). The percentage for less studied plants including cereals could be expected to be much higher. Even when certain protein is known and established to be upregulated by abiotic stresses, the exact function remains unknown (Zagorchev et al. 2014).

5.4.8 Identification of Induced Cell Wall Proteins

Studying plant cell wall proteins appeared to be a tough task. The obtaining of a pure fraction is usually complicated due to the complexity of the cell wall structure, the manner of binding of the proteins, and the probable contamination from intracellular proteins. Furthermore, the great variation of posttranslational modifications and especially the abundant glycosylation makes the plant cell wall protein fraction extremely complex and difficult to investigate (Jamet et al. 2008). Another issue is the comparatively low percentage of cell wall proteins from the whole cell proteome that impede the isolation of large amount of material needed for a thorough investigation especially of low abundant proteins.

The establishment of an exhaustive classification of CWP is not an easier task than the choice of an appropriate experimental methodology. The first attempts for comprehensive reviews in the mid- and late 1990s were focused mainly on hydroxyproline- and glycine-rich proteins such as extensins, arabinogalactan proteins, and lectins (Showalter 1993; Cassab 1998). Further in the investigation of CWP, a number of proteins with enzymatic or signaling activity emerged. The most recent classification proposed subdivision of CWP into nine different classes (Albenne et al. 2013) including proteins related to the lipid and carbohydrate metabolism, proteins with oxido-reductive and proteolytic enzymatic activities, proteins involved in signaling and molecular interactions, structural proteins, miscellaneous proteins, and protein with unknown functions. Classification of CWP is, however, further rendered difficult due to the multiple functions of some of them. For example, AGPs are currently regarded as signaling proteoglycans but also as molecules that link the cell wall with the plasma membrane and the cytoskeleton, putatively structural function (Ellis et al. 2010). In another class of structural proteins, the extensins are essential for the primary cell wall architecture (Lamport et al. 2011) but belong to the same class as AGPs, the hydroxyproline-rich proteins, and several authors reported chimeric AGPs that share similar structural components as extensins (Showalter et al. 2010; Poon et al. 2012; Zagorchev et al. 2014).

The role of the plant cell wall in every substantial process in plant growth and development is emerging in recent years, and the involvement of the cell wall proteins as the functionally active component gains great attention. It is clear that those proteins are important in stress response and tolerance and along with the thoroughly investigated intracellular mechanisms could provide the necessary knowledge to overcome the negative impact of the environment on agriculture. Therefore, more and more studies on the changes in the structure and function of the cell wall are conducted. But does the increasing quantity of evidences leads to improved quality of our understanding on the intimate mechanisms of stress response? It seems that since now the available data is posing more questions than answers given. Nevertheless, the fast-developing technology is able to provide more and more powerful tools for investigation. A complex investigation on the transcriptome, proteome, and metabolome of halophytes, relative to commercial crops, is needed for a thorough understanding of mechanisms underlying salinity tolerance. Plant cell wall proteome is an inevitable player in the molecular events, following the initial response and further adaptation to stress. The perception of stress signals along with downward signaling events should be of special interest in future studies (Zagorchev et al. 2014).

5.5 Salt Stress-Responsive Genes

There are changes in gene expression in plants on an exposure to salt. A number of salt-responsive genes have been isolated and characterized during the last decade. A wide range of tolerance to salinity exists within crop species from very high (*Beta vulgaris*) to extremely low (*Citrus* spp.), and salt-tolerant and salt-sensitive varieties exist within a given crop species also. To isolate genes involved in conferring salt tolerance, salt-tolerant species have been frequently used which gives an insight of the mechanisms that distinguish them from their salt-sensitive counterparts (Sairam and Tyagi 2004).

To isolate genes whose expression is influenced by salinity in plants, a variety of approaches are used which involved screening of cDNA libraries constructed from mRNA populations isolated from salt-stressed plants or cells. To isolate salt-responsive cDNA clones from these libraries, the most common and successful methods are differential screening with probes derived from mRNA isolated from salt-stressed and non-stressed plant tissue. The mRNA which corresponds to genes preferentially expressed in the salt-affected plants is isolated and characterized.

By using salt-sensitive mutants of yeast that are rescued following transformation with plant cDNA libraries, cDNAs that encode proteins to enhance salt tolerance have been selected. Based on their ability to complement yeast cells deficient in the phosphoprotein phosphatase calcineurin, two cDNA clones of *A. thaliana* were isolated (Lippuner et al. 1996). Genes were also isolated on the basis of their enhanced expression in response to other environmental stresses, especially water deficit, that are salt responsive, as both salinity and water deficit confer an osmotic stress upon plants.

After a salt treatment, changes in gene expression at the transcriptional and post-transcriptional levels occur which were demonstrated by the analysis of protein profiles elicited in plants. There are both qualitative and quantitative changes in the pattern of polypeptide synthesis following salt treatment (Ericscon and Alfinito 1984, Singh et al. 1989, Ramgopal 1987, Chen and Tabaerzadeh 1991, Moons et al. 1997). Analysis of mRNA changes effected by salinity as assessed by in vitro translation of mRNA populations was also done which indicate changes in relative abundance of mRNA under salt (Gulick and Dvorak 1987, Robinson et al. 1990). There is accumulation of unique mRNA also in the roots of a tolerant barley genotype (Ramgopal 1987).

Salt-sensory pathways can be linked to many tolerance responses through transcription factors. Transcription factor (TF) family genes like basic leucine zipper (bZIP) (Yang et al. 2009), WRKY (Jiang and Deyholos 2009), APETALA2/ETHYLENE RESPONSE FACTOR (AP2/ERF) (Kasuga et al. 1999), MYB (Cui et al. 2013), basic helix–loop–helix (bHLH) (Jiang et al. 2009), and NAC (Tran et al. 2004) are differentially expressed in response to elevated external salinity (Golldack et al. 2011). The expression levels of various genes that may ultimately influence the level of salt tolerance of plants are in turn regulated by these transcriptional factors. Genes relevant for inorganic ion uptake and osmolyte synthesis are upregulated to counteract the water potential decrease resulting from the osmotic component of enhanced salinity (Geng et al. 2013). Dynamic changes in hormone biosynthesis transcriptionally regulate these stress-response genes in plants (Geng et al. 2013; Dinneny et al. 2008). An initial quiescence period is followed by a growth recovery phase after stress induction. Both of these correlate with changes in the levels of the plant hormones abscisic acid (ABA), jasmonate (JA), gibberellic acid (GA), and brassinosteroid (BR). A secondary signaling network that controls plant growth after salt stress has been revealed by mining of data from the At Gen Express consortium (Geng et al. 2013; Kilian et al. 2007). Most stress-induced transcriptional changes occur approximately 3 h after application of salt stress (Geng et al. 2013). The expression of 5590 genes was salt regulated in the roots of *A. thaliana* seedlings (Chen et al. 2002), and fluorescence-activated cell sorting (FACS) has revealed that root cortex cells were the most transcriptionally active (Geng et al. 2013). According to molecular analyses, the root endodermis is the main cell layer in the context of lateral root development under salt stress conditions. Lateral root elongation into surrounding media with high salt concentrations is prevented by ABA (Duan et al. 2013). Plants try to circumvent highly saline media by altering the direction of root growth (Galvan-Ampudia et al. 2013). This phenotype is known as halotropism and is induced by salt-triggered auxin responses, but not by osmotic stress (Galvan-Ampudia et al. 2013). Our understanding of plant salinity tolerance strategies may be improved by further exploring molecular mechanisms behind halotropism. There are significant knowledge gaps that need to be filled, although several key components of the plant salt stress response network have been identified in recent years. Ethylene was recently shown to confer plant salt tolerance in soil-grown *Arabidopsis* plants by improving the Na^+/K^+ ratio in shoots (Jiang et al. 2013). Knock out of ETHYLENE OVERPRODUCER1 (ETO1) resulted in elevated ethylene levels, which stimulated root stele reactive oxygen species (ROS) production by the respiratory burst oxidase homologue F (RBOHF). The increase in stele ROS accumulation led to reduced net Na^+ influx in roots, decreased Na^+ xylem loading and to root K^+ retention and subsequent enhanced salinity tolerance (Jiang et al. 2013; Deinlein et al. 2014).

Microarray methods have been employed to catalogue all the changes in gene expression in response to drought, cold, and high salt stress conditions over time in recent studies. The potential function of approximately 130 genes of *A. thaliana* which have been shown to be upregulated points to signaling events, detoxification, and other functions involved in cellular response to stress (Bray 2002). According to Seki et al. (2002), there is about five times increase in transcripts of 53, 277 and 194 genes after cold, drought, and salinity treatments, respectively.

5.5.1 Genes Encoding Proteins Involved in Regulating Gene Expression

Some salt-responsive genes encode proteins that are involved in the regulation of other salt-responsive genes. These salt-responsive regulatory genes are mostly transacting factors (Urao et al. 1993) and protein kinases (Mizoguchi et al. 1996). In response to salinity and ABA, the expression of receptor-like protein kinase gene is induced in *Arabidopsis thaliana* (Hong et al. 1997). A receptor protein kinase cDNA was also isolated from rice (Naot et al. 1995) that transduces an extracellular signal across the membrane to activate cellular signal transduction pathways. In *Arabidopsis thaliana*, a gene was reported to encode components of signal transduction and found to be mitogen-activated protein kinase (MAPK) by Mizoguchi et al. (1995, 1996). MAPKK kinase and a ribosomal S6 kinase are other similar types of genes which also function in the MAPK cascade. It was found that expression of these genes increased under salt stress.

5.5.2 Salinity-Resistant Genotypes/Varieties Developed by Conventional Breeding

Although development of genotypes/varieties resistant to salt stress has not been successful, some genotypes have been developed and released for commercial cultivation in salinity-prone areas.

The salinity-resistant rice varieties have been developed by crossing with Pokkali, a salinity-resistant local land race from Kerala. Some moderately tolerant to tolerant varieties have been developed in the wheat also. In other crops, development of salinity-tolerant cultivars has not been successful and has not been attempted with the same zeal. In rice and wheat, breeders were also lucky to have local land races (Pokkali and Kharchia) tolerant to salinity.

In plant stress biology, cloned plant genes and transgenic plants have become a standard tool which mainly have applied to model systems and extended our knowledge of tolerance mechanism. At all levels of organization like morphological, physiological, biochemical, and molecular, changes in plant processes occur through various abiotic stresses. Attention has been focused on alterations in gene expression in recent years. There is rapid increase in the number of genes that are transcriptionally upregulated in response to stress. The functions of some of these polypeptides are being identified, and their likely role in stress physiology is also being determined. Different ways in which plants can be utilized will be expanded by our understanding of mechanisms that regulate gene expression and the ability to transfer genes from other organism into plants. Cloned genes can be exploited to alter the function of gene products in transgenic plants to assess their biological role in a stress response.

A large collection of characterized genes can be built by the molecular analysis of stress responses. Qualitative trait loci identification for abiotic stress resistance is an effective analytical tool, which is promising, considering that saturated DNA marker maps are now available for both genetic model plants and crop plants. The use of novel approaches combining genetic, physiological, and molecular techniques should provide excellent results in the near future. Conventional breeding has been of some success only in the case of wheat and rice.

5.5.3 Marker-Assisted Selection

The process of variety development by conventional breeding methods can be complemented through marker-assisted selection (MAS) by accurate genotyping and targeting on the desired region of the genome that result in transfer on only the desired gene(s) and avoids any undesirable linkage drags. A closely linked marker may be used to screen large number of samples for rapid identification of progeny that carry desirable characteristics, by mapping the traits into chromosomes. MAS is a form of indirect selection and most widely used application of DNA markers and one that plant breeders have been quick to embrace.

To increase the precision of selection for best trial combinations, molecular markers have been used by breeders. MAS developed variety are not considered genetically modified organisms (GMOs) and accepted by local and international market. Selection methods assisted by molecular marker have resulted in significant improvement in breeding efficiency by reducing trial and error aspect of the breeding process and by allowing for time and cost savings. Molecular marker systems have benefitted from the constant increase in the integration of biotechnological production of segregating populations like homozygous double haploids in wheat breeding cycles since the major requirement of being codominant for molecular markers will disappear. An opportunity to select desirable lines based on genotype rather than phenotype has been offered by marker-assisted selection which is an invaluable tool for gene pyramiding (bringing genes from different individuals together in one individual) and has been fairly successful for combining single gene traits.

5.5.4 Importance of QTL Mapping for MAS

The identification of genes and quantitative trait loci (QTLs) and DNA markers that are linked to them is done by QTL mapping experiments which are the foundation of the development of markers for MAS. It was generally assumed that markers could be directly used in MAS, but many factors such as population size and type, level of replication of phenotypic data, environmental effects, and genotyping errors influence the accuracy of QTL mapping. They are very important for more complex quantitative traits with many QTLs each with relatively small effects like drought tolerance and yield. These days it has become widely accepted that QTL confirmation, validation, and/or additional marker testing steps may be required after QTL mapping and prior to MAS. These steps are:

(i) Marker conversion for reliability improvement
(ii) QTL confirmation for testing the accuracy of results from the primary QTL mapping study
(iii) QTL validation for verifying that a QTL is effective in different genetic backgrounds
(iv) Marker validation for testing the level of polymorphism of most tightly linked markers within a narrow window (say 5–10 cM) spanning a target locus and also testing the reliability of markers to predict phenotype.

5.6 Resistance to Abiotic Stress

As most of stress tolerance traits are quantitatively linked traits (QTLs), traditional approaches at transferring resistance to crop plants are limited because of their complexity. A convenient alternative and a rapid approach for the improvement of stress tolerance are the direct introduction of a small number of genes. These gene products provide some protection either directly or indirectly against environmental stresses, although present engineering strategies are dependent on the transfer of one or several genes that encode biochemical pathways or endpoints of signaling pathways. Salinity stress is a major abiotic factor that limits crop productivity, thereby causing enormous loss.

5.6.1 Marker-Assisted Breeding

There is significant influence of environmental factors on the response of plants to salinity which limits the direct selection of superior salt-tolerant genotypes under field conditions (Richards 1996). Salt tolerance is a complex trait involving the function of many genes (Foolad 2004; Flowers 2004) and in plants appears to be developmentally regulated process. The tolerance of the plants at one stage of development is not always correlated with tolerance at other stages (Foolad 2004; Flowers 2004; Greenway and Munns 1980; Tal and Shannon 1983) like in tomato, barley, corn, rice, and wheat; salt tolerance tends to increase with the age of the plant (Foolad 2004). QTLs associated with salt tolerance at the early stage of growth were different from those QTLs associated with salt tolerance at the germination stage in barley (Mano and Takeda 1997), tomato (Foolad et al. 1999), and *Arabidopsis* (Quesada et al. 2002). The plants selected by their ability to germinate at high salinity did not display similar salt tolerance during vegetative growth.

Development of DNA markers that can be used to identify QTLs is done by the development of molecular biology techniques. The efficiency of selection, in particular for those traits that are controlled by several genes and are highly influenced by environmental factors, has been improved by the use of QTLs (Flowers 2004). There are several advantages of QTLs and marker-assisted selection over direct phenotypic screening, particularly because the time needed to screen individuals and the impact of environmental effects on the trait under study are reduced by the PCR-based methodologies which are used to detect the markers. Salt tolerance and its sub-traits are determined by multiple QTLs, and both additive and dominance effects are important in the inheritance of many of the traits associated with salt tolerance (Foolad 2004; Flowers 2004; Gregorio et al. 2002). Improvement in crop salt tolerance can be achieved through the development of high-density DNA maps that incorporate microsatellite markers, RFLP (restriction fragment length polymorphisms), and AFLP (amplified fragment length polymorphisms), and advances in marker-assisted selection techniques will facilitate pyramiding traits of interest to attain (Yamaguchi and Blumwald 2005).

There is physiologically linkage of drought and salt stress because both limit the crop's physiological access to water. Yadav et al. (2011) tested whether the major DT-QTL mapping to LG 2 of pearl millet would have a pleiotropic effect on salt tolerance under a range of salinity and alkalinity stress treatments. They have tested drought-tolerant (PRLT 2/89-33) and drought-sensitive (H 77/833-2) parents, and NILs introgressed with the DT-QTL for a range of traits under salinity and alkalinity stress treatments. They have evaluated these genotypes for germination and seedling emergence and for vegetative growth and yield-related traits. In these treatments, Na^+ and K^+ accumulation, their compartmentation in different plant parts, and their effects on growth and/or yield parameters were also measured. It was observed that reduction in seedling emergence was much higher (80%) in the hybrid of the drought-sensitive parent H 77/833-2 than in the hybrids of the drought-tolerant parent PRLT 2/89-33 (50%) and QTL-NIL ICMR 01029 (64%). The drought-tolerant parent and the QTL-NIL also recorded greater shoot growth than the drought-sensitive parent under both salinity and alkalinity stress treatments at the vegetative stage (24 days after sowing). There are lesser reductions in yield and yield-related parameters, in the drought-tolerant parent and in the QTL-NIL in comparison to the sensitive parent. The drought-sensitive parent H 77/833-2 accumulated significantly higher concentrations of toxic Na^+ in shoot parts than the drought-tolerant parent PRLT 2/89-33 and QTL-NIL ICMR 1029 in all salinity and alkalinity stress treatments irrespective of the growth stage.

5.7 Strategies to Improve Stress Tolerance

A new opportunity for understanding the genetics of stress-resistance genes and their contribution to plant performance under stress has been developed by recent advances in plant genome mapping and molecular biology techniques, which will also provide new tools for breeding in stress environment. For major crop plants, including rice, wheat, maize, barley, sorghum, and potato, molecular genetic maps have been developed which made it possible for scientists to tag desirable traits using known DNA landmarks. Genetic loci controlling stress resistance can be tracked by breeders through molecular genetic markers without having to measure the phenotype, which reduces the need for extensive field testing over time and space. By using molecular tags, gene pyramiding or introgression can be done more precisely which together with molecular genetic markers offers a new strategy known as marker-assisted selection. Genetic engineering of selected genes into elite breeding lines is another molecular strategy which depends on gene cloning and plant transformation technology. In genetic engineering experiments, the following three inputs are needed:

(i) The gene of interest
(ii) An effective technique for transferring the desired gene from one species to another
(iii) Promoter sequences for regulated expression of that gene

Among all of these, the first one is actually a rate-limiting factor. A lot of stress-induced genes have been isolated (Bartels et al. 1993) which can be analyzed through targeted or nontargeted strategy. The targeted approach depends on the availability of relevant biochemical information in terms of defined enzyme, protein, a biochemical reaction, or a physiological phenomenon, while the nontargeted strategy to obtain a desired gene is indirect and includes differential hybridization and shotgun cloning. The list of those genes, whose transcription is upregulated in response to stress, is rapidly increasing. Understanding of the mechanisms, which regulate gene expression and the ability to transfer genes from other organisms into plants, will expand the ways in which plants can be utilized. It is essential that the knowledge is applied to agriculturally and ecologically important plant species to exploit the full potential of these approaches.

5.8 Conclusion

Salt stress is a well-studied stress which utilized proteomic-based techniques as a powerful tool to reveal the molecular mechanisms of salt stress responses. Several salt-responsive proteins have been identified for crops by using these techniques. It will be possible to change salt-sensitive crops to salt-tolerant crops in the near future by using these proteins and their corresponding genes. In the past few years, great progress has been made toward understanding plant salt stress responses and tolerance mechanisms, but still many challenges lie ahead. Both molecular breeding and advanced biotechnology methods should help scientists to develop crops with enhanced salt tolerance. For this, relevant genes for enhancing salinity tolerance could be combined like Na^+ transporters, such as HKTs and NHXs, ROS scavengers, and other traits which play major roles in salt homeostasis and positively influence the capability of the plant to deal with elevated salinity.

References

Abbasi FM, Komatsu S (2004) A proteomic approach to analyze salt-responsive proteins in rice sheath. Proteomics 4:2072–2081

Abrol IP, Yadav JSP, Massoud FI (1988) Salt-affected soils and their management. FAO Soils

Acquaah G (2007) Principles of plant genetics and breeding. Blackwell, Oxford, p 385

Aghaei K, Komatsu S (2013) Crop and medicinal plants proteomics in response to salt stress. Front Plant Sci 4:8. http://dx.doi.org/10.3389/fpls.2013.00008

Aghaei K, Ehsanpour AA, Komatsu S (2008a) Proteome analysis of potato under salt stress. J Proteome Res 7:4858–4868

Aghaei K, Ehsanpour AA, Shah AH, Komatsu S (2008b) Proteome analysis of soybean hypocotyls and root under salt stress. Amino Acids 36:91–98

Aghaei K, Ehsanpour AA, Komatsu S (2009) Potato responds to salt stress by increased activity of antioxidant enzymes. J Integr Plant Biol 51:1095–1103

Ahmad P (2010) Growth and antioxidant responses in mustard (Brassica juncea L.) plants subjected to combined effect of gibberellic acid and salinity. Arch Agron Soil Sci 56:575–588

Ahmad P, Prasad MNV (2012a) Abiotic stress response in plants: metabolism, productivity and sustainability. Springer, New York

Ahmad P, Prasad MNV (2012b) Environmental adaptations and stress tolerance in plants in the era of climate change. Springer Science + Business Media, New York

Ahmad P, Hakeem KR, Kumar A, Ashraf M, Akram NA (2012) Salt-induced changes in photosynthetic activity and oxidative defense system of three cultivars of mustard (Brassica juncea. L). Afr J Biotechnol 11:2694–2703

Albenne C, Canut H, Jamet E (2013) Plant cell wall proteomics: the leadership of Arabidopsis thaliana. Front Plant Sci 4:111

Albert R (1975) Salt regulation in halophytes. Oceologia 21:57–71

Amini F, Ehsanpour AA, Hoang QT, Shin JS (2007) Protein pattern changes in tomato under in vitro salt stress. Russ J Plant Physiol 54:464–471

Ashraf M (2004) Some important physiological selection criteria for salt tolerance in plants. Flora 199:361–376

Ashraf M, Harris PJC (2004) Potential biochemical indicators of salinity tolerance in plants. Plant Sci 166:3–16

Askari H, Edqvist J, Hajheidari M, Kafi M, Salekdeh GH (2006) Effects of salinity levels on proteome of *Suaeda aegyptiaca* leaves. Proteomics 6:2542–2554

Azevedo Neto AD, Prisco JT, Eneas-Filho J, Abreu CEB, Filho EG (2006) Effect of salt stress on antioxidative enzymes and lipid peroxidation in leaves and roots of salt-tolerant and salt-sensitive maize genotypes. Environ Exp Bot 56:87–94

Bartels D et al (1993) Dessication related gene products analysed in a resurrection plant and in barley embryos. In: Close TJ, Bray EA (eds) Plant responses to cellular dehydration during environmental stress: current trends in plant physiology. An American Society of Plant Physiologist Series, Rockville, pp 119–127

Bayuelo Jimenez JS, Debouk DG, Lynch JP (2002) Salinity tolerance in phaseolus species during early vegetative growth. Crop Sci 42:2184–2192

Beltagi MS, Ismail MA, Mohamed FH (2006) Induced salt tolerance in common bean (Phaseolus vulgaris L.) by gamma irradiation. Pak J Biol Sci 6:1143–1148

Bethke PC, Drew MC (1992) Stomatal and nonstomatal components to inhibition of photosynthesis in leaves of Capsicum annum during progressive exposure to NaCl salinity. Plant Physiol 99:219–228

Bhatnagar – Mathur P, Vadez V, Sharma KK (2008) Transgenic approaches for abiotic stress tolerance in plants: retrospect and prospects. Plant Cell Rep 27:411–424

Blackwood GC, Miflin BJ (1976) The effects of salts on NADH malate dehydrogenase activity in maize and barley. Plant Sci Lett 7:435–446

Bliss RD, Platt-Aloia KA, Thomson WW (1984) Effects of salt on cell membranes of germinating seeds. Calif Agric:24–25

Bohnert HJ, Nebson DW, Jensen RG (1995) Adaptation to environmental stress. Plant Cell 7:1099–1111

Bonales-Alatorre E, Shabala S, Chen ZH, Pottosin I (2013) Reduced tonoplast fast-activating and slow-activating channel activity is essential for conferring salinity tolerance in a facultative halophyte, Quinoa. Plant Physiol 162:940–952

Bonhert HJ, Jensen RG (1996) Strategies for engineering water-stress tolerance in plants. Trends Biotechnol 14:89–97

Bose J, Rodrigo-Moreno A, Shabala S (2014) ROS homeostasis in halophytes in the context of salinity stress tolerance. J Exp Bot 65:1241–1257

Boyer JS (2009) Cell wall biosynthesis and the molecular mechanism of plant enlargement. Funct Plant Biol 36:383–394

Brady CJ, Gibson TS, Barlow EWR, Speirs J, Wyn Jones RG (1984) Salt-tolerance in plants. I. Ions, compatible organic solutes and the stability of plant ribosomes. Plant Cell Environ 7:571–578

Bray EA (2002) Classification of genes differentially expressed during water-deficit stress in *Arabidopsis thaliana*: an analysis using micro-array and differential expression data. Ann Bot 89:803–811

Bresler E, McNeal BL, Carter DL (1982) Saline and sodic soils. Principles-dynamics-modeling, Advanced Series in Agricultural Sciences, vol 10. Springer, Berlin

Britto DT, Kronzucker HJ (2006) Futile cycling at the plasma membrane: a hallmark of low-affinity nutrient transport. Trends Plant Sci 11:529–534

Brownell PF (1965) Sodium as an essential micronutrient element for a higher plant (Atriplex vesicaria). Plant Physiol 40:460–468

Brownell PF, Crossland (1972) The requirement for sodium as a micronutrient by species having C4 dicarboxylic photosynthetic pathway. Plant Physiol 49:794–797

Brugnoli E, Björkman O (1992) Growth of cotton under continuous salinity stress: influence on allocation pattern, stomatal and non-stomatal components of photosynthesis and dissipation of excess light energy. Planta 187:335–347

Bulletin 39 (n.d.) Food and Agriculture Organization of the United Nations, Rome, Italy

Caruso G, Cavaliere C, Guarino C, Gubbiotti R, Foglia P, Lagana A (2008) Identification of changes in *Triticum durum* L. leaf proteome in response to salt stress by two-dimensional electrophoresis and MALDI-TOF mass spectrometry. Anal Bioanal Chem 391:381–390

Cassab GI (1998) Plant cell wall proteins. Annu Rev Plant Biol 49:281–309

Cheeseman JM (2013) The integration of activity in saline environments: problems and perspectives. Funct Plant Biol 40:759–774

Chen T, Murata N (2002) Enhancement of tolerance of abiotic stress by metabolic engineering of betaines and other compatible solutes. Curr Opin Plant Biol 5:250–257

Chen RD, Tabaerzadeh Z (1991) Alteration of gene expression in tomato plants (*Lycopersicon esculentum*) by drought and salt stress. Genome 35:385–391

Chen WQ et al (2002) Expression profile matrix of Arabidopsis transcription factor genes suggests their putative functions in response to environmental stresses. Plant Cell 14:559–574

Chen C, Tao C, Peng H, Ding Y (2007) Genetic analysis of salt stress responses in asparagus bean (Vigna unguiculata L. ssp. Sesquipedalis verdc.) J Hered 98(7):655–665

Chinnusamy V, Zhu J, Zhu JK (2006) Salt stress signaling and mechanism of plant salt tolerance. Genet Eng 27:141–177

Cho WK, Chen XY, Chu H et al (2009) Proteomic analysis of the secretome of rice calli. Physiol Plant 135(4):331–341

Chutipaijit S, Cha-um S, Sompornpailin K (2011) High contents of proline and anthocyanin increase protective response to salinity in *Oryza sativa* L. spp. indica. Aust J Crop Sci 5:1191–1198

Colmer TD, Fan Teresa W-M, Läuchli A, Higashi RM (1996) Interactive effects of salinity, nitrogen and sulphur on the organic solutes in leaf blades. J Exp Bot 47(3):369–375

Crosbie TM, Pearce RB (1982) Effects of recurrent phenotypic selection for high and low photosynthesis on agronomic traits in two maize populations. Crop Sci 22:809–813

Cui MH et al (2013) An Arabidopsis R2R3-MYB transcription factor, AtMYB20, negatively regulates type 2C serine/threonine protein phosphatases to enhance salt tolerance. FEBS Lett 587:1773–1778

Dani V, Simon WJ, Duranti M, Croy RRD (2005) Changes in the tobacco leaf apoplast proteome in response to salt stress. Proteomics 5:737–745

Deinlein U, Stephan AB, Horie T, Luo W, Xu G, Schroeder JI (2014) Plant salt-tolerance mechanisms. Trends Plant Sci 19(6):371–379

Dinneny JR et al (2008) Cell identity mediates the response of Arabidopsis roots to abiotic stress. Science 320:942–945

Dixon M, Webb E (1964) Enzymes. Longmans, London

Dooki AD, Mayer-Posner FJ, Askari H, Zaiee A, Salekdeh GH (2006) Proteomic responses of rice young panicles to salinity. Proteomics 6:6498–6507

Duan L et al (2013) Endodermal ABA signaling promotes lateral root quiescence during salt stress in Arabidopsis seedlings. Plant Cell 25:324–341

Dubey RS (2005) Photosynthesis in plants under stressful conditions. In: Pessarakli M (ed) Handbook of photosynthesis, 2nd edn. CRC Press, Boca Raton, pp 479–497

DuPont FM (1992) Salt-induced changes in ion transport: regulation of primary pumps and secondary transporters. In: DT Cooke, DT Clarkson (eds) Transport and receptor proteins of plant membranes. Plenum Press, New York, pp 91–100

Ellis M, Egelund J, Schultz CJ, Bacic A (2010) Arabinogalactan-proteins: key regulators at the cell surface? Plant Physiol 153(2):403–419

English JP, Colmer TD (2013) Tolerance of extreme salinity in two stem-succulent halophytes (*Tecticornia* species). Funct Plant Biol 40:897–912

Ericscon MC, Alfinito SH (1984) Proteins produced during salt stress in tobacco cell culture. Plant Physiol 74:506–509

Ewing WSB (1981) The effects of salinity on the morphological and anatomical characteristics of Atriplex triangularis willd. Master's thesis. Department of Botany, Ohio University, Athens, OH. Excess Light Energy Planta 187: 335–347

Fagerberg WR (1984) Cytological changes in palisade cells of developing sunflower leaves. Protoplasma 119:21–30

FAO (2005) Global network on integrated soil management for sustainable use of salt affected FAO land and plant nutrition management service. Rome

FAO (2008) Land and plant nutrition management service. https:// www.fao.org/ag/agl/agll/spush

FAO (2009) High level expert forum – how to feed the world in 2050. Economic and social development department. Food and Agricultural Organization of the United Nations, Rome

Farooq M, Hussain M, Wakeel M, Siddique KHM (2015) Salt stress in maize : effects, resistance mechanism and management: a review. Agron Sustain Dev 35(2):461–481. Springer-verlag/ EDP Sciences/INRA

Fisarakis I, Chartzoulakis K, Stavrakas D (2001) Response of Sultana vines (*V. vinifera* L.) on six rootstocks to NaCl salinity exposure and recovery. Agric Water Manag 51:13–27

Flowers TJ (1972a) The effect of sodium chloride on enzyme activities from four halophyte species of Chenopodiaceae. Phytochemistry 11:1881–1886

Flowers TJ (1972b) Salt tolerance in *Suaeda maritima* (L) Dum: effect of sodium-chloride on growth, respiration, and soluble enzymes in a comparative study with *Pisum sativum* L. J Exp Bot 23:310–321

Flowers TJ (1985) Physiology of halophytes. Plant Soil 89:41–56

Flowers TJ (2004) Improving crop salt tolerance. J Exp Bot 55:307–319

Flowers TJ, Colmer TD (2008) Salinity tolerance in halophytes. New Phytol 179:945–963

Flowers TJ, Dalmond D (1992) Protein-synthesis in halophytes – the influence of potassium, sodium and magnesium in vitro. Plant Soil 146:153–161

Flowers TJ, Yeo AR (1986) Ion relations of plants under drought and salinity. Aust J Plant Physiol 13:75–91

Flowers TJ, Hall JL, Ward ME (1976a) Salt tolerance in halophyte *Suaeda maritima* – further properties of the enzyme malate dehydrogenase. Phytochemistry 15:1231–1234

Flowers TJ, Ward ME, Hall JL (1976b) Salt tolerance in the halophyte *Suaeda maritima*: some properties of malate dehydrogenase. Philos Trans R Soc B Biol Sci 273:523–540

Flowers TJ, Troke PF, Yeo AR (1977) The mechanism of salt tolerance in halophytes. Annu Rev Plant Physiol 28:89–112

Flowers TJ, Galal HK, Bromham L (2010) Evolution of halophytes: multiple origins of salt tolerance in land plants. Funct Plant Biol 37:604–612

Flowers TJ, Munns R, Colmer TD (2015) Sodium chloride toxicity and the cellular basis of salt tolerance in halophytes. Ann Bot 115(3):419–431

Foolad MR (2004) Recent advances in genetics of salt tolerance in tomato. Plant Cell Tissue Organ Cult 76:101–119

Foolad MR et al (1999) Comparison of QTLs for seed germination under non-stress, cold stress and salt stress in tomato. Plant Breed 118:167–173

Galvan-Ampudia CS et al (2013) Halotropism is a response of plant roots to avoid a saline environment. Curr Biol 23:2044–2050

Gama PBS, Inanaga S, Tanaka K, Nakazawa R (2007) Physiological response of common bean (Phaseolus vulgaris L.) seedlings to salinity stress. Afr J Biotechnol 6(2):79–88

Gama PBS, Tanaka K, Eneji A, Eltayeb AE, Elsiddig K (2009) Salt induced stress effects on biomass, photosynthetic rate and reactive oxygen species scavenging enzyme accumulation in common bean. J Plant Nutr 32(5):837–854

Geng Y et al (2013) A spatio-temporal understanding of growth regulation during the salt stress response in Arabidopsis. Plant Cell 25:2132–2154

Ghoulam C, Foursy A, Fares K (2002) Effects of salt stress on growth, inorganic ions and proline accumulation in relation to osmotic adjustment in five sugar beet cultivars. Environ Exp Bot 47(1):39–50

Gibson TS, Speirs J, Brady CJ (1984) Salt-tolerance in plants. II. In vivo translation of m-RNAs from salt-tolerant and salt-sensitive plants on wheat germ ribosomes. Responses to ions and compatible organic solutes. Plant Cell Environ 7:579–587

Gil R, Boscaiu M, Lull C, Bautista I, Lidon A, Vicente O (2013) Are soluble carbohydrates ecologically relevant for salt tolerance in halophytes? Funct Plant Biol 40:805–818

Golldack D, Lüking I, Yang O (2011) Plant tolerance to drought and salinity: stress regulating transcription factors and their functional significance in the cellular transcriptional network. Plant Cell Rep 30:1383–1391. https://doi.org/10.1007/s00299-011-1068-0

Gorham J, Mcdonnell E, Budrewicz E, Wyn Jones RG (1985) Salt Tolerance in the Triticeae: growth and solute accumulation in leaves of Thinopyrum bessarabicum. J Exp Bot 36(7):1021–1031

Gossett DR, Millhollon EP, Lucas MC (1994) Antioxidant response to NaCl stress in salt tolerant and salt-sensitive cultivars of cotton. Crop Sci 34:706–714

Greenway H, Munns R (1980) Mechanism of salt tolerance in nonhalophytes. Annu Rev Plant Physiol 31:149–190

Greenway H, Munns R (1983) Interactions between growth and uptake of Cl^- and Na^+, and water relations of plants in saline environments. Plant Cell Environ 6:575–589

Greenway H, Osmond CB (1972) Salt responses of enzymes from species differing in salt tolerance. Plant Physiol 49:256–259

Gregorio GB et al (2002) Progress in breeding for salinity tolerance and associated abiotic stresses in rice. Field Crop Res 76:91–101

Grime JP (1979) Plant strategies and vegetation process. Wiley, New York

Gulick PJ, Dvorak J (1987) Gene induction and repression by salt treatment in roots of the salinity sensitive Chinese Spring Wheat and the salinity-tolerant Chinese Spring x Elytrigia elongata amphiploid. Proc Natl Acad Sci U S A 84:99–103

Ha E, Ikhajiagba B, Bamidele JF, Ogic-odia E (2008) Salinity effects on young healthy seedling of kyllingia peruviana collected from escravos, Delta state. Glob J Environ Res 2(2):74–88

Hagege D, Kevers C, Boucaud J, Gasper T (1988) Activities, peroxydasiques, production d' ethylene, lignification et limitation de croissance chez Suaeda maritima cultive en l'absence de NaCl. Plant Physiol Biochem 26:609–614

Hajheidari M, Abdollahian-Noghabi M, Askari H (2005) Proteome analysis of sugar beet leaves under drought stress. Proteomics 5:950–960

Hajibagheri MA, Hall JL, Flowers TJ (1984) Stereological analysis of leaf cells of the halophyte *Suaeda maritima* (L.) Dum. J Exp Bot 35:1547–1557

Hasanuzzaman M, Nahar K, Fujita M (2013) Plant response to salt stress and role of exogenous protectants to mitigate salt-induced damages. In: Ahmad et al. (eds) Ecophysiology and responses of plants under salt stress. doi:https://doi.org/10.1007/978–1–4614-4747-4_2, © Springer Science+Business Media, LLC

Hasegawa PM (2013) Sodium (Na^+) homeostasis and salt tolerance of plants. Environ Exp Bot 92:19–31

Hasegewa PM, Bressan RA, Zhu JK, Bohnert HJ (2000) Plant cellular and molecular responses to high salinity. Annu Rev Plant Physiol Mol Biol 51:463–499

Hassanein RA, Bassuony FM, Baraka DM, Khalil RR (2009) Physiological effects of Nicotinamide and Ascorbic acid on ZeaMays plants grown under salinity stress. I. Changes in growth, some relevant metabolic activities and oxidative defense systems. Res J Agric Biol Sci 5(1):72–81

Hawkins HJ, Lewis OAM (1993) Combination effect of NaCl salinity, nitrogen form and calcium concentration on the growth and ionic content and gaseous properties of *Triticum aestivum* L. cv. Gamtoos. New Phytol 124:161–170

Hématy K, Cherk C, Somerville S (2009) Host-pathogen warfare at the plant cell wall. Curr Opin Plant Biol 12(4):406–413

Hernandez JA, Olmas E, Copras FJ, Sevilla F, del Rio LA (1995) Salt induced oxidative stress in chloroplast of pea plants. Plant Sci 105:151–167

Holmström K-O, Somersalo S, Mandal A, Palva TE, Welin B (2000) Improved tolerance to salinity and low temperature in transgenic tobacco producing glycine betaine. J Exp Bot 51(343):177–185

Hillel D (2000) Salinity management for sustainable irrigation. The World Bank, Washington, DC

Hong SW, Jon JH, Kwak JM, Nam HG (1997) Identification of a receptor-like protein kinase gene rapidly induced by absciscic acid, dehydration, high salt and cold treatment in *Arabidopsis thaliana*. Plant Physiol 13:1203–1212

Houimli SIM, Denden M, Elhadj SB (2008) Induction of salt tolerance in pepper (B0 or Hbt 10mmtt1) by 24-epibrassinolide. Eurasia J Biol Sci 2:83–90

Huang Z, Zhang Z, Zhang X, Zhang H, Huang D, Huang R (2004) Tomato TERF1 modulates ethylene response and enhances osmotic stress tolerance by activating expression of downstream genes. FEBS Lett 573:110–116

Hurkman WJ (1992) Effect of salt stress on gene expression: a review. Plant Soil 146:145–151

Iyengar ERR, Reddy MP (1996) Photosynthesis in highly salt-tolerant plants. In: Pessaraki M (ed) Handbook of photosynthesis. Marcel Dekker, New York, pp 897–909

James RA, Rivelli AR, Munns R, Caemmerer SV (2002) Factors affecting CO_2 assimilation, leaf injury and growth in salt-stressed durum wheat. Funct Plant Biol 29:1393–1403

Jamet E, Albenne C, Boudart G, Irshad M, Canut H, Pont-Lezica R (2008) Recent advances in plant cell wall proteomics. Proteomics 8(4):893–908
Jamil M, Lee CC, Rehman SU, Lee DB, Ashraf M, Rha ES (2005) Salinity (NaCl) tolerance of brassica species at germination and early seedling growth. Electron J Environ Agric Food Chem 4:970–976. ISSN: 1579–4377
Jamil M, Rehman SU, Lee KJ, Kim JM, Rha HK (2007a) Salinity reduced growth ps2 photochemistry and chlorophyll content in radish. Sci Agric (Piracicaba, Braz) 64(2):111–118
Jamil M, Rehman S, Rha ES (2007b) Salinity effect on plant growth, ps11 photochemistry and chlorophyll content in sugar beet (*Beta vulgaris* L.) and cabbage (*Brassica oleracea capitata* L.) Pak J Bot 39(3):753–760
Jiang Y, Deyholos MK (2009) Functional characterization of Arabidopsis NaCl-inducible WRKY25 and WRKY33 transcription factors in abiotic stresses. Plant Mol Biol 69:91–105
Jiang Y et al (2009) Functional characterization of the Arabidopsis bHLH92 transcription factor in abiotic stress. Mol Gen Genomics 282:503–516
Jiang C et al (2013) An Arabidopsis soil-salinity-tolerance mutation confers ethylene-mediated enhancement of sodium/potassium homeostasis. Plant Cell 25:3535–3352
Johnson MK, Johnson EJ, MacElroy RD, Speer HL, Bruff BS (1968) Effects of salts on the halophilic alga *Dunaliella viridis*. J Bacteriol 95:1461–1468
Kalir A, Flowers TJ (1982) The effect of salts on malate dehydrogenase from leaves of *Zea mays*. Phytochemistry 21:2189–2193
Kasuga M et al (1999) Improving plant drought, salt, and freezing tolerance by gene transfer of a single stress-inducible transcription factor. Nat Biotechnol 17:287–291
Katerji N, van Hoorn JW, Hamdy A, Mastrorilli M, Mou Karzel E (1997) Osmotic adjustment of sugar beets in response to soil salinity and its influence on stomatal conductance, growth and yield. Agric Water Manag 34(1):57–69
Kaur G, Kumar S, Nayyar H, Upadhyaya HD (2008) Cold stress injury during the pod-filling phase in chickpea (Cicer arietinum L.) effects on quantitative and qualitative components of seeds. J Agron Crop Sci 194:457–464
Kaymakanova M, Stoeva N (2008) Physiological reaction of bean plants (Phaseolus vulgares L.) to salinity stress. Gen Appl Plant Physiol 34(3–4):177–188
Kaymakanova M, Stoeva N, Mincheva T (2008) Salinity and its effect on physiological response of bean (Phaseolus vulgares L.) Cent Eur Agric 9(4):749–756
Kilian J et al (2007) The At Gen Express global stress expression data set: protocols, evaluation and model data analysis of UV-B light, drought and cold stress responses. Plant J 50:347–363
Kosová K, Prášil IT, Vítámvás P (2013) Protein contribution to plant salinity response and tolerance acquisition. Int J Mol Sci 14:6757–6789. https://doi.org/10.3390/ijms14046757
Koyro H-W, Zörb C, Debez A, Huchzermeyer B (2013) The effect of hyper-osmotic salinity on protein pattern and enzyme activities of halophytes. Funct Plant Biol 40(9):787–804
Kurek I, Dulberger R, Azem A, Tzvi BB, Sudhakar D, Christou P et al (2002) Deletion of the C-terminal 138 amino acids of the wheat FKBP73 abrogates calmodulin binding, dimerization and male fertility in transgenic rice. Plant Mol Biol 48:369–381
Kurth E, Cramer GR, Lauchli A, Epstein E (1986) Effects of NaCl and CaCl2 on cell production in cotton roots. Plant Physiol 82:1102–1106
Kwon T, Abe T, Sasahara T (1995) Enhanced saline stress resistance in threonine and methionine overproducing mutant cell line from protoplast culture of rice (Oryza sativa L.) J Plant Physiol 145:551–556
Lamport DTA, Kieliszewski MJ, Chen Y, Cannon MC (2011) Role of the extension superfamily in primary cell wall architecture. Plant Physiol 156(1):11–19
Lauchli A, Schubert S (1989) The role of calcium in the regulation of membrane and cellular growth processes under salt stress. In: Cherry JH (ed) Environmental stress in plants, NATO ASI Series, vol 19. Springer, Berlin, pp 131–138
Leach RP, Wheeler KP, Flowers TJ, Yeo AR (1990a) Molecular markers for ion compartmentation in cells of higher plants. I. Isolation of vacuoles of high purity. J Exp Bot 41:1079–1087
Leach RP, Wheeler KP, Flowers TJ, Yeo AR (1990b) Molecular markers for ion compartmentation in cells of higher plants. II. Lipid composition of the tonoplast of the halophyte *Suaeda maritima* (L.) Dum. J Exp Bot 41:1089–1094

Lippuner V, Cyert MS, Gasser CS (1996) Two classes of plant cDNA clones differentially complement yeast calcineurin mutants and increase salt tolerance of wild-type yeast. J Biol Chem 271:12589–12866

Liu Y, Du H, He X, Huang B, Wang Z (2012) Identification of differentially expressed salt-responsive proteins in roots of two perennial grass species contrasting in salinity tolerance. J Plant Physiol 169:117–126

Lodeyro AP, Carrillo N (2015) Salt stress in higher plants: mechanisms of toxicity and defensive responses. In: Tripathi BN, Miller M (eds) Stress responses in plants. Springer International Publishing, Cham, pp 1–33

Maathuis FJM, Flowers TJ, Yeo AR (1992) Sodium chloride compartmentation in leaf vacuoles of the halophyte *Suaeda maratima* (L.) Dum. and its relation to tonoplast permeability. J Exp Bot 43:1219–1223

Manchanda G, Garg N (2008) Salinity and its effects on the functional biology of legumes. Acta Physiol Plant 30(5):595–618

Manetas Y (1990) A reexamination of NaCl effects on phosphoenolpyruvate carboxylase at high (physiological) enzyme concentrations. Physiol Plant 78:225–229

Mano Y, Takeda K (1997) Mapping quantitative trait loci for salt tolerance at germination and the seedling stage in barley (Hordeum vulgare L.) Euphytica 94:263–272

Mansour MMF, Salama KHA, Al-Mutawa MM (2003) Transport proteins and salt tolerance in plants. Plant Sci 164:891–900

Mantri N, Patade V, Penna S, Ford R, Pang E (2012) Abiotic stress responses in plants: present and future. In: Ahmad P, Prasad MNV (eds) Abiotic stress response in plants : metabolism, productivity and sustainability. Springer, New York, pp 1–9

Marschner H (1995) Mineral nutrition of higher plants. Academic, London

Mathur N, Singh J, Bohra S, Bohra A, Vyas A (2006) Biomass production, productivity and physiological changes in moth bean genotypes at different salinity levels. Am J Plant Physiol 1(2):210–213

Maxwell K, Johnson GN (2000) Chlorophyll fluorescence – a practical guide. J Exp Bot 51:659–668

Memon SA, Hou X, Wang LJ (2010) Morphological analysis of salt stress response of pak Choi. EJEAF Che 9(1):248–254

Mizoguchi T, Hayashida N, Yamaguchi-shinazaki K, Kamada H, Shinozaki K (1995) Two genes that encode ribosomal protein S6 kinase homologs are induced by cold or salinity stress in *Arabidopsis thaliana*. FEBS Lett 358:199–204

Mizoguchi T, Irie K, Hirayama T, Hayashida N, Yamaguchi-shinazaki K, Matsumoto K, Shinozaki K (1996) A gene encoding a mitogen-activated protein kinase and an S6 ribosomal protein kinase by touch, cold and water stress in *Arabidopsis thaliana*. Proc Natl Acad Sci U S A 93:765–769

Moons A, Bauw G, Prinsen E, Van Montagu M, Van Der Straeten D (1995) Molecular and physiological responses to abscisic acid and salts in roots of salt-sensitive and salt-tolerant indica rice varieties. Plant Physiol 107(1):177–186

Moons A, Gielen J, Vandekerckerckhove J, VanDer SD, Gneysen G, Van Montagu M (1997) An abscisic acid and salt stress responsive rice cDNA from a novel plant gene family. Planta 202:443–454

Munns R (1993) Physiological processes limiting plant growth in saline soils: some dogmas and hypotheses. Plant Cell Environ 16:15–24

Munns R (2002a) Comparative physiology of salt and water stress. Plant Cell Environ 25:239–250

Munns R (2002b) Salinity, growth and phytohormones. In: Läuchli A, Lüttge U (eds) Salinity: environment – plants – molecules. Kluwer, Amsterdam, pp 271–290

Munns R (2005) Genes and salt tolerance: bringing them together. New Phytol 167:645–663

Munns R, Tester M (2008) Mechanisms of salinity tolerance. Annu Rev Plant Biol 59:651–681

Munns R, James RA, Xu B, Athman A, Conn SJ, Jordans C, Byrt CS, Hare RA, Tyerman SD, Tester M (2012) Wheat grain yield on saline soils is improved by an ancestral Na^+ transporter gene. Nat Biotechnol 30:360–364

Mustard J, Renault S (2006) Response of red-osier dogwood (Cornus sericea) seedling to NaCl during the onset of bud break. Can J Bot 84(5):844–851

Naot D, Ben-Hayyim G, Eshdat Y, Holland D (1995) Drought, heat and salt stress induce the expression of a citrus homologue of an atypical late-embryogenesis Lea5 gene. Plant Mol Biol 27(3):619–622
Nawaz K, Hussain K, Majeed A, Khan F, Afghan S, Ali K (2010) Fatality of salt stress to plants: morphological, physiological and biochemical aspects. Afr J Biotechnol 9(34):5475–5480
Nazir N, Ashraf M, Rasul E (2001) Genomic relationships in oilseed Brassica with respect to salt tolerance-photosynthetic capacity and ion relations. Pak J Bot 33:483–501
Netondo GW, Onyango JC, Beck E (2004) Crop physiology and metabolism Sorghum and salinity II – gas exchange and chlorophyll fluorescence of sorghum under salt stress. Crop Sci 44(3):806–811
Neubert AB, Zorb C, Schubert S (2005) Expression of vacuolar Na^+/H^+ antiporters (ZmNHX) and Na^+ exclusion in roots of maize (*Zea mays* L.) genotypes with improved salt resistance. In: Li CJ, Zhang FS, Dobermann A, Hinsinger P, Lambers H, Li XL et al (eds) Plant nutrition for food security, human health and environmental protection. Tsinghua University Press, Beijing, pp 544–545
Neumann P (1997) Salinity resistance and plant growth revisited. Plant Cell Environ 20:1193–1198
Ngara R, Ndimba R, Jensen JB, Jensen ON, Ndimb B (2012) Identification and profiling of salinity stress-responsive proteins in Sorghum bicolor seedlings. J Proteome 75:4139–4150
Niaz BH, Athar M, Salim M, Rozema J (2005) Growth and ionic relations of fodder beet and sea beet under saline. CEERS 2(2):113–120
Osmond CB, Greenway H (1972) Salt responses of carboxylation enzymes from species differing in salt tolerance. Plant Physiol 49(2):260–263
Ouyang B, Yang T, Li H, Zhang L, Zhang Y, Zhang J et al (2007) Identification of early salt stress response genes in tomato root by suppression subtractive hybridization and microarray analysis. J Exp Bot 58:507–520
Parida AK, Das AB, Mohanty P (2004) Investigations on the antioxidative defense responses to NaCl stress in a mangrove, *Bruguiera parviflora*: differential regulations of isoforms of some antioxidative enzymes. Plant Growth Regul 42:213–226
Pettigrew WT, Meredith WR (1994) Leaf gas exchange parameters vary among cotton genotypes. Crop Sci 34:700–705
Pitman MG, Läuchli A (2002) Global impact of salinity and agricultural ecosystems. In: Läuchli A, Lüttge U (eds) Salinity: environment – plants – molecules. Kluwer, Dordrecht, pp 3–20
Pollard A, Wyn Jones WG (1979) Enzyme activities in concentrated solutions of glycinebetaine and other solutes. Planta 144:291–298
Pons LJ (1973) Outline of genesis, characteristics, classification and improvement of acid sulfate populations of *Trifolium repens* L. differing in salt tolerance. Plant Soil 146:131–136
Poon S, Heath RL, Clarke AE (2012) A chimeric arabinogalactan protein promotes somatic embryogenesis in cotton cell culture. Plant Physiol 160(2):684–695
Qados AMS (2011) Effect of salt stress on plant growth and metabolism of bean plant Vicia faba (L.) J Saudi Soc Agric Sci 10:7–15
Quesada V et al (2002) Genetic architecture of NaCl tolerance in Arabidopsis. Plant Physiol 130:951–963
Qureshi MI, Qadir S, Zoll L (2007) Proteomics-based dissection of stress-responsive pathways in plants. J Plant Physiol 164:1239–1260
Ramgopal S (1987) Differential mRNA transcription during salinity stress in barley. Proc Natl Acad Sci U S A 84:94–98
Raul L, Andres O, Armado L, Bernardo M, Enrique T (2003) Response to salinity of three grain legumes for potential cultivation in arid areas (plant nutrition). Soil Sci Plant Nutr 49(3):329–336
Raza SH, Athar HR, Ashraf M, Hameed A (2007) Glycinebetaine-induced modulation of antioxidant enzymes activities and ion accumulation in two wheat cultivars differing in salt tolerance. Environ Exp Bot. https://doi.org/10.1016/j.envexpbot.2006.12.009
Rea PA, Poole RJ (1993) Vacuolar H^+-translocating pyrophosphatase. Annu Rev Plant Physiol Plant Mol Biol 44:157–180
Reddy VS, Reddy ASN (2004) Proteomics of calcium-signaling components in plants. Phytochemistry 65:1745–1776
Rengasamy P (2006) World salinization with emphasis on Australia. J Exp Bot 57:1071–1023
Rengel Z (1992) The role of calcium in salt toxicity. Plant Cell Environ 15:625–632
Reuter DJ, Robinson JB (1986) Plant analysis: an interpretation manual. Inkata Press, Melbourne

Richards RA (1996) Defining selection criteria to improve yield under drought. Plant Growth Regul 20:157–166

Robinson NL, Tanaka CK, Hurkman WJ (1990) Time dependent changes in polypeptide and translatable mRNA levels caused by NaCl in barley roots. Physiol Plant 78:128–134

Rodriguez P, Amico JD, Morales D, Blanco MJS, Alarcon JJ (1997) Effects of salinity on growth, shoot water relations and root hydraulic conductivity in tomato plants. J Agric Sci 128:439–444

Rodríguez M, Canales E, Borrás- Hidalgo O (2005) Molecular aspect of abiotic stress in plants. Biotecnol Apl 22:1–10

Rogers ME, Noble CL (1992) Variation in growth and ion accumulation between two selected populations of *Trifolium repens* L. differing in salt tolerance. Plant Soil 146:131–136

Romero-Aranda R, Soria T, Cuartero J (2001) Tomato plant-water uptake and plant-water relationships under saline growth conditions. Plant Sci 160(2):265–272

Römheld V (2012) Diagnosis of deficiency and toxicity of nutrients. In: Marschner P (ed) Marschner's mineral nutrition of higher plants, 3rd edn. Elsevier/Academic, San Diego, pp 299–312

Roy SJ, Negrão S, Tester M (2014) Salt resistant crop plants. Curr Opin Biotechnol 26:115–124. https://doi.org/10.1016/j.copbio.2013.12.004

Rui L, Wei S, Mu-xiang C, Cheng-jun J, Min W, Bo-ping Y (2009) Leaf anatomical changes of Bruguiera gymnorrhiza seedlings under salt stress. J Trop Subtrop Bot 17(2):169–175

Sachs MM, Ho TD (1986) Alteration of gene expression during environmental stress in plants. Annu Rev Plant Physiol 37:363–376

Saffan SE (2008) Effect of salinity and osmotic stresses on some economic plants. Res J Agric Biol Sci 4(2):159–166

Sairam RK, Tyagi A (2004) Physiology and molecular biology of salinity stress tolerance in plants. Curr Sci 86(3):407–421

Sakamoto A, Murata N (2000) Genetic engineering of glycine betaine synthesis in plants: current status and implications for enhancement of stress tolerance. J Exp Bot 51:81–88

Salekdeh GH, Siopongco J, Wade LJ, Ghareyazie B, Bennett J (2002) Proteomic analysis of rice leaves during drought stress and recovery. Proteomics 2:1131–1145

Saqib M, Zorb C, Schubert S (2006) Salt resistant and salt-sensitive wheat genotypes show similar biochemical reaction at protein level in the first phase of salt stress. J Plant Nutr Soil Sci 169:542–548

Seeman JR, Sharkey TD (1986) Salinity and nitrogen effects on photosynthesis, Ribulose-1,5-Biphosphate carboxylase and metabolite pool sizes in Phaseolus vulgaris L. Plant Physiol 82:555–560

Seemann JR, Critchley C (1985) Effects of salt stress on the growth, ion content, stomatal behaviour and photosynthetic capacity of a salt sensitive species. Phaseolus vulagris L. Plant Physiol 82:555–560

Seki M, Narusaka M, Ishida J, Nanjo T, Fujita M, Oono Y, Kamiya A, Nakajima M, Enju A, Sakurai T, Satou M, Akiyama K, Taji T, Yamaguchi-Shinozaki K, Carninci P, Kawai J, Hayashizaki Y, Shinozaki K (2002) Monitoring the expression profiles of 7000 Arabidopsis genes under drought, cold and high-salinity stresses using a full-length cDNA microarray. Plant J 31(3):279–292

Shabala S (2013) Learning from halophytes: physiological basis and strategies to improve abiotic stress tolerance in crops. Ann Bot 112:1209–1221

Shanker A, Venkateswarlu B (2011) Abiotic stress in plants – mechanism and adaptations. InTech, Rijeka, p ix

Shininger TL (1979) The control of vascular development. Annu Rev Plant Physiol 30:313–337

Showalter AM (1993) Structure and function of plant cell wall proteins. Plant Cell 5(1):9–23

Showalter AM, Keppler B, Lichtenberg J, Gu D, Welch LR (2010) A bioinformatics approach to the identification, classification, and analysis of hydroxyproline-rich glycoproteins. Plant Physiol 153(2):485–513

Singh NK, LaRosa PC, Nelson D, Iraki N, Carpita NC, Hasegawa PM, Bressan RA (1989) In: Cherry JH (ed) Reduced growth rate and changes in cell wall proteins of plant cells adapted to NaCl, Environmental Stress in Plants. NATO ASI Series, vol G19. Springer, Berlin, pp 173–194

Sobhanian H, Razavizadeh R, Nanjo Y, Ehsanpour AA, Rastgar Jazii F, Motamed N et al (2010) Proteome analysis of soybean leaves, hypocotyls and roots under salt stress. Proteome Sci 8:1–15

Steudle E (2000) Water uptake by roots: effects of water deficit. J Exp Bot 51(350):1531–1542

Sudhir P, Murthy SDS (2004) Effects of salt stress on basic processes of photosynthesis. Photosynthetica 42:481–486

Szabolcs I (1974) Salt affected soils in Europe. Martinus Nijhoff, The Hague. 63 p

Taffouo VD, Kouamou JK, Ngalangue LMT, Ndjeudji BAN, Akoa A (2009) Effects of salinity stress on growth, ions partitioning and yield of some cowpea (*Vigna ungiuculata* L., walp) cultivars. Int J Bot 5(2):135–143

Taffouo VD, Wamba OF, Yombi E, Nono GV, Akoe A (2010) Growth, yield, water status and ionic distribution response of three bambara groundnut (Vigna subterranean (L.) verdc.) land races grown under saline conditions. Int J Bot 6(1):53–58

Tal M (1984) Physiological genetics of salt resistance in higher plants, studies on the level of the whole plant and isolated organs, tissues and cells. In: Staples RC, Toenniessen (eds) Salinity tolerance in plants, strategies for crop improvement. Wiley, New York, p 443

Tal M, Shannon MC (1983) Salt tolerance in the wild relatives of the cultivated tomato: responses of Lycopersicon esculentum, Lycopersicon cheesmani, Lycopersicon peruvianum, Solanum pennelli, and F1 hybrids to high salinity. Aust J Plant Physiol 10:109–117

Tavakkoli E, Rengasamy P, McDonald GK (2010) High concentrations of Na^+ and Cl^- ions in soil solution have simultaneous detrimental effects on growth of faba bean under salinity stress. J Exp Bot 61:4449–4459

Thakur P, Kumar S, Malik JA, Berger JD, Nayyar H (2010) Cold stress effects on reproductive development in grain crops. An overview. Environ Exp Bot 67:429–443

Tran LS et al (2004) Isolation and functional analysis of Arabidopsis stress-inducible NAC transcription factors that bind to a drought-responsive cis-element in the early responsive to dehydration stress 1 promoter. Plant Cell 16:2481–2498

Treichel JC, Hettfleisch H, Eilhardt S, Faist K, Kluge H (1988) A possible induction of CAM by NaCl-stress in heterotrophic cell suspension cultures of Mesembryanthemum crystallinum. J Plant Physiol 133:419–424

Turan MA, Kalkat V, Taban S (2007) Salinity-induced stomatal resistance, proline, chlorophyll and ion concentrations of bean. Int J Agric Res 2(5):483–488

Ul Haq T, Akhtar J, Steele KA, Munns R, Gorham J (2013) Reliability of ion accumulation and growth components for selecting salt tolerant lines in large populations of rice. Funct Plant Biol 41:379–390

Urao T, Yamaguchi-Shinozaki K, Urao S, Shinozaki K (1993) An *Arabidopsis* myb homolog is induced by dehydration stress and its gene product binds to the conserved MYB recognition sequence. Plant Cell 5:1529–1539

Vítámvás P, Prášil IT, Kosová K, Planchon S, Renaut J (2012) Analysis of proteome and frost tolerance in chromosome 5A and 5B reciprocal substitution lines between two winter wheats during long-term cold acclimation. Proteomics 12:68–85

Von Willert DJ, Kirst GO, Treichel S, Von Willert E (1976) The effect of leaf age and salt stress on malate accumulation and phosphoenolpyruvate carboxylase activity in Mesembryanthemum crystallinum. Plant Sci Lett 7:341346

Vorasoot N, Songsri P, Akkasaeng C, Jogloy S, Patanothai A (2003) Effect of water stress on yield and agronomic characters of peanut. Songklanakarin J Sci Technol 25:283–288

Vysotskaya L, Hedley PE, Sharipova G, Veselov D, Kudoyarova G, Morris J, Jones HG (2010) Effect of salinity on water relations of wild barley plants differing in salt tolerance. AoB Plant. https://doi.org/10.1093/aobpla/plq006

Wakeel A, Asif AR, Pitann B, Schubert S (2011) Proteome analysis of sugar beet (*Beta vulgaris* L.) elucidates constitutive adaptation during the first phase of salt stress. *J*. Plant Physiol 168:519–526

Walker RR, Törökfalvy E, Scott NSS, Kriedemann PE (1981) An analysis of photosynthetic response to salt treatment in Vitis vinifera. Aust J Plant Physiol 8:359–374

Wang L, Showalter AM, Ungar IA (1997) Effect of salinity on growth, ion content and cell wall chemistry in Atriplex prostrata (Chenopodiaceae). Am J Bot 84(9):1247–1255

Wang M, Dan G, Liu T, Wang Z, Guo X, Hou W, Bai Y, Chen X, Wang G (2007) Overexpression of a putative maize calcineurin B-like protein in Arabidopsis confers salt tolerance. Plant Mol Biol 65(6):733–746

Wang W, Vinocur B, Altman A (2003) Plant responses to drought, salinity and extreme temperatures: towards genetic engineering for stress tolerance. Planta 218:1–14

Wang W, Vinocur B, Shoseyov O, Altman A (2004) Role of plant heat-shock proteins and molecular chaperones in the abiotic stress response. Trends Plant Sci 9:244–252
Wang X, Yang P, Gao Q, Liu X, Kuang T, Shen S, He Y (2008) Proteomic analysis of the response to high-salinity stress in Physcomitrella patens. Planta 228:167–177
Ward JM, Hirschi KD, Sze H (2003) Plants pass the salt. Trends Plant Sci 8:200–201
White PJ, Broadley MR (2001) Chloride in soils and its uptake and movement within the plant: a review. Ann Bot 88:967–988
Wyn Jones RG (1981) Salt tolerance. In: Johnson CB (ed) Physiological processes limiting plant productivity. Butterworths, London, pp 271–292
Wyn Jones R, Storey R, Leigh RA, Ahmad N, Pollard A (1977) A hypothesis on cytoplasmic osmoregulation. In: Marre E, Ciferri O (eds) Regulation of cell membrane activities in plants. Elsevier/North Holland, Amsterdam, pp 121–136
Xu G, Magen H, Tarchitzky J, Kafkafi U (2000) Advances in chloride nutrition of plants. Adv Agron 68:97–150
Yadav RS, Yadav, Sehgal D, Vadez V (2011) Using genetic mapping and genomics approaches in understanding and improving drought tolerance in pearl millet. J Exp Bot 62(2):397–408
Yamaguchi T, Blumwald E (2005) Developing salt-tolerant crop plants: challenges and opportunities. Trends Plant Sci 10(12):615–620
Yancey PH, Clark ME, Hand SC, Bowlus RD, Somero GN (1982) Living with water stress: evolution of osmolyte systems. Science 217:1214–1222
Yang O et al (2009) The Arabidopsis basic leucine zipper transcription factor AtbZIP24 regulates complex transcriptional networks involved in abiotic stress resistance. Gene 436:45–55
Yeo AR (1981) Salt tolerance in the halophyte *Suaeda maritima* (L.) Dum.: intracellular compartmentation of ions. J Exp Bot 32:487–497
Yeo AR (1983) Salinity resistance: physiologies and prices. Physiol Plant 58:214–222
Yeo AR (1998) Molecular biology of salt tolerance in the context of whole plant physiology. J Exp Bot 49:915–929
Yeo AR, Flowers TJ (1980) Salt tolerance in the halophyte *Suaeda maritima* (L.) Dum.: evaluation of the effect of salinity upon growth. J Exp Bot 31:1171–1183
Yeo AR, Flowers TJ (1983) Varietal differences in the toxicity of sodium ions in rice leaves. Physiol Plant 59:189–195
Yilmaz H, Kina A (2008) The influence of NaCl salinity on some vegetative and chemical changes of strawberries (Fragaria x ananassa L.) Afr J Biotechnol 7(18):3299–3305
Zagorchev L, Kamenova P, Odjakova M (2014) The role of plant cell wall proteins in response to salt stress. Sci World J Volume Article ID 764089, 9 pages. https://doi.org/10.1155/2014/764089
Zhao GQ, Ma BL, Ren CZ (2007) Growth, gas exchange, chlorophyll fluorescence and ion content of naked oat in response to salinity. Crop Sci 47(1):123–131
Zhu JK (2001) Overexpression of a delta-pyrroline-5-carboxylate synthetase gene and analysis of tolerance to water and salt stress in transgenic rice. Trends Plant Sci 6:66–72
Zhu JK (2002) Salt and drought stress signal transduction in plants. Annu Rev Plant Biol 53:247–273
Zhu JK (2003) Regulation of ion homeostasis under salt stress. Curr Opin Plant Biol 6:441–445
Zhu J, Chen S, Alvarez S et al (2006) Cell wall proteome in the maize primary root elongation zone.I. Extraction and identification of water-soluble and lightly ionically bound proteins. Plant Physiol 140(1):311–325

Biochar Mitigates Salinity Stress in Plants

6

Anju Patel, Puja Khare, and D.D. Patra

Abstract

Salinity is a major problem in India. It not only hampers the growth of plant and productivity but also decreases the soil productivity. Some plants develop several mechanisms to cope up Salinity stress like ion regulation by Na/H antiporter; synthesis of amino acids like valine, aspartic acid, and proline; etc. Various antioxidants play crucial role in combating salinity stress. Biochar (product obtained after pyrolysis of any organic compound) enhances the fertility of soil as it improves the soil cation exchange capacity and water holding capacity. That in turn improves the nutrient capacity of the soil. A biochar property is also dependent on the type of material and pyrolysis temperature. After biochar amendment to the saline soil, its physicochemical parameters improve like organic carbon, CEC, available phosphorous, etc. Thus, biochar not only enhances the crop productivity but also improves soil enzymatic activity.

Keywords

Biochar · Salinity · Antioxidants · Soil enzymes

A. Patel (✉)
School of Earth Sciences, Banasthali Vidyapith, Banasthali, Rajasthan, India
e-mail: patel7anju@gmail.com

P. Khare
Agronomy and Soil Science Division, CSIR-CIMAP, Lucknow, India

D.D. Patra
Bidhan Chandra Krishi Viswa Vidyalaya, Nadia, West Bengal, India

V. Shukla et al. (eds.), *Plant Adaptation Strategies in Changing Environment*,
https://doi.org/10.1007/978-981-10-6744-0_6

6.1 Introduction

The genesis of biochar is linked to the ancient Amerindian populations of the Amazon region, regionally known as Terra Preta de Indio, where dark earth was created through the use of slash-and-char techniques (Lehmann 2009; Lehmann and Joseph 2009a). Research on Terra Preta soils (hortic anthrosols) discloses various effects of biochars on the functionality of soils in the Amazonia. Generally, biochar ameliorates with the soil for maintaining soil fertility and sustainability. Biochar is also recognized as a very crucial tool for the management of the environment (Lehmann and Joseph 2009b). Biochar is a newly fabricated scientific term, which means a carbon (C)-rich product is obtained when organic substance like agricultural residues. Wood or manure is heated in a closed chamber in absence of air or little air (Lehmann and Joseph 2009a). Further, Shackley et al. (2012) defined biochar more precisely as "the porous carbonaceous solid material generated by the thermochemical conversion of organic materials in an anaerobic atmosphere that owns physicochemical properties suitable for safe and long-term storage of carbon in the environment." The International Biochar Initiative (IBI) standardized its definition as "a solid material obtained from the thermochemical conversion of biomass in an oxygen-limited environment" (IBI 2012). All of these definitions are directly or indirectly related to the biochar production condition and its application to soil. Lehmann and Joseph (2009b) differentiate biochar operationally from charcoal. Chiefly, the difference lies in the end use of biochar and charcoal. Charcoal is used for producing fuel and energy, and basically it is a charred organic matter, whereas biochar is usually applied for carbon sequestration and environmental management. The term hydrochar is closely related to biochar; however, it is distinguished by different conditions like the hydrothermal carbonization of biomass (Libra et al. 2011). Broadly, biochar is generated by dry carbonization or pyrolysis and gasification of biomass, whereas hydrochar is produced as slurry in water by hydrothermal carbonization of biomass under pressure. These two chars varied widely in physical as well as chemical properties (Bargmann et al. 2013). Four major areas where biochar is being used in environmental management are (1) soil reclamation, (2) waste management, (3) climate change mitigation, and (4) energy generation (Lehmann and Joseph 2009a).

6.2 Biochar Properties

6.2.1 Biomass Pyrolysis

Biomass resources may be limited for the biochar production in a sustainable way. For example, biomass obtained from agricultural crops and trees (forests) may cause diminution of forest areas and increase soil erosion, hence decreasing soil fertility (Cowie et al. 2012). To overcome this situation, Brick (2010) categorized feedstocks into two groups: (1) primarily produced biomass as a resource of bioenergy and biochar and (2) by-products as waste biomass. Nowadays, waste biomass

Table 6.1 Pyrolysis processes and its end products

Process	Residence time	Products (%)		
		Liquid (bio-oil)	Solid (biochar)	Gas (syngas)
Fast (temperature around 300–1000 °C)	Less than 2 s	75%	12%	13%
Intermediate (temperature near about 500 °C)	Moderate residence time	50%	25%	25%
Slow (temperature around 100–1000 °C)	Long residence time	30%	35%	35%
Gasification (temperature more than 800 °C)	Moderate residence time	5% tar	10%	85%

has been used extensively for production of biochar because of its cost-effectiveness and food security advantages compared to other types of biomass (Brick 2010). Biochar is usually made by thermochemical decomposition of biomass at temperatures between 200 and 900 °C in anaerobic condition, which is commonly known as pyrolysis (Demirbas and Arin 2002). Pyrolysis is further categorized into fast, intermediate, and slow depending on the residence time and temperature (Table 6.1; Mohan et al. 2006). Fast pyrolysis process have a short residence time (less than 2 s) and usually generate about 75% of bio-oil from biomass (Mohan et al. 2006). Biochar production around 25–35% comes from slow and intermediate pyrolysis procedure which has a residence time between a few minutes and several hours (Brown 2009).

Gasification is different from general pyrolysis process. In gasification, organic biomass is transformed into gases which are usually rich in carbon monoxide and hydrogen; this is obtained when biomass is burned at high temperature (more than 800 °C) in a controlled aerobic environment. The resulting gas mixture is known as synthetic gas or syngas (Mohan et al. 2006; Sohi et al. 2009).

6.2.2 Factors Affecting Biochar Properties

There are various factors that affect biochar properties, and they are feedstock type, pyrolysis temperature, and heating rate. Table 6.2 showed various types of feedstock and biochar properties.

Commonly, animal litter and solid waste produced a large amount of biochar compared to the biochar obtained from agricultural fields and wood biomasses (Enders et al. 2012). The high yield is associated with the higher inorganic constituents of the raw materials, as indicated by their high ash content. Biochar derived from animal litter contains inherent metals which protect the biochar material from the loss of volatile matter present in biochar by charging the bond dissociation energies of organic and inorganic C bonds (Cantrell et al. 2012). This finding supported by Raveendran et al. (1995) also suggested that high-yield biochar was obtained from rice husk, groundnut shells, coir pith, and wheat straw due to the higher levels of K and Zn. Usually, materials with high lignin percent produce good amount of

Table 6.2 Characteristics of biochar derived from different feedstock

Feedstock	Pyrolysis temperature °C	Heating rate °C min^{-1}	Yield (%)	Ash (%)	C (%)	H (%)	O (%)	N (%)	Surface area (m^2 g^{-1})	References
Rapeseed	500	5.0	35.6	12.9	75	2.62	7.79	1.41	15.7	Karaosmanoğlu et al. (2000)
Rapeseed	600	5.0	32.2	13.9	78.48	1.88	3.94	1.53	17.6	Karaosmanoğlu et al. (2000)
Rapeseed	700	5.0	29.6	14.4	79.48	1.20	3.29	1.35	19.3	Karaosmanoğlu et al. (2000)
Pine needles	300	–	48.6	1.9	68.87	4.31	25.74	1.08	19.9	Chen et al. (2008)
Pine needles	500	–	26.1	2.8	81.67	2.26	14.96	1.11	236.4	Chen et al. (2008)
Pine needles	700	–	14	2.2	86.51	1.28	11.08	1.13	490.8	Chen et al. (2008)
Swine solid	620	13	43.49	44.7	50.70	1.9	0.01	3.26	–	Ro et al. (2010)
Boiler litter	350	–	–	–	45.60	4	18.30	4.50	60	Uchimiya et al. (2010)
Boiler litter	700	–	–	–	46	1.42	7.40	2.82	94	Uchimiya et al. (2010)
Soybean straw	400	20	24.7	–	44.10	–	–	2.38	–	Tong et al. (2011)
Swine solid	700	8.3	36.4	52.9	44.06	0.74	4.03	2.61	4.1	Cantrell et al. (2012)
Wheat straw	400	8.0	34	9.7	65.70	4.05	–	1.05	4.8	Kloss et al. (2012)
Wheat straw	525	8.0	34	9.7	74.40	2.83		1.04	14.2	Kloss et al. (2012)

biochar (Sohi et al. 2010a). Increased heating rate from 5 to 15 °C min^{-1} showed a slight decrease in biochar yield (Karaosmanoğlu et al. 2000). Pyrolysis temperature plays an influential role in determining biochar characteristics. In a study on cottonseed hulls, Uchimiya et al. (2011a) revealed that the biochar yield was affected by the pyrolysis temperatures. A swift reduction in biochar generation was found at more than 400 °C because of the loss of volatile matter and noncondensable gases (CO_2, CO, H_2, and CH_4). The biochar yield was stable at less than 400 °C in cottonseed hulls due to low lignin content. A quick decline in biochar yields was observed at more than 300 °C due to initial degradation reactions, in the case of biochar obtained from grass and wood materials (Keiluweit et al. 2010). However, lignin content of grass is less than the lignin content of wood which causes thermal breakdown at low temperature (200–400 °C).

An enhanced temperature during pyrolysis increases the carbon content, whereas hydrogen and oxygen contents were decreased (Table 6.2). This results in lower molar H/C and O/C ratios, thereby indicating dehydration and deoxygenation of the biomass (Keiluweit et al. 2010). Biochar derived from sewage sludge and poultry manure does not undergo depolymerization due to the absence of lignocellulosic compounds. They are generally rich in N content. Regarding other elements, insufficient database is available for S and P contents of biochar. Usually, no significant change was observed in pyrolysis temperature of various feedstocks (Table 6.2) related to N content of biochar obtained from various sources. S containing functional groups (present in biochar) increases ammonia retention on the surface of char due to ammonium sulfate salts (Petit et al. 2010). Hence, the functional groups in biochar affect the nutrient cycling in soil. Pyrolysis temperature highly influences the morphology and surface structure in biochar (Liu et al. 2010; Uchimiya et al. 2011a). In general, increase in pyrolysis temperature enhances the surface area of char. Although at 700 °C a reduction in surface area has also been reported (Uchimiya et al. 2011b). At high pyrolysis temperature, the aliphatic alkyl and ester groups are disrupted or dissolved which exposes the aromatic lignin core which may be responsible for an increase in surface area (Chen and Chen 2009). Downie et al. (2009) suggested that the pore size distribution is a key factor responsible for the increase in surface area in biochar. Biochars obtained from animal litter and solid waste displayed lower surface areas, whereas biochars obtained from crop residue and wood biomass showed higher surface area even at high temperatures (Table 6.2). This may be due to the low C content and high molar H/C and O/C ratios in the latter biomass samples, leading to the formation of extensive cross-linkages (Bourke et al. 2007; Ahmad et al. 2014a).

6.3 Physiological and Biochemical Response of Plant Under Salt Stress

Salinity is the most important environmental factor that limits plant growth and productivity (Allakhverdiev et al. 2000). The harmful effects of high salt content on plants can be observed such as reduced growth or the death of plants. Many plants create different ways to eliminate salt from their cells or to agglomerate the salt in the cells. During the initiation and development of the salt stress within a plant, all crucial processes such as photosynthesis, protein synthesis, and energy and lipid metabolism are strongly affected. The prompt response is to reduce the speed of blade surface expansion, followed by a termination of expansion as the stress increases. The growth of the plant increases again when the stress is relieved. The photosynthetic rate is generally lower for plants under salt stress.

6.3.1 Salt Tolerance of Plants

Plants which are salt tolerant have a capability to grow and complete their life cycle on a medium containing high amount of salts. Plants that sustain high concentrations of salt in their rhizospheric zone and grow well are known as halophytes. Depending on their salt tolerance, halophytes are obligate or facultative. Obligate halophytes are characterized by low morphology and taxonomic diversity with relatively higher growth rates. In facultative halophytes, they are found in less healthy habitats along the border between saline and nonsaline lands and characterized by a greater physiological diversity that allows them to cope with saline conditions and nonsaline situations.

6.3.2 Mechanism Adopted by the Plant Against Salt Tolerance

Plants follow a series of molecular and biochemical strategies to deal with salt stress. Biochemical processes leading to various products which in turn stimulate various processes that help the plants to cope up with the salinity stress (Iyengar and Reddy 1996). Biochemical mechanism includes (1) selective agglomeration or elimination of ions, (2) selective ion uptake by the roots and transfer it to the leaves, (3) compartmentalization and fixation of ions at the cellular levels, (4) production of various compatible solutes, (5) alterations in photosynthetic pathway, (6) alteration in membrane structure, (7) stimulation of various enzymatic and nonenzymatic antioxidants, and (8) induction of plant hormones. Salt tolerance mechanisms operated inside the plants are generally low complex or high complex. Low-complex mechanism involves alterations in various biochemical pathways. High-complex mechanisms involve changes in crucial process like chromosomal structural changes, viz., DNA methylation and DNA elimination (Walbot and Cullis 1985), and in photosynthesis and respiration processes, i.e., water use efficiency and plasma membrane–cell wall interactions (Botella et al. 1994). It is usually found that high-complex mechanism works in coordination with low-complex mechanism (Bohnert et al. 1995).

6.3.2.1 Ion Modulation and Compartmentalization

Plant generally restricts salt uptake under saline condition or compartmentalizes the salts within various tissues (Adams et al. 1992). Plants, whether glycophyte or halophyte, cannot sustain high concentration of salts within the cytoplasm, and thus they are confined within the vacuole or fix the ions in various tissues to promote normal metabolic functions of the cell (Reddy et al. 1992; Iyengar and Reddy 1996; Zhu 2003). Cheeseman (1988) reported that glycophytes restrict Na uptake or store them in eldest tissues which will eventually be sacrificed (Cheeseman 1988). Salt-inducible enzymes expel sodium from the cytoplasm or confined within the vacuoles and this is usually done by a Na^+/H^+ antiporter (Apse et al. 1999). Two types of electrogenic H+ pumps exist. The first one is vacuolar-type H^+-ATPase (V-ATPase) and the second one is vacuolar pyrophosphatase (V-PPase). Dietz et al. (2001) reported that H^+ pumps work in a synchronized way and exist at the membranes of secretory pathways of plants. V-ATPase is the dominant H^+ pump found at the endomembranes of most of the plant cells. Under normal conditions, the V-ATPase is imperative for plant growth because it provides energy to secondary transport, maintains solute homeostasis, and also enhances vesicle fusion. During abiotic stress conditions like salinity, drought, anoxia, excess heavy metals, etc. in the soil, survival of the plant cells largely resides on retaining the activity of the V-ATPase. Modulation of gene expression and its activity based on the V-ATPase adapting capability lies on long- and short-term bases. A study by Otoch et al. (2001) on salt stress in hypocotyls of *Vigna unguiculata* seedlings revealed that tonoplast regulates salt by H^+-pumping, V-ATPase, and H^+-pyrophosphatase. During salt stress, V-ATPase activity was enhanced, whereas V-PPase activity was restricted (Otoch et al. 2001). The crucial mechanism in halophytes (*Suaeda salsa*) against salinity stress is the modulation of V-ATPase activity, which in turn activates the tonoplast to facilitate ion uptake inside the vacuole; however, V-PPase plays a secondary role (Wang et al. 2001). When the plant is under salt stress, it maintains high concentrations of K^+ and lower concentrations of Na^+ in the cytosol. This is maintained by K^+ and Na^+ transporters, and the driving force for transport is generated by H^+ pumps (Zhu et al. 1993). Some salt stress sensors have been identified which help sustain the plant during stress. Zhu et al. (1993) demonstrated that calcium signal activates the myristoylated calcium-binding protein SOS3 and the serine/threonine protein kinase SOS2. This protein kinase complex then phosphorylates and triggers various ion transporters, such as the plasma membrane Na^+/H^+ antiporter SOS1. *Arabidopsis thaliana* (AtNHX1 gene) encodes a vacuolar Na^+/H^+ antiporter which is crucial for salt tolerance. A similar study by Shi and Zhu (2002) reported modulation of AtNHX1 expression along with ABA production during salt stress. Experimental evidence revealed that salts like NaCl, KCl, or ABA modulate the level of AtNHX1 transcript. AtNHX1 promoter (GUS) analysis in transgenic *Arabidopsis* displayed that AtNHX1 is expressed in all plant tissues except the root tip. High GUS expression was observed in guard cells which further disclosed that AtNHX1 played crucial role in regulation of K+ homeostasis into the specialized cells. NaCl, KCl, or ABA regulates the expression of AtNHX1 at the transcriptional level. AtNHX1 store Na^+ in the enlarged vacuoles of the root hair cells.

Studies (Liu and Zhu 1997; Lauchli and Schubert 1989) showed that Ca^{2+} plays a major role in salt adaptation by the plants. It also stimulates K^+/Na^+ transporters. High salinity also induces high cytosolic Ca^{2+} which is shifted from the intracellular compartments and apoplast (Knight et al. 1997). The resultant transient Ca^{2+}

increase potentiates stress signal transduction and leads to salt adaptation (Mendoza et al. 1994; Knight et al. 1997). Other strategies of salt modulation are salt secretion and selective salt agglomeration or elimination. Salt secretion takes place with the help of unique cellular structure known as salt glands. Salts are expelled from these glands from the leaf surface and maintain ion concentration inside the cell (Hogarth 1999). Salt ejection occurs through the roots in many halophytes (Levitt 1980). Selective agglomeration of ions enables the plants to make osmotic adjustments which results in enhanced water retention and Na exclusion.

6.3.2.2 Induced Biosynthesis of Compatible Solutes

The cytoplasm cumulates low molecular mass compounds known as compatible solutes to harmonize ionic balance in vacuoles, as it does not disrupt the normal metabolic reactions rather it replaces water from the biochemical reactions (Yancey et al. 1982; Ford 1984; Ashihara et al. 1997; Hasegawa et al. 2000; Zhifang and Loescher 2003). Osmolytes maintain the osmotic balance of the tissues by water influx (or reduced efflux) to protect plant structure from salt stress. Compatible solutes contain mainly proline (Khatkar and Kuhad 2000; Singh et al. 2000), glycine betaine (GB) (Rhodes and Hanson 1993; Khan et al. 2000; Wang and Nil 2000), sugars (Kerepesi and Galiba 2000; Bohnert and Jensen 1996; Pilon-Smits et al. 1995), and polyols (Ford 1984; Popp et al. 1985; Orthen et al. 1994; Bohnert et al. 1995). Polyols have various functions, viz., as low-molecular-weight chaperones and also act as scavengers of oxygen radicals (stress-induced) (Smirnoff and Cumbes 1989; Bohnert et al. 1995). There are two types of polyols: acyclic (e.g., mannitol) and cyclic (e.g., pinitol). Mannitol serves as a compatible solute to deal with salt stress, as it is produced through the action of a mannose-6-phosphate reductase (M6PR) in celery. A bacterial gene is used to engineer mannitol biosynthesis in plants to make up the plants to cope salt stress. For example, *A. thaliana* (nonmannitol-producer) has been introduced with M6PR gene under control of the CaMV 35S promotor. After transformation, *Arabidopsis* M6PR transformants started accumulating mannitol throughout the plants, and it ranges between 0.5 and 6 mmol g-1 fresh weight. A unique compound (i.e. mannitol), neither found in celery or *Arabidopsis*. In the absence of NaCl, all transformants are phenotypically the same as the wild type; however, in the presence of NaCl, mature transgenic plants show a high level of salt tolerance (Zhifang and Loescher 2003). Salt stress enhances various reducing sugars like glucose, fructose, sucrose, and fructans in various plants (Kerepesi and Galiba 2000; Khatkar and Kuhad 2000; Singh et al. 2000). However, Gadallah (1999) reported decreased soluble and hydrolyzable sugars in *Vicia faba* due to salinity. Alamgir and Ali (1999) revealed that enhanced sugar content was observed in some genotypes of rice; however, reduction in sugar content was also observed in some genotypes under salinity stress. During salinity stress, Parida et al. (2002) reported that a decrease in starch content vis-a-vis an enhanced content of both reducing and nonreducing sugar has been found in the leaves of *Bruguiera parviflora*. Other studies (Khavarinejad and Mostofi 1998) illustrated that the content of soluble sugars and total saccharides in tomato was increased, but the starch content was not affected by NaCl treatment. Similarly, Gao et al. (1998) have reported upraised sucrose content along with sucrose phosphate in tomato (*Lycopersicon esculentum* L.), but reduction in acid invertase activity was observed due to salinity stress. A number of nitrogen-containing compounds (NCC) are agglomerated in the plants when subjected to salinity stress. The most common NCC is polyamines, amides, quaternary ammonium compounds, imino acids, and proteins. The specific NCC accumulation varies from plant to plant. Wang and Nil (2000) found that during salt stress, glycine and betaine content increase in various plants. Parida et al. (2002) also confirmed that the proline content is also known to increase under saline condition in the leaves of *B. parviflora*.

Amino acid like cysteine, arginine, and methionine contents were reduced in wheat plants due to NaCl treatments. However, proline, isoleucine, aspartic acid, and valine contents were enhanced in the salinity stress condition (Elshintinawy and Elshourbagy 2001).

6.3.2.3 Induction of Antioxidative Enzymes

Salt stress introduces water paucity due to osmotic effects on various metabolic activities (Greenway and Munns 1980; Cheeseman 1988). This condition leads to the formation of reactive oxygen species (ROS) such as superoxide ($O_2^{\cdot-}$), hydrogen peroxide (H_2O_2), hydroxyl radical (OH) (Halliwell and Gutteridge 1985), and singlet oxygen ($^1O_2$) (Elstner 1987). These reactive oxygen species (ROS) can cause severe injury to the normal metabolism via lipid peroxidation (Fridovich 1986; Wise and Naylor 1987) and damage to protein and nucleic acids (Fridovich 1986; Imlay and Linn 1988). O_2 concentration is very high during photosynthesis; thus, the chloroplast is more likely to generate ROS (Asada and Takahashi 1987). Superoxide ($O_2^{\cdot-}$) dismutates, either enzymatically or nonenzymatically, to produce H_2O_2 and O_2. Further, H_2O_2 may interact with certain metal ions or metal chelates to produce highly reactive $^{\cdot}OH$ (Imlay and Linn 1988). Plants are equipped with a series of antioxidants to protect themselves from reactive oxygen species. The metalloenzyme superoxide dismutase (SOD) converts $O_2^{\cdot-}$ to H_2O_2. Catalase and a variety of peroxidases (Chang et al. 1984) activate the catalysis of H_2O_2. However, catalase is absent in the chloroplast; H_2O_2 can be detoxified by an ascorbate-specific peroxidase (through the ascorbate–glutathione cycle) that is present in high concentration in this organelle (Chen and Asada 1989; Halliwell and Gutteridge 1986; Asada 1992). Plants that have high levels of antioxidants (either constitutive or induced) have been reported to cope up better against the oxidative damage. The enhanced activities of various antioxidative enzymes such as catalase (CAT), ascorbate peroxidase (APX), guaiacol peroxidase (POD), glutathione reductase (GR), and superoxide dismutase are in correlation with the high levels of salinity stress (Benavides et al. 2000; Lee et al. 2001; Mittova et al. 2003). Comba et al. (1998) revealed that soybean root nodules under salt stress showed decline of ascorbate peroxidase and glutathione reductase activities, whereas superoxide dismutase and reduced glutathione content were increased, and malondialdehyde (MDA) and total protein content remain unchanged. Willkens et al. (1997) reported that tobacco plant displayed enhanced sensitivity against salinity stress. Studies (Orr and Sohal 1992; Allen et al. 1997; Noctor and Foyer 1998) revealed that transgenic plants showed overexpression of genes which leads to enhanced activities of Mn-SOD, Fe-SOD, chloroplastic Cu/Zn-SOD, bacterial catalase, and glutathione S-transferase (GST)/glutathione peroxidase (GPX) under salinity stress (Roxas et al. 2000). Takemura et al. (2002) observed that salt stress causes stimulation of superoxide in the cytosol which in turn builds up the tolerance capacity of plants (*B. gymnorrhiza*).

6.3.2.4 Stimulation of Plant Hormones

A huge amount of salt stimulate plant hormones like ABA and cytokinins (Vaidyanathan et al. 1999). Abscisic acid is liable for the changes in salt stress-induced genes (De-Bruxelles et al. 1996). Gupta et al. (1998) suggested that in rice, ABA-induced genes play critical role in salt tolerance. An enhanced level of ABA, aminocyclopropane-1-carboxylic acid, and ethylene was observed in *Citrus sinensis* (GomezCadenas et al. 1998) during salt stress. ABA lightens the stress generated due to NaCl (Popova et al. 2002). ABA favors stomatal closure by swiftly changing ion fluxes in guard cells under stress conditions. ABA is also involved in modifications of gene expression and the diversity of potential cis-acting regulatory elements. Chen et al. (2001) demonstrated the experimental evidence which showed that the enhanced Ca^{2+} uptake is correlated with high levels of ABA under salt stress which further maintain membrane integrity. GomezCadenas et al. (2002) reported that ABA caused diminution of ethylene release and leaf absorption in citrus under stress. This is probably due to the decrease in the accumulation of Cl^- ion in the leaves. Salt tolerance of facultative halophytic (*Lophopyrum elongatum)* and less salt-tolerant wheat (*T. aestivum* L.) increased in due course of time, when they are gradually acclimatized with the salt condition rather than suddenly shocked (Noaman et al. 2002). This acclimation to salt stress is regulated by ABA.

6.3.2.5 Change in Photosynthetic Pathway

A high salt concentration inhibits the photosynthesis by reducing the water potential. To cope up from this, plants enhance their water use efficiency under salt stress. Facultative halophytic plants such as *M. crystallinum* change their C3 mode to CAM photosynthesis, so as to reduce their water loss by facilitating the opening of stomata at night (Cushman et al. 1989). Plants like *Atriplex lentiformis* change their mode of photosynthesis from C3 to C4 pathway in response against salinity stress (Zhu and Meinzer 1999).

6.3.2.6 Molecular Mechanism of Salt Tolerance

Metabolic adaptation at cellular level makes the plant to sustain against salt stress, and a large number of genes have also been identified that copes the plant against stress (Ingram and Bartels 1996; Bray 1997; Shinozaki et al. 1998). Salt tolerance is a multigenic trait, and various genes having different functional groups are responsible for encoding salt stress proteins: (1) genes for photosynthetic enzymes, (2) genes for synthesis of compatible solutes, (3) genes for vacuolar-sequestering enzymes, and (4) genes for radical scavenging enzymes. Majority of the genes in the functional groups have been classified as salt inducible under stress conditions. Wu et al. (1996) identified mutants by a salt hypersensitivity assay in *Arabidopsis*, which caused K uptake due to salt stress. Kawasaki et al. (2001) investigated the transcript regulation in rice-tolerant variety Pokkali with microarrays. Adjustment of microorganisms to particular environmental stress is deeply related to the expression of various genes present in the microorganism. Kanesaki et al. (2002) found that hyperosmotic stress showed different effects on the cytoplasmic volume and gene expression in *Synechocystis* sp. PCC 6803. DNA microarray analysis revealed that salt stress stimulates genes for some ribosomal proteins. However, hyperosmotic stress also strongly triggers the genes for 3-ketoacyl-acyl carrier protein reductase and rare lipoprotein A. Each kind of stress stimulates a number of genes for proteins of unknown function.

6.4 How Biochar Mitigates Salinity Stress in Plants

The world population is increasing at an alarming rate and is expected to reach 9.6 billion by 2050 (FAO 2009). Increased population puts pressure for food requirement, and generation of more food in turn creates pressure on natural resources. Agricultural crops are usually subjected to abiotic stresses like salinity, drought, and heavy metal stress (Osakabe et al. 2014; Parihar et al. 2015; Rizwan et al. 2016a). Among the abiotic stresses, salinity and drought are the most serious threats to agricultural production. According to Wicke et al. (2011), more than 1100 million hectares (1128 Mha) of land surface is affected by salinity. Due to salinity problem, approximately an annual economic loss of 27.2 billion USD in terms of crop loss in irrigated agriculture land (Qadir et al. 2014). Loss in revenue increased day by day, and it will reach up to 69% if no precautionary measure is taken to mitigate the deteriorated land. However, the C emanation from degraded lands enhances the cost of reclamation. Plants under salinity have to deal with two types of stresses: one is osmotic stress and the other is ionic stress. The osmotic stress raises the salt level in the soil solution which surrounds the roots that results in water uptake hindrance and interferes in lateral bud development (Munns and Tester 2008). On the other hand, in ionic stress, when Na^+ concentration is more than the threshold, it leads to leaf mortality, chlorosis, necrosis, and inhibition of photosynthesis (Glenn et al. 1999; Panuccio et al. 2014). Biochar is a charcoal-like material gathered after heating any organic material in anaerobic condition (process known as pyrolysis). It has a great demand in the agricultural sector as it enhances the soil physicochemical properties (e.g., soil water holding capacity, aggregate stability, aeration, bulk density, nutrient holding capacity, EC, pH, surface area, and CEC) when amended with degraded soil (Lehmann and Joseph 2009a; Sohi et al. 2010b; Andrenelli et al. 2016; Bamminger et al. 2016). The overall mean increase in crop productivity due to increased plant growth and biomass reported in the literature with biochar amendment was about 10–12% (Jeffery et al. 2011; Haider et al. 2015; Kim et al. 2016). Some studies have shown negative effects on crop productivity when biochar was added in the soil, which may be due to specific types of biochar (Liu et al. 2013). Biochar consists of a large degree of recalcitrant carbon (C) (Cheng et al. 2008), which may remain in the soil for 100–1000 years, and thus, biochar could be very

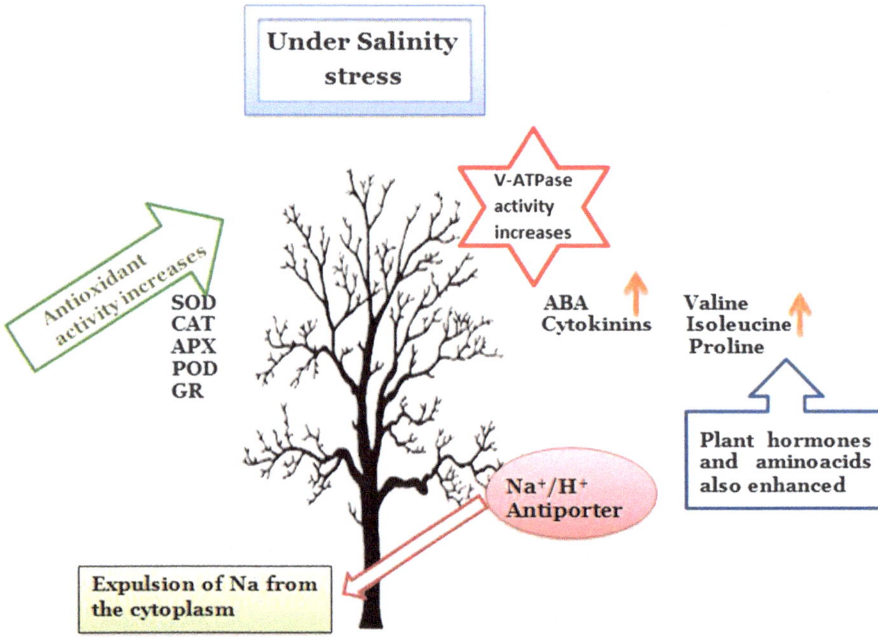

Fig. 6.1 General strategies used by the plants to cope up from salinity stress

effective in fixing the carbon in the soil and hence reducing global greenhouse gas (GHG) emission (Lehmann 2007; Sohi et al. 2010a; Chowdhury et al. 2014). There are plenty of reports which illustrate the remediation of toxic/contaminated soils (Beesley et al. 2011; Uchimiya et al. 2012; Zhang et al. 2013; Ahmad et al. 2014b; Samsuri et al. 2014). The crucial factor or mechanism behind the reclamation of contaminated soil with biochar was due to its high adsorption capacity (Samsuri et al. 2014; Zhang et al. 2013). Thomas et al. (2013) also reported high salt sorption potential of biochar, and therefore, it can alter the negative consequences of salinity by reducing Na^+ uptake or by eliminating Na^+ from the plant cells. The adsorption capacity of biochar is mainly governed by the characteristics of biochar (Lehmann and Joseph 2009a) and the type of feedstock and also the pyrolysis conditions under which the biochar is produced (Chen et al. 2011). The determining feature of biochar which makes it jack of all trades is the high adsorption capacity which includes its high surface area and cation exchange properties (Fig. 6.1).

6.4.1 Functional Properties of Biochar

Adsorption capacity of biochar is mainly because of the presence of functional groups generated during the pyrolysis process which in turn depends on raw material (feedstock) and the pyrolysis temperature (Chun et al. 2004; Ahmad et al. 2012). During pyrolysis at temperature around 300 °C, organic material loses various degradable compounds such as cellulose and lignin. As the temperature is intensified, the aromatic compounds become more condensed which leads to enhanced surface area of biochar (Inyang et al. 2016). The functional properties of biochar which are derived from agricultural residues can be assessed through FTIR, Boehm titrations, and scanning electron microscopy (Zhang and Luo 2014). Other studies by Zhang et al. (2014) found that the functional properties like carboxyl, lactones, and phenolic groups were observed more in wood-derived biochar than the biochar obtained from bamboo, rice husk, and rice husk ash. A similar study by Qayyum et al. (2012) illustrated high aromaticity in biochar derived from wood as compared to biochar derived from sewage sludge. Biochar derived from pine needles contains various functional groups (Ahmad et al. 2013). The presence of these and other functional groups in biochar makes them a suitable and economical choice for the adsorption of various salts present in the soils, thus mitigating the salinity of the soil (Ahmad et al. 2014a; Rajapaksha et al. 2016).

6.4.2 Salt Stress Effect on Soil Properties

Salt stress negatively affects the soil properties, plant growth, and overall productivity of plants (Ohashi et al. 2014; Rath and Rousk 2015). Commonly, salt-affected soils are further divided into saline, sodic, or saline–sodic which are mainly based on their electrical conductivity (EC), sodium adsorption ratio (SAR), and exchangeable sodium percentage (ESP) of the saturated paste extracts (Richards 1954). Among these three, the saline–sodic soils are highly deteriorated and deprived of nutrient which in turn corresponds to least productivity (Rengasamy and Olsson 1991). Salinity is inversely proportional to the soil properties such as organic matter and C:N ratio (Morrissey et al. 2014). Due to salinity and sodicity, the microbial biomass and microbial activity alter in the soil and gradually decline (Yan et al. 2015). A similar study was done by Rath and Rousk (2015); they found that soil respiration and soil microbial enzyme activities were restricted due to short- or long-term salinity and it also affects the C and nutrient cycling in the soil. However, the net respiration also depends on the residue properties. The adverse consequences of salinity are more prompt in the degraded or infertile soil having materials that are less degradable (Hasbullah and Marschner 2014).

6.4.3 Biochar Effect on Soil Properties Under Salt Stress

Biochar is well known to enhance the plant tolerance against salt stress. Amelioration of soil with the help of biochar was studied extensively; it improves the physicochemical properties of soil. Studies (Chaganti et al. (2015), Diacono and Montemurro (2015), Sun et al. (2016)) revealed that biochar not only improves the soil characteristics but also related to sodium removal as sodium leaching and EC. Biochar also rectifies the physical, chemical, and biological properties of soil under abiotic stresses (Rizwan et al. 2016b). Lu et al. (2015) reported during maize cultivation, when biochar derived from poultry manure compost (BPC) amended in saline soil showed increased microbial biomass carbon content and along with the enhanced activities of urease, invertase, and phosphatase in rhizospheric soils. Similar studies were done by Bhaduri et al. (2016); they illustrated that saline soil mitigation by biochar depends upon the amount of biochar, incubation time, biochar material, and type of soil enzymes. However, an organic amendment also improves the physico-chemical properties of saline soil (Wang et al. 2014a, b). Not much literature is available on the effect of biochar on the saline soil properties. Thomas et al. (2013) showed that application of biochar (30 g m^{-2}) on salt-affected soil did not alter the pH, although increase in EC was observed over the control. Similarly, enhanced SOC and CEC were observed after furfural biochar was applied in saline soil (Wu et al. 2014). Biochar enhanced the soil organic matter along with the CEC and inhibited the exchangeable Na (Luo et al. 2017). Aforesaid studies showed that addition of biochar in saline soils mitigates the salinity in soil and also enhances the soil microbiota.

6.4.4 Regulation of Stomatal Conductance and Reduction in Oxidative Stress

Various studies (Akhtar et al. 2015a; Lashari et al. 2015) revealed that biochar reduces salt stress in plants by lower production of phytohormones. Akhtar et al. (2015a) reported that under salinity stress, biochar restricts the ABA content in leaf and xylem sap of potato. When biochar is amended in saline soil with endophytic bacteria, it inhibits the ABA concentration in xylem of wheat and maize as compared to the unamended controls (Akhtar et al. 2015b, c). Stomatal conductance is known to be enhanced in plants grown in biochar-mediated soil under salt stress. Many recent studies (Thomas et al. 2013; Akhtar et al. 2015b, c) have revealed that biochar application to sodic/saline soils ameliorates the stomatal density and conductance in various herbaceous plants like wheat and tomato. Biochar not only improves soil properties but also enhances the soil moisture and sodium-binding capacity to biochar-amended soil which in turn reduces the root reactivity to osmotic stress (Akhtar et al. 2015c). Thus, the generation of ABA declines in the root which in turn enhances the stomatal conductance and leaf growth.

Salinity induces oxidative stress in plants by the excessive production of reactive oxygen species (Parihar et al. 2015; Farhangi-Abriz and Torabian 2017). Tartoura et al. (2014) showed that organic amendments lessen the salinity stress in plants by modulating the synthesis of antioxidant enzymes. A few studies reported that biochar mitigates the oxidative stress in plants grown in saline soils. Kim et al. (2016) found that maize under salinity stress generates high amount of ROS, but after amendment with biochar, the ascorbate peroxidase (APX) and glutathione reductase (GR) activities were declined as compared to control. Similarly, study by Lashari et al. (2015) revealed that compost of poultry manure plus diluted pyroligneous solution when applied on maize field under salt stress decreases MDA content in leaf sap. A very recent study (Farhangi-Abriz and Torabian 2017) reported that biochar when amended in saline soil decreases the antioxidant enzyme activities of plants and releases the oxidative stress generated in bean seedlings. These studies formulated that biochar could improve plant growth and biomass under salt stress by reducing oxidative stress.

6.4.5 Effects on Plant Growth, Biomass, and Photosynthesis

Employment of biochar in salt-affected soils is proclaimed to augment the soil properties which in turn promote plant growth and photosynthesis. For example, biochar derived from wood, when amended in saline soil, enhanced the shoot biomass, tuber yield, and Pn in potato (Akhtar et al. 2015a). Another study by Kim et al. (2016) restores the tidal land (which contained high concentration of exchangeable sodium) with the help of biochar derived from rice hull, which further promotes the maize growth. Biochar addition enhances the tomato growth and biomass under saline condition (3.6 dS m^{-1}) (Usman et al. 2016). A study by Thomas et al. (2013) reported that biochar (pyrolyzed at 378 °C and applied at the rate of 50 t ha^{-1}) enhances the growth of two herbaceous plant (*Abutilon theophrasti* and *Prunella vulgaris*) under salt stress; however, photosynthetic carbon gains and chlorophyll fluorescence (Fv/Fm) value have not been affected by the presence of biochar.

Overall, the biochar response under salt stress varied from plant species to species. During salinity stress, biochar derived from compost when amended in soil could also promote the plant growth and biomass. For example, Lashari et al. (2013) reported that employment of biochar poultry manure compost (BPC) along with pyroligneous solution (PS) spray for 6 weeks in the saline soil enhanced the wheat grain production as compared to the control. In another study in maize field, BPC and PS treatments in salt-affected soil promote plant height, leaf area index, and photosynthetic pigments along with the increased grain yield (Lashari et al. 2015). However, when composted biochar was applied on two halophyte species under saline conditions, similar result was reported (Luo et al. 2017). During saline conditions, biochar application with the suitable microbial inoculants like plant growth-promoting rhizobacteria (PGPR) further improved the growth of plant as compared to the control and biochar-only treatments (Nadeem et al. 2013; Fazal and Bano 2016; Akhtar et al. 2015b, c). Another study by Hammer et al. (2015) found that arbuscular mycorrhizal (AM) fungi when incorporated with biochar under salinity stress showed increased growth of lettuce as compared to the treatments alone. The beneficial effect of biochar on ion homeostasis under salinity stress could be further enhanced by co-application of biochar with endophytic bacteria (Akhtar et al. 2015a, b). However, the combined application of wheat straw-derived biochar and P increased the phosphate precipitation/sorption in the saline–sodic soil and decreased P concentrations in plants (Xu et al. 2016). These studies illustrated that biochar might be very effective in mitigating salinity stress from the soil and also inhibits Na^{+} uptake by plants grown in saline soils which in turn enhanced the mineral nutrients in plants. For example, the biochar and AM application promotes the P and Mn content in lettuce plants (Hammer et al. 2015). However, further research is needed to establish the mechanisms of biochar-mediated mineral uptake by plants under saline conditions both at the soil and plant levels.

6.5 Effect on Soil Properties After Addition of Biochar

6.5.1 Modification of the Soil Habitat by Biochar

The material properties of biochar are very diverse from that of the uncharred organic matter in the soil (Schmidt and Noack 2000) and are evidently changed over a period of time due to weathering, interactions with soil mineral and organic matter, and oxidation by microorganisms in soil (Lehmann et al. 2005; Cheng et al. 2008; Cheng and Lehmann 2009; Nguyen et al. 2010). The mechanism of physical and chemical characteristics of biochar and its interaction with soil biota and microorganism are poorly understood.

6.5.1.1 Basic Properties: Organic and Inorganic Composition

Biochar composition can be divided into recalcitrant C, labile or leachable C, and ash. The major chemical difference between biochar and other organic matter is the much larger proportion of aromatic C and, specifically, the occurrence of fused aromatic C structures, in contrast to other aromatic structures of soil organic matter such as lignin (Schmidt and Noack 2000). The fused aromatic structure of biochars, when obtained at lower temperature, includes amorphous C, whereas when obtained at higher temperature, turbostratic C was formed (Keiluweit et al. 2010; Nguyen et al. 2010). The stability of biochar depends largely on these C structures (Nguyen et al. 2010). However, exact mechanism of the stability to the aromatic C structures in soil is not yet clear. The microbiota present in soil does not readily absorb the C (from the biochar) although a fraction of C may be leached and therefore mineralizable (Lehmann et al. 2009), and in some studies, it has been shown that biochar stimulates the microbial activity and hence abundance (Steiner et al. 2008). At present, such fractions may be quantified by incubation studies and are frequently referred to as "volatile matter" or the labile fraction. Volatile matter refers to an ASTM standard methodology that was developed to evaluate the quality of coals as fuels and determine the stability of biochar (Deenik et al. 2010; Zimmerman 2010). However, such quantified volatile matter (5–37% of C in the study by Zimmerman 2010) is typically much larger than the corresponding mineralization (2–18% of C over 1 year). This may indicate that the mineralizable fraction is imperfectly captured by volatile matter. The next main component consists of minerals present as ash inclusions in biochar. These minerals include several significant macro- and micronutrients for biological absorption and thus the concept of valuable resources in the soil food chain. The presence of these elements during pyrolysis stimulates organo-metal reactions which are thermodynamically favorable at high temperatures. For example, N may replace one or two C atoms in aromatic compounds (Leinweber et al. 2007) with largely unknown effects on biochar behavior in soil. Iron (Fe)-rich biochars made from peat and investigated by ^{57}Fe Mossbauer spectroscopy show the formation of Fe_3C bonds and small ferromagnetic iron clusters at pyrolysis temperatures over 600 °C (Freitas et al. 2002). Grass and a variety of common raw materials (rice shells, sludge from purification plants, etc.) also contain significant amounts of amorphous silica (>2 wt%). The Si-C bonds usually participate in cross-links between aromatic domains and crystallites (Freitas et al. 2000). At temperatures of 400–600 °C, pyrolysis alters the chemical structure of biosilicates, with a continuous increase of SiO_4 relative to SiO_{2-3} with increasing heat treatment temperature (Freitas et al. 2000). Silicates can absorb a significant proportion (>14% for corn cobs and 88% for rice shells) of biochar pore volume (Bourke et al. 2007; Freitas et al. 2000).

6.5.2 Responses of the Soil Biota to Biochar

The use of biochar as a targeted strategy for controlling soil biota is an issue of increasing interest, and unintended changes to the soil's biota as a result of biochar application are of the same concern. This line of research is important as the health and diversity of soil microbial populations are crucial for soil functional and ecosystem services, which in turn affect the soil structure and stability, nutrient cycle, aeration, water utilization efficiency, disease resistance, and C storage capacity (Brussaard 1997). Incorporation of various organic matters to the soil is one of the most important means of managing biodiversity in soils (Brussaard et al. 2007). The distributions of organic amendments, quantity, and quality affect the trophic structure of the soil food web (Moore et al. 2004). Therefore, all three of these aspects should be considered in the use of biochar as a soil management tool. In the following section, we discuss how the biochar affects the:

I. Growth of microorganism
II. Nutrient transformation

In the following sections, we consider how biochar affects soil biota on several trophic levels, including root dynamics, and discuss the reasons behind observed changes with respect to different biochar properties.

6.5.3 Abundance of Microorganisms

Microbial abundance has been assured in biochar-amended soil by various methods like total genomic DNA extraction (Jin 2010), culturing and plate counting (Jackson 1958), substrate-induced respiration (Steiner et al. 2004; Kolb et al. 2009), fumigation extraction (Jin 2010; Liang et al. 2010), phospholipid fatty acid (PLFA) extraction (Birk et al. 2009), staining, and direct observation of individual biochar particles (Jackson 1958; Pietikäinen et al. 2000; Warnock et al. 2007; Jin 2010; Fig. 6.2). The microbial reproduction rate has been uplifted in some biochar-amended soils (Steiner et al. 2004) and in wastewater (Koch et al. 1991). In case of bio-digesters

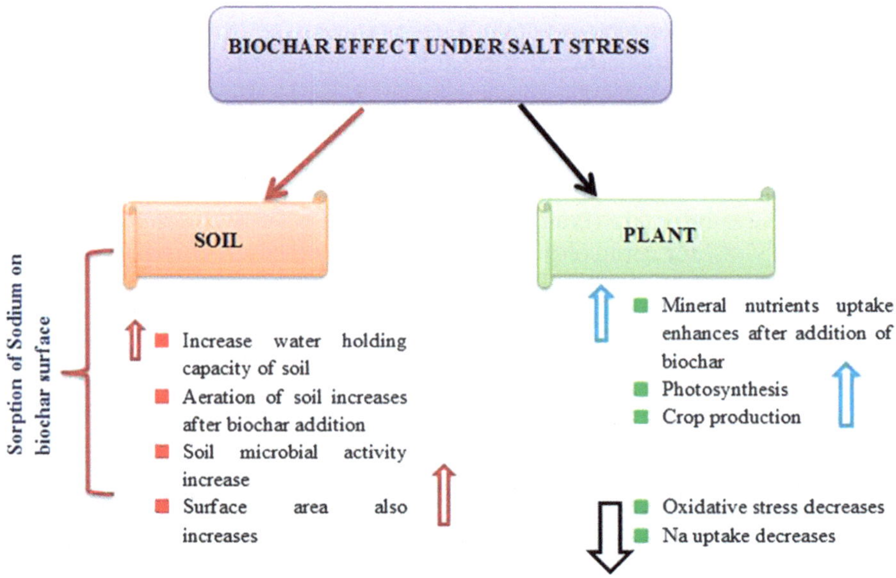

Fig. 6.2 Biochar effects under salt stress

which are used to evolve methane (CH_4) (as an energy source) after the addition of biochar led to an increase in anaerobic and cellulose-hydrolyzing bacteria (Kumar et al. 1987). The changes in microbial abundance vary within the groups of microorganism. In case of mycorrhizal fungi (arbuscular and ectomycorrhizal), enhanced growth was observed (Warnock et al. 2007). Mycorrhizal response in the host plant is most commonly measured by measuring the root colonization. Both formation rate and tip number of EM infection of larch seedling roots were increased by 19–157% with biochar additions (Makoto et al. 2010). Likewise, AM colonization of wheat roots was found to increase to 20–40% 2 years after *Eucalyptus* wood biochar additions of 0.6–6 t ha^{-1}, in comparison to a colonization rate of 5–20% in unamended controls (Solaiman et al. 2010). Biochar derived through hydrothermal carbonization stimulates the spore germination of AM fungi, which further enhances the populations of these symbionts in the soil (Rillig et al. 2010). Although, some studies (Gaur and Adholeya 2000; Birk et al. 2009; Warnock et al. 2010) revealed that after biochar addition AM abundance decreases. The reasons behind the decline in population are still not clear; however, it could be because of (1) a reduced requirement for mycorrhizal symbiosis due to increased in nutrient and water availability to plants which cause decreases in mycorrhizal abundance, for example, with greater P availability in soil (Gryndler et al. 2006); (2) changes in soil conditions, e.g., due to modifications of pH or water relations (discussed below); (3) direct negative effects from high contents of mineral elements or organic compounds detrimental to the fungi, such as high salt or heavy metal contents (Killham and Firestone 1984; Killham 1985); and (iv) sorption of organic C and organically bound nutrients may influence their availability (Pietikäinen et al. 2000; Chan and Xu 2009).

The nutrient and C availability change the increase or decrease in microbial biomass, depending on (1) the existing nutrient and C availability in soil, (2) the magnitude of change, and (3) the microorganism group (Warnock et al. 2010). Bacteria may sorb to biochar surfaces and make them less susceptible to leaching in soil (Pietikäinen et al. 2000). Thus, it increases bacterial abundance although it has no effect on fungal abundance. The ability of biochars to retain bacteria will vary greatly depending on the biochar properties including the ash content, pore size, and volatile content that are highly variable (Bond 2010). Formation of surfactants by microorganisms (Ron and Rosenberg 2001) may additionally facilitate adhesion to biochars.

6.5.4 Effect of Biochar on Nutrient Transformation

Biochar can have influential effects on microbially mediated transformation of nutrients in soil. In forest soils, nitrification has been enhanced after the amelioration of soil with the biochar (MacKenzie and DeLuca 2006; Ball et al. 2010) which in turn increases sorption of phenolic that would otherwise inhibit nitrification (Zackrisson et al. 1996; DeLuca et al. 2006) and an increase in ammonia-oxidizing bacteria (Ball et al. 2010). However, biochar additions to agricultural soil show no change in net N mineralization (DeLuca et al. 2006) and less N availability to the plants (Lehmann et al. 2003). The higher the mineralizable fraction of biochar (often quantified and described as volatile matter), the greater the N immobilization with resultant decreases in N uptake and growth of crops (Deenik et al. 2010). A high microbial biomass content was observed after biochar additions. An enhanced activity of alkaline phosphatase, aminopeptidase, and *N*-acetylglucosaminidase was found to increase with biochar applications (Bailey et al. 2010; Jin 2010). Alkaline phosphatase increased by 615% and aminopeptidase by 15% with increasing rates of corn biochar application to an alfisol (Jin 2010). Biochar triggers the growth of fine roots and root hairs which in turn raised the production of organic N and P mineralization enzymes (Bailey et al. 2010). Biochar induces changes in the bacterial community similar to rhizosphere effects.

6.5.5 Biochar and Plant Roots

Various studies (Breazeale 1906; Nutman 1952) showed that the biochar materials have been reported to stimulate the root growth (Breazeale 1906; Nutman 1952). The different properties of biochar in comparison to the soil cause the improved root growth; however, roots may grow into the biochar pores (Lehmann et al. 2003; Joseph et al. 2010). After forest fire, a layer of char enhances the root biomass (47%) and also root tip number (64%) (Makoto et al. 2010). The root length of rice was also increased with biochar additions (Noguera et al. 2010). Germination and rooting of fir embryos (*Abies numidica*) significantly increased from 10% to 20% without additions to 32–80% of embryos when activated carbon was added to various growth media (Vookova and Kormutak 2001). Therefore, not only abundance but also growth behavior of roots may change in response to the presence of biochar. When the soils are amended with biochar, it improves the physicochemical properties of the soil like water availability, pH, and aeration which likely enhance root growth and also enhance shoot to root ratio (Wilson 1988). A study by Breazeale (1906) and Dachnowski (1908) have illustrated the pronounced increase in root growth after additions of carbon black (soot) to soil with sorption of allelopathic compounds that were phytotoxic. Inderjit and Callaway (2003) found that activated carbon neutralizes the phytotoxic compounds present in soil. Although these results have been challenged by Lau et al. (2008), they revealed that nutrients leaching from the activated carbons and that the addition of carbonaceous adsorbents may have multiple effects on soil. No studies have been published which showed the toxic effect of biochar on plant growth, i.e., plant growth decreased while shoot to root ratio increased.

6.6 Conclusion and Future Perspectives

Crop growth and yields have seriously been affected due to salt stress. Studies reported that the biochar application enhances the plant growth and biomass during salt stress. Biochar when applied in soil enhances photosynthesis and nutritional uptake and modified gas exchange properties in plants. Biochar inhibits Na^+ uptake while increased K^+ uptake in salt-stressed plants. The biochar amelioration increases the capability of the plant to tolerate high salt condition by the reduction of Na^+ uptake, accumulation of minerals, and regulation of stomatal conductance and plant hormones. Overall, this chapter develops a better understanding of the biochar-mediated tolerance mechanisms in plants under salt stress. However, more research is needed to establish a different mechanism for biochar-mediated mineral uptake at plant and soil. The effect of biochar on soil also depends on the type of biochar (raw material), pyrolysis temperature, and the type of soil.

References

Adams P, Thomas JC, Vernon DM, Bohnert HJ, Jensen RG (1992) Distinct cellular and organismic responses to salt stress. Plant Cell Physiol 33:1215–1223

Ahmad M, Lee SS, Dou X, Mohan D, Sung JK, Yang JE, Ok YS (2012) Effects of pyrolysis temperature on soybean stover- and peanut shell derived biochar properties and TCE adsorption in water. Bioresour Technol 118:536–544

Ahmad M, Lee SS, Rajapaksha AU, Vithanage M, Zhang M, Cho JS, Lee SE, Ok YS (2013) Trichloroethylene adsorption by pine needle biochars produced at various pyrolysis temperatures. Bioresour Technol 143:615–622

Ahmad M, Rajapaksha AU, Lim JE, Zhang M, Bolan N, Mohan D, Vithanage M, Lee SS, Ok YS (2014a) Biochar as a sorbent for contaminant management in soil and water: a review. Chemosphere 99:19–33

Ahmad M, Lee SS, Lim JE, Lee SE, Cho JS, Moon DH, Hashimoto Y, Ok YS (2014b) Speciation and phytoavailability of lead and antimony in a small arms range soil amended with mussel shell, cow bone and biochar: EXAFS spectroscopy and chemical extractions. Chemosphere 95:433–441

Akhtar SS, Andersen MN, Liu F (2015a) Biochar mitigates salinity stress in potato. J Agron Crop Sci 201:368–378

Akhtar SS, Andersen MN, Liu F (2015b) Residual effects of biochar on improving growth, physiology and yield of wheat under salt stress. Agric Water Manag 158:61–68

Akhtar SS, Andersen MN, Naveed M, Zahir ZA, Liu F (2015c) Interactive effect of biochar and plant growth-promoting bacterial endophytes on ameliorating salinity stress in maize. Funct Plant Biol 42:770–781

Alamgir ANM, Ali MY (1999) Effect of salinity on leaf pigments, sugar and protein concentrations and chloroplast ATPase activity of rice (Oryza sativa L.) Bangladesh J Bot 28:145–149

Allakhverdiev SI, Sakamoto A, Nishiyama Y, Inaba M, Murata N (2000) Ionic and osmotic effects of NaCl-induced inactivation of photosystems I and II in *Synechococcus* sp. Plant Physiol 123:1047–1056

Allen RD, Webb RP, Schake SA (1997) Use of transgenic plants to study antioxidant defenses. Free Radic Biol Med 23:473–479

Andrenelli MC, Maienza A, Genesio L, Miglietta F, Pellegrini S, Vaccari FP, Vignozzi N (2016) Field application of pelletized biochar: short term effect on the hydrological properties of a silty clay loam soil. Agric Water Manag 163:190–196

Apse MP, Aharon GS, Snedden WA, Blumwald E (1999) Salt tolerance conferred by overexpression of a vacuolar Na+/H+ antiport in Arabidopsis. Science 285:1256–1258

Asada K (1992) Ascorbate peroxidase—a hydrogen peroxide scavenging enzyme in plants. Physiol Plant 85:235–241

Asada K, Takahashi M (1987) Production and scavenging of active oxygen radicals in photosynthesis. In: Kyle DJ, Osmond CB, Arntzen CJ (eds) Photoinhibition, vol 9. Elsivier, Amsterdam, pp 227–288

Ashihara H, Adachi K, Otawa M, Yasumoto E, Fukushima Y, Kato M, Sano H, Sasamoto H, Baba S (1997) Compatible solutes and inorganic ions in the mangrove plant *Avicennia marina* and their effects on the activities of enzymes. Z Naturforsh 52:433–440

Bailey VL, Fansler SJ, Smith JL, Bolton JH (2010) Reconciling apparent variability in effects of biochar amendment on soil enzyme activities by assay optimization. Soil Biol Biochem 43:296–301

Ball PN, MacKenzie MD, DeLuca TH, Holben WE (2010) Wildfire and charcoal enhance nitrification and ammonium-oxidizing bacteria abundance in dry montane forest soils. J Environ Qual 39:1243–1253

Bamminger C, Poll C, Sixt C, Högy P, Wüst D, Kandeler E, Marhan S (2016) Short-term response of soil microorganisms to biochar addition in a temperate agroecosystem under soil warming. Agric Ecosyst Environ 233:308–317

Bargmann I, Rilling MC, Buss W, Kruse A, Kuecke M (2013) Hydrochar and biochar effects on germination of spring barley. J Agron Crop Sci 199:360–373

Beesley L, Moreno-Jimenez E, Gomez-Eyles JL, Harris E, Robinson B, Sizmur T (2011) A review of biochars' potential role in the remediation, revegetation and restoration of contaminated soils. Environ Pollut 159:3269–3282

Benavides MP, Marconi PL, Gallego SM, Comba ME, Tomaro ML (2000) Relationship between antioxidant defence systems and salt tolerance in Solanum tuberosum. Aust J Plant Physiol 27:273–278

Bhaduri D, Saha A, Desai D, Meena HN (2016) Restoration of carbon and microbial activity in salt-induced soil by application of peanut shell biochar during short-term incubation study. Chemosphere 148:86–98

Birk JJ, Steiner C, Teixeira WC, Zech W, Glaser B (2009) Microbial response to charcoal amendments and fertilization of a highly weathered tropical soil. In: Woods WI, Teixeira WG,

Lehmann J, Steiner C, WinklerPrins AMGA, Rebellato L (eds) Amazonian dark earths: Wim Sombroek's vision. Springer, Berlin, pp 309–324
Bohnert HJ, Jensen RG (1996) Strategies for engineering waterstress tolerance in plants. Trends Biotechnol 14:89–97
Bohnert HJ, Nelson DE, Jensen RG (1995) Adaptations to environmental stresses. Plant Cell 7:1099–1111
Bond DR (2010) Electrodes as electron acceptors, and the bacteria who love them. In: Barton LL, Mandl M, Loy A (eds) Geomicrobiology: molecular and environmental perspective. Springer, Berlin, pp 385–399
Botella MA, Quesada MA, Kononowicz AK, Bressan RA, Pliego F, Hasegawa PM, Valpuesta V (1994) Characterization and in-situ localization of a salt-induced tomato peroxidase messenger-RNA. Plant Mol Biol 25:105–114
Bourke J, Harris MM, Fushimi C, Dowaki K, Nunoura T, Antal MJ Jr (2007) Do all carbonized charcoals have the same chemical structure? 2. A model of the chemical structure of carbonized charcoal. Ind Eng Chem Res 46:5954–5967
Bray EA (1997) Plant responses to water deficit. Trends Plant Sci 2:48–54
Breazeale JF (1906) Effect of certain solids upon the growth of seedlings in water cultures. Bot Gaz 41:54–63
Brick S (2010) Biochar: assessing the promise and risks to guide US policy. Natural Resource Defense Council, New York
Brown R (2009) Biochar production technology. In: Lehmann J, Joseph S (eds) Biochar for environmental management science and technology. Earthscans, London, pp 127–146
Brussaard L (1997) Biodiversity and ecosystem functioning in soil. Ambio 26:563–570
Brussaard L, de Ruiter PC, Brown GG (2007) Soil biodiversity for agricultural sustainability. Agric Ecosyst Environ 121:233–244
Cantrell KB, Hunt PG, Uchimiya M, Novak JM, Ro KS (2012) Impact of pyrolysis temperature and manure source on physicochemical characteristics of biochar. Bioresour Technol 107:419–428
Chaganti VN, Crohn DM, Šimůnek J (2015) Leaching and reclamation of a biochar and compost amended saline–sodic soil with moderate SAR reclaimed water. Agric Water Manag 158:255–265
Chan KY, Xu Z (2009) Biochar: nutrient properties and their enhancement. In: Lehmann J, Joseph S (eds) Biochar for environmental management: science and technology. Earthscan, London, pp 67–84
Chang H, Siegel BZ, Siegel SM (1984) Salinity-induced changes in isoperoxidases in taro Colocasia esculenta. Phytochemistry 23(2):233–235
Cheeseman JM (1988) Mechanism of salinity tolerance in plants. Plant Physiol 87:547–550
Chen G, Asada K (1989) Ascorbate peroxidase in tea leaves: occurrence of two isozymes and the differences in their enzymatic and molecular properties. Plant Cell Physiol 30:987–998
Chen B, Chen Z (2009) Sorption of naphthalene and 1-naphthol by biochars of orange peels with different pyrolytic temperatures. Chemosphere 76:127–133
Chen S, Li J, Wang S, Huttermann A, Altman A (2001) Salt, nutrient uptake and transport, and ABA of Populus euphratica; a hybrid in response to increasing soil NaCl. Trees Struct Funct 15:186–194
Chen B, Zhou D, Zhu L (2008) Transitional adsorption and partition on nonpolar and polar aromatic contaminants by biochars of pine needles with different pyrolytic temperatures. Environ Sci Technol 42:5137–5143
Chen X, Chen G, Chen L, Chen Y, Lehmann J, McBride MB, Hay AG (2011) Adsorption of copper and zinc by biochars produced from pyrolysis of hardwood and corn straw in aqueous solution. Bioresour Technol 102(19):8877–8884
Cheng CH, Lehmann J (2009) Ageing of black carbon along a temperature gradient. Chemosphere 75:1021–1027
Cheng CH, Lehmann J, Thies JE, Burton SD (2008) Stability of black carbon in soils across a climatic gradient. J Geophys Res 113:20–27

Chowdhury MA, de Neergaard A, Jensen LS (2014) Potential of aeration flow rate and biochar addition to reduce greenhouse gas and ammonia emissions during manure composting. Chemosphere 97:16–25
Chun Y, Sheng G, Chiou CT, Xing B (2004) Compositions and sorptive properties of crop residue-derived chars. Environ Sci Technol 38:4649–4655
Comba ME, Benavides MP, Tomaro ML (1998) Effect of salt stress on antioxidant defence system in soybean root nodules. Aust J Plant Physiol 25:665–671
Cowie AL, Downie AE, George BH, Singh B, Zweiten LV, O'Connell D (2012) Is sustainability certification for biochar the answer to environmental risks? Pesqui Agrop Brasil 47:637–648
Cushman JC, Meyer G, Michalowski CB, Schmitt JM, Bohnert HJ (1989) Salt stress leads to differential expression of two isogenes of PEP case during CAM induction in the common ice plant. Plant Cell 1:715–725
Dachnowski A (1908) The toxic property of bog water and bog soil. Bot Gaz 46:130–143
De-Bruxelles GL, Peacock WJ, Dennies ES, Dolferus R (1996) Abscisic acid induces the alcohol dehydrogenase gene in Arabidopsis. Plant Physiol 111:381–391
Deenik JL, McClellan T, Uehara G, Antal NJ, Campbell S (2010) Charcoal volatile matter content influences plant growth and soil nitrogen transformations. Soil Sci Soc Am J 74:1259–1270
DeLuca TH, MacKenzie MD, Gundale MJ, Holben WE (2006) Wildfire-produced charcoal directly influences nitrogen cycling in ponderosa pine forests. Soil Sci Soc Am J 70:448–453
Demirbas A, Arin G (2002) An overview of biomass pyrolysis. Energy Sources 24(5):471–482
Diacono M, Montemurro F (2015) Effectiveness of organic wastes as fertilizers and amendments in salt-affected soils. Agriculture 5:221–230
Dietz KJ, Tavakoli N, Kluge C, Mimura T, Sharma SS, Harris GC, Chardonnens AN, Golldack D (2001) Significance of the V-type ATPase for the adaptation to stressful growth conditions and its regulation on the molecular and biochemical level. J Exp Bot 52:1969–1980
Downie A, Crosky A, Munroe P (2009) Physical properties of biochar. In: Lehmann J, Joseph S (eds) Biochar for environmental management science and technology. Earthscans, London, pp 13–32
Elshintinawy F, Elshourbagy MN (2001) Alleviation of changes in protein metabolism in NaCl-stressed wheat seedlings by thiamine. Biol Plant 44:541–545
Elstner EF (1987) Metabolism of activated oxygen species. In: Davies DD (ed) The biochemistry of plants. Vol. II, biochemistry of metabolism. Academic, San Diego, pp 252–315
Enders A, Hanley K, Whitman T, Joseph S, Lehmann J (2012) Characterization of biochars to evaluate recalcitrance and agronomic performance. Bioresour Technol 114:644–653
FAO (2009) www.fao.org/How to feed the world in 2050
Farhangi-Abriz S, Torabian S (2017) Antioxidant enzyme and osmotic adjustment changes in bean seedlings as affected by biochar under salt stress. Ecotoxicol Environ Saf 137:64–70
Fazal A, Bano A (2016) Role of plant growth-promoting rhizobacteria (pgpr), biochar, and chemical fertilizer under salinity stress. Commun Soil Sci Plant Anal 47:1985–1993
Ford CW (1984) Accumulation of low molecular solutes in water stress tropical legumes. Phytochemistry 23:1007–1015
Freitas JCC, Emmerich FG, Bonagamba TJ (2000) High-resolution solid-state NMR study of the occurrence and thermal transformations of silicon containing species in biomass materials. Chem Mater 12:711–718
Freitas JCC, Passamani EC, Orlando MTD, Emmerich FG, Garcia F, Sampaio LC, Bonagamba TJ (2002) Effects of ferromagnetic inclusions on 13C MAS NMR spectra of heat-treated peat samples. Energy Fuel 16:1068–1075
Fridovich I (1986) Biological effects of the superoxide radical. Arch Biochem Biophys 247:1–11
Gadallah MAA (1999) Effects of proline and glycinebetaine on Vicia faba response to salt stress. Biol Plant 42:249–257
Gao ZF, Sagi M, Lips SH (1998) Carbohydrate metabolism in leaves and assimilate partitioning in fruits of tomato (*Lycopersicon esculentum* L.) as affected by salinity. Plant Sci 135:149–159
Gaur A, Adholeya A (2000) Effects of the particle size of soil-less substrates upon AM fungus inoculum production. Mycorrhiza 10:43–48

Glenn EP, Brown JJ, Blumwald E (1999) Salt tolerance and crop potential of halophytes. Crit Rev Plant Sci 18:227–255
GomezCadenas A, Tadeo FR, PrimoMillo E, Talon M (1998) Involvement of abscisic acid and ethylene in the responses of citrus seedlings to salt shock. Plant Physiol 103:475–484
GomezCadenas A, Arbona V, Jacas J, PrimoMillo E, Talon M (2002) Abscisic acid reduces leaf abscission and increases salt tolerance in citrus plants. J Plant Growth Regul 21:234–240
Greenway H, Munns R (1980) Mechanisms of salt tolerance in nonhalophytes. Annu Rev Plant Physiol 31:149–190
Gryndler M, Larsen J, Hrselova H, Rezaccova V, Gryndlerova H, Kuba J (2006) Organic and mineral fertilization, respectively, increase and decrease the development of external mycelium of arbuscular mycorrhizal fungi in a longterm field experiment. Mycorrhiza 16:159–166
Gupta S, Chattopadhyay MK, Chatterjee P, Ghosh B, SenGupta DN (1998) Expression of abscisic acid-responsive elementbinding protein in salt tolerant indica rice (Oryza sativa L. cv. Pokkali). Plant Mol Biol 137:629–637
Haider G, Koyro HW, Azam F, Steffens D, Müller C, Kammann C (2015) Biochar but not humic acid product amendment affected maize yields via improving plant-soil moisture relations. Plant Soil 395:141–157
Halliwell B, Gutteridge JMC (1985) Free radicals in biology and medicine. Clarendon Press, Oxford
Halliwell B, Gutteridge JMC (1986) Free radicals in biology and medicine. Oxford University Press, London
Hammer EC, Forstreuter M, Rillig MC, Kohler J (2015) Biochar increases arbuscular mycorrhizal plant growth enhancement and ameliorates salinity stress. Appl Soil Ecol 96:114–121
Hasbullah H, Marschner P (2014) Residue properties influence the impact of salinity on soil respiration. Biol Fertil Soils 51:99–111
Hasegawa PM, Bressan RA, Zhu JK, Bohnert HJ (2000) Plant cellular and molecular responses to high salinity. Annu Rev Plant Physiol Plant Mol Biol 51:463–499
Hogarth PJ (1999) The biology of mangroves. Oxford University Press, New York
IBI (2012) Standardized product definition and product testing guidelines for biochar that is used in soil. International Biochar Initiative April 2012
Imlay JA, Linn S (1988) DNA damage and oxygen radical toxicity. Science 240:1302–1309
Inderjit, Callaway RM (2003) Experimental designs for the study of allelopathy. Plant Soil 256:1–11
Ingram J, Bartels D (1996) The molecular basis of dehydration tolerance in plants. Annu Rev Plant Physiol Plant Mol Biol 47:377–403
Inyang MI, Gao B, Yao Y, Xue Y, Zimmerman A, Mosa A, Pullammanappallil P, Ok YS, Cao X (2016) A review of biochar as a low-cost adsorbent for aqueous heavy metal removal. Crit Rev Environ Sci Technol 46:406–433
Iyengar ERR, Reddy MP (1996) Photosynthesis in highly salt tolerant plants. In: Pesserkali M (ed) Handbook of photosynthesis. Marshal Dekar, Baten Rose, USA, pp 897–909
Jackson RM (1958) The ecology of fungi in the soil with special reference to fungistasis. PhD thesis, University of London, Rothamsted Experimental Station, Harpenden, UK
Jeffery S, Verheijen GAM, Vander-Velde V, Bastos AC (2011) A quantitative review of the effects of biochar application to soils on crop productivity using meta-analysis. Agric Ecosyst Environ 144:175–187
Jin H (2010) Characterization of microbial life colonizing biochar and biochar amended soils. PhD. dissertation, Cornell University, Ithaca, NY
Joseph SD, Camps-Arbestain M, Lin Y, Munroe P, Chia CH, Hook J, van Zwieten L, Kimber S, Cowie A, Singh BP, Lehmann J, Foidl N, Smernik RJ, Amonette JE (2010) An investigation into the reactions of biochar in soil. Aust J Soil Res 48:501–515
Kanesaki Y, Suzuki I, Allakhverdiev SI, Mikami K, Murata N (2002) Salt stress and hyperosmotic stress regulate the expression of different sets of genes in Synechocystis sp. PCC 6803. Biochem Biophys Res Commun 290:339–348

Karaosmanoğlu F, Işığıgür-Ergüdenler A, Sever A (2000) Biochar from the straw-stalk of rapeseed plant. Energy Fuel 14(2):336–339

Kawasaki S, Borchert C, Deyholos M, Wang H, Brazille S, Kawai K, Galbraith D, Bohnert HJ (2001) Gene expression profiles during the initial phase of salt stress in rice. Plant Cell 13:889–905

Keiluweit M, Nico PS, Johnson MG, Kleber M (2010) Dynamic molecular structure of plant biomass-derived black carbon (biochar). Environ Sci Technol 44:1247–1253

Kerepesi I, Galiba G (2000) Osmotic and salt stress-induced alteration in soluble carbohydrate content in wheat seedlings. Crop Sci 40:482–487

Khan MA, Ungar IA, Showalter AM (2000) Effects of sodium chloride treatments on growth and ion accumulation of the halophyte *Haloxylon recurvum*. Commun Soil Sci Plant Anal 31:2763–2774

Khatkar D, Kuhad MS (2000) Short-term salinity induced changes in two wheat cultivars at different growth stages. Biol Plant 43:629–632

Khavarinejad RA, Mostofi Y (1998) Effects of NaCl on photosynthetic pigments, saccharides, and chloroplast ultrastructure in leaves of tomato cultivars. Photosynthetica 35:151–154

Killham K (1985) A physiological determination of the impact of environmental stress on the activity of microbial biomass. Environ Pollut Ser A 38:283–294

Killham K, Firestone MK (1984) Salt stress control of intracellular solutes in *Streptomyces* indigenous to saline soils. Appl Environ Microbiol 47:301–306

Kim HS, Kim KR, Yang JE, Ok YS, Owens G, Nehls T, Wessolek G, Kim KH (2016) Effect of biochar on reclaimed tidal land soil properties and maize (Zea mays L.) response. Chemosphere 142:153–159

Kloss S, Zehetner F, Dellantonio A, Hamid R, Ottner F, Liedtke V, Schwanninger M, Gerzabek MH, Soja G (2012) Characterization of slow pyrolysis biochars: effects of feedstocks and pyrolysis temperature on biochar properties. J Environ Qual 41:990–1000

Knight H, Trewavas AJ, Knight MR (1997) Calcium signaling in Arabidopsis thaliana responding to drought and salinity. Plant J 12:1067–1078

Koch B, Ostermann M, Höke H, Hempel DC (1991) Sand and activated carbon as biofilm carriers for microbial degradation of phenols and nitrogen-containing aromatic compounds. Water Res 25:1–8

Kolb SE, Fermanich KJ, Dornbush ME (2009) Effect of charcoal quantity on microbial biomass and activity in temperate soils. Soil Sci Soc Am J 73:1173–1181

Kumar S, Jain MC, Chhonkar PK (1987) A note on the stimulation of biogas production from cattle dung by addition of charcoal. Biol Wastes 20:1209–1215

Lashari MS, Liu Y, Li L, Pan W, Fu J, Pan G, Zheng J, Zheng J, Zhang X, Yu X (2013) Effects of amendment of biochar-manure compost in conjunction with pyroligneous solution on soil quality and wheat yield of a salt-stressed cropland from Central China Great Plain. Field Crop Res 144:113–118

Lashari MS, Ye Y, Ji H, Li L, Kibue GW, Lu H, Zheng J, Pan G (2015) Biochar–manure compost in conjunction with pyroligneous solution alleviated salt stress and improved leaf bioactivity of maize in a saline soil from central China: a 2-year field experiment. J Sci Food Agric 95:1321–1327

Lau JA, Puliafico KP, Kopshever JA, Steltzer H, Jarvis EP, Schwarzländer M, Strauss SY, Hufbauer RA (2008) Inference of allelopathy is complicated by effects of activated carbon on plant growth. New Phytol 178:412–423

Lauchli A, Schubert S (1989) The role of calcium in the regulation of membrane and cellular growth processes under salt stress. NATO ASI Ser G19:131–137

Lee DH, Kim YS, Lee CB (2001) The inductive responses of the antioxidant enzymes by salt stress in the rice (Oryza sativa L.) J Plant Physiol 158:737–745

Lehmann J (2007) A handful of carbon. Nature 447:143–144

Lehmann J (2009) Terra preta Nova – where to from here? In: Woods WI, Teixeira WG, Lehmann J, Steiner C, Prins AW (eds) Amazonian dark earths: Wim Sombroek's vision. Springer, Berlin, pp 473–486

Lehmann J, Joseph S (2009a) Biochar for environmental management: science and technology. Earthscan, London

Lehmann J, Joseph S (2009b) Biochar for environmental management: an introduction. In: Lehmann J, Joseph S (eds) Biochar for environmental management science and technology. Earthscans, London, pp 1–12

Lehmann J, da Silva Jr JP, Steiner C, Nehls T, Zech W, Glaser B (2003) Nutrient availability and leaching in an archaeological Anthrosol and a Ferralsol of the Central Amazon basin: fertilizer, manure and charcoal amendments. Plant Soil 249:343–357

Lehmann J, Liang B, Solomon D, Lerotic M, Luizão F, Kinyangi F, Schäfer T, Wirick S, Jacobsen C (2005) Near-edge X-ray absorption fine structure (NEXAFS) spectroscopy for mapping nano-scale distribution of organic carbon forms in soil: application to black carbon particles. Glob Biogeochem Cycles 19:10–13

Lehmann J, Czimczik C, Laird D, Sohi S (2009) Stability of biochar in soil. In: Lehmann J, Joseph S (eds) Biochar for environmental management: science and technology. Earthscan, London, pp 183–205

Leinweber P, Kruse J, Walley FL, Gillespie A, Eckardt K-U, Blyth R, Regier T (2007) Nitrogen K-edge XANES an overview of reference compounds used to identify 'unknown' organic nitrogen in environmental samples. J Synchrotron Radiat 14:500–511

Levitt J (1980) Responses of plant to environmental stress chilling, freezing, and high temperature stresses, 2nd edn. Academic, New York

Liang B, Lehmann J, Sohi SP, Thies JE, O'Neill B, Trujillo L, Gaunt J, Solomon D, Grossman J, Neves EG, Luizão FJ (2010) Black carbon affects the cycling of non-black carbon in soil. Org Geochem 41(2):206–213

Libra JA, Ro KS, Kammann C, Funke A, Berge ND, Neubauer Y, Titirici MM, Fühner C, Bens O, Kern J, Emmerich KH (2011) Hydrothermal carbonization of biomass residuals: a comparative review of the chemistry, processes and applications of wet and dry pyrolysis. Biofuels 2:89–124

Liu J, Zhu JK (1997) An Arabidopsis mutant that requires increased calcium for potassium nutrition and salt tolerance. Proc Natl Acad Sci U S A 94:14960–14964

Liu Z, Zhang FS, Wu J (2010) Characterization and application of chars produced from pinewood pyrolysis and hydrothermal treatment. Fuel 89:510–514

Liu X, Zhang A, Ji C, Joseph S, Bian R, Li L, Pan G, Paz-Ferreiro J (2013) Biochar's effect on crop productivity and the dependence on experimental conditions—a meta-analysis of literature data. Plant Soil 373:583–594

Lu W, Kang C, Wang Y, Xie Z (2015) Influence of biochar on the moisture of dark brown soil and yield of maize in Northern China. Int J Agric Biol 17:1007–1012

Luo X, Liu G, Xia Y, Chen L, Jiang Z, Zheng H, Wang Z (2017) Use of biochar-compost to improve properties and productivity of the degraded coastal soil in the Yellow River Delta, China. J Soils Sediments 17:780–789

MacKenzie MD, DeLuca TH (2006) Charcoal and shrubs modify soil processes in ponderosa pine forests of western Montana. Plant Soil 287:257–267

Makoto K, Tamai Y, Kim YS, Koike T (2010) Buried charcoal layer and ectomycorrhizae cooperatively promote the growth of Larix gmelinii seedlings. Plant Soil 327:143–152

Mendoza I, Rubio F, Rodriguez-Navarro A, Prado JM (1994) The protein phosphatase calcineurin is essential for NaCl tolerance of Saccharomyces cerevisiae. J Biol Chem 269:8792–8796

Mittova V, Tal M, Volokita M, Guy M (2003) Up-regulation of the leaf mitochondrial and peroxisomal antioxidative systems in response to salt-induced oxidative stress in the wild salt-tolerant tomato species Lycopersicon pennellii. Plant Cell Environ 26:845–856

Mohan D, Pittman CU, Steele PH (2006) Pyrolysis of wood/biomass for bio-oil: a critical review. Energy Fuel 20:848–889

Moore JC, Berlow EL, Coleman DC, de Ruiter PC, Dong Q, Hastings A, Johnson NC, McCann KS, Melville K, Morin PJ, Nadelhoffer K, Rosemond AD, Post DM, Sabo JL, Scow KM, Vanni MJ, Wall DH (2004) Detritus, trophic dynamics and biodiversity. Ecol Lett 7:584–600

Morrissey EM, Gillespie JL, Morina JC, Franklin RB (2014) Salinity affects microbial activity and soil organic matter content in tidal wetlands. Glob Chang Biol 20:1351–1362

Munns R, Tester M (2008) Mechanisms of salinity tolerance. Annu Rev Plant Biol 59:651–681

Nadeem S, Zahir Z, Naveed M, Nawaz S (2013) Mitigation of salinityinduced negative impact on the growth and yield of wheat by plant growth-promoting rhizobacteria in naturally saline conditions. Ann Microbiol 63:225–232

Nguyen B, Lehmann J, Hockaday WC, Joseph S, Masiello CA (2010) Temperature sensitivity of black carbon decomposition and oxidation. Environ Sci Technol 44:3324–3331

Noaman MM, Dvorak J, Dong JM (2002) Genes inducing salt tolerance in wheat, *Lophopyrum elongatum* and amphiploid and their responses to ABA under salt stress. Prospects Saline Agric (Ser: Tasks Veg Sci) 37:139–144

Noctor G, Foyer CH (1998) Ascorbate and glutathione: keeping active oxygen under control. Annu Rev Plant Physiol Plant Mol Biol 49:249–279

Noguera D, Rondon M, Laossi K-R, Hoyos V, Lavelle P, de Carvalho MHC, Barot S (2010) Contrasted effect of biochar and earthworms on rice growth and resource allocation in different soils. Soil Biol Biochem 42:1017–1027

Nutman PS (1952) Host factors influencing infection and nodule development in leguminous plants. Proc R Soc Lond Ser B Biol Sci 139:176–185

Ohashi M, Kume T, Yoshifuji N, Kho LK, Nakagawa M, Nakashizuka T (2014) The effects of an induced short-term drought period on the spatial variations in soil respiration measured around emergent trees in a typical Bornean tropical forest, Malaysia. Plant Soil 387:337–349

Orr WC, Sohal RS (1992) The effects of catalase gene overexpression on life span and resistance to oxidative stress in transgenic *Drosophila melanogaster*. Arch Biochem Biophys 297:35–41

Orthen B, Popp M, Smirnoff N (1994) Hydroxyl radical scavenging properties of cyclitols. Proc R Soc Edinburg Sect B 102:269–272

Osakabe Y, Osakabe K, Shinozaki K, Tran LP (2014) Response of plants to water stress. Front Plant Sci 5:1–19

Otoch MDL, Sobreira ACM, deAragao MEF, Orellano EG, Lima MDS, deMelo DF (2001) Salt modulation of vacuolar H+-ATPase and H+-pyrophosphatase activities in *Vigna unguiculata*. J Plant Physiol 158:545–551

Panuccio MR, Jacobsen SE, Akhtar SS, Muscolo A (2014) Effect of saline water irrigation on seed germination and early seedling growth of the halophyte quinoa. AoB Plants 6:plu047

Parida A, Das AB, Das P (2002) NaCl stress causes changes in photosynthetic pigments, proteins and other metabolic components in the leaves of a true mangrove, *Bruguiera parviflora*, in hydroponic cultures. J Plant Biol 45:28–36

Parihar P, Singh S, Singh R, Singh VP, Prasad SM (2015) Effect of salinity stress on plants and its tolerance strategies: a review. Environ Sci Pollut Res 22:4056–4075

Petit C, Kante K, Bandosz TJ (2010) The role of sulfur-containing groups in ammonia retention on activated carbons. Carbon 48:654–667

Pietikäinen J, Kiikkilä O, Fritze H (2000) Charcoal as a habitat for microbes and its effects on the microbial community of the underlying humus. Oikos 89:231–242

Pilon-Smits EAH, Ebskamp MJM, Paul MJ, Jeuken MJW, Weisbeek PJ, Smeekens SCM (1995) Improved performance of transgenic fructan-accumulating tobacco under drought stress. Plant Physiol 107:125–130

Popova OV, Ismailov SF, Popova TN, Dietz KJ, Golldack D (2002) Salt-induced expression of NADP-dependent isocitrate dehydrogenase and ferredoxin-dependent glutamate synthase in Mesembryanthemum crystallinum. Planta 215:906–913

Popp M, Larther F, Weigel P (1985) Osmotic adaptation in Australian mangroves. Vegetation 61:247–254

Qadir M, Quillérou E, Nangia V, Murtaza G, Singh M, Thomas RJ, Drechsel P, Noble AD (2014) Economics of salt-induced land degradation and restoration. Nat Res Forum 38:282–295

Qayyum MF, Steffens D, Reisenauer HP, Schubert S (2012) Kinetics of carbon mineralization of biochars compared with wheat straw in three soils. J Environ Qual 41:1210–1220

Rajapaksha AU, Chen SS, Tsang DC, Zhang M, Vithanage M, Mandal S, Gao B, Bolan NS, Ok YS (2016) Engineered/designer biochar for contaminant removal/immobilization from soil and water: potential and implication of biochar modification. Chemosphere 148:276–291

Rath KM, Rousk J (2015) Salt effects on the soil microbial decomposer community and their role in organic carbon cycling: a review. Soil Biol Biochem 8:108–123

Raveendran K, Ganesh A, Khilar KC (1995) Influence of mineral matter on biomass pyrolysis characteristics. Fuel 74:1812–1822

Reddy MP, Sanish S, Iyengar ERR (1992) Photosynthetic studies and compartmentation of ions in different tissues of Salicornia brachiata Roxb. under saline conditions. Photosynthetica 26:173–179

Rengasamy P, Olsson K (1991) Sodicity and soil structure. Aust J Soil Res 29:935–952

Rhodes D, Hanson AD (1993) Quaternary ammonium and tertiary sulphonium compounds in higher plants. Annu Rev Plant Physiol Plant Mol Biol 44:357–384

Richards LA (1954) Diagnosis and improvement of saline and alkali soils. In: Agricultural handbook No. 60. U. S. Salinity Laboratory, Riverside, California

Rillig MC, Wagner M, Salem M, Antunes PM, George C, Ramke HG, Titirici MM, Antonietti M (2010) Material derived from hydrothermal carbonization: effects on plant growth and arbuscular mycorrhiza. Appl Soil Ecol 45:238–242

Rizwan M, Ali S, Qayyum MF, Ibrahim M, Rehman MZ, Abbas T, Ok YS (2016a) Mechanisms of biochar-mediated alleviation of toxicity of trace elements in plants: a critical review. Environ Sci Pollut Res 23:2230–2248

Rizwan M, Meunier JD, Davidian JC, Pokrovsky OS, Bovet N, Keller C (2016b) Silicon alleviates Cd stress of wheat seedlings (*Triticum turgidum* L. cv. Claudio) grown in hydroponics. Environ Sci Pollut Res 23:1414–1427

Ro KS, Cantrell KB, Hunt PG (2010) High-temperature pyrolysis of blended animal manures for producing renewable energy and value-added biochar. Ind Eng Chem Res 49:10125–10131

Ron EZ, Rosenberg E (2001) Natural roles of biosurfactants. Environ Microbiol 3:229–236

Roxas VP, Lodhi SA, Garrett DK, Mahan JR, Allen RD (2000) Stress tolerance in transgenic tobacco seedlings that overexpress glutathione S-transferase/glutathione peroxidase. Plant Cell Physiol 41:1229–1234

Samsuri AW, Sadegh-Zadeh F, Seh-Bardan BJ (2014) Characterization of biochars produced from oil palm and rice husks and their adsorption capacities for heavy metals. Int J Environ Sci Technol 11:967–976

Schmidt MWI, Noack AG (2000) Black carbon in soils and sediments: analysis, distribution, implications, and current challenges. Glob Biogeochem Cycles 14:777–793

Shackley S, Carter S, Knowles T, Middelink E, Haefele S, Sohi S, Cross A, Haszeldine S (2012) Sustainable gasification-biochar systems? A case-study of rice-husk gasification in Cambodia, Part 1: Context, chemical properties, environmental and health and safety issues. Energ Policy 42:49–58

Shi HZ, Zhu JK (2002) Regulation of expression of the vacuolar Na+/H+ antiporter gene AtNHX1 by salt stress and abscisic acid. Plant Mol Biol 50:543–550

Shinozaki K, Yamaguchi-Shinozaki K, Mizoguchi T, Uraro T, Katagiri T, Nakashima K, Abe H, Ichimura K, Liu QA, Nanjyo T, Uno Y, Luchi S, Srki M, Lto T, Hirayama T, Mikami K (1998) Molecular responses to water stress in *Arabidopsis thaliana*. J Plant Res 111:345–351

Singh SK, Sharma HC, Goswami AM, Datta SP, Singh SP (2000) In vitro growth and leaf composition of grapevine cultivars as affected by sodium chloride. Biol Plant 43:283–286

Smirnoff N, Cumbes QJ (1989) Hydroxyl radical scavenging activity of compatible solutes. Phyotochemistry 28:1057–1060

Sohi S, Loez-Capel S, Krull E, Bol R (2009) Biochar's roles in soil and climate change: a review of research needs. CSIRO Land Water Sci Rep 5(9):64

Sohi SP, Krull E, Lopez-Capel E, Bol R (2010a) A review of biochar and its use and function in soil. Adv Agron 105:47–82

Sohi SP, Krull E, Lopez-Capel E, Bol R (2010b) A review of biochar and its use and function in soil. In: Sparks DL (ed) Advances in agronomy. Academic, Burlington, pp 47–82

Solaiman ZM, Blackwell P, Abbott LK, Storer P (2010) Direct and residual effect of biochar application on mycorrhizal colonization, growth and nutrition of wheat. Aust J Soil Res 48:546–554
Steiner C, Teixeira WG, Lehmann J, Zech W (2004) Microbial response to charcoal amendments of highly weathered soils and Amazonian dark earths in Central Amazonia – preliminary results. In: Glaser B, Woods WI (eds) Amazonian dark earths: explorations in time and space. Springer, Berlin, pp 195–212
Steiner C, Das KC, Garcia M, Förster B, Zech W (2008) Charcoal and smoke extract stimulate the soil microbial community in a highly weathered xanthic ferralsol. Pedobiologia Int J Soil Biol 51:359–366
Sun J, He F, Shao H, Zhang Z, Xu G (2016) Effects of biochar application on Suaedasalsa growth and saline soil properties. Environ Earth Sci 75:1–6
Takemura T, Hanagata N, Dubinsky Z, Karube I (2002) Molecular characterization and response to salt stress of mRNAs encoding cytosolic Cu/Zn superoxide dismutase and catalase from Bruguiera gymnorrhiza. Trees Struct Funct 16:94–99
Tartoura KA, Youssef SA, Tartoura ESA (2014) Compost alleviates the negative effects of salinity via up-regulation of antioxidants in *Solanum lycopersicum* L. plants. Plant Growth Regul 74:299–310
Thomas SC, Frye S, Gale N, Garmon M, Launchbury R, Machado N, Melamed S, Murray J, Petroff A, Winsborough (2013) Biochar mitigates negative effects of salt additions on two herbaceous plant species. J Environ Manag 129:62–68
Tong SJ, Li JY, Yuan JH, Xu RK (2011) Adsorption of Cu(II) by biochars generated from three crop straws. Chem Eng J 172:828–834
Uchimiya M, Wartelle LH, Lima IM, Klasson KT (2010) Sorption of deisopropylatrazine on broiler litter biochars. J Agric Food Chem 58:12350–12356
Uchimiya M, Chang S, Klasson KT (2011a) Screening biochars for heavy metal retention in soil: role of oxygen functional groups. J Hazard Mater 190:432–441
Uchimiya M, Klasson KT, Wartelle LH, Lima IM (2011b) Influence of soil properties on heavy metal questration by biochar amendment: 1. Copper sorption isotherms and the release of cations. Chemosphere 82:1431–1437
Uchimiya M, Bannon DI, Wartelle LH (2012) Retention of heavy metals by carboxyl functional groups of biochars in small arms range soil. J Agric Food Chem 60:1798–1809
Usman ARA, Al-Wabel MI, Abdulaziz AH, Mahmoud WA, El-Naggar AH, Ahmad M, Al-Faraj AAO (2016) Conocarpus biochar induces changes in soil nutrient availability and tomato growth under saline irrigation. Pedosphere 26:27–38
Vaidyanathan R, Kuruvilla S, Thomas G (1999) Characterization and expression pattern of an abscisic acid and osmotic stress responsive gene from rice. Plant Sci 140:21–30
Vookova B, Kormutak A (2001) Effect of sucrose concentration, charcoal, and indole-3-butyric acid on germination of Abies numidica somatic embryos. Biol Plant 44:181–184
Walbot V, Cullis CA (1985) Rapid genomic change in higher plants. Annu Rev Plant Physiol Plant Mol Biol 36:367–396
Wang Y, Nil N (2000) Changes in chlorophyll, ribulose biphosphate carboxylase–oxygenase, glycine betaine content, photosynthesis and transpiration in Amaranthus tricolor leaves during salt stress. J Hortic Sci Biotechnol 75:623–627
Wang BS, Luttge U, Ratajczak R (2001) Effects of salt treatment and osmotic stress on V-ATPase and V-PPase in leaves of the halophyte Suaeda salsa. J Exp Bot 52:2355–2365
Wang L, Chen W, Zhou W (2014a) Assessment of future drought in Southwest China based on CMIP5 multimodel projections. Adv Atmos Sci 31:1035–1050
Wang L, Sun X, Li S, Zhang T, Zhang W, Zhai P (2014b) Application of organic amendments to a coastal saline soil in north China: effects on soil physical and chemical properties and tree growth. PLoS One 9:1–9
Warnock DD, Lehmann J, Kuyper TW, Rillig MC (2007) Mycorrhizal responses to biochar in soil e concepts and mechanisms. Plant Soil 300:9–20

Warnock DD, Mummey DL, McBride B, Major J, Lehmann J, Rillig MC (2010) Influences of non-herbaceous biochar on arbuscular mycorrhizal fungal abundances in roots and soils: results from growth-chamber and field experiments. Appl Soil Ecol 46:450–456
Wicke B, Smeets E, Dornburg V, Vashev B, Gaiser T, Turkenburg W, Faaij A (2011) The global technical and economic potential of bioenergy from salt-affected soils. Energy Environ Sci 4:2669–2681
Willkens H, Chamnogopol S, Davey M, Schraudner M, Langebartels C (1997) Catalase is a sink for H2O2 and is indispensable for stress defense in C3 plants. EMBO J 16:4806–4816
Wilson JB (1988) A review of the evidence on the control of shoot: root ratio in relation to models. Ann Bot 61:433–449
Wise RR, Naylor AW (1987) Chilling-enhanced photooxidation: evidence for the role of singlet oxygen and endogenous antioxidants. Plant Physiol 83:278–282
Wu S, Ding L, Zhu J (1996) SOS1, a genetic locus essential for salt tolerance and potassium acquisition. Plant Cell 8:617–627
Wu Y, Xu G, Shao HB (2014) Furfural and its biochar improve the general properties of a saline soil. Solid Earth 5:665–671
Xu G, Zhang Y, Sun J, Shao H (2016) Negative interactive effects between biochar and phosphorus fertilization on phosphorus availability and plant yield in saline sodic soil. Sci Total Environ 568:910–915
Yan N, Marschner P, Cao W, Zuo C, Qin W (2015) Influence of salinity and water content on soil microorganisms. Int Soil Water Conserv Res 3:316–323
Yancey P, Clark ME, Had SC, Bowlus RD, Somero GN (1982) Living with water stress: evolution of osmolyte system. Science 217:1214–1222
Zackrisson O, Nilsson M-C, Wardle DA (1996) Key ecological function of charcoal from wildfire in the Boreal forest. Oikos 77(1):10
Zhang Y, Luo W (2014) Adsorptive removal of heavy metal from acidic wastewater with biochar produced from anaerobically digested residues: kinetics and surface complexation modeling. Biol Res 9:2484–2499
Zhang J, Lü F, Luo C, Shao L, He P (2014) Humification characterization of biochar and its potential as a composting amendment. J Environ Sci 26:390–397
Zhang X, Wang H, He L, Kouping L, Sarmah A, Li J, Bolan NS, Pei J, Huang H (2013) Using biochar for remediation of soils contaminated with heavy metals and organic pollutants. Environ Sci Pollut Res 20(12):8472–8483
Zhifang G, Loescher WH (2003) Expression of a celery mannose 6- phosphate reductase in *Arabidopsis thaliana* enhances salt tolerance and induces biosynthesis of both mannitol and a glucosyl-mannitol dimmer. Plant Cell Environ 26:275–283
Zhu JK (2003) Regulation of ion homeostasis under salt stress. Curr Opin Plant Biol 6:441–445
Zhu J, Meinzer FC (1999) Efficiency of C-4 photosynthesis in Atriplex lentiformis under salinity stress. Aust J Plant Physiol 26:79–86
Zhu J-K, Shi J, Singh U, Wyatt SE, Bressan RA, Hasegawa PM, Carpita NC (1993) Enrichment of vitronectinand fibronectin-like proteins in NaCl-adapted plant cells and evidence for their involvement in plasma membrane-cell wall adhesion. Plant J 3(5):637–646
Zimmerman A (2010) Abiotic and microbial oxidation of laboratory-produced black carbon (biochar). Environ Sci Technol 44:1295–1301

Fly Ash-Induced Metabolic Adaptations in Three Ferns

7

Alka Kumari

Abstract

Three fern species, namely, *Pteris vittata* L, *Ampelopteris prolifera* (Retz.) Copel., and *Diplazium esculentum* (Retz.) Sw., were grown on three different amendments of fly ash (FA) with garden soil (GS), viz., 100% GS as control, 50% FA+50% GS, and 100% FA. Their growth, metal accumulation, and response to antioxidants were evaluated. It was observed that all of these species accumulated significant amount of metals in their fronds and rhizomes (including rhizoids), while the amount of metal being accumulated by each fern varied. Results revealed that there was a significant increase in their biomass and photosynthetic pigments, for all the test species grown on 50% FA-amended GS in comparison to control; however, it further decreased in ferns grown on 100% FA, indicating that 50% FA amendment did not generate oxidative stress in ferns as well as it seems favorable substratum for fern growth.

Furthermore, while the activity of antioxidant enzymes such as melanoaldehydes (MDA), superoxide dismutase (SOD), ascorbate peroxidase (APX), and guaiacol peroxidase (GPX) increased to a considerable extent in 50% FA amendment, it was found to be maximum in the case of 100% FA amendment. In all the species, the fronds accumulated more metals than rhizomes; they also experienced more oxidative stress as the activities of antioxidant enzymes were observed to be higher in frond's biomass. Overall, the results of the experiment showed fly ash-induced metabolic adaptation in these ferns and further utility of these species in phytoremediation of toxic metals from fly ash as well as ecorestoration of fly ash landfills with the same species.

A. Kumari (✉)
Department of Botany, University of Lucknow, Lucknow, UP, India
e-mail: kumarialkasanjay@gmail.com; kumarialkasanjay@rediffmail.com

V. Shukla et al. (eds.), *Plant Adaptation Strategies in Changing Environment*,
https://doi.org/10.1007/978-981-10-6744-0_7

Keywords
Fly ash · Fern · Metals · Photosynthetic pigments · Oxidative stress · Antioxidative enzymes

7.1 Introduction

Coal is one of the chief sources of electricity generation in India. A large number of power-generating stations in India are coal based (Khan and Khan 1996), wherein a huge quantity of fly ash (FA) is generated in the form of coal combustion residue. The global generation of FA is estimated to be above 600 million tons per year (Singh and Siddiqui 2003), and India contributes to the number by generating 120 million tons FA every year from its 82 power plants; this amount is likely to exceed to 150 million tons by 2020. This huge quantity of FA requires enough space and money for storage/dumping. Physicochemical studies on FA have been revealed that it contains significant amount of noxious metals and metalloids, which percolated from FA and resulted a hazard to receiving ecosystem. Management of such contaminated sites becomes imperative in view of ecological health and hygiene (Kumari et al. 2013; Pandey and Singh 2010). Although a big proportion of the FA has been utilized in building materials and pit filling in mines as well as other landfilling purposes, even then a huge quantity of the FA need environmental friendly uses. Its utilization for agriculture and cultivation of nonagricultural crops like ferns would be an alternative for its disposal/utilization.

Consequently, the revegetation of the FA landfill area by the FA-tolerant plants is the only lucrative and eco-friendly solution/alternative for the management of FA as well as to maintain an aesthetic and pollution-free pleasant landscape (Pandey 2012; Rai et al. 2004; Vajpayee et al. 2000; Wong and Wong 1990). A number of leguminous plant species have been proven suitable in ecorestoration of FA dikes (Gupta et al. 2004; Jambhulkar and Juwarkar 2009; Rai et al. 2004; Ram et al. 2008), and amalgamation of multiple chemical constituents of FA as well as combination of N_2-fixing microbes has been found useful in enhancing growth of the plants. Some rhizospheric bacteria also enhance metal mobilization in FA and consequently uptake in plants (Tiwari et al. 2008, 2010). Besides, some fern species have also been reported as hyperaccumulator of metals including arsenic (Ma et al. 2001a, b; Mehrag 2002; Singh et al. 2006), which could be also used for phytoremediation of FA (Kumari et al. 2011, 2013).

Plants growing on FA dikes experience a variety of physiological and biochemical stresses due to the presence of toxic metals. Due to the presence of high pH; low levels of N, P, and K; and high concentration of many toxic elements like Cu, Pb, Hg, Cd, Ni, Al, and As in FA (Haynes 2009; Mehra et al. 1998; Pandey and Singh 2010), some plants showed visual toxicity symptoms, but the common fern species, studied during present work, namely, *Pteris vittata*, *Ampelopteris prolifera*, and *Diplazium esculentum*, did not show any visible toxicity symptoms. It could be due to some internal defense mechanism in these plants against toxicity of heavy metals.

It is known that plants developed internal defense mechanism in the presence of antioxidative enzymes to protect the damage caused by free radicals. Metal compounds, upon entry into the plant cells, produce reactive oxygen species either by direct electron transfer involving metal cations or as a consequence of metal-mediated metabolic reactions. However, to cope up with reactive oxygen species, plant has scavenging system mediated by nonenzymatic antioxidant and enzymatic antioxidant systems like melanoaldehyde (MDA), superoxide dismutase (SOD), ascorbate peroxidase (APX), guaiacol peroxidase (GPX), and glutathione (GSH) (Sinha and Gupta 2005; Sinha et al. 2005). Such plants under stress condition also synthesize metal-binding peptides (phytochelatins) showing high level of nonprotein thiols (NPSH) and cysteine (Grill et al. 1987; Zenk 1996). Further, arsenic-induced biochemical changes in *P. vittata* have been reported (Cao et al. 2004; Srivastava et al. 2005), whereas FA-induced biochemical studies are reported in both non-nodulated species *Cassia siamea* Lamk. (Kumar et al. 2002) and nodulated species chickpea of leguminous family (Pandey et al. 2010). FA-induced biochemical changes in *Prosopis juliflora* L. have been done by Sinha et al. (2005).

However, such studies have been conducted on many crop plants and vegetation with varying degree of success, but such studies on the ferns experiencing FA stress are limited, and a single report has been made by Kumari et al. (2013) to explain antioxidant systems in the fern *Thelypteris dentata* growing on varied level of FA amendments. Therefore, the present study was planned to pursue the objectives on how these species could establish themselves in the vicinity of FA, even in the presence of toxic metals and lack of N, P, and K. In this context, the present study aimed to elucidate the internal defense mechanism by analyzing antioxidative enzymes in target ferns generated to combat with metal stress of FA. Thus, the present study is also an attempt to eco-friendly management of FA landfills with nonedible, ornamental, and potential metal accumulator ferns which may prove a useful alternative for FA utilization.

7.2 Material and Methods

7.2.1 Sample Collection and Preparation

FA used in this experiment was collected randomly from dumping sites of National Thermal Power Corporation, Kanti, Muzaffarpur (Bihar), India in large plastic bags and brought to the garden field near experimental fern house. The garden soil (GS) was collected from CSIR-Institute of Himalayan Bioresource Technology, Palampur, India. Before preparation of various amendments, both FA and GS were air dried for 7 days and sieved through 2 mm mesh. These amendments were denoted as 100% FA, 50% FA (50% FA+50% GS), and 100% GS alone served as control. Sporophylls of *P. vittata*, *A. prolifera*, and *D. esculentum* were collected from the plants growing naturally in FA dikes of NTPC Kanti, Muzaffarpur (Bihar), India, in spore packets made by brown papers.

7.2.2 Experimental Setup

Collected sporophylls of all the three species with mature spores were kept in desiccators for 4 weeks. Spores from sporophyll of each fern were released on the tissue paper automatically and by destructing the sporangia with the help of needle. Spores were surface sterilized by 2% sodium hypochlorite for 2 min and rinsed four times with sterilized deionized water. Spores were sprinkled on petri plates containing 25 ml P&T medium solidified with addition of 1% agar as stock culture. Cultures were placed in culture room (temperature, 22 ± 2°; photoperiod, 16:8 h; a light intensity of 350 μ mol m^{-2} s^{-1}). After 1 week of sowing, germination of spores started which was observed at weekly interval under microscopes till complete gametophyte formation. Nine earthen pots were taken of 2 kg air-dried FA capacity. Three sets of pot were prepared by filling FA with GS: first set containing only 100% FA, second set containing 50:50 ratio of FA and GS, and third set containing only 100% GS which was treated as control. Each set of experiment was performed in triplicates.

Similarly, nine earthen pots (1 kg capacity) were prepared for each species, and experiments were set up in randomized block design. Each pot was transplanted with three healthy gametophytes (5 weeks old) of *P. vittata*, *A. prolifera*, and *D. esculentum*, from stock culture, respectively. Pots were arranged in three plastic trays, one tray for each species, and covered by acclimatization hood and put randomly inside experimental fern house of IHBT, Palampur. A dish was placed under each pot to avoid leaching during the experiments. For healthy growth of gametophyte, moisture was maintained by spraying tap water daily or as and when necessary, and it is allowed to grow till sporophyte forms five to six fronds. The plants were grown for approximately 90 days after inoculation in pots and then harvested and washed in tap water. Harvested plants were washed repeatedly, blotted dry, and separated into the aboveground (fronds) and belowground (rhizomes including roots) portions weighed (fresh weight (FW) basis). The uptakes of metals as well as various growth, biochemical parameters, and antioxidative enzymes were studied in harvested samples.

7.2.3 Physicochemical Characterization of Different Amendments of FA and GS

FA and GS prepared for pot experiment were analyzed for their physicochemical characterization. The pH and electrical conductivity were determined through pH meter and conductivity meter by adding distilled water in FA in 1:5 ratio (Piper 1966), total organic carbon (Walkely and Black 1934), total nitrogen (micro-Kjeldahl digestion: Nelson and Sommer 1982), available phosphorus (Oleson and Sommers 1982), and cation exchange capacity (CEC) extracted with ammonium acetate (Allen et al. 1974).

7.2.4 Plant Growth Parameters (Biomass and Photosynthetic Pigments)

Plants were harvested after 90 days of growth in different amendments and repeatedly washed with double-distilled water. Plants were blotted dry, and the biomass (fresh weight) of belowground part and aboveground parts was measured on fresh weight basis. Biomass of plants growing on various amendments of FA was observed by recording the fresh weight of both parts of the plant (data not shown separately) and was expressed as total biomass of each fern on various amendments on fresh weight (FW) basis. For estimation of photosynthetic pigments, leaf frond (300 mg) was ground in chilled 80% acetone in dark. After centrifugation at 10,000×g for 10 min. at 4 °C, absorbance of the supernatants was taken at 480, 510, 645, and 663 nm. Chlorophyll and carotenoids contents were calculated using the formula given by Arnon (1949) and Dexbury and Yentch (1956), respectively.

7.2.5 Metal Estimation in Fronds and Root Parts

The harvested fern sample was washed and separated in aboveground (fronds) and belowground parts (Rhizome with roots) and then oven-dried at 80 °C for 24 h. The oven-dried plant tissue (frond and root) of treated and control plants was grinded at room temperature and digested with 0.1 g of sample in glass digestion tube of 250 mL along with mixture of concentrated nitric acid (HNO_3) and perchloric acid ($HClO_4$) (V/V 3:1) at 140 °C. After digestion the solution was cooled, filtered, and made up to 50 mL with distilled water for heavy metal analysis. The heavy metal measurements in the foliar and root samples were performed with Flame Atomic Absorption Spectrophotometer (Perkin Elmer, Model A Analyst 300) following Tripathi et al. (2008).

7.2.6 Lipid Peroxidation

The level of lipid peroxidation in roots and leaves was measured in terms of melanoaldehyde (MDA) content as estimated by thiobarbituric acid reaction following the method of Heath and Packer (1968) with slight modification. Plant tissues (500 mg) were homogenized in 5 ml of 0.1% TCA. The homogenate was centrifuged at 10,000× g for 15 min. For every 1 ml of aliquot, 4 ml of 20% TCA containing 0.5% thiobarbituric acid was added. Mixture was heated at 95 °C for 30 min and then cooled quickly on ice bath. The resulting mixture was centrifuged at 10,000-×g for 15 min, and the absorbance of the supernatants was taken at 532 and 600 nm. The non-specific absorbance at 600 nm was subtracted from the absorbance at 532 nm. The concentration of MDA was calculated and expressed by total thiobarbituric acid reaction substrates (TBARS) in term of μmole g^{-1} by using an extinction coefficient of 155 mM^{-1} cm^{-1}.

7.2.7 Analysis of Antioxidative Enzymes

Plant material (500 mg, leaves or roots) was homogenized in 100 mM potassium phosphate buffer (pH 7.0) containing 0.1 mm EDTA and 1% polyvinylchloride (w/v) at 4 °C. Homogenate was filtered through four layers of cheesecloth and centrifuged at 15,000×g for 15 min at 4 °C. Supernatant was used to measure the activities of enzymes.

7.2.7.1 Superoxide Dismutase (SOD)

The activity of SOD was assayed by measuring its ability to inhibit the photochemical reduction of nitrobluetetrazolium according to the method of Beauchamp and Fridovich (1971). The 3 ml reaction mixture contained 40 mM phosphate buffer (pH 7.8), 13 mM methionine, 75 mM nitrobluetetrazolium, 2 μm riboflavin, and 0.1 μm EDTA, and a suitable aliquot of enzyme extract riboflavin was added; at the end the test tubes were shaken and placed 30 cm below the light source consisting of 15 watt fluorescent lamp. Switching on the light started the reaction, and after 30 min, switching off the light stopped the reaction. A tube containing a protein kept in the dark served as blank, while the control tube was without the enzyme and kept in the light. The absorbance of the solution was taken at 560 nm. Activity of SOD is the measure of NBT reduction in light without protein minus NBT reduction with protein. One unit of activity is the amount of protein required to inhibit 50% initial reduction of NBT under light.

7.2.7.2 Ascorbate Peroxidase (APX)

The activity of APX was measured according to the method of Nakano and Asada (1981) by estimating the rate of ascorbate oxidation (extinction coefficient 2.8 mM^{-1} cm^{-1}). The 3 ml reaction mixture contains 50 mM phosphate buffer (pH, 7.0), 0.1 mM H_2O_2, 0.5 mM sodium ascorbate, 0.1 mM EDTA, and a suitable aliquot of enzyme extract. The change in absorbance was monitored at 290 nm, and enzyme activity was expressed as μmole of ascorbate oxidized min^{-1} g^{-1} fw.

7.2.7.3 Guaiacol Peroxidase (GPX)

GPX activity was assayed according to the method of Hemeda and Klein (1990). A 100 ml of reaction mixture was prepared by adding 10 ml of 1% guaiacol (v/v), 10 ml of 0.3% H_2O_2, and 80 ml of 50 mM phosphate buffer (pH, 6.6). 75 μl of enzyme extract was added to reaction mixture with a final volume of 3 ml. The increase in absorbance due to oxidation of guaiacol (extinction coefficient 26.6 mM^{-1} cm^{-1}) was monitored at 470 nm. Enzyme activity was expressed as μmole of guaiacol oxidized min^{-1} g^{-1} fw.

7.2.8 Statistical Analysis

All the values are presented as mean ± SD ($n = 3$) in figures and tables. The data was subjected to analysis of variance (ANOVA) to test the significant differences by using software package SYSTAT-9.0.

7.3 Results and Discussion

All the three fern species were grown on three different amendments of FA and GS, viz., 100% GS, 50% FA: 50% GS, 100% FA) for 90 days. Physicochemical studies were carried out to assess the physical status and metal toxicity present in the above amendments used during pot experiment. Oxidative stress created by the metals accumulated by plants was measured by changes in the biomass, photosynthetic pigments, TBARS, SOD, and antioxidant enzymes. Further, we tested the hypothesis that antioxidant molecules APX and GPX were involved in the detoxification of FA-induced stress in ferns. Since aboveground part (especially frond) contains more metal contents in comparison to belowground parts, aboveground part experienced more oxidative stress. Present study evaluated the extent of FA-induced oxidative stress and associated metal detoxification systems in three target fern species.

7.3.1 Physicochemical Characterization of Different Substratum

Analysis of physicochemical properties of different substratum used during present study is depicted in Table 7.1. Result reveals that FA was alkaline but GS used during experiment was acidic in nature. The pH and electrical conductivity (EC) of FA were 9.3 ± 0.03 and 7.51 ± 0.52 dS m^{-1}, while the pH and EC of GS were 6.4 ± 0.02 and 1.61 ± 0.03 dS m^{-1}, respectively. Cation exchange capacity (CEC) of GS was significantly higher in comparison to 100% FA, but it was found in moderate

Table 7.1 Physicochemical properties of FA amended with GS in different ratio

Parameter	100% GS (control)	50% FA+50% GS	100% FA
PH	6.4 ± 0.02	8.4 ± 0.17	9.33 ± 0.03
EC	1.61 ± 0.03	2.05 ± 0.06	7.51 ± 0.05
CEC	1.89 ± 0.05	1.05 ± 0.02	0.93 ± 0.03
Total nitrogen (%)	1.54 ± 0.02	0.23 ± 0.01	0.01 ± 0.01
Total phosphorous (%)	1.35 ± 0.03	0.27 ± 0.03	0.02 ± 0.02
Organic carbon (%)	2.43 ± 0.04	2.01 ± 0.01	1.73 ± 0.01
WHC (%)	35.5 ± 0.15	43.5 ± 0.35	48.5 ± 0.32
Metals (μg/g)			
Zn	34.5 ± 2.3	63 ± 3.4	79 ± 4.7
Fe	1135 ± 69	2397 ± 93	4536 ± 113
Si	21.5 ± 1.5	2118 ± 114	4977 ± 175
Ni	25.5 ± 2.1	145 ± 5.8	237 ± 11.7
Mn	47 ± 0.15	67 ± 5.7	86 ± 6.5
Cu	36 ± 2.3	49 ± 4.3	69 ± 3.7
Cd	ND	23 ± 0.03	34 ± 1.5
Pb	ND	21 ± 0.05	35 ± 2.3
B	4.9 ± 0.1	19 ± 0.02	27 ± 0.3
Al	5.9 ± 0.3	169 ± 3.9	379 ± 23

Mean ± SD (n = 3), student's t-test significant ($p < 0.05$) as compared to control content at 50:50 ratios of FA and GS. However, the presence of total N and P in FA was almost negligible (0.01 and 0.02, respectively) in comparison to GS but having better water holding capacity, which retains moisture inside. The concentration of trace elements in different substrates such as GS-, FA-, and FA-treated soils is also given in Table 7.1. The level of trace elements (Fe, Si, Ni, Al, Mg, Cu Cd, Pb, and As) in 100% FA was significantly high (student's t-test significant at $p < 0.05$) as compared to GS (control), while the concentration of As, Cd, and Pb in GS was below detectable limit (BDL). Results of the present study reveal that FA contains many plant growth essential micronutrient elements that are used in normal plant metabolism because they are essential constituents of various coenzymes and food substances like carbohydrate, fat, and protein, but the role of heavy metals is not known in plant metabolism, and they pose significant health risks to the receiving environment. The results of the present study are in support of earlier studies (Haynes 2009; Mehra et al. 1998; Pandey and Singh 2010). Chemically, all naturally existing elements can be found in FA (Klien et al. 1995), which is also subsequently enriched in trace elements compared with the parent coal. Among the elements enriched in ashes were Zn, Fe, Ni, Mn, Cu, Cd, Pb, Cr, B, Al, Si, and As (Mehra et al. 1998).

7.3.2 Metal Accumulation in Ferns Growing on FA Amendments

All the three fern species grown on different amendments of FA accumulated significant amount of Fe, Cu, Al, Ni, Si, Pb, Cr, and Cd in their fronds and rhizomes (Fig. 7.1). However, the concentration of metals in different plant parts (fronds and rhizome including roots) and order of preference for uptake and accumulation of metals in the plant tissues varied in each fern species. In the present study, *P. vittata* accumulated maximum iron followed by Al, Ni, Si, Cu, Cr, Pb, and Cd in frond part, but in rhizome metals accumulated in the order of Al>Fe>Cu>Si>Cr>Pb>Ni>Cd in 100% as well as 50% FA, but in 100% GS the patterns changed slightly as Fe>Al>Cu>Ni>Si>Pb>Cr>Cd. In each amendment, *A. prolifera* accumulated metals in Fe>Al>Si>Cu>Ni>Cr>Pb>Cd pattern. In fronds of *D. esculantum*, the metal was accumulated in the order of Fe>Al>Si>Ni>Cu>Cr>Pb>Cd in 100% FA and in 50% FA but negligible silicon in fronds grown on 100% GS. In *D. esculantum*, accumulation of metals in rhizome was found in the order of Fe>Al>Si>Ni>Cu>Cr>Pb>Cd at various amendments. Although metal accumulation was almost equal in other two species *A. prolifera* and *D. esculentum*, *P. vittata* accumulated about two times more heavy metals (Fe, Al, Cr, Pb, and Cd) than other two species in both plant parts. Further, all the three species accumulated significant amount of all the metals tested in this study in fronds than rhizomes including roots (Fig. 7.1a–c). Results showed potential of all the three fern species to grow on FA without any visible phytotoxic symptoms.

Many plants are known to accumulate high quantity of toxic metals in their tissues, and hence they are being used for phytoremediation of contaminated sites (Baker and Brooks 1989; Wenzel and Jockwer 1999). Among such plants, ferns have been demonstrated to accumulate and resist high metal toxicity growing on contaminated soil

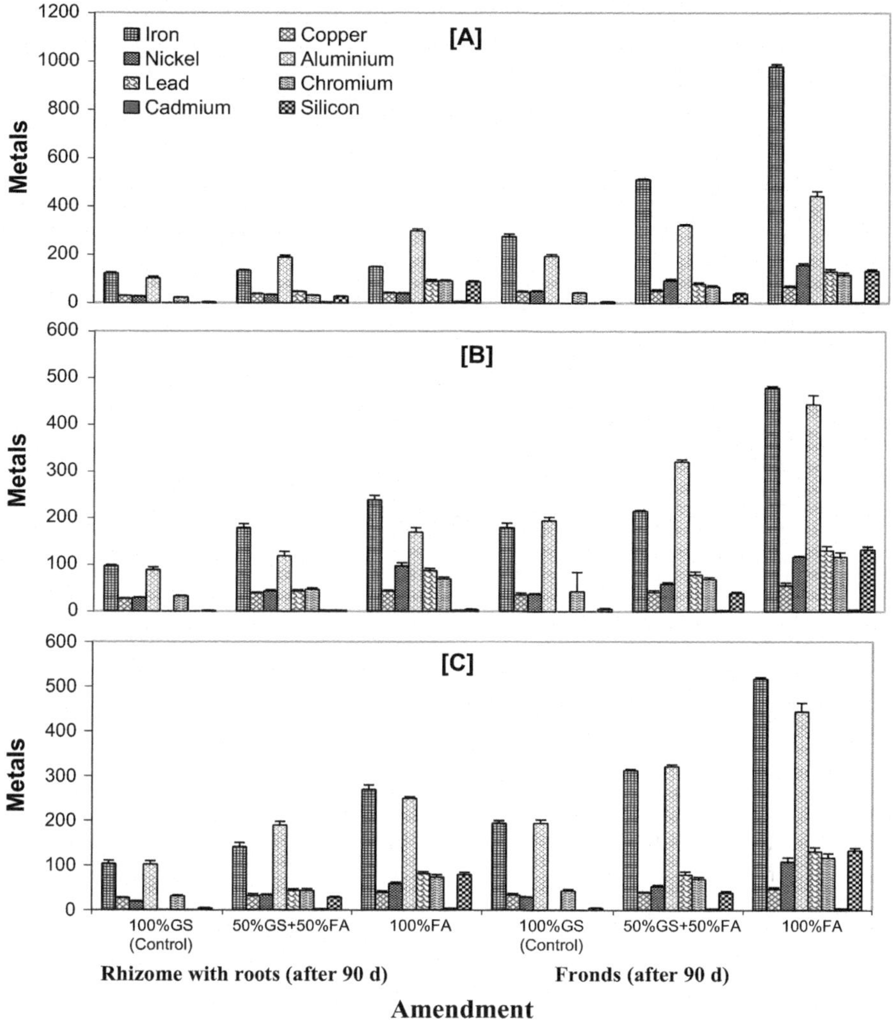

Fig. 7.1 Effect of different amendments of fly ash on metal accumulation of three fern species (**a**) *P. vittata*, (**b**) *A. prolifera*, and (**c**) *D. esculentum* on rhizome with roots and fronds after 90 days of treatment

(Honjo et al. 1980; Ma et al. 2001a, b). Despite of this fact, these species have never been utilized for removal of metals from contaminated wastes like FA. Results obtained during the present study clearly demonstrated a very high potential of *P. vittata*, *A. prolifera*, and *D. esculentum* in removal of metals from FA-amended soil. Further, it was also observed that fronds contain higher amount of heavy metals than rhizomes of the plants in all the three test species. It indicates that accumulated metal content was translocated and deposited inside frond which is reported as a good symbol of efficient hyperaccumulator, because suitability of a plant for phytoremediation could be determined by its ability to produce high aboveground biomass and high bioconcentration and transfer factors of metals (Feyiga et al. 2004). Several studies reported the suitability of *P. vittata* for phytoremediation of As-contaminated lands (Feyiga et al. 2004; Singh et al. 2006) and also recently reported in phytoremediation of FA-contaminated lands (Kumari et al. 2011). Application of *A. prolifera* and *D. esculentum* for phytoremediation of FA has first time reported in this paper. Our results showed that iron accumulation was maximum in all the three species which could be attributed to the high requirements of iron for plants for various metabolic activities as constituent of many enzymes and cytochromes of certain porphyrins in plant growth (Hewitt 1998). These findings confirm earlier reports on terrestrial plants growing on FA-amended soil (Gupta et al. 2007; Kumari et al. 2011; Rai et al. 2004).

7.3.3 Effect of FA Amendments on Biomass and Photosynthetic Pigments of Ferns

The comparative growth parameters of all the three species were observed weekly, and their rate of germination and growth rate were studied on P&T medium solidified with 1% agar under laboratory conditions (Table 7.2). The percentage rate of germination and growth was maximum in *P. vittata* L. in comparison to other two species, i.e., *A. prolifera* and *D. esculentum*. Both species took approximately equal time for germination and possessed almost equal growth rate. Sex ontogeny and differential growth stages were also appeared delayed in comparison to *P. vittata*.

Ferns grown on various amendments showed variation in their growth patterns; hence, they possessed varied biomass. Most significant increase in biomass was observed on 50% concentration of FA in all the three test species. *P. vittata* performed maximum increase by 40% in total biomass growing on 50% FA, and it decreases on 100% FA by denoting only 20% increase in total biomass. *A. prolifera* showed 20% increase in total biomass on 50% amendments, but there was no significant increase found on 100% FA as compared to control. Similarly *D. esculentum* performed 15% increase in total biomass on 50% amendments but decreased biomass on 100% FA amendments in comparison to control. At 100% FA, all the test species experienced toxicity in some extent, and consequently biomass decreases in comparison to 50% FA (Fig. 7.2). However, there were no visible toxicity symptoms observed in all the three species.

Table 7.2 Comparative growth of fern species on FA

Name of species	% germination	Gametophyte formation	Sex ontogeny	Sporophyte formation
Pteris vittata L.	95–100	After 4 weeks	Initiated after 4 weeks	After 6 weeks
Ampelopteris prolifera Copel.	60–70	After 5–6 weeks	Initiated after 5 weeks	After 7–8 weeks
Diplazium esculentum (Retz.) Sw.	70–80	After 4–5 weeks	Initiated after 5 weeks	After7–8 weeks

As exposure of 100% FA possessed metal toxicity in ferns, the photosynthetic pigments also exerted by the same and consequently decreased chlorophyll a and b and total chlorophyll (Fig. 7.2), but ferns growing on 50% FA enhanced chlorophyll a and b and total chlorophyll by 20–25% in *P. vittata*. Ferns grown on 50% FA amendments showed always enhanced photosynthetic pigments in comparison to control in all species. However, chlorophyll content was decreased at 100% FA, but

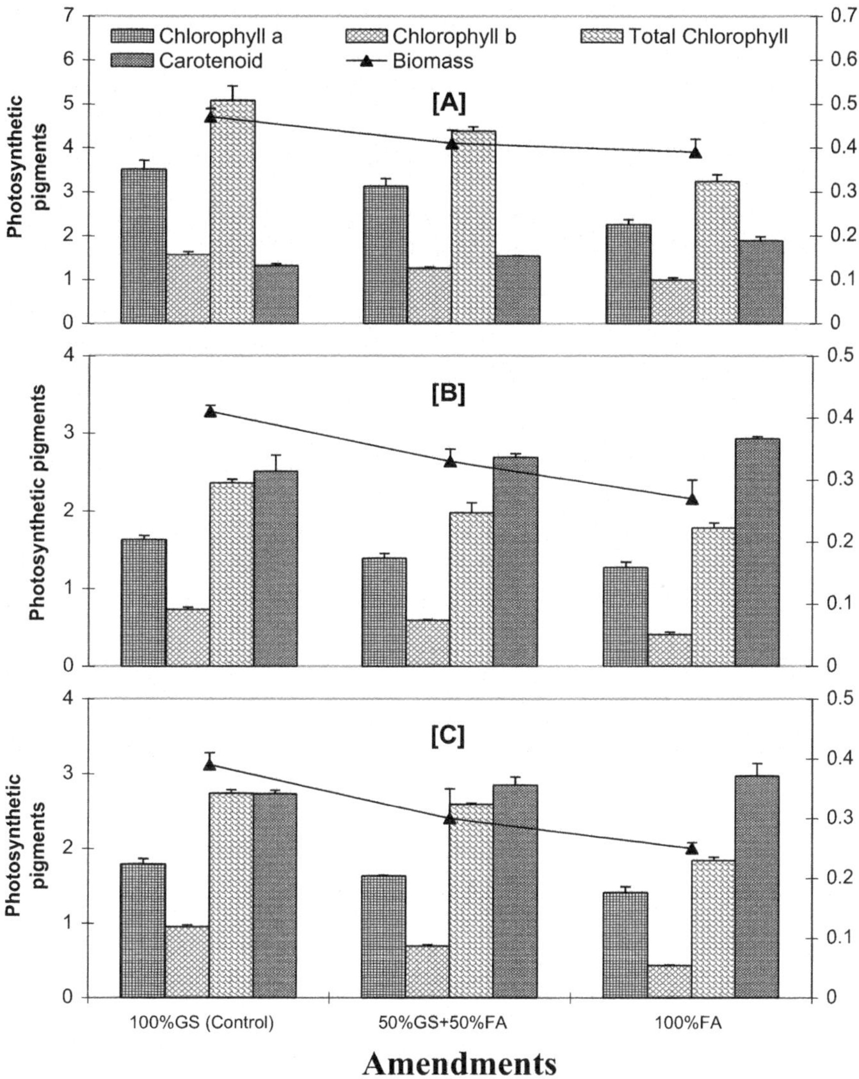

Fig. 7.2 Effect of different amendments of fly ash on total biomass and photosynthetic pigments of three fern species (**a**) *P. vittata*, (**b**) *A. prolifera*, and (**c**) *D. esculentum* on fronds after 90 days of treatment

carotenoid contents were enhanced in all the three species. It showed young sporophytes raised on maximum dose of FA loss in photosynthetic pigments due to metal toxicity present in FA. Thus, the result obtained by study of biomass and photosynthetic pigments suggests that 50% FA amendment do not generate oxidative stress in ferns and seem to be suitable substratum for healthy fern growth (Fig. 7.2).

Carotenoid is a class of natural fat-soluble pigments found in plants and plays an important role in photosynthetic process as well as membrane-associated antioxidant activity. The decline in chlorophyll concentration implied heavy metal-induced stress due to FA in all the three species. Earlier *Pteris vittata* has been shown to have reduced chlorophyll content in higher-dose exposure of arsenic, but in lesser doses it showed enhancement in chlorophyll content (Singh et al. 2006). Similar result was obtained during present study on various FA amendments. The same trend was followed in other two species *D. esculantum* and *A. prolifera*. The chlorophyll data suggest that a combination of antioxidant compounds and enzymes resulted in a greater protection of the photosynthetic system of metal-tolerant *Pteris vittata* and other hyperaccumulator ferns against ROS (Singh et al. 2006).

7.3.4 Effect of FA Amendments on Lipid Peroxidation and Antioxidant Enzymes

7.3.4.1 Lipid Peroxidation (MDA)

The effect of FA toxicity on lipid peroxidation was determined by evaluating MDA content in the form of TBARS content of the fern tissues. It was found maximum in more toxic site, i.e., in 100% FA and minimum in control (100% GS) site. A significant increase by 47%, 31%, and 27% in TBARS was observed in frond tissues of *P. vittata*, *A. prolifera*, and *D. esculentum*, respectively, as compared to ferns grown on control. Similarly, MDA content was enhanced in rhizome tissues of *P. vittata*, *A. prolifera*, and *D. esculentum* by 41%, 25%, and 28%, respectively. As metal content in fronds was found higher in comparison to rhizome tissues, the level of MDA in frond tissues was also found higher. MDA level increased by increasing ratio of FA amendments in GS. Lipid peroxidation in terms of MDA was found maximum in the case of sporophytes of *P. vittata*, grown in 100% FA as compared to other treatments showing its lower scavenging ability. Therefore, high levels of lipid peroxidation in fern tissues are also indirect indicator of metal stress provided by FA (Fig. 7.3).

The formation of TBARS in plants exposed to adverse environmental conditions is an indicator of free-radical formation in the tissues, and it may be used as an index of lipid peroxidation in biological systems (Heath and Packer 1968). There is considerable evidence that inorganic exposure results in the generation of ROS in plants (Hartley-Whitaker et al. 2001). ROS induced by the ferns growing in FA-amended soil showed an enhanced level of melanoaldehyde compounds, which is degreedative product of a variety of biologically important compounds like amino acid, protein, and carbohydrate; therefore, increased concentration of MDA in these ferns following FA stress is an indicator of metal-induced oxidative damage as reported in earlier studies (Kumari et al. 2013). This indicated that all these target species

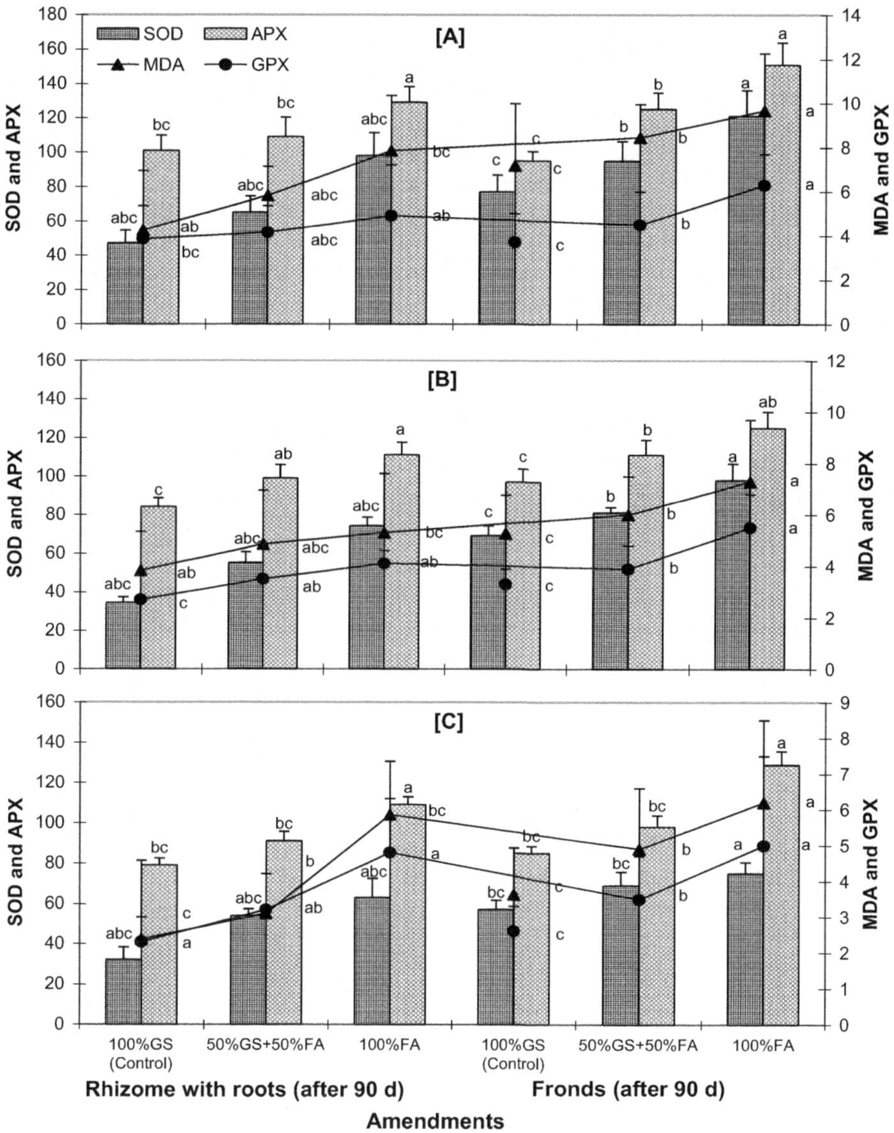

Fig. 7.3 Effect of different amendments of fly ash on SOD (μ mg^{-1} fw), APX (min^{-1} g^{-1}fw), MDA content (μ mol g^{-1} fw), and GPX (μ mol g^{-1} fw) of three fern species (**a**) *P. vittata*, (**b**) *A. prolifera*, and (**c**) *D. esculentum* on fronds and rhizomes including roots after 90 days. Similar letters are not significant ($P < 0.5$ level) a/c to Duncan's multiple range test

possessed a defense system to protect themselves against oxidative damages as evidenced by increased activity of antioxidant enzymes.

7.3.4.2 Superoxide Dismutase (SOD)

SOD is an essential component of plant's antioxidative defense system. The SOD activities of the frond tissue in all the three species were found significantly higher in 100% FA-grown ferns. It was found enhanced by 57%, 27%, and 34% in frond part of *P. vittata*, *A. prolifera*, and *D. esculentum*, respectively, on ferns grown on 100% FA. Similarly, in belowground part, the level of SOD was found higher in FA-grown ferns by 42%, 24%, and 26%. It was further observed that fronds maintained higher SOD activity than roots and rhizomes, and it increased by increasing level of FA stress (Fig. 7.3).

Antioxidant enzymes are considered to be an important defense system of plants against oxidative stress caused by metals (Weckx and Clijsters 1997). The activities of SOD, APX, and GPX in general showed simultaneous induction and decline, which may be due to their co-regulation (Shigeoka et al. 2002). Significant increase in the level of SOD in all three ferns was found in this study. However, the highest level of SOD was found in *P. vittata* fronds grown on 100% FA indicating its greater scavenging or acclimatization capacity to FA stress by producing rapidly antioxidative defense system. SOD plays an important role in dismutation of free hydroxyl radicals by the formation of hydrogen peroxides. However, during well-organized functioning of SOD, which prevents oxygen-driven cell damage and converting it to H_2O_2 which is then reduced to water and molecular oxygen by the action of various enzymes like APX and GPX working at different location of cell.

7.3.4.3 Ascorbate Peroxidase (APX)

In comparison to fronds, the root tissues contained higher APX concentration in all the three test species. In 100% FA, fronds of *P. vittata* showed 18% increase in APX activity, but it increased by 22% in belowground part. *A. prolifera* and *D. esculentum* showed approximately similar APX activity. But both species showed maximum APX content in belowground part (rhizomes and roots) by 28% and 24% increase than control. In frond part of both these species, APX activity increased when treated with maximum FA toxicity by 18% and 14%, respectively (Fig. 7.3).

Ascorbate is an essential compound in plant tissues and has a major role in relation to enzymatic and nonenzymatic oxidation reactions in the biological systems (Gupta et al. 1999; Smirnoff 1996). It can react directly by reducing superoxide, hydrogen peroxide, and hydroxyl radical or quenching singlet oxygen and functions as co-substrate of plant oxidases, such as the ascorbate peroxidase system, which produces dehydroascorbate (Halliwell 1982). Though APX play key role in functioning of chloroplast in ascorbate glutathione cycle, but in present study the role of APX during metal detoxification seems not much significant, as it shown slight increase in ferns growing on 100% FA in comparison to GPX, which is basically a cell-bound enzyme found in cytoplasm play better response to combat metal toxicity in ferns.

7.3.4.4 Guaiacol Peroxidase (GPX)

GPX activities perform remarkably higher in FA-grown ferns than control, but it also showed similarity with APX and was found higher in root tissues than fronds. When treated with 100% FA, it showed maximum increase by 71% in *P. vittata* fronds and 60% in rhizome parts. Other two species *A. prolifera* and *D. esculentum* showed approximately similar increase by 52% and 50% in fronds and by 50% and 55% in rhizomes, respectively. Almost similar GPX activity was performed by both species, and no significant changes in GPX activity of both plant parts were observed (Fig. 7.3).

In present study, GPX activity in both fronds and rhizomes showed remarkable better response as compared to APX. GPX thus seemed to play a lead role in metal detoxification. GPX is thought to be stress marker enzymes (Castillo 1986), and its higher induction may indicate stress exerted by heavy metals, which can be correlated with amount of accumulated metal.

7.4 Conclusions

The present study concludes feasibility of using *P. vittata*, *D. esculantum*, and *A. prolifera* in phytoremediation of metals from FA. Metal-accumulating ferns, despite showing slight reduction in the biomass and photosynthetic pigments, showed enhanced level of antioxidative enzymes; thus, these plants could scavenge oxygen radicals due to stimulated activities of SOD and GPX. However, contribution of APX found is not much significant, and other mechanisms of thio-metabolism and phytochelatin induction could be ruled out. The result of present study also suggests that 50% amendment ratio is most suitable substratum for healthy and luxuriant growth of ferns, which address a major problem of FA management in agriculture of ornamental ferns. Further studies were aimed to analyze both GSH and PC biosynthesis in these potential ferns under FA stress. Besides these, molecular mechanism underlying hyperaccumulation in ferns is also prerequisite for better understanding of the complete mechanism of metal detoxification.

Acknowledgment We thank Directors, CSIR-Institute of Himalayan Bioresource Technology, Palampur, H.P., and CSIR-National Botanical Research Institute, Lucknow, UP, India, for motivation of collaborating research to exchange required research facilities. Alka Kumari is grateful to DST for providing financial support under WOS-A scheme (SR/WOS-A/LS-117/2008).

References

Allen SE, Grimsha WHM, Parkinson JA, Quarnby C (1974) Chemical analysis of ecological materials. Blackwell Scientific Publisher, Oxford

Arnon DI (1949) Copper enzymes in isolated chloroplast, polyphenol/oxidase in *Beta vulgaris*. Plant Physiol 24:1–15

Baker AJM, Brooks RR (1989) Terrestrial higher plants which hyperaccumulate metallic elements – a review of their distribution, ecology and phytochemistry. Biorecovery 1:81–126

Beauchamp C, Fridovich I (1971) Superoxide dismutase: improved assays and an assay applicable to acrylamide gels. Anal Biochem 44 (1):276–287
Cao X, Ma LQ, Tu C (2004) Antioxidant responses to arsenic in the arsenic hyperaccumulator Chinese brake fern (*Pteris vittata* L.) Environ Pollut 128:463–468
Castillo FJ (1986) Extracellular peroxidases as markers of stress? In: Grepin H, Penel C, Gaspar T (eds) Molecular and physiological aspects of plant peroxidases. University of Geneva Press, Geneva, pp 419–426
Dexbury AC, Yentch CS (1956) Plankton pigment monograph. J Mar Res 5:93–101
Feyiga OA, Ma LQ, Xinde C, Rathinasabapathi B (2004) Effects of heavy metals on growth and arsenic accumulation in the arsenic hyperaccumulator *Pteris vittata* L. Environ Pollut 132:289–296
Grill E, Winnacker EL, Zenk MH (1987) Phytochelatins, a class of heavy metal binding peptides from plants, are functionally analogous to metallothioneins. Proc Natl Acad Sci U S A 84:439–443
Gupta M, Cuypers H, Vangronsveld H, Clijsters (1999) Copper effects the enzymes of the ascorbate-glutathione cycle and its related metabolites in the roots of *Phaseolus vulgaris*. Physiologiqua Plant 106:262–267
Gupta DK, Rai UN, Sinha S, Tripathi RD, Nautiyal BD, Rai P, Inouhe M (2004) Role of rhizobium (CA-1) inoculation in increasing growth and metal accumulation in *Cicer arietinum* L. growing under fly ash stress condition. Bull Environ Contam Toxicol 73:424–431
Gupta AK, Dwivedi S et al (2007) Metal accumulation and growth performance of *Phaseolus vulgaris* grown in fly ash amended soil. Bioresour Technol 98:3404–3407
Halliwell H (1982) Ascorbic acid and the illuminated chloroplast. In: Seib PA, Tolbert BM (eds) Ascorbic acid: chemistry, metabolism and uses. American Chemical Society, Washington, DC, pp 263–274
Hartley-Whitakar J, Ainsworth G, Vooijs R, Ten WB, Schat H, Mehrag AA (2001) Phytochelatins are involved in differential arsenate tolerance in *Holcus lanatus*. Plant Physiol 126:299–306
Haynes RJ (2009) Reclamation and revegetation of fly ash disposal sites – challenges and research needs. J Environ Manag 90:43–53
Heath RL, Packer L (1968) Photoperoxidation in isolated chloroplast 1 kinetics and stoichiometry of fatty acid peroxidation. Arch Biochem Biophys 125:189–198
Hemeda HM, Klein BP (1990) Effects of naturally occurring anti-oxidants on peroxidase activity of vegetable extracts. J Food Sci 55:184–185
Hewitt EJ (1998) The role of mineral elements in the activity of plant enzymes. In: Ruhl W (ed) Hand buch der pfflanzen physiologie, vol IV. Springer, Berlin, p 427
Honjo T, Suganuma H, Satomi N (1980) The vegetation of the pollution areas caused by the lead title in Kanazawa Castle. J Phytogeogr Taxon 27:70–73
Jambhulkar HP, Juwarkar AA (2009) Assessment of bioaccumulation of heavy metals by different plant species grown on fly ash dump. Ecotoxicol Environ Saf 72:1122–1128
Khan MR, Khan MW (1996) The effect of fly ash on plant growth and yield of tomato. Environ Pollut 92:105–111
Klien DH, Andew AW et al (1995) Pathways of thirty seven trace elements through coal fired power plants. Environ Sci Technol 9:973–979
Kumar A, Vajpayee P, Ali MB, Tripathi RD, Singh N, Rai UN, Singh SN (2002) Biochemical responses of *Cassia siamea* Lamk. grown on coal combustion residue (fly ash). Bull Environ Contam Toxicol 68:675–683
Kumari A, Lal B, Pakade YB, Chand P (2011) Assessment of bioaccumulation of heavy metals by *Pteris vittata* L. growing in the vicinity of fly ash. Int J Phytoremediation 13:779–787
Kumari A, Pandey VC, Rai UN (2013) Feasibility of fern *Thelypteris dentata* for revegetation of coal fly ash landfills. J Geochem Explor 128:147–152
Ma LQ, Komar KM, Tu C, Zhang W, Cai Y (2001a) A fern that hyperaccumulates arsenic. Nature 409:579
Ma LQ, Komar KM, Tu C, Zhang W, Cai Y (2001b) A fern that hyperaccumulates arsenic-addendum. Nature 410:411–438
Mehra A, Farago ME, Banerjee DK (1998) Impact of fly ash from coal fired station in Delhi, with particular reference to metal contamination. Environ Monit Assess 50:15–35

Mehrag AA (2002) Variation in arsenic accumulation/hyperaccumulation in ferns and their allies. New Phytol 157:25–31

Nakano Y, Asada K (1981) Hydrogen peroxide is scavenged by ascorbate-specific peroxidase in spinach chloroplast. Plant Cell Physiol 22:867–880

Nelson DW, Sommer LE (1982) Total carbon, organic carbon, and organic matter. In: Page AL (ed) Methods of soil analysis, ASA Monogr 9 (2), 2nd edn. American Society of Agronomy, Madison, pp 539–579

Oleson SR, Sommers LE (1982) Phosphorus. In: Page AL, Miller RH, Keene DR (eds) Methods of soil analysis, Part 2 Chemical and Microbiological. Properties ASA, Madison, pp 403–427

Pandey VC (2012) Invasive species based efficient green technology for phytoremediation of fly ash deposits. J Geochem Explor 123:13–18

Pandey VC, Singh N (2010) Impact of fly ash incorporation in soil systems. Agric Ecosyst Environ 136:16–27

Pandey VC, Singh JS, Kumar A, Tewari DD (2010) Accumulation of heavy metals by chickpea grown in FA treated soil: effect on antioxidants. Clean (Weinh) 38:1116–1123

Piper CS (1966) Soil and plant analysis. Inter Science, New York

Rai UN, Pandey K, Sinha S, Singh A, Saxena R, Gupta DK (2004) Revegetating fly-ash landfills with *Prosopis juliflora* L. impact of different amendments and rhizobium inoculation. Environ Int 30:293–300

Ram LC, Jha SK, Tripathi RC, Masto RE, Selvi VA (2008) Remediation of fly ash landfills through plantation. Remediation 18:71–90

Shigeoka S, Ishikawa T, Tamoi M, Miyagawa Y, Takeda T, Yabuta Y, Yoshimura K (2002) Regulation and function of ascorbate peroxidase iso-enzyme. J Exp Bot 53:1305–1319

Singh IP, Siddiqui ZA (2003) Effects of fly ash and *Helminthosporium oryzae* on growth and yield of three cultivars of rice. Bioresour Technol 86:73–78

Singh N, Ma LQ, Srivastava M, Rathinasabapathi B (2006) Metabolic adaptations to arsenic-induced oxidative stress in *P vittata* L. and *Pteris ensiformis* L. Plant Sci 170:274–282

Sinha S, Gupta AK (2005) Translocation of metals from fly ash amended soil in the plant of *Sesbania cannabina* L Ritz: effect on antioxidants. Chemosphere 61:1204–1214

Sinha S, Rai UN, Bhatt K, Pandey K, Gupta AK (2005) Fly ash induced oxidative stress and tolerance in *Prosopis juliflora* L. grown on different amended substrates. Environ Monit Assess 102:447–457

Smirnoff N (1996) The function and metabolism of ascorbic acid in plants. Ann Bot 78:661–669

Srivastava M, Ma LQ et al (2005) Antioxidant responses of hyper-accumulator and sensitive fern species to arsenic. J Exp Bot 56:1335–1342

Tiwari S, Kumari B, Singh SN (2008) Microbe-induced changes in metal extractability from fly ash. Chemosphere 71:1284–1294

Tiwari S, Kumari B, Singh SN (2010) Evaluation of metal mobility/immobility in fly ash induced by bacterial strains isolated from the rhizospheric zone of *Typha latifolia* growing on fly ash dumps. Bioresour Technol 99:1305–1310

Tripathi RD, Dwivedi S et al (2008) Role of blue green algae biofertilizer in ameliorating the nitrogen demand and fly-ash stress to the growth and yield of rice (*Oryza sativa* L.) plants. Chemosphere 70:1919–1929

Vajpayee P, Rai UN et al (2000) Management of fly ash landfills with *Cassia surattensis* Burm. – a case study. Bull Environ Contam Toxicol 65:675–682

Walkely YA, Black IA (1934) An examination of the Degtjareff method for determining soil organic matter and a proposed modification of the chromic acid titration method. Soil Sci 37:29–38

Weckx JEJ, Chlisters HMM (1997) Zn phytotoxicity induces oxidative stress in primary leaves of *Phaseolus vulgaris*. Plant Physiol Biochem 35:405–410

Wenzel WW, Jockwer F (1999) Accumulation of heavy metals in plants grown on mineralized soils of the Austrian Alps. Environ Pollut 104:145–155

Wong JWC, Wong MH (1990) Effects of fly ash on yields and elemental composition of two vegetables, *Brassica parachinensis* and *Brassica chinensis*. Agric Ecosyst Environ 30:254–264

Zenk MH (1996) Heavy metals detoxification in higher plants – a review. Gene 179:21–30

The Multiple Properties of Some of the Lichenized Ascomycetes: Biological Activity and Active Metabolites

8

Valery M. Dembitsky

Abstract

The main objective of this chapter was to describe the physicochemical and biological characteristics of selected lichenized ascomycetes and the influence of their physiologically active compounds on human health, through scientifically proven information. The chapter presents the biologically active metabolites derived from lichen species (polyphenols, volatile compounds, lipids, phospholipids, fatty acids, and organic acids). Lichens and their metabolites have been demonstrated to possess numerous biological activities, including antiviral, antibacterial, antitumor, and enzyme inhibitory activity. The influence of environmental factors on the lipid and fatty acid composition of some lichen species has also been reported. Lichens are easily exposed to halogens, which are present in polluted air. We present chlorinated metabolites, which are isolated from various species of lichens. Chlorine is one of the main pollutants in nature. The structures of about 50 chlorinated metabolites of phenolic nature generated by lichenized ascomycetes are considered. Some active lichen substances are used in the pharmaceutical industry.

Keywords

Lichens · Lichenized ascomycetes · Lipids · Fatty acids · Glycosides · Polar lipids · Activity · Chlorinated metabolites · Depsides · Depsidones

V.M. Dembitsky (✉)
National Scientific Center of Marine Biology, Vladivostok, Russia
e-mail: valeryde@imb.dvo.ru

V. Shukla et al. (eds.), *Plant Adaptation Strategies in Changing Environment*,
https://doi.org/10.1007/978-981-10-6744-0_8

8.1 Introduction

Lichen is a living organism, formed by symbiosis of the fungus and algae. Algae can be green algae or blue-green algae. Blue-green algae are actually bacteria; they are called cyanobacteria (Dembitsky 2003a, b; Dembitsky and Tolstikov 2003a, 2005; Huneck 2001). Lichens first appeared about 400 million years ago. Some individual species, such as *Hypogymnia physodes*, have been around for at least 25 million years and maybe for as long as 70 million years. Colors of lichens are grayish, greenish-gray, and light or dark brown, less often yellow, orange, white, and black. The coloring is due to the pigments that are found in the hyphae of the fungus. There are five groups of pigments: green, blue, violet, red, and brown. The color of lichens can also depend on the coloring of lichen acids, which are deposited in the form of crystals or grains on the surface of the hyphae (Dembitsky 2003a; Dembitsky and Tolstikov 2003a, 2005; Purvis 2010; Hawksworth and Hill 1984).

The number of different species of lichens is about 25,000 species. Lichens are found in all continents of the Earth, even in Antarctica. Lichens are found everywhere, and people have used them since ancient times for various purposes (as pet food, as medicine and food, for dyeing fabrics). The symbiosis of algae and fungus makes it possible to live lichen in a variety of environments that are not adapted to life. Lichens can grow on rocks, walls of houses, desert, and tundra. And, of course, they are found everywhere in the forests. However, lichens are very sensitive to contamination. If the air is smoggy, there are harmful gases in it, and then lichens die. Therefore, lichens can serve as indicators of the purity of the environment. The complexity of lichen partnerships has caused lichens to be described as "small ecosystems" (Huneck 1996; Dembitsky and Tolstikov 2005).

Symbiosis of the fungus and algae in the body of the lichen is very tight, which as a result gives a single organism. In the thallus the hyphae of the fungus are intertwined; between them are cells of green algae or cyanobacteria. These cells can be located either singly or in groups. Thus, lichen combines two very different organisms. The fungus feeds heterotrophically (it absorbs ready-made organic substances), and the algae autotrophically (synthesizes organic substances from inorganic substances). You can draw an analogy. Mycorrhiza is a symbiosis between higher plants and fungi, and lichen is a symbiosis between lower plants and fungi. However, in a lichen, the symbiosis is much closer. After all, the species of fungi that are part of lichens cannot exist at all without algae, although most algae lichens are found in nature separately (Purvis 2010; Hawksworth and Hill 1984; Dembitsky and Tolstikov 2005).

Lichens are the symbiotic phenotype of nutritionally specialized fungi that acquire, in an ecologically obligate symbiosis, fixed carbon from a population of green algal or cyanobacterial cells (Dembitsky 2003a; Dembitsky and Tolstikov 2005). Lichens produce more than 1200 different metabolites that are not found in other organisms. These organisms have both algal and fungal properties and produce n-alkane (Torres et al. 2003; Enk et al. 2002), unusual betaine ether glycerolipids (Dembitsky 1996; Dembitsky et al. 1993; Temina et al. 2010), and saturated, unsaturated, branched, and halogenated fatty acids (Dembitsky 1992, 1996;

Dembitsky et al. 1991a, b, c, 1992a, b, c, d, e, 1993a, b, c, 1994a, b; Torres et al. 2003; Rezanka and Dembitsky 2001, 2003; Rezanka et al. 2003; Dembitsky and Srebnik 2002). Many different bioactive secondary metabolites have also been isolated from lichen species (Torres et al. 2004; Enk et al. 2005) which have been used in pharmaceutical sciences (Hawksworth and Rossman 1997). Arsenolipids (Dembitsky and Levitsky 2004; Dembitsky and Rezanka 2003), boron-containing fatty acid derivatives (Dembitsky et al. 2002, 2011), natural surfactants (Dembitsky 2004a, b, 2005a, b, c, d, e, 2006a, b), and allenic fatty acids (Dembitsky and Maoka 2007) have also been described.

In this chapter, we first of all present the results of our own research, which we studied for the last 30 years.

8.2 Chemical Constituents of the Lichenized Ascomycetes Belonging to the Genus *Collema*

Collema is a genus of lichens in the family Collemataceae. The photobiont is the cyanobacterium genus *Nostoc*. Several species of lichens from this genus were investigated.

8.2.1 A UV-B Absorbing Mycosporine with Photoprotective Activity from the Lichenized Ascomycetes Belonging to the Genus *Collema*

The sun radiation reaching the Earth spans from the shortwaved UV (UV-C), which is absorbed in the ozone layer, through the UV-B (280–315 nm), the UV-A (315–400 nm), and the visible range (400–800 nm), to the infrared. The shorter the wavelength, the more energetic and potentially harmful is the radiation. UV-B radiation was recognized long ago as the cause for skin erythema (sunburn), and accumulated exposure results in DNA damage and immunosuppression, eventually leading to skin cancer (Dembitsky and Tolstikov 2005). Most commercial sunscreens are designed to prevent sunburn on the assumption that this activity will also prevent skin cancer. However, the dramatic increase in skin cancer incidence rates demonstrates the inadequacy of traditional sun protection agents and emphasizes the urgent need to look for new, alternative molecules.

UV spectra of lichen and mycobiont have the same absorption maximum, at 311 nm (Fig. 8.1). According to this, a molar extinction coefficient (ε) of 34,000 m^{-1} cm^{-1} was determined. The high ε-value gives the isolated compound a huge advantage over common commercial sunscreens by having a great potential to protect from UV-B radiation. Collemin A is a new mycosporine, which incorporates a novel pyrrolidine ring. The UV-B absorbance ability of this compound is probably of fungal origin. Isolation of the dominant fungus (one out of four) from the lichen gave a UV spectrum with a similar λ_{max} when compared to the lichen itself (λ max = 311 nm).

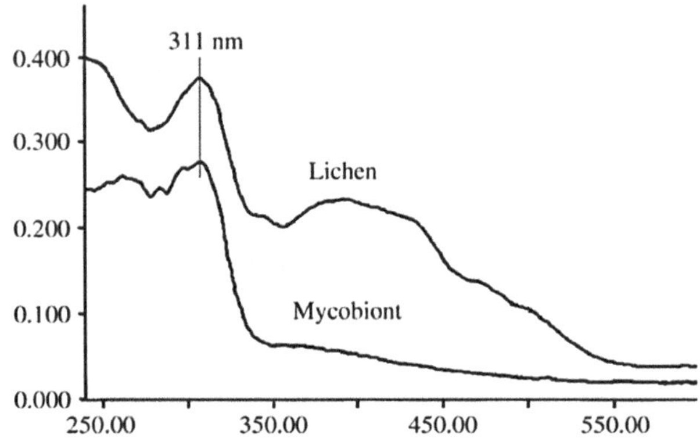

Fig. 8.1 Comparative UV spectra of the fungus (mycobiont) and lichen *Collema cristatum*. The mycobiont metabolite has a maximum at 311 nm.

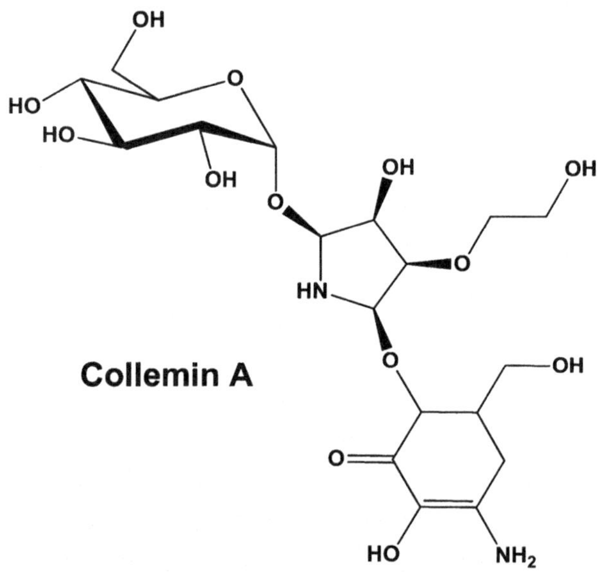

Fig. 8.2 Structure of a photoprotective mycosporine from the lichenized ascomycete *Collema cristatum*

A photoprotective mycosporine, which we have named collemin A (Fig. 8.2), was isolated from the lichenized ascomycete *Collema cristatum* (Torres et al. 2004; Rezanka et al. 2004a; Enk et al. 2002, 2003, 2004, 2005) Biological activity was measured in terms of protection against UV-B-induced membrane destruction and pyrimidine dimer formation in cultured human keratinocytes and prevention of UV-B-induced erythema. It was found that the pure isolated compound prevented UV-B-induced cell destruction in a dose-dependent manner and that the compound partially prevented pyrimidine dimer formation and completely prevented UV-B-induced erythema when applied to the skin prior to irradiation.

A known photoprotective mycosporine, collemin A (Torres et al. 2004; Enk et al. 2004, 2005), was also isolated from the *C. cristatum* and *C. callopumum* by HPLC, both lichens growing on calcareous rocks (Temina et al. 2010). It was found that the

Table 8.1 Nitrogen-containing compounds detected in the genus *Collema*

Lichen metabolites	*C. cristatum*	*C. callopumum*	*C. flaccidum*	*C. fuscovirens*
Collemin A	114.3	29.8	ND	ND
Indole-3-acetic acid	49.9	22.1	28.9	33.7
Bufotenine	ND	ND	14.2	2.9
5-Hydroxy-*N*-methyltryptamine	ND	ND	4.8	18.3
5-Hydroxytryptophan	ND	ND	11.3	6.9
Serotonin	ND	4.9	6.9	ND
Tryptamine	5.2	3.2	ND	ND
Tryptophan	2.6	3.9	11.7	17.9

μg/100 g dry wt
ND not detected

pure isolated compound prevents UV-B-induced cell destruction in a dose-dependent manner, partially prevents pyrimidine dimer formation, and completely prevents UV-B-induced erythema when applied to the skin prior to irradiation. Species of *C. flaccidum* and *C. fuscovirens* growing on trees did not produce this metabolite. Other nitrogen-containing compounds were also detected in lichens (Table 8.1).

Indole-3-acetic acid was detected in many lichen species (*Canoparmelia caroliniana*, *C. crozalsiana*, *C. texana*, *Parmotrema sancti-angeli*, and *P. tinctoria*, *Dermatocarpon intestiniforme*, *Flavoparmelia caperata*, *Lecanora muralis*, *Neofuscelia pulla*, *Rhizocarpon geographicum*, *Tephromela atra*, and *Xanthoria elegans*) and also in the genera *Cladonia*, *Parmelia*, *Peltigera*, and *Usnea*. Tryptophan was isolated from *Schizophyllum commune* (Temina et al. 2010).

8.2.2 Chiral Bianthraquinone Glycosides with Antitumor Activity

Anthraquinones are widely distributed pigments mainly in higher plants, fungi, and lichens (Dembitsky and Tolstikov 2005). Rhodocladonic acid was identified in lichen *Cladonia* sp.; it is also the intermediate of synthetic product, i.e., paeciloquinone C. This compound, the common anthraquinone structurally comparable to our monomer compound, was also isolated from fungus *Paecilomyces carneus*.

Colleflaccinosides A and B, two chiral bianthraquinone glycosides (Fig. 8.3) from the two geographical varieties of lichen *Collema flaccidum* collected in Russia and Israel, have been isolated as new natural products. The colleflaccinosides B had significant antitumor activity in the crown gall tumor inhibition test (Řezanka and Dembitsky 2006b).

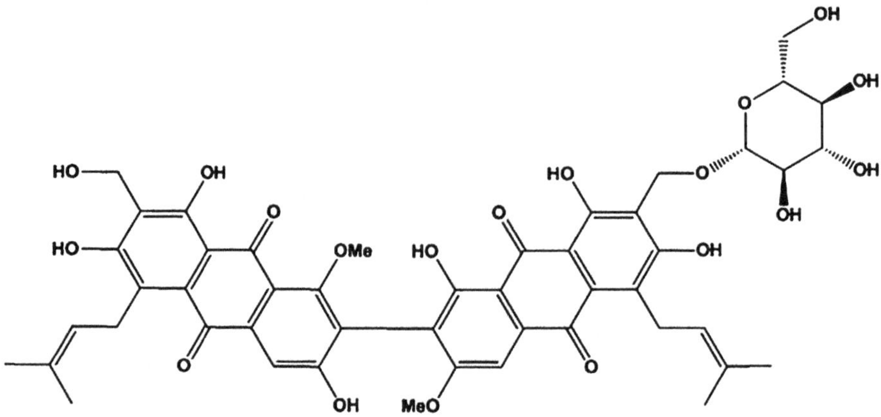

Colleflaccinoside A

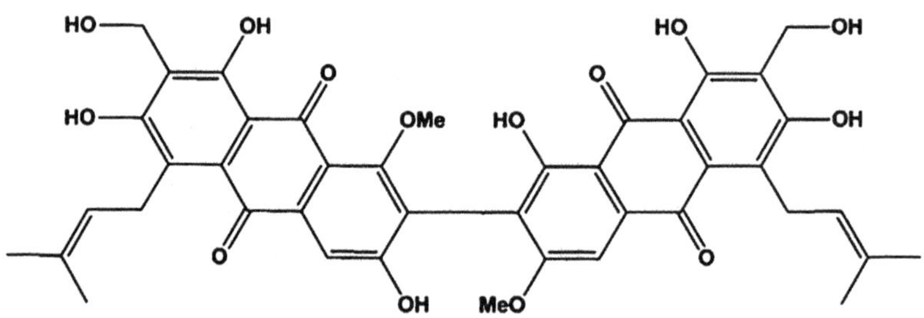

Colleflaccinoside B

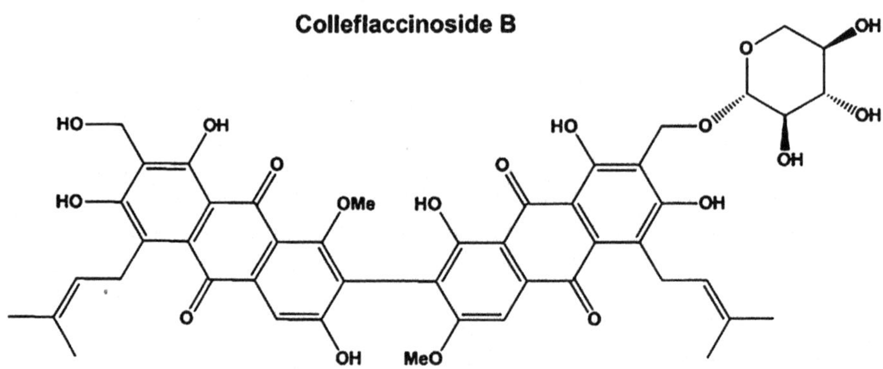

Bianthraquinone glycoside

Fig. 8.3 Colleflaccinosides *A* and *B*, chiral bianthraquinone glycosides isolated from lichen *Collema flaccidum*

8.2.3 Phospholipids and Fatty Acids

Four epiphytic and lithophilic species belonging to the genus *Collema* (family Collemataceae) were chosen for a comparative examination and to identify some chemicals in these groups of their fatty acid, polar lipids, and aromatic compounds. The GC-MS analysis of fatty acids in Collema, which grows on calcareous rocks and trees, revealed a high polyenoic content in species which grows on the Palestine oak (*Quercus calliprinos*) (62.1% and 61.2%; Table 8.2), but much lower amounts of such acids in two other *Collema* species, viz., 22.5% and 19.2%, respectively. The amounts of trienoic acids, however, were higher in the samples from tree-growing species, 43.9% (No 3) and 42.7% (No 4) (Temina et al. 2010).

Strong discrepancy of saturated fatty acids between species growing on rocks and trees was identified with n-16:0 as a major one (56–58% and 12–13%, respectively). All species studied had almost identical monoenoic acid contents. Dienoic and trienoic acids had also a strong discrepancy, as well as the saturated acids.

Tetraenoic and pentaenoic acids have not been found in *Collema* species growing on rocks, and they are present in species growing on the trees (Table 8.2). Among other interesting acids, 4,7,10,13-hexadecatetraenoic acid [16:4(n-3)], the presence of which is characteristic for green marine macrophytic and microscopic algae (Dembitsky and Rezanka 2005; Rezanka and Dembitsky 2006a), and stearidonic acid [18:4(n-3)], characteristic for brown marine algae (Dembitsky et al. 1990, 1991), were detected. It seems likely that the lichen photobiont synthesizes these

Table 8.2 Main fatty acids in four lichens of the genus *Collema*

Fatty acids	No 1	No 2	No 3	No 4
Saturated	56.8	58.9	13.6	13.0
16:0	34.1	38.3	6.9	6.0
18:0	4.0	2.5	1.3	1.9
Others	18.7	18.1	5.4	5.1
Monoenes	14.8	14.9	13.9	11.6
18:1(n-9)	11.0	11.4	6.4	5.8
Others	3.8	3.5	7.5	5.8
Dienes	5.9	5.6	10.4	14.2
18:2(n-6)	4.8	4.2	9.1	10.3
Others	1.1	1.4	1.3	3.9
Polyenes	22.5	19.2	62.1	61.2
18:3(n-6)	2.2	1.9	4.1	4.0
18:3(n-3)	17.5	16.6	33.3	30.7
16:4(n-3)	ND	ND	4.3	5.1
18:4(n-3)	ND	ND	4.2	4.7
20:4(n-3)	ND	ND	1.9	2.1
20:5(n-3)	ND	ND	4.3	3.7
Others	2.8	0.7	10.0	10.9

ND not detected. Lichen names: 1, *C. cristatum*; 2, *C. callopismum*; 3, *C. fuscovirens*; and 4, *C. flaccidum*

acids, though two species (No 1 and No 2) do not synthesize these acids. The total level of isomers of arachidonic acid [20:4(n-6) and 20:4(n-3)] in two species was considerably higher than that from all of the examined species, reaching over 5%. These amounts of arachidonic acid isomers are the highest known in the lichen literature with regard to the total lipid extract, although still higher levels have been found in the individual lipid classes, for example, *Peltigera aphthosa* (Dembitsky et al. 1992a, b, c, d, e).

Total lipid content has been studied in all collected lichen species, having different climatic peculiarities and substrates. Table 8.3 shows total lipid compositions in lichens collected during January (No 1) and in July (Nos 2, 3, and 4); the total lipid content in such lichens shows variations from 28.8 to 66.8 mg/g dry wt. Neutral lipids make up the highest percentage among the total lipids and vary from 15.8 to 36.6 mg/g dry wt.

Study of the lichen polar lipids, including betaine lipids and phospholipids, showed DGTA (diacylglyceryltrimethylalanine), DTGS (diacylglyceryltrimethylhomoserine), and PC (phosphatidylcholine) to be the major polar lipids with concentrations varying from 71.5% to 83.9% of total polar lipids (Table 8.3). The PC content in various fungal species is known to vary from 19% to 50% (Dembitsky 1996; Dembitsky and Reozentsvet 1993), whereas PC contents in red algal species vary up to 78%. PE was detected in all lichen species studied; its level was low in

Table 8.3 Lipid composition of the *Ramalina lacera* growing on different substrates

Lipid classes	No 1	No 2	No 3	No 4
Total lipids (mg/g dry wt)	28.8	29.4	66.8	64.4
Neutral lipids (mg/g dry wt)	17.5	15.8	36.6	30.2
Free fatty acids	TR	0.9	2.9	2.9
Free sterols	2.6	1.6	4.6	2.6
Diacylglycerols	1.1	1.9	3.1	2.1
Triacylglycerols	8.3	8.0	12.3	14.0
Steryl esters	5.5	2.2	6.2	5.4
Wax esters	TR	1.2	3.0	3.2
Glycolipids (mg/g dry wt)	6.9	8.0	21.1	26.6
MGDG	2.0	2.2	3.2	10.4
DGDG	4.3	5.3	7.2	12.9
SQDG	0.6	0.5	0.9	3.3
Polar lipids (% of total polar lipids	4.4	5.6	9.1	7.6
DGTA	27.1	28.3	2.8	1.9
DGTS	19.9	16.8	8.1	3.3
PC	27.8	29.4	62.7	66.3
PE	3.8	5.2	10.3	11.8
PI	14.8	11.2	16.7	26.7
PA	ND	1.9	ND	ND
X	6.6	7.2	ND	ND

ND Not detected, *TR* trace, *MGDG* monogalactosyl diacylglycerol, *DGDG* digalactosyl diacylglycerol, *SQDG* sulfoquinovosyl diacylglycerol, *DGTA* diacylglyceryltrimethylalanine, *DGTS* diacylglyceryltrimethylhomoserine, *PC*, phosphatidylcholine, *PE* phosphatidylethanolamine, *PI* phosphatidylinositol, *PA* phosphatidic acid, *X* non-identified polar lipid the first two species (ca 3–5%), but reached 11–12% in species No 3 and No 4. PI was also found in all species studied; its level was highest in No 4 (26.7%).

Biodiversity of secondary lichen metabolites by the epiphytic and lithophilic lichens of the genus *Collema* was reported. The most abundant fatty acids were a-linolenic acid (18:3n-3), oleic acid (18:1n-9), and palmitic acid (16:0), but a great variation of the ester composition from one to another was found.

A comparison of neutral lipids, glycolipids, polar lipids, and fatty acid composition of four species was done. DGTS, DGTA, PC, and PI were found as major components among polar lipids. Several indole alkaloids and other nitrogen-containing metabolites have been isolated from the genus *Collema*: *C. cristatum*, *C. callopismum*, *C. fuscovirens*, and *C. flaccidum* (Temina et al. 2010).

8.3 Chemical Constituents of the Lichenized Ascomycetes Belonging to the Genus *Ramalina*

8.3.1 Lipids in the Epiphytic Lichenized Ascomycete *Ramalina lacera* Grown on Difference Substrates

The identification of lichen metabolite production by the epiphytic lichenized ascomycete *Ramalina lacera* collected from different substrates: *Crataegus sinaicus*, *Pinus halepensis*, and *Quercus calliprinos*. The most abundant fatty acids were α-linolenic acid, oleic acid, and palmitic acid, but a considerable variability of the ester composition from one to another was found. A comparison of neutral lipids, glycolipids, polar lipids, and fatty acid composition of the tree-growing lichen *Ramalina lacera* was done. Diacylglyceryl-*N,N,N*-trimethylhomoserine (DGTS), diaclyglycerylhydroxymethyl-*N,N,N*-trimethyl-*β*-alanine (DGTA), phosphatidylcholine (PC), and phosphatidylinositol (PI) were found as major components among polar lipids. Diffractaic, lecanoric, norstictic, protocetraric, and usnic acids were isolated as major aromatic compounds in all samples of *R. lacera* (Hanuš et al. 2008b, c).

Ramalina lacera is a moderately xeric epiphytic fruticose lichen that grows in the Mediterranean areas on different shrubs and trees. These epiphytic species belonging to the family Ramalinaceae were chosen for a comparative examination of their fatty acid, polar lipids, and aromatic compounds. The GC-MS analysis of fatty acids in *R. lacera*, which grows on *Crataegus sinaica*, *Pinus halepensis*, and *Quercus calliprinos*, revealed a high polyenoic content in species which grows on the Palestine oak *Q. calliprinos* (54.28%, Table 8.4), but much lower amounts of such acids in two other ones, viz., 49.16% and 47.84%, respectively. The amounts of trienoic acids, however, were higher in the samples from tree-growing species, 42.75% (sample 1), 37.95% (sample 2), and 38.62% (sample 3). A total of 11 saturated fatty acids were identified, with n-16:0 and n-18:0 as major ones (8.47–12.94% and 3.29–5.13%, respectively). Total saturation in various species varied from 18.36 (sample 3) to 26% (sample 1). All three species studied had almost identical monoenoic and dienoic acid contents. Among other interesting acids, 16:4(n-3), the

Table 8.4 Fatty acid composition of the *Ramalina lacera* growing on different substrates

Fatty acids	*Crataegus sinaica* (sample 1)	*Pinus halepensis* (sample 2)	*Quercus calliprinos* (sample 3)
Saturated	26.00	29.38	18.36
16:0	12.94	11.88	8.47
18:0	5.13	4.96	3.29
Other acids	7.93	12.54	6.60
Monoenes	17.63	15.39	17.73
16:1(n-9)	5.14	2.15	4.96
18:1(n-9)	6.98	7.23	8.14
Other acids			
Dienes	7.21	7.39	9.62
18:2(−6)	5.94	6.16	8.57
Other acids			
Polyenes	49.16	47.84	54.28
16:3(n-3)	2.98	2.87	3.11
18:3(n-6)	35,87	30.19	28.70
16:4(n-3)	3.14	2.98	3.49
18:4(n-3)	1.85	1.70	2.98
20:4(n-6)	0.96	1.21	2.25
20:4(n-3)	1.34	2.11	3.66
20:5(n-3)	1.12	1.89	3.28
Other acids			

presence of which is characteristic for green marine algae27, and 18:4(n-3), characteristic for brown marine algae (Dembitsky et al. 1990, 1991, 1993; Go et al. 2002; Dembitsky and Rozentsvet 1989), were detected. It seems likely that the lichen photobiont synthesizes these acids. In sample 3, the total level of isomers of 20:4(n-6) and 20:4(n-3) was considerably higher than that from the rest of the examined species, reaching 5.91%. These amounts of arachidonic acid isomers are the highest known in the lichen literature with regard to the total lipid extract, although still higher levels have been found in individual lipid classes, for example, *Peltigera aphthosa*. Total lipid content was studied in all collected lichen species, having common climatic peculiarities. Table 8.5 shows total lipid compositions in lichens collected during July; the total lipid content in such lichens shows variations from 36.9 to 51.3 mg/g dry wt. Neutral lipids make up the highest percentage among of total lipids (Table 8.5) and vary from 22.1 to 32.1 mg/g dry wt. Examination of neutral lipids using HR-TLC revealed the domination of TAG and diacylglycerols over the rest of neutral lipids, thereby representing more than 50% in the majority of lichen species. Free fatty acids, free sterols, and its esters were also detected. The amount of glycolipid is comparatively lower than that of neutral lipids and varies between 8% and 11%. Examination of the lichen polar lipids, including betaine lipids (DGTA and DGTS) and phospholipids, showed PC to be the major phospholipids with concentrations varying from 34.9% to 44.2% of total polar lipids (Table 8.5). The PC content in various fungal species is known to vary from 20% to

Table 8.5 Lipid composition of the *Ramalina lacera* growing on different substrates

Lipid classes	*Crataegus sinaica* (sample 1)	*Pinus halepensis* (sample 2)	*Quercus calliprinos* (sample 3)
Total lipids (mg/g dry wt)	36.9	42.4	51.3
Neutral lipids (mg/g dry wt)	22.1	26.8	32.1
Free fatty acids[a]	2.6 ± 0.2	1.6 ± 0.1	2.9 ± 0.4
Free sterols	3.9 ± 0.4	6.6 ± 0.3	4.5 ± 0.2
Diacylglycerols	1.5 ± 0.1	2.3 ± 0.2	3.1 ± 0.3
Triacylglycerols	10.1 ± 0.6	8.5 ± 0.7	12.3 ± 0.9
Steryl esters	2.9 ± 0.2	4.2 ± 0.2	6.2 ± 0.5
Wax esters	1.6 ± 0.1	3.2 ± 0.3	3.0 ± 0.6
Glycolipids (mg/g dry wt)	9.0	8.0	11.3
MGDG	3.2 ± 0.6	2.6 ± 0.4	3.2 ± 0.3
DGDG	5.0 ± 0.8	4.7 ± 0.6	7.2 ± 0.5
SQDG	0.8 ± 0.2	0.7 ± 0.1	0.9 ± 0.1
Polar lipids (% of total polar lipids	5.8	7.6	7.9
DGTA	14.2 ± 0.8	12.6 ± 0.9	8.9 ± 0.6
DGTS	18.6 ± 0.9	21.3 ± 1.3	25.2 ± 1.9
PC	35.6 ± 1.4	44.2 ± 3.7	34.9 ± 2.8
PE	6.8 ± 0.4	7.4 ± 0.6	14.3 ± 0.9
PI	15.3 ± 0.7	11.2 ± 0.8	16.7 ± 0.7
PA	2.2 ± 0.2	0.9 ± 0.1	
X	7.3 ± 0.4	2.4 ± 0.2	

MGDG monogalactosyl diacylglycerol, *DGDG* digalactosyl diacylglycerol, *SQDG* sulfoquinovosyl diacylglycerol, *DGTA* diacylglyceryltrimethylalanine, *DGTS* diacylglyceryltrimethylhomoserine, *PC* phosphatidylcholine, *PE* phosphatidylethanolamine, *PI* phosphatidylinositol, *PA* phosphatidic acid, *X* non-identified polar lipid

[a]Mean ± standard deviation

55%, whereas PC contained in red algal species varies from 61.6% to 77.8%. Phosphatidylethanolamine (PE) was detected in all lichen species studied; its level was low in the first two species (ca 6.8%), but reached 14.3% in sample 3. Phosphatidylinositol was also found in all species studied; its level was highest in sample 3 (16.7%). Both betaine lipids were detected in all samples; their level varies from 8.9% to 14.2% for DGTA, and 18.6–25.2% for DGTS.

Thus, the analyses of fatty acids, lipids, and aromatic compounds from three samples of the *R. lacera* showed the presence of rare fatty acids: 16:4(n-3) and 18:4(n-3). Such acids were found in some marine algal species. As the photobiont component may form the main part of lichens, the presence of the above acids is to be expected for lichens. Hexadeca-4,7,10,13-tetraenoic acid 16:4(n-3), octadeca-6,9,12,15-tetraenoic acid (stearidonic acid), 18:4(n-3), and α-linolenic acid were isolated from marine green alga *Ulva fasciata* (family Ulvaceae). These polyunsaturated fatty acids (PUFAs) showed potent algicidal activity against microscopic highly toxic alga *Heterosigma akashiwo* (LC_{50} 1.35 μg/mL, 0.83 μg/mL, and 1.13 μg/mL for 16:4, 18:4, and α-linolenic acid, respectively), and the result

Table 8.6 Aromatic compounds of the *Ramalina lacera* growing on different substrates

Aromatic compounds (μg/100 g dry wt)	*Crataegus sinaica* (sample 1)	*Pinus halepensis* (sample 2)	*Quercus calliprinos* (sample 3)
Orsellinic acid	11.4	12.5	19.2
Lecanoric acid	22.5	26.2	21.7
Protocetraric acid	27.4	32.8	41.2
Diffractaic acid	28.7	33.7	38.9
Homosekikaic acid	4.9	5.9	6.2
Usnic acid	26.8	25.3	29.4
Norstictic acid	23.5	29.9	22.8
Total compounds	145.2	166.3	179.4

demonstrated the potential of these PUFAs for practical harmful algal bloom control. α-Linolenic acid [18:3(n-3)] and oleic acid [18:1(n-9)] were found as major unsaturated fatty acids. *α*-Linolenic acid showed anti-inflammatory activity. Among polar lipids, both betaine ether lipids (DGTA and DGTS) and PC and PI were found as major lipid components.

All samples of *R. lacera* produced the same aromatic acids but in different amounts. Some biological activities of isolated aromatic compounds from *R. lacera* have also been reported. Thus, we recently reported that orsellinic, lecanoric, diffractaic, protocetraric, usnic, and norstictic acids (Table 8.6; for structures, see Fig. 8.4) from *R. lacera* possess antibacterial and antifungal activities. Diffractaic acid exhibited antifungal activity against the phytopathogenic fungus *Cladosporium sphaerospermum* (Table 8.7). Potent antiproliferative agents, usnic and diffractaic acids, showed inhibitory activities against the human keratinocyte cell line HaCaT, with IC_{50} values of 2.1 and 2.6 μM, respectively. Diffractaic acid also showed strong inhibitory activity against tumor promoter-induced Epstein-Barr virus. Orsellinic acid revealed antibacterial activity against *Escherichia coli*, *Ralstonia solanacearum*, *Staphylococcus aureus*, and *Xanthomonas campestris vesicatoria*. Protocetraric acid showed activity against yeasts *Candida albicans* and *C. glabrata*, and norstictic acid was active against *Aeromonas hydrophila*, *Bacillus subtilis*, *Listeria monocytogenes*, *Proteus vulgaris*, *Staphylococcus aureus*, *Streptococcus faecalis*, *Candida albicans*, and *C. glabrata*. (+)-Usnic acid and (−)-usnic acid isolated from the lichen *Ramalina farinacea* showed cytotoxic and genotoxic activities against V-79 (Chinese hamster lung fibroblast-like) and A549 (human lung carcinoma epithelial-like) cell lines.

8.3.2 Antibacterial and Antifungal Activities of Some Phenolic Metabolites Isolated from the Lichenized Ascomycete *Ramalina lacera*

The antibacterial and antifungal activities and the MIC values of the six phenolic acids from ethanol-water extract of lichenized ascomycete *Ramalina lacera* have been investigated against some microorganisms (Hanus et al. 2006, 2007). Nearly

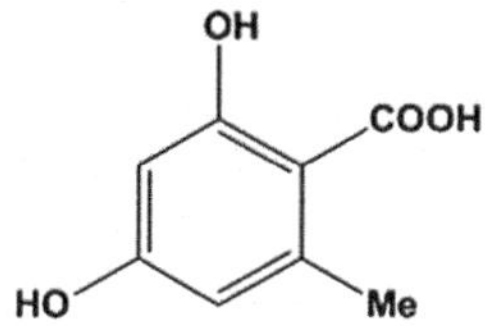

1 Orsellinic acid

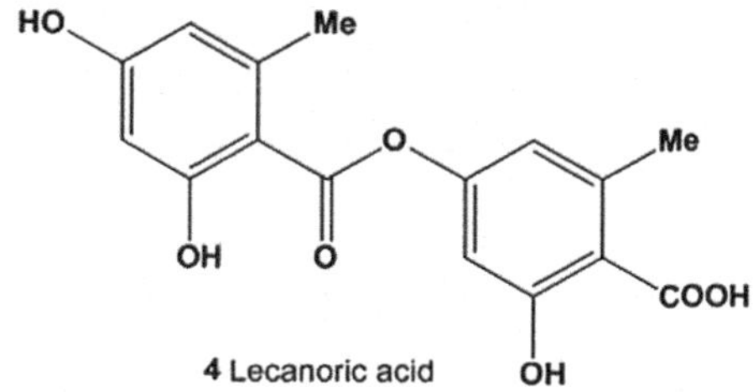

4 Lecanoric acid

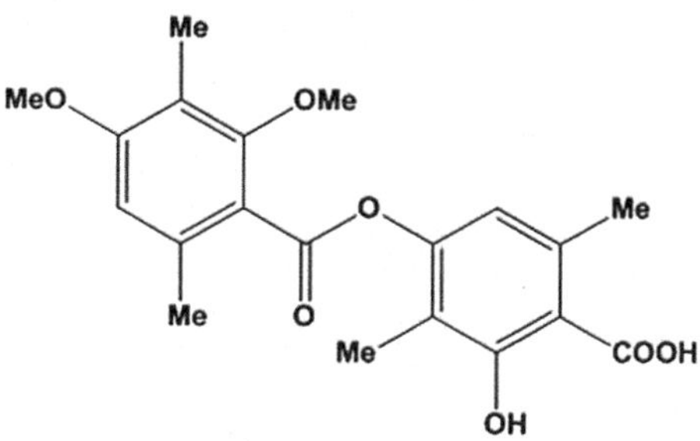

2 Diffractaic acid

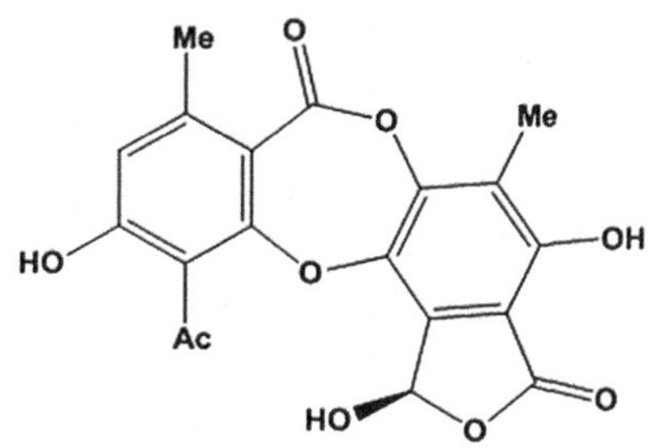

5 Norstictic acid

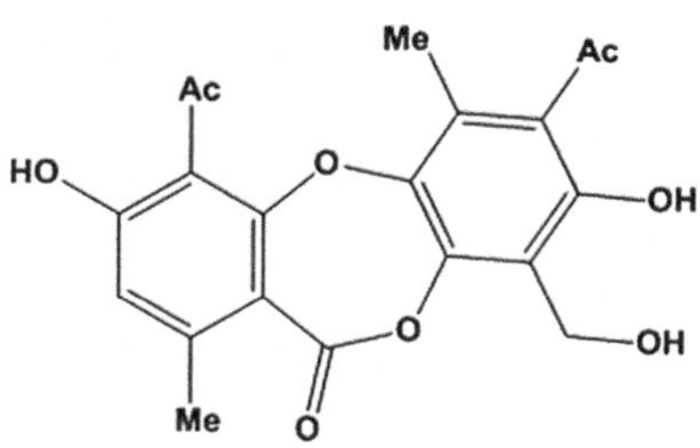

3 Protocetraric acid

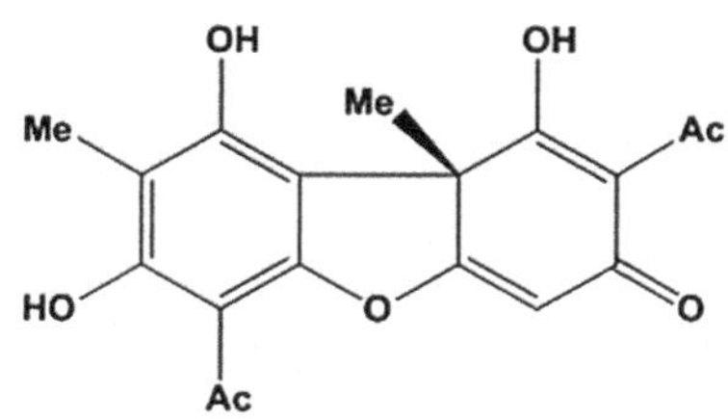

6 Usnic acid

Fig. 8.4 Orsellinic, lecanoric, diffractaic, protocetraric, usnic, and norstictic acids isolated from *R. lacera*

all tested compounds showed antibacterial and antifungal activities against *Bacillus cereus*, *B. megaterium*, *B. subtilis Escherichia coli*, *Pseudomonas aeruginosa*, *Salmonella choleraesuis*, *S. enterica typhi*, *S. enterica typhimurium*, *Staphylococcus aureus*, *Sarcina lutea*, *Alternaria brassicicola*, *Aspergillus flavus*, *A. nidulans*, *A. niger*, *A. ochraceus*, *A. sojae*, *Candida albicans*, *C. glabrata*, *C. krusei*, *C. tropicalis*, *Fusarium moniliforme*, *Trichoderma viride*, and *Trichophyton mentagrophytes*. All these substances were tested against 10 bacteria and 13 fungus species by a microdilution method. *Salmonella enterica typhimurium*, *Pseudomonas aeruginosa*, and *Aspergillus nidulans* were the most sensitive to diffractaic, protocetraric, and lecanoric acids with MIC values ranging from 3.8 to 5.2 μg/mL. These results showed that some lichen phenolic metabolites may lead to compounds with more pronounced activities (Table 8.7).

Table 8.7 Antibacterial and antifungal activities (MIC values are expressed in μg/mL, ST) of the phenolic compounds (1–6) isolated from *Ramalina lacera*

Bacteria	**(1)**	**(2)**	**(3)**	**(4)**	**(5)**	**(6)**	**TTC**
Bacillus cereus	11.3	18.4	12.3	10.3	16.4	6.9	20.6
Bacillus megaterium	9.6	10.9	8.6	7.7	10.1	7.7	19.4
Bacillus subtilis	10.4	12.3	5.6	6.9	9.4	6.3	15.2
Escherichia coli	9.2	15.6	10.3	3.3	5.1	4.9	2.4
Pseudomonas aeruginosa	10.1	11.9	8.7	9.4	12.3	5.2	3.5
Salmonella choleraesuis	21.4	19.7	11.4	10.1	8.6	6.3	8.9
Salmonella enterica typhi	19.2	10.8	5.2	4.9	3.6	8.2	3.2
Salmonella enterica typhimurium	18.6	12.3	6.1	3.8	5.2	5.4	6.9
Staphylococcus aureus	11.1	14.6	8.6	11.4	12.6	8.3	13.2
Sarcina lutea	16.9	23.4	20.8	16.3	19.4	19.1	20.3
Fungi	**(1)**	**(2)**	**(3)**	**(4)**	**(5)**	**(6)**	**CTM**
Alternaria brassicicola	11.4	14.6	8.6	11.4	12.6	8.3	13.2
Aspergillus flavus	9.8	8.3	14.1	12.6	13.8	20.1	18.1
Aspergillus nidulans	6.7	9.1	11.6	10.3	6.4	14.3	15.6
Aspergillus niger	12.6	7.4	6.9	3.9	8.6	5.8	19.3
Aspergillus ochraceus	14.5	5.2	7.4	18.2	13.8	11.3	20.4
Aspergillus sojae	7.8	5.9	9.3	11.9	12.4	6.4	16.2
Candida albicans	38.1	24.6	20.1	15.2	13.1	21.4	31.6
Candida glabrata	20.6	22.4	18.4	16.8	16.2	29.2	33.3
Candida krusei	19.9	18.8	13.4	18.1	15.3	30.1	34.5
Candida tropicalis	14.8	29.3	15.6	18.4	15.6	20.9	29.9
Fusarium moniliforme	15.1	16.3	12.6	14.8	16.1	18.6	18.8
Trichoderma viride	19.1	20.2	11.4	24.6	15.1	14.2	22.3
Trichophyton mentagrophytes	18.3	22.4	23.4	16.9	17.8	18.0	24.6

ST standard, *TTC* tetracycline, *CTM* clotrimazole. Tests were done in ten replicates
orsellinic (1), diffractaic (2), protocetraric (3), lecanoric (4), norstictic (5), and usnic (6) acids

8.4 Brominated Metabolites in Some Lichen Species

Halogenated fatty acids which are one of the most interesting groups among the naturally occurring halogen compounds are insufficiently sighted in the literature (Dembitsky and Srebnik 2002). References are introduced, which encompass different groups of organisms from microorganisms to the highest plants and lichen species (Dembitsky 2002; Dembitsky and Tolstikov 2000, 2001, 2002a, b, c, d, 2003a, b, c, d, e, f, g, h, 2004a, b).

The composition of novel brominated aliphatic compounds (Fig. 8.5 and Table 8.8) from lichens, *Acarospora gobiensis*, *Cladonia furcate*, *Lecanora fructulosa*, *Leptogium saturninum*, *Peltigera canina*, *Rhizoplaca peltata*, *Xanthoparmelia camtschadalis*, *X. tinctina*, and *Xanthoria elegans*, collected during summer from stones around the lake Issyk-Kul (Central Asia) was described (Rezanka and Dembitsky 1999a, b, 2001; Dembitsky and Srebnik 2002). The compounds, predominantly fatty acid derivatives with unique groups (bromine, cyclopropane,

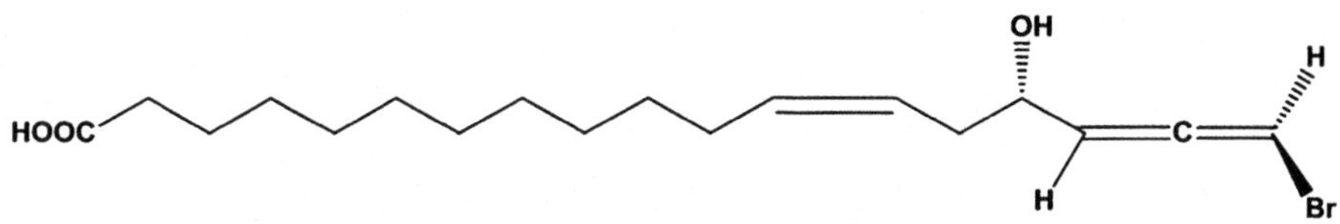

(15*S*,17*S*,*Z*)-18-bromo-15-hydroxyoctadeca-12,16,17-trienoic acid

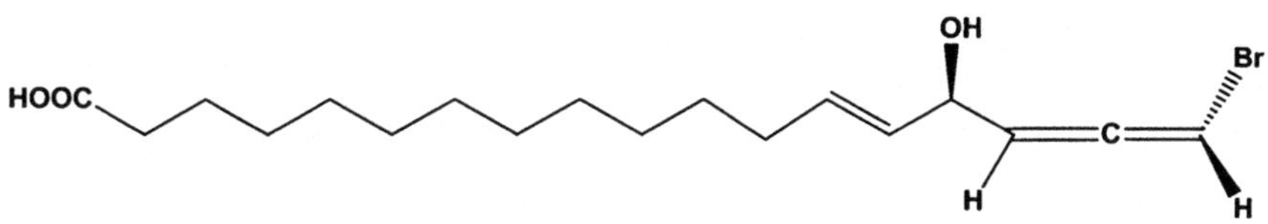

(15*R*,17*R*,*E*)-18-bromo-15-hydroxyoctadeca-13,16,17-trienoic acid

Fig. 8.5 Bromoallenic acids (*A* and *B*) isolated from some lichens

Table 8.8 Occurrence of bromoallenic acids from some lichen species

Name of lichens	Fatty acid, A	Fatty acid, B
Acarospora gobiensis	2.8	0.7
Cladonia furcate	0.1	1,9
Lecanora fructulosa	3.2	ND
Leptogium saturninum	2.5	ND
Peltigera canina	ND	1.8
Rhizoplaca peltata	ND	ND
Xanthoparmelia camtschadalis	ND	3.6
X. tinctina	ND	0.7
Xanthoria elegans	1.5	ND

mg/100 g dry wt

oxirane) and conjugated double and triple bonds, were identified by means of 1H and 13C-NMR, MS, IR, and UV spectra.

Unusual brominated fatty acids, (5*E*,17*E*)-methyl 18-bromooctadeca-5,17-dien-15-ynoate (**A**), methyl 18-bromooctadeca-5,7,17-triynoate (**B**), (15*E*,17*Z*)-methyl 16,18-dibromooctadeca-15,17-dien-5,7-diynoate (**C**), methyl 18,18-dibromooctade ca-17-en-5,7-diynoate (D), (5*E*,17*Z*)-methyl 18-bromooctadeca-5,17-dien-15-ynoate (**E**), (4*E*,14*Z*)-methyl 5-bromoheptadeca-4,14-dien-10,12,16-triynoate (**F**), (*R*,*E*)-methyl 11-((2*S*,3*S*)-3-((Z)-6-bromohex-3-en-5-yn-1-yl)oxiran-2-yl)-9-hydroxyundec-10-enoate (**G**), methyl 4-((1*S*,2*S*)-2-((3*E*,5*E*)-13-bromo-6,10-dimethyltrideca-3,5-dien-12-yn-2-yl)cyclopropyl)-butanoate (**H**), have been isolated from several lichen species (Fig. 8.6 and Table 8.9).

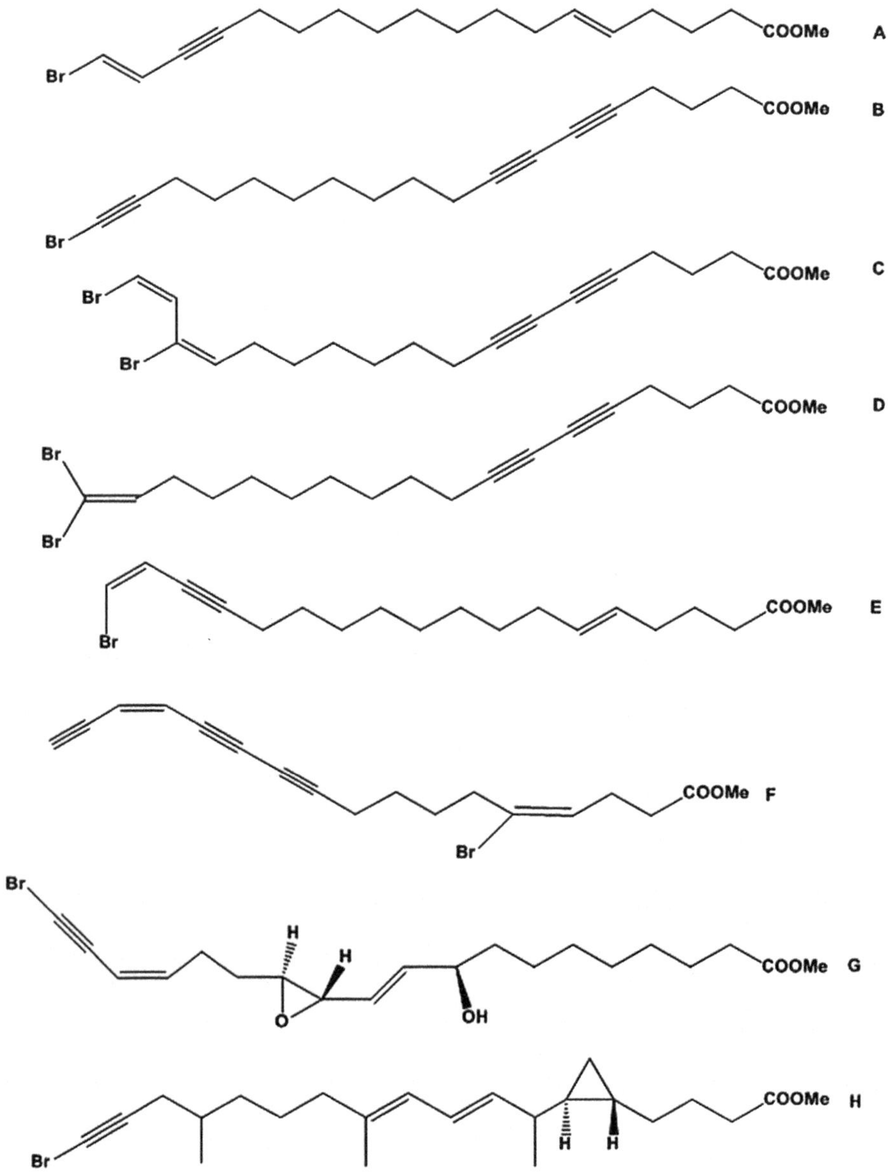

Fig. 8.6 Unusual brominated fatty acids were isolated from lichen species

Table 8.9 Occurrence of brominated fatty acids from some lichen species (mg/50 g dry wt)

Lichens	(A)	(B)	(C)	(D)	(E)	(F)	(G)
Acarospora gobiensis	13.1	8.8	0.1	ND	ND	ND	ND
Cladonia furcate	1.7	2.4	ND	3.3	0.4	ND	ND
Lecanora fructulosa	ND	ND	3.2	ND	ND	ND	ND
Leptogium saturninum	0.4	4.2	ND	1.9	ND	ND	5.1
Peltigera linctina	ND	ND	6.3	ND	2.8	ND	ND
Peltigera canina	0.9	ND	ND	0.3	ND	9.4	4.8
Xanthoparmelia camtschadalis	2.2	2.5	0.3	5.8	ND	3.0	ND
Xanthoria sp.	ND	0.8	2.4	ND	5.7	ND	ND

8.5 Bioactive Furanone Derivatives Have Been Isolated from the Lichenized Ascomycete *Tornabea scutellifera*

Tornabeatins A, B, C, and D (Fig. 8.7) have been isolated as new natural products from the lichenized ascomycete *Tornabea scutellifera* and their structures elucidated using UV, IR, MS, 1D, and 2D NMR spectral data and chemical degradation (Rezanka et al. 2004a, b).

Unusual tetrahydro-2-furanone and χ-lactonic acid derivatives are widely distributed among lichen species and were reviewed in part in recently published books (Huneck and Yoshimura 1996; Huneck 2001) and some papers (Dembitsky 1992; Dembitsky and Tolstikov 2005). 3-Methylidene-tetrahydro-2-furanones (α-methylene-χ-butyrolactones) constitute an important group of natural products and possess wide-ranging biological activities. The 3-methylenetetrahydro-2-furanone ring is an internal building block of the sesquiterpene lactones, such as vernolepin, aromaticin, and others, which were isolated from some lichen and plant species. All these compounds have cytotoxic, antitumor, and often bactericidal properties. The cytotoxic activity of 3-methylidene-tetrahydro-2-furanones is mainly associated with the exocyclic, conjugated double bond, which acts as an alkylating agent in a Michael-type reaction with thiol-rich bionucleophiles. In addition, the 3-methylene-tetrahydro-2-furanone moiety is a characteristic component of a large number of biologically active natural products, especially the sesquiterpene lactones and even the parent 3-methylene-tetrahydro-2-furanone have significant pharmacological activities. Tetrahydro-2-furanone diacylglycerol analogues are known to have activated protein kinase C, stimulated phosphorylation of the R-pseudosubstrate peptide, and inhibited binding of epidermal growth factor with an ED_{50} in primary mouse keratinocytes. The fruticose epiphytic lichen *Tornabea scutellifera* is widely distributed in the forest of north Israel and Cyprus (Temina et al. 2010) and has a large biomass. Up to now, its secondary metabolites have not been studied.

Goniothalamusin and acetogenins A and B isolated from *M. elutina* exhibited significant antibacterial and cytotoxic activities against Gram-positive (*Bacillus cereus*, *Staphylococcus aureus*, *Streptomyces haemolyticus*) and Gram-negative (*Salmonella typhi*, *Shigella flexneri*, *Shigella dysenteriae*) bacterial species (Dembitsky and Tolstikov 2005). We discovered new cyclopentenones, namely,

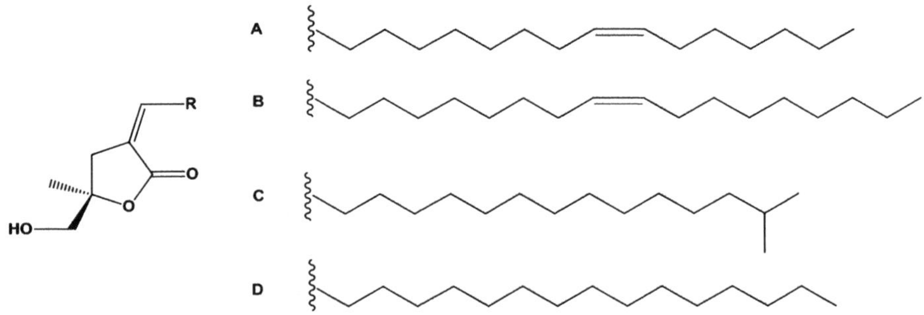

Fig. 8.7 Tornabeatins *A*, *B*, *C*, and *D* were isolated from the lichenized ascomycete *Tornabea scutellifera*

Table 8.10 Bioactivities of cyclopentenones (μg/mL)

Test organism	1	2	3	4
Staphylococcus aureus	28	35	21	24
Bacillus subtilis	32	24	27	20
Escherichia coli	0	0	0	0
S. cerevisiae	14	17	5	6
A. salina	2	2	4	5
P. lividus	2	3	2	2
A. tumefaciens	35	71	49	53

A-D tornabions, which show moderate activity against various microorganisms (see Table 8.10). They are active against *S. aureus* and *Bacillus subtilis* but inactive against the Gram-negative bacterium *Escherichia coli* (Rezanka et al. 2004).

8.6 Fatty Acids of the Lichenized Ascomycetes Belonging to the Genus *Cladonia*

Fatty acids of nine lichen species belonging to the genus *Cladonia* were examined by GC-MS. Twenty-two saturated, twenty-three monoene, five diene, nine triene, and eight tetra-, penta-, and hexane fatty acids were identified. Unusually for lichens, very long-chain fatty acids, 25:1, 26:0, 26:1, 26:3, 28:0, 28:1, and 30:1, were detected (Dembitsky et al. 1991; Rezanka and Dembitsky 1999a, b). Major fatty acids are shown in Table 8.11, and compositions of phospho- and polar lipids are shown in Table 8.12. Polar and phospholipids were detected using TLC chromatography method (Dembitsky 1988; Hanuš et al. 2008).

Table 8.11 Major fatty acids isolated from the *Cladonia* species

Fatty acids	1	2	3	4	5	6	7	8	9
16:0	9.3	8.5	6.9	6.9	4.0	2.0	7.2	1.3	6.8
18:0	5.1	5.2	3.3	4.8	7.1	0.3	0.8	0.3	5.5
Total saturated	25.4	18.7.	17.6	17.3	17.6	3.8	16.2	3.2	18.8
16:1(n-9)	1.8	7.0	5.2	8.6	4.4	1.3	0.7	0.5	5.0
16:1(n-7)	2.7	5.2	0.9	3.2	7.0	0.5	0.1	1.4	1.0
18:1(n-11)	2.5	1.0	0.3	1.3	3.1	0.2	1.4	0.0	0.7
18:1(n-9)	6.8	7.9	16.4	9.1	2.4	15.5	0.8	1.4	11.4
18:1(n-7)	1.7	1.5	1.1	2.2	3.6	1.2	15.2	0.8	4.1
Total monoenes	24.8	29.3	30.0	28.5	26.1	19.8	24.4	5.5	28.1
18:2(n-6)	8.1	5.8	7.3	16.2	10.4	6.2	6.6	8.0	10.2
Total dienes	10.1	7.2	9.1	18.5	13.0	7.0	7.4	9.0	15.3
18:3(n-6)	3.1	1.5	0.6	1.0	1.4	20.3	14.4	0.5	1.1
18:3(n-3)	13.4	17.7	17.7	13.5	19.0	1.3	0.8	18.2	9.5
Total trienes	24.0	26.4	29.3	23.6	28.3	24.2	18.6	18.6	15.3
18:4(n-3)	1.7	1.7	2.9	7.0	3.2	4.9	17.9	5.4	1.4
20:4(n-6)	1.8	0.9	1.4	1.0	1.4	4.4	2.3	9.8	2.5
20:4(n-3)	2.7	0.9	2.0	0.4	0.8	7.1	0.9	1.1	1.6
20:5(n-3)	3.0	3.5	2.3	1.5	2.0	8.9	4.4	8.3	6.1
Total polyenes	15.7	18.4	14.0	12.1	15.0	45.2	33.4	54.9	24.6

Studied lichen species: *1 C. deformis*, *2 C. cervicornis* subsp. *verticillata*, *3 C. cornuta*, *4 C. pyxidata*, *5 C. phyllophora*, *6 C. subulata*, *7 C. fimbriata*, *8 C. anomea*, *9 C. furcata*

Table 8.12 Phospholipid composition of some lichen species from the Volga river basin

Lichen species	PI	PS	PC	PE	PG	X	TL
Cladonia arbuscula			77.1	4.9	18.0		27.5
C. rangiferina	20.6		61.8	10.0	7.6		22.7
Cladonia sp. 1			85.5	9.7	4.8		33.6
Cladonia sp. 2			61.4	16.8	20.0	1.8	28.4
Cladonia sp. 3	20.6		44.1	22.1	13.2		30.2
Cladonia sp. 3		9.9	56.2	23.1	10.8		22.1
Cladonia sp. 4			73.9	12.3	13.8		24.8
Cladonia sp. 5			72.2	11.1	16.7		34.2
Cladonia sp. 6			80.8	9.2	10.0		20.9
Cladonia sp. 7	8.6		68.0	15.5	6.9		18.8
Cladonia sp. 8			70.2	20.0	9.8		22.4
Parmelia scortea		10.3	89.7				28.0
P. sulcata	21.4		57.1	14.3	7.2		26.3
Peltigera sp. 1	18.4		51.7	15.9	10.0	4.0	30.2
Peltigera sp. 2	8.0	1.8	56.6	13.3	20.3		23.5
Placolecanora sp.		11.8	60.5	13.3	14.4		27.9
Pseudevernia furfuracea	18.4	7.3	52.3	4.0	5.2	12.8	25.3

TL total lipids. Mg/g^{-1} dry wt

Phospholipids: *PC* phosphatidylcholine, *PE* phosphatidylethanolamine, *PI* phosphatidylinositol, *PS* phosphatidylserine, *PG* phosphatidylglycerol, *X* non-identified polar lipid

8.7 Composition of Surface N-Alkanes and Fatty Acids of the Epiphytic Lichen *Xanthoria parietina*, Its Photobiont a Green Alga *Trebouxia* sp., and Its Mycobiont

Lichens have both algal and fungal properties and produce n-alkane, unusual betaine ether glycerolipids, and saturated, unsaturated, branched, and halogenated fatty acids (Dembitsky and Tolstikov 2005). One of the main questions in lichen biology and chemistry is which compounds are synthesized by which symbiotic lichen partner? As fungi are often rich in secondary products (Dembitsky 2014; Dembitsky et al. 1993a, b, c, d; Hanuš et al. 2008; Dembitsky et al. 2010), it is not surprising that many typical lichen substances are synthesized by the mycobiont. It is known that experimentally produced combinations of a mycobiont with foreign photobionts generate the same lichen products as the mycobiont in a natural thallus with its usual partner.

Surface alkanes and fatty acids from the thalli of the lichen *Xanthoria parietina*, its photobiont *Trebouxia* sp., and its mycobiont were analyzed by GC-MS. The green alga *Trebouxia* sp. synthesized mainly unsaturated fatty acids such as (*Z*,*Z*,*Z*)-9,12,15–18:3 (*Z*,*Z*)-9,12-18:2 and (*Z*)-9-18:1 (Fig. 8.8) and light alkanes C8–C15 (up to 83% of total n-alkanes) (Fig. 8.8, Torres et al. 2003).

However, the mycobiont contained mainly saturated fatty acids such as hexadecanoic (16:0) and octadecanoic acid (18:0) and also very long-chain n-alkanes C22–C34. Dehydroabietic acid was found in both lichen and mycobiont. The occurrence of different amounts of n-alkanes and fatty acids in the photobionts and mycobionts of *X. parietina* was shown for the first time. Lichens collected from different locations in the Jerusalem hills contained n-alkanes ranging in concentration from 187 to 211 mg/g dry wt; n-alkane concentrations in the photobiont and mycobiont were 17–24 and 215–262 mg/g dry wt, respectively (Torres et al. 2003).

Obtained results have shown that *Trebouxia* sp. predominantly produces the following unsaturated fatty acids: 9,12,15–18:3, 9,12–18:2, and 9–18:1, as well as light n-alkanes C8–C15. The major saturated fatty acids obtained from the mycobiont are tetradecanoic (14:0), hexadecanoic (16:0), and octadecanoic acid (18:0) and very long-chain n-alkanes C22–C34.

8.8 Seasonal Variability of Lipids and Fatty Acids in Some Tree-Growing Lichens

8.8.1 Seasonal Examination of Lipids of the Tree-Growing Lichen of Genus *Physcia*

Lipids, including phospholipids and fatty acids from two lichen species, *Physcia dubia* and *P. stellaris*, collected between March and May were examined. The seasonal changes occurring in lipid composition due to temperature were identified. Each species had a characteristic temperature-based lipid pattern. *Physcia stellaris*

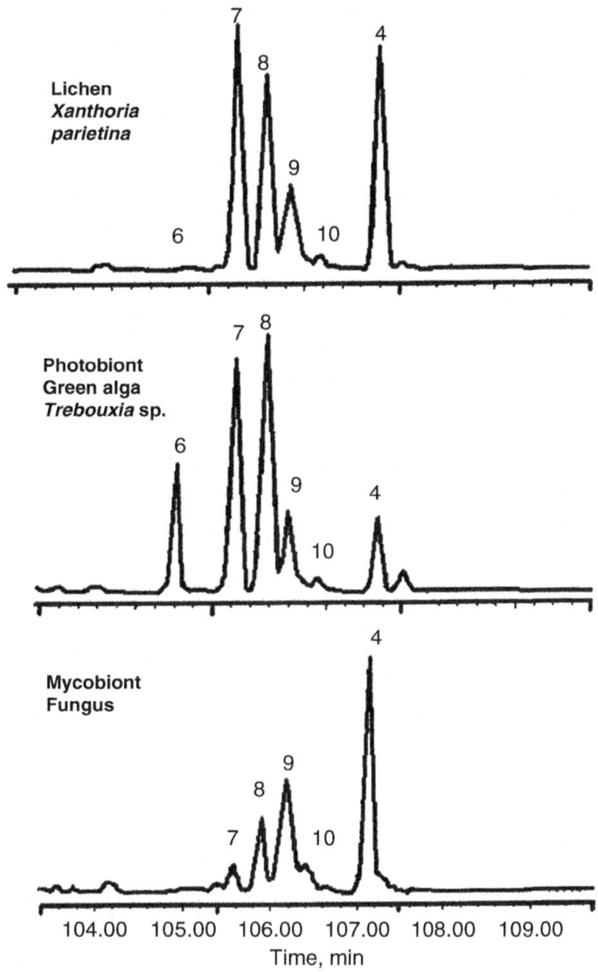

Fig. 8.8 The part of gas chromatogram (full run is 162 min) of C18 family methyl ester of fatty acids of lichen *Xanthoria parietina*, photobiont *Trebouxia* sp., and mycobiont. Peaks identified by GC-MS: 6, (Z,Z,Z)-9,12,15–18:3; 7, (Z,Z)-9,12–18:2; 8, (Z)-9–18:1; 9, 10–18:1; 10, 12–18:1; 4, 18:0

was found to increase its total lipid content from 24.8 in March to 37.5 mg/g dry wt in May, whereas the lipid content decreased from 63.4 to 51.8 mg/g dry wt in *P. dubia* from March to May. Both species possessed high neutral lipid contents in March, 70.1% and 56%, respectively, and showed an increase in their glycolipid contents during the same period. Polar and phospholipid contents were also determined, and certain patterns for phosphatidylcholine and diacylglyceryltrimethylhomoserine proportions were observed. Capillary GC-MS demonstrated that seasonal changes also occurred in the fatty acid composition of the two species. Hydroxy acid dynamics were also studied (Dembitsky et al. 1994).

The total lipid contents in each species changed differently. *Physcia stellaris* was found to increase its total lipids from 24.8 to 37.5 mg/g dry wt (Table 8.13). Both *Physcia* species decreased their neutral lipid amounts but increased those of glycolipids. The amount of phospholipids in *P. stellaris* increased from 7% to 17.4%,

Table 8.13 Composition of total, neutral, glyco-, and phospholipids of two lichen species collected during March to May 1992

Lichen *Physcia stellaris*									
Collection date	Weather conditions during sampling	TL	NL	GL	PL	TL	NL	GL	PL
8 March	Freezing, snow (−7 °C)	63.4	56.0	27.5	16.5	24.8	70.1	22.9	7.0
23 April	Cool, wet (+7 °C)	54.9	57.6	25.8	16.6	28.8	56.4	28.7	14.9
28 May	Warm, sunny (+23 °C)	51.8	53.8	34.8	11.4	37.5	28.2	54.4	17.4

TL total lipids, mg/g dry wt, *NL* neutral lipids, *GL* glycolipids, *PL* polar lipids as % of TL

Table 8.14 Composition of *DGTS* and phospholipids in *P. dubia*

Lichen *Physcia dubia*									
Collection date	Weather conditions during sampling	DGTS	PC	PG	PI	PS	PE	LPE	X
8 March	Freezing, snow (−7 °C)	17.8	29.3	9.9	7.7	7.2	20.0	7.1	0.1
23 April	Cool, wet (+7 °C)	19.7	22.3	11.5	8.1	8.7	20.4	8.9	0.4
28 May	Warm, sunny (+23 °C)	23.0	20.4	9.2	5.6	8.1	20.6	13.1	–

DGTS diacylglyceryltrimethylhomoserine, *PC* phosphatidylcholine, *PE* phosphatidylethanolamine, *PI* phosphatidylinositol, *PA* phosphatidic acid, *X* non-identified polar lipid

Table 8.15 Composition of DGTS and phospholipids in *P. stellaris*

Lichen *Physcia stellaris*									
Collection date	Weather conditions during sampling	DGTS	PC	PG	PI	PS	PE	LPE	X
8 March	Freezing, snow (−7 °C)	13.5	23.3	14.5	9.1	11.6	19.3	8.5	0.2
23 April	Cool, wet (+7 °C)	19.3	24.3	9.9	5.5	11.0	201	6.9	3.0
28 May	Warm, sunny (+23 °C)	21.7	16.6	11.1	5.6	7.9	23.9	13.2	–

Values are means + s.d. ($n = 3$–5). For abbreviations, see Table 8.2.

while in *P. dubia*, it remained unchanged for March and April with a decrease in May due to sunlight. The majority of phospholipid classes underwent quantitative changes; the phosphatidylcholine (PC) content in *P. dubia* and *P. stellaris* decreased by 9.1% and 6.7% during the period from March to May, respectively.

Similar changes were observed for phosphatidylglycerol (PG) and phosphatidylinositol (PI) (Tables 8.14 and 8.15). The greatest quantitative changes occurred in PC and diacylglyceryltrimethylhomoserine (DGTS). It is noteworthy that the March sample of *P. dubia* contained the greatest amount of DGTS among the polar lipids, including phospholipids, except PC and phosphatidylethanolamine (PE), namely, 17.8%, which increased toward May to make up 9.1%. At the same time,

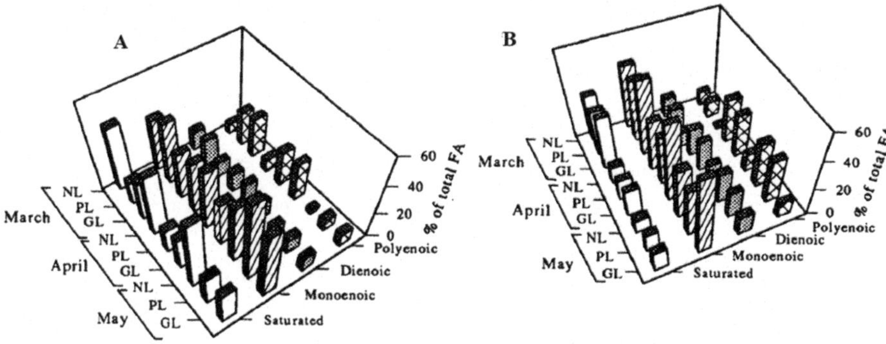

Fig. 8.9 Seasonal distribution of major fatty acids in different lipid fractions in *Physcia dubia* (**a**) and *P. stellaris* (**b**) collected from March to May

the PC content was observed to decrease from 29.3% to 20.4% during the same period. These results suggest that there is a correlation between the levels of PC and DGTS in lichens. The amounts of saturated fatty acids in the neutral lipid fraction increased from 31.8% to 37% in *P. stellaris* and from 44.5% to 49.9% in *P. dubia* (Table 8.14).

Monoenoic acids were found to increase in a similar way, from 44.9% to 52.4% in *P. stellaris* and from 36.1% to 40.6% in *P. dubia*. Dienoic and polyenoic acids were observed to decrease.

The major fatty acids of the neutral lipids were saturated 14:0, 16:0, 18:0, monoenoic 16:1 (n-9) and 16:1(n-7) and 18:1(n-9) and 18:1(n-7), dienoic 18:2 (n-6), and polyenoic 18:3 (n-6) and 18:3 (n-3). Besides the major fatty acids, we also found iso-, anteiso-, prystanic, phytanic, and small amounts of long-chain acids (24:0–28:0) (Fig. 8.9).

8.8.2 Seasonal Variability of Lipids of Lipids and Fatty Acids in the Tree-Growing Lichen *Xanthoria parietina*

The composition of lipids, PLs, and FAs in the tree-growing lichen *Xanthoria parietina*, collected during the period from March to May, was studied. The major polar lipids found, including phospholipids, were DGTS, PC, PE, and PG. Polar and PL contents were also identified, and certain trends in the changing proportions of PC (increasing from 17.8% to 50.1%) and DGTS (decreasing from 27.1% to 12.6%) were determined. The fatty acid composition was examined using capillary GC-MS in the NL, GL and PL fractions. Hydroxy acids were detected only in the glycolipid fraction; their seasonal dynamics were also studied. The seasonal changes occurring in lipid composition due to the temperature factor were identified. It was found that *X. parietina* had a characteristic temperature-based lipid pattern, increasing in its neutral lipid content from 40.6% in March to 52.7% in May but decreasing in glycolipid from 39.0% in March to 27.0% in May (Dembitsky et al. 1994) (Tables 8.16 and 8.17).

Table 8.16 Composition of total, neutral, glyco-, and phospholipids of lichen *Xanthoria parietina* collected during March to May 1992

Collection date	Weather conditions during sampling	TL	NL	GL	PL
8 March	Freezing, snow (−7 °C)	41.2	40.6	39.0	20.4
23 April	Cool, wet (+7 °C)	42.7	57.8	26.9	15.3
28 May	Warm, sunny (+23 °C)	39.5	52.7	27.0	20.3

TL total lipids, mg/g dry wt, *NL* neutral lipids, *GL* glycolipids, *PL* polar lipids as % of TL

Table 8.17 Composition of DGTS and phospholipids in *Xanthoria parietina*

Lichen *Xanthoria parietina*									
Collection date	Weather conditions during sampling	DGTS	PC	PG	PI	PS	PE	LPE	X
8 March	Freezing, snow (−7 °C)	26.1	17.8	12.7	11.2	9.9	13.9	7.4	1.0
23 April	Cool, wet (+7 °C)	23.3	32.0	9.0	6.9	6.1	15.4	6.6	0.7
28 May	Warm, sunny (+23 °C)	12.6	50.1	7.1	6.0	4.1	14.2	5.9	–

DGTS diacylglyceryltrimethylhomoserine, *PC* phosphatidylcholine, *PE* phosphatidylethanolamine, *PI* phosphatidylinositol, *PA* phosphatidic acid, *X* non-identified polar lipid
Values are means ± s.d. ($n = 3$–5)

8.9 Prenylated Xanthone Glucosides from Ural's Lichen *Umbilicaria proboscidea*

Two new compounds, umbilicaxanthoside A and B (Fig. 8.10), were isolated from an extract of Central Asian lichen *Umbilicaria proboscidea*. The isolated glucosides had mono- and di-prenylated xanthones as the aglycones and a saccharide moiety from the two glucoses linked at C-7 (Rezanka and Dembitsky 2003; Rezanka et al. 2003). According to the literature data, only a few phytochemical research studies were carried out on lichens. As a constitution of our search for compounds from lichens, we examined constituents of the lichen *Umbilicaria proboscidea*, collected in Ural Mountains, Muslimovo Village, 50 km west from Chelyabinsk (Russia). Many lichen species have been studied, and xanthones and other colored compounds have been identified as their constituents. In contrast, only the depsides and tridepsides from the genus of *Umbilicaria* (family Umbilicariaceae) were described in the literature (Huneck 2001; Dembitsky and Tolstikov 2005).

8.10 Polyisoprenoid Alcohols from Lichens

Lower plants, i.e., mosses, liverworts, algae, and lichens, were found to contain several classes of polyisoprenoid alcohols (Fig. 8.11) in the range C40–C130 (Rezanka and Dembitsky 1993). These were isolated by means of reverse phase HPLC and identified by EI-mass spectrometry, field desorption-mass spectrometry, IR spectroscopy, and 1H and 13C NMR spectroscopy. Mosses contain predominantly

Fig. 8.10 Two umbilicaxanthosides *A* and *B* were isolated from lichen *Umbilicaria proboscidea*

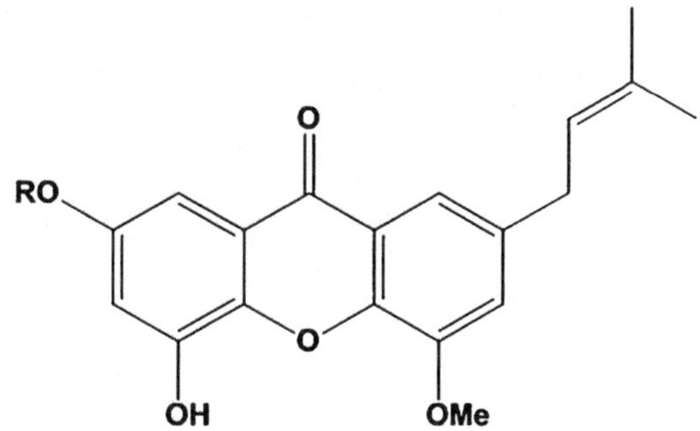

Umbilicaxanthoside A

R = β-D-Glcp
R = H

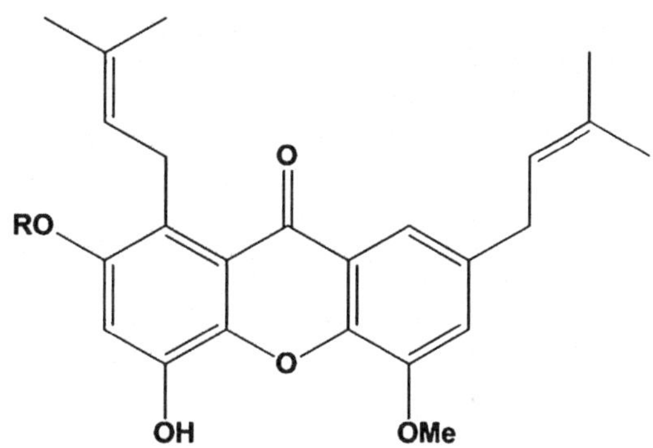

Umbilicaxanthoside B

R = β-D-Glcp-(1->)-β-D-Glcp
R = H

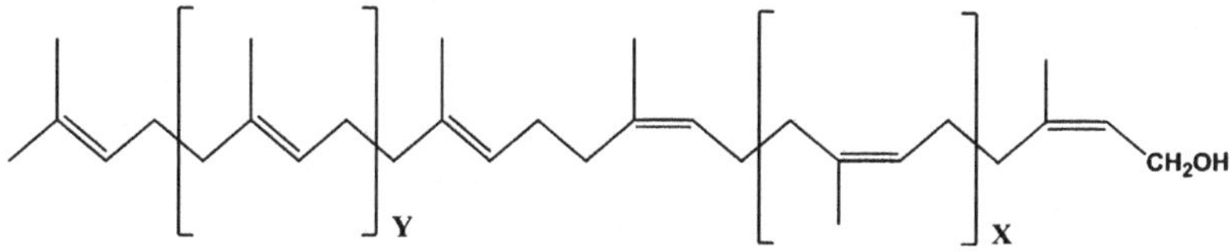

Fig. 8.11 The structure of prenols ($X = 7$, $Y = 1$ betulaprenol C_{60}; $X = 8$, $Y = 2$, ficaprenol C_{70}) which isolated from lichens

betulaprenols (polyprenols with two *trans*- and the rest *cis*-isoprene residues) with the average chain length of C60. The liverwort examined had two maxima of chain lengths, the first at C70 and the second at C115. It contained predominantly ficaprenols (three *trans*- and the rest *cis*-isoprenes) with minor amounts (up to 5%) of some betulaprenols. Algae have a similar composition to mosses; only the average chain length is moved to higher values (approximately C70). Lichens were different in average chain length and in composition of long-chain alcohols. They contain ficaprenols with two maxima, the first at C80 and the second at C105–110, and dolichols (two *trans*-, many *cis*-, and one saturated isoprene residue; Table 8.18).

8.11 Chlorinated Metabolites in Lichenized Ascomycetes

Lichens generate a number of biologically active molecules, among which a noticeable role is played by chlorinated phenol compounds. Lichens are sensitive to chlorine ions, as their absorption systems are active, which leads to the accumulation of chlorine when exposed to high levels of chlorine contamination. Chlorine is absorbed by mycobionts and metabolizes them into the embedding of chlorine in various lichen metabolites and especially in depsides and depsidones (Dembitsky and Tolstikov 2003a, b; Hunek and Yoshimura 1996; Huneck 2001).

8.11.1 Lichen Depsides and Depsidones

Depsides are esters formed by the derivatives of *p*-oxybenzoic acid and phenols (see structures, Figs. 8.12, 8.13, and 8.14). They comprise a small group of polyketides and are generated mainly by lichens. For example, the most widespread chlorine-containing depside chloroatranorin (**1**) was discovered in more than 40 lichen species: *Evernia prunastri, Pseudevernia furfuracea, P. intens, Parmelia perlata, P. pseudoreticulata, P. olivetorum, P. cryptochlorophaea, P. tinctoria, P. pseudofatiscens, P. horrescens, P. damaziana, P. furfuracea, Buellia canescens, Ramalina siliquosa, R. druidarum, Lecidea carpathica, Lecidea* sp., *Phiscia picta, Cetralia cetrarioides, C. japonica, Usnea canariensis, Hypogymnia physodes, H. billardieri, H. enteromorpha, H. lugubris, H. subphysodes, Anaptychia neoleucomelaena, Menegazzia asahinae, M. terebrata, M. dispora, Platismatia glauca, Parmotrema demethylmicrophyllinicum, P. praesorediosum, Lacanora braccha, L. epibryon, L. rupicola*, and *L. gangaleoides* (Dembitsky and Tolstikov 2003a, 2005; Huneck 2001).

Tumidulin (**2**), a less widely spread lichen metabolite, was discovered only in *Ramalina* genus: *R. ceruchis, R. flaccescens, R. tumidula, R. inanis, R. cactacearuum, R. peruviana*, and *R. chilensis*. The *Erioderma wrightii* lichen generates wrightin (**3**), (**4–10**) metabolites were detected in *Erioderma* sp., and depside (**11**) was isolated from extracts of *Pseudocyphellaria pickeringii*. Metabolite (**12**) was isolated from the extracts of *Lecanora sulphurella* lichen; its structure was confirmed by synthesis (Dembitsky and Tolstikov 2003a, b). 3-Chlorodivaricatinic (**13**), 3-chlorostenosporic (**14**), and 3-chloroperlatolic (**15**) acids were discovered in *Thelomma mammosum* and *Dimelaena* sp. lichens (Huneck 2001).

Table 8.18 Composition of alcohols from lichens (content of prenologues)

	10	11	12	13	14	15	16	17	18	19	20	21	22	23	24	25	26
M	698	766	834	902	970	1038	1106	1174	1242	1310	1378	1446	1514	1582	1650	1718	1786
A		0.5	2.5	5.4	9.4	12.8	22.2	13.3	5.4	2.5	3.9	8.9	5.9	3.8	2.0	1.0	0.5
B	0.4	1.3	3.9	9.9	13.8	17.7	22.8	7.3	2.6	0.9	1.3	3.9	6.9	3.4	2.2	1.3	0.4
C	0.8	1.6	3.3	3.7	6.1	10.2	19.2	13.5	8.6	3.3	3.7	9.5	7.8	4.7	2.4	1.2	0.4
D			0.6	1.7	5.0	12.6	19.5	10.7	2.8	1.7	4.4	8.0	12.6	10.6	5.8	3.5	0.5
E	0.4	1.7	3.4	8.0	11.0	12.2	17.3	10.5	6.8	3.4	1.3	2.5	8.0	6.3	5.1	1.7	0.4

M, FD-MS. Lichen names: A, *Anaptychia ciliaris*; B, *Aspicilia transbaicalica*; C, *Evernia mesomorpha*; D, *Hypogymnia physodes*; E, *Cladonia* sp.

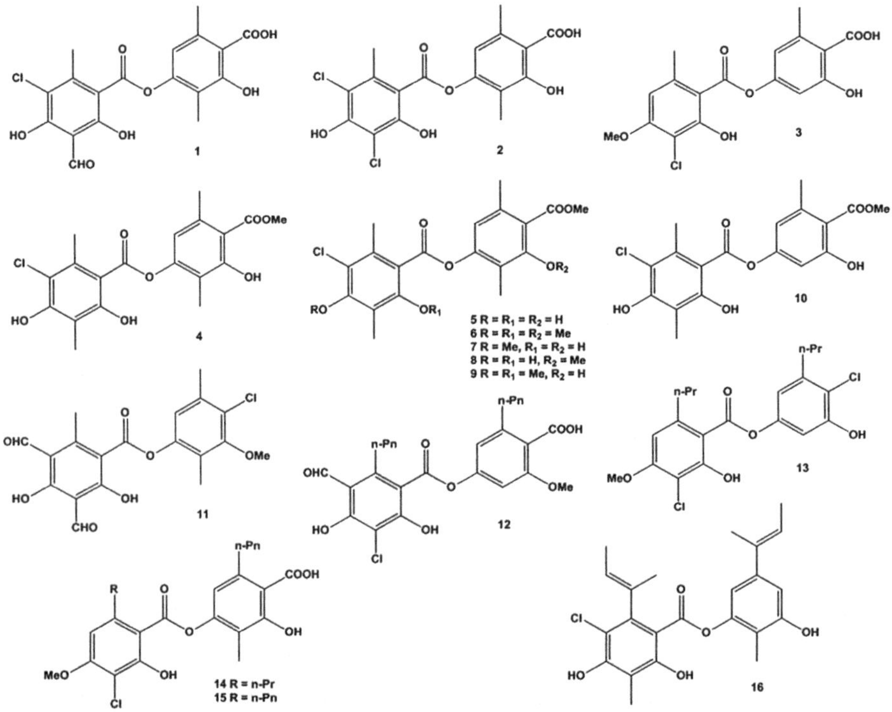

Fig. 8.12 Chlorinated depsides were discovered from lichens

Depsidones are the derivatives of diphenyloxide containing the heterocyclic system of 1,4-dioxacycloheptanone-7 as a result of the formation of intramolecular ester bond. The first representative of depsidones, namely, diploicin (**17**), was isolated in 1904 by Zopf from a lichen of undetermined species (Hunek and Yoshimura 1996). Subsequent research showed that (**17**) is present in *Buellia canescens* and Lecidea cargaleoides lichens (Dembitsky and Tolstikov 2003a).

Though gangaleoidin (**18**) was discovered more than 65 years ago in *Lecanora gangaleoides* lichen, its structure was confirmed only after 30 years (Hunek and Yoshimura 1996). Pannarin (**19**) was initially isolated from the lichens of *Pannaria* genus: *P. lanuginosa*, *P. fulvescens*, and *P. lurida*, *P. pityrea*, *P. rubiginosa*, and later from *Lecanora hercynica*, *Bombyliospora japonica*, and *Eriodeerma chilense* (Huneck 2001).

The structure of pannarin (**19**) was additionally investigated by other authors, while nidulin (**20**) was also discovered in *Emericella unguis* lichen [Dembitsky and Tolstikov 2003b]. Vicanicin (**23**), which was initially discovered in *Teloschistes flavicans*, was later isolated from other species: *Caloplaca* sp., *Psoroma sphinctrinum*, *P. allorhizum*, and *Erioderma chilense*. Norvicanicin (**24**) and isovicanicin (**25**) are metabolites in *Psoroma athrophyllum*, while *O*-methylvicanicin (**26**) is a metabolite of *Erioderma* sp. The *Caloplaca* sp. generates caloploicin (**27**); its structure was confirmed by synthesis (Hunek and Yoshimura 1996).

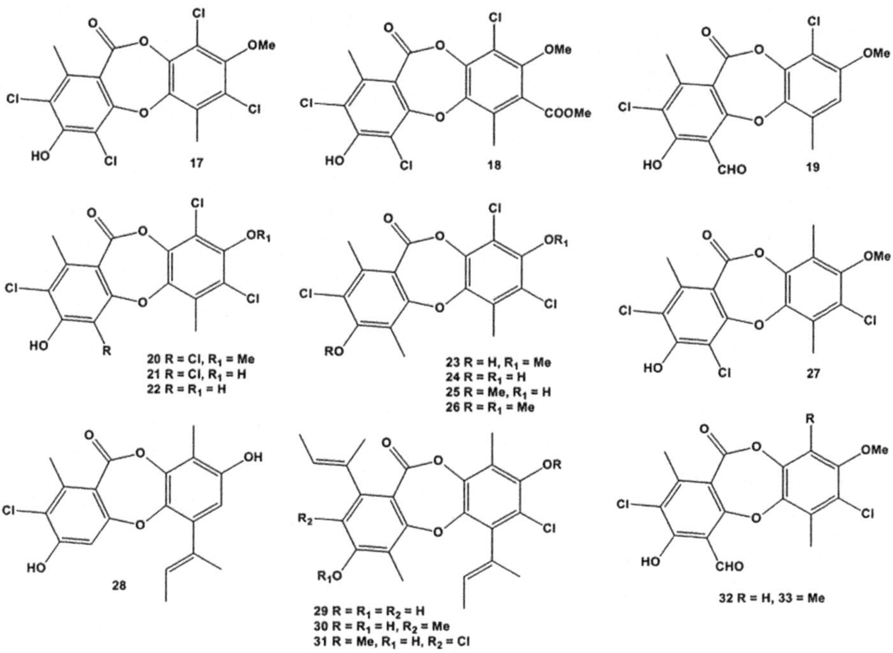

Fig. 8.13 Chlorinated depsides were discovered from lichens

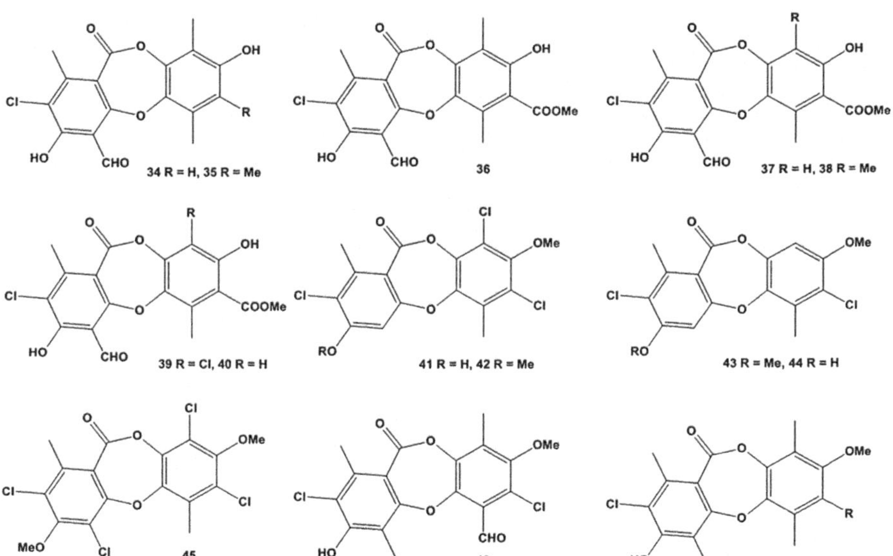

Fig. 8.14 Chlorinated depsides were discovered from lichens

Structurally similar 2-chlorounguinol (**28**), as well as the depsidones emeguisins A–C (**29–31**) were detected in *Emericella unguis*. The metabolite 1-chloropannarin (**32**) known also as argopsin, with its structure established by chlorination of pannarin (**19**), was isolated from *Agropsis megalospora*. This metabolite was also discovered in such lichens as *Agropsis friesiana* (otherwise called *Agropsis megalospora*), *Erioderma chilense*, *E. phaeorhizym*, and *E. tomentosum*. Eriodermin (**33**) was discovered in *Erioderma physcioides* and *E. phaeorhizum*. The *E. chilense* generates norpannarin (**34**) and norargopsin (**35**) (Dembitsky and Tolstikov 2005).

Physcosporin (**36**), a new chlorine-containing depsidones, was isolated from water-methanol extracts of *Pseudocyphellaria physciospora* lichen growing in Canada and in Europe. It was also discovered among metabolites of *P. granulata* and *P. flaveolata*. Physcosporin (**36**), hypophyscosporin (**37**), and 3-*O*-methylphyscosporin (**38**) are present among the extractives from *Erioderma phaeorhizum*. The Australian lichen *Lecidea* sp. generates lecideoidin (**39**) and its dechlorinated analogue 3′-dechlorolecideoidin (**40**). Depsidones, namely, dechlorolecideoidins (**41**) and (**42**), along with scencidin (**43**) and compound (**44**), were discovered in *Buellia canescens*. 3-*O*-demethylscencidin (**44**) and *O*-methyldi ploicin (**45**) were isolated from Australian lichen *Diploicia canescens*. Metabolite allorhizin (**46**) was found in *Psoroma allorhizum* lichen from New Zealand; two depsidones, i.e., phyllopsorin (**47**) and chlorophyllopsorins (**48–53**), are vital products of the Australian lichen *Phyllopsora coralline* (Dembitsky and Tolstikov 2005).

The *Lecanora gangaleoides* lichen synthesized leonidin (**54**), while *Fulgensia* sp. generates metabolite fulgidin (**55**), and components (**56**) and (**57**) are discovered in the Australian lichen *Buellia canescens* (Dembitsky and Tolstikov 2005) (see structures, Fig. 8.15).

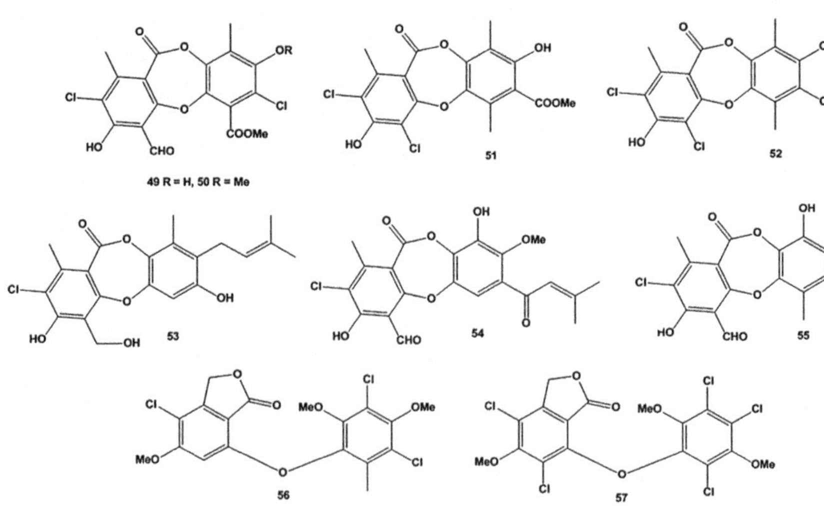

Fig. 8.15 Structures of chlorinated depsides and depsidones from lichens

8.12 Conclusion Remarks

More than 1200 secondary metabolites, including hydrocarbons, fatty and carboxylic acids, halogenated compounds and other components, have been isolated from lichens. Many of the isolated lichen metabolites, as investigated by many scientists, have shown anticancer, antibacterial, anti-inflammatory, antifungal and other activities. Our own studies of lichens over the past 30 years have shown that they have valued more than 300 new secondary metabolites, many of which have shown antifungicidal, antibacterial, and anticancer properties.

Acknowledgments The author would like to thank two anonymous reviewers for their helpful comments that significantly improved the manuscript as well as Ms. Jenny Dembitsky for her valuable assistance.

References

Dembitsky VM (1988) Quantification of plasmalogen, alkyl-acyl- and diacyl-glycerolipids by micro-thin-layer chromatography. J Chromatogr 436:467–473
Dembitsky VM (1992) Lipids of lichens. Prog Lipid Res 31:373–397
Dembitsky VM (1996) Betaine ether-linked glycerolipids: chemistry and biology. Prog Lipid Res 35:1–51
Dembitsky VM (2002) Bromo and iodo containing alkaloids of marine microorganisms and sponges. Russ Bioorg Chem 28:196–208
Dembitsky VM (2003a) Lichens: nature's pioneers. INFORM, Amer Oil Chem Soc 14:30–34
Dembitsky VM (2003b) Oxidation, epoxidation, and sulfoxidation reactions catalyzed by haloperoxidases. Tetrahedron 59:4701–4720
Dembitsky VM (2004a) Chemistry and biodiversity of biologically active natural glycosides. Chem Biodivers 1:673–781
Dembitsky VM (2004b) Astonishing diversity of natural surfactants. 1. Glycosides of fatty acids and alcohols. Lipids 39:933–953
Dembitsky VM (2005a) Astonishing diversity of natural surfactants. 2. Polyether glycosidic ionophores and macrocyclic glycosides. Lipids 40:219–248
Dembitsky VM (2005b) Astonishing diversity of natural surfactants. 3. Carotenoid glycosides and isoprenoid glycolipids. Lipids 40:535–557
Dembitsky VM (2005c) Astonishing diversity of natural surfactants. 4. Fatty acid amide glycosides, their analogues and derivatives. Lipids 40:641–660
Dembitsky VM (2005d) Astonishing diversity of natural surfactants. 5. Biological active glycosides of aromatic metabolites. Lipids 40:869–900
Dembitsky VM (2005e) Astonishing diversity of natural surfactants. 6. Biological active marine and terrestrial alkaloid glycosides. Lipids 40:1081–1105
Dembitsky VM (2006a) Natural neo acids and neo alkanes: their analogues and derivatives. Lipids 41:309–340
Dembitsky VM (2006b) Anticancer activity of natural and synthetic acetylenic lipids. Lipids 41:883–924
Dembitsky VM, Levitsky DO (2004) Arsenolipids. Prog Lipid Res 43:403–448
Dembitsky VM, Maoka T (2007) Allenic and cumulenic lipids. Prog Lipid Res 46:328–375
Dembitsky VM, Rezanka T (2003) Natural occurrence of arseno compounds in plants, lichens, fungi, algal species, and microorganisms. Plant Sci 165:1177–1192
Dembitsky VM, Rezanka T (2005) Metabolites produced by nitrogen-fixing Nostoc species. Folia Microbiol 50:363–391

Dembitsky VM, Rozentsvet OA (1989) Diacylglyceryltrimethylhomoserines and phospholipids of some green marine macrophytes. Phytochemistry 28:3341–3343
Dembitsky VM, Rozentsvet OA (1993) Phospholipid composition of some marine red algae. Phytochemistry 29:3149–3152
Dembitsky VM, Srebnik M (2002) Natural halogenated fatty acids: their analogues and derivatives. Prog Lipid Res 41:315–367
Dembitsky VM, Tolstikov GA (2000) Natural halogen carbonic acids. Chem Sustain Dev 8:623–643
Dembitsky VM, Tolstikov GA (2001) Natural chloro containing alkaloids. Chem Sustain Dev 9:169–181
Dembitsky VM, Tolstikov GA (2002a) Natural halogenated diterpenes. Chem Sustain Dev 3:269–280
Dembitsky VM, Tolstikov GA (2002b) Chloro-containing sesquiterpenes of plants. Chem Sustain Dev 4:383–390
Dembitsky VM, Tolstikov GA (2002c) Natural halogenated sesquiterpenes of marine organisms. Chem Sustain Dev 10:383–390
Dembitsky VM, Tolstikov GA (2002d) Natural halogenated C_{15} acetogenines of marine organisms. Chem Sustain Dev 11:329–340
Dembitsky VM, Tolstikov GA (2003a) Natural organic halogenated compounds. Geo-Science Publishing House, Novosibirsk, p 367
Dembitsky VM, Tolstikov GA (2003b) Natural halogenated furanones, terpenes and sterols. Chem Sustain Dev 11:699–705
Dembitsky VM, Tolstikov GA (2003c) Natural halogenated polyacrylamides. Chem Sustain Dev 11:341–348
Dembitsky VM, Tolstikov GA (2003d) Natural halogenated alkaloids. Chem Sustain Dev 11:461–476
Dembitsky VM, Tolstikov GA (2003e) Natural halogenated complex phenols. Chem Sustain Dev 11:819–830
Dembitsky VM, Tolstikov GA (2003f) Halogenated phenols of lichens and fungi. Chem Sustain Dev 11:569–578
Dembitsky VM, Tolstikov GA (2003g) Natural halogenated simple phenols and derivatives. Chem Sustain Dev 11:579–587
Dembitsky VM, Tolstikov GA (2003h) Natural halogenated macrolides, tetracyclines and quinones. Chem Sustain Dev 11:689–698
Dembitsky VM, Tolstikov GA (2004a) Natural halogenated sesquiterpenes marine organisms. Chem Sustain Dev 12:1–12
Dembitsky VM, Tolstikov GA (2004b) Natural halogenated xantones. Chem Sustain Dev 12:13–18
Dembitsky VM, Tolstikov GA (2005) Organic metabolites of lichens. Geo-Science, Novosibirsk, p 135
Dembitsky VM, Rozentsvet OA, Pechenkina EE (1990) Glycolipids, phospholipids and fatty acids of brown algae species. Phytochemistry 29:3417–3421
Dembitsky VM, Pechenkina-Shubina EE, Rozentsvet OA (1991a) Glycolipids and fatty acids of some seaweeds and marine grasses from the Black Sea. Phytochemistry 30:2279–2283
Dembitsky VM, Rezanka T, Bychek IA, Shustov MV (1991b) Identification of fatty acids from Cladonia lichens. Phytochemistry 30:4015–4018
Dembitsky VM, Shustov MV, Rozentsvet OA, Bychek IA (1991c) Phospholipid and fatty acid compositions of some lichen species from the Volga river basin. Phytochemistry 30:837–839
Dembitsky VM, Bychek IA, Rozentsvet OA, Shustov MV (1992a) Fatty acid and phospholipid compositions of some lichen species from the Volga river basin. Cryptogam Bot 2:391–394
Dembitsky VM, Bychek IA, Kashin AG (1992b) Chemical constituents of some lichen species. J Hattori Bot Lab 71:255–262
Dembitsky VM, Rezanka T, Bychek IA, Shustov MV (1992c) Fatty acid composition of Parmelia lichens. Phytochemistry 31:841–843

Dembitsky VM, Rezanka T, Bychek IA (1992d) Lipid composition of some lichens. Phytochemistry 31:1617–1620
Dembitsky VM, Rezanka T, Bychek IA (1992e) Fatty acid and phospholipids from lichens of the order Leconorales. Phytochemistry 31:851–853
Dembitsky VM, Bychek IA, Rozentsvet OA (1993a) Diacylglyceryltrimethyl-homoserine and phospholipid composition of some lichen species. Phytochemistry 34:1535–1536
Dembitsky VM, Rezanka T, Rozentsvet OA (1993b) Lipid composition of three macrophytes from the Caspian Sea. Phytochemistry 33:1015–1019
Dembitsky VM, Rezanka T, Shubina EE (1993c) Unusual hydroxy fatty acids from some fungi. Phytochemistry 34:1057–1059
Dembitsky VM, Rezanka T, Shubina EE (1993d) Chemical constituents of some higher fungi. 2. Fatty acid composition of ascomycetes. Cryptogam Bot 3:378–381
Dembitsky VM, Rezanka T, Bychek IA (1994a) Seasonal variability of lipids of lipids and fatty acids in the tree-growing lichen Xanthoria parietina. J Exp Bot 45:403–408
Dembitsky VM, Rezanka T, Bychek IA (1994b) Seasonal examination of lipids of the tree-growing lichen of genus Physcia. Phytochemistry 36:601–608
Dembitsky VM, Smoum R, Al-Quntar AA, Abu Ali H, Pergament I, Srebnik M (2002) Natural occurrence of boron-containing compounds in plants, algae and microorganisms. Plant Sci 163:931–942
Dembitsky VM, Terent'ev AO, Levitsky DO (2010) Amino and fatty acids of wild edible mushrooms of the genus Boletus. Rec Nat Prod 4:218–223
Dembitsky VM, Quntar A, Srebnik M (2011) Natural and synthetic small boron-containing molecules as potential inhibitors of bacterial and fungal quorum sensing. Chem Rev 111:209–237
Enk DC, Hochberg M, Torres A, Lev O, Dor I, Niddam V, Dembitsky VM, Srebnik M (2002) Natural UV filters derived from pigments of lichens (Sept. 03), Israel Patent Application: IL02/00725
Enk DC, Srebnik M, Lev O, Hochberg M, Dor I, Torres-Kerner A, Dembitsky VM (2003) Verwendung Von Natü Rlichen Pigmenten Aus Flechten, Cyanobakterien, Pilzen Und Pflanzen Zum Sonnenschutz. WO 2003/020236 2003/11 13.03.2003. German Patent
Enk CD, Hochberg M, Torres A, Dor I, Lev O, Srebnik M, Dembitsky VM (2004) Photoprotection by Cichorium endivia extracts: prevention of UVB-induced erythema, pyrimidine dimer formation and IL-6 expression. Skin Pharmacol Appl Ski Physiol 17:19–28
Enk CD, Srebnik M, Lev O, Hochberg M, Dor I, Torres-Kerner A, Dembitsky VM (2005) Utilization of natural pigments from lichens, cyanobacteria, fungi and plants for sun protection, US Patent #20050129630, 06.15.2005, Application 424059000 (USPTO), A61K007/42 (Intl Class)
Go JV, Řezanka T, Dembitsky VM (2002) Variability of fatty acid components of marine and freshwater gastropod species from the littoral zone of the Red Sea, Mediterranean Sea, and Sea of Galilee. Biochem Syst Ecol 30:819–835
Hanus LO, Dembitsky VM, Moussaieff A (2006) Comparative studies of the volatile compounds from fresh fruits of Mandragora autumnalis. Acta Chromatogr 17:151–160
Hanus LO, Moussaieff A, Rezanka T, Abu-Lafi S, Dembitsky VM (2007) Phytochemical analysis and comparison for differentiation of Boswellia carterii and B. serrata. Nat Prod Commun 2:139–142
Hanuš LO, Shkrob I, Dembitsky VM (2008a) Lipids and fatty acids of wild edible mushrooms of the genus Boletus. J Food Lipids 15:370–383
Hanuš LO, Temina M, Dembitsky VM (2008b) Antibacterial and antifungal activities of phenolic metabolites from lichenized ascomycete Ramalina lacera. Nat Prod Commun 3:233–236
Hanuš LO, Temina M, Dembitsky VM (2008c) Biodiversity of chemical constituents by the epiphytic lichenized ascomycete Ramalina lacera grows on Crataegus sinaicus, Pinus halepensis, and Quercus calliprinos. Biomed Papers 152:203–208
Hawksworth D, Hill D (1984) The lichen-forming fungi. Blackie. 158
Hawksworth DL, Rossman AY (1997) Where are all the undescribed fungi? Phytopathology 87(9):888–891

Huneck S (2001) New results on the chemistry of lichen substances. Progress in the chemistry of organic natural products, vol 81. Springer, Vienna, p 313
Huneck S, Yoshimura I (1996) Identification of lichen substances. Springer, Berlin
Purvis W (2010) Lichens. Natural history museum. Softcover, p 112
Rezanka T, Dembitsky VM (1993) Occurrence of C_{40}-C_{130} polyisoprenoid alcohols in lower plants. Phytochemistry 34:1335–1339
Rezanka T, Dembitsky VM (1999a) Bromo-containing fatty acids of Acorospora gobiensis lichen. Phytochemistry 50:97–99
Rezanka T, Dembitsky VM (1999b) Novel brominated lipidic compounds from lichen of Central Asia. Phytochemistry 51:963–968
Rezanka T, Dembitsky VM (2001) Bromoallenic lipid compounds from lichens of Central Asia. Phytochemistry 56:869–874
Rezanka T, Dembitsky VM (2003) Identification of acylated xanthone glycosides by liquid chromatography-atmospheric pressure chemical ionization mass spectrometry in positive and negative modes from the lichen Umbilicaria proboscidea. J Chromatogr A 995:109–118
Rezanka T, Dembitsky VM (2006a) Metabolites produced by cyanobacteria belonging to several species of the family Nostocaceae. Folia Microbiol 51:159–182
Řezanka T, Dembitsky VM (2006b) The colleflaccinosides, two chiral bianthraquinone glycosides with antitumor activity from the lichen Collema flaccidum collected in Israel and Russia. Nat Prod Res 20:969–980
Rezanka T, Jáchymová J, Dembitsky VM (2003) Prenylated xanthone glucosides from Ural's lichen Umbilicaria proboscidea. Phytochemistry 62:607–612
Rezanka T, Temina M, Tolstikov AG, Dembitsky VM (2004a) Natural microbial UV radiation filters – mycosporine-like amino acids. Folia Microbiol 49:339–352
Rezanka T, Temina M, Hanuš L, Dembitsky VM (2004b) Tornabeatins, four new tetrahydro-2-furanone derivatives, from the lichenized ascomycete Tornabea scutellifera (With.) J.R. Laundon. Phytochemistry 65:2605–2612
Temina M, Levitsky DO, Dembitsky VM (2010) Chemical constituents of the epiphytic and Lithophilic lichens of the genus *Collema*. Rec Nat Prod 4:79–86
Torres A, Dor I, Rottem E, Srebnik M, Dembitsky VM (2003) n-Alkane and fatty acid variations in the lichen Xanthoria parietina, their photobiont Trebouxia sp. and mycobiont, from Jerusalem hills. Eur J Biochem 270:2120–2125
Torres A, Hochberg M, Pergament I, Smoum R, Niddam V, Dembitsky VM, Temina M, Dor I, Lev O, Srebnik M, Enk CD (2004) A new UV-B absorbing mycosporine with photo protective activity from the lichenized ascomycete Collema cristatum. Eur J Biochem 271:780–784

Metabolic Profiling and Its Plausible Environmental Significance in a Common Himalayan Lichen

9

Vertika Shukla, Ankita Asati, Devendra K. Patel, Manoj Semwal, Narendra Kumar, and Dalip K. Upreti

Abstract

Metabolomics is an important technique that detects change in the quantitative profile of the organism to observe its response towards diverse genetic alterations and environmental conditions. In the present study, metabolite profiling of *Heterodermia diademata*, a common foliose lichen in Garhwal Himalayas, has been carried out in different environmental niches within an altitudinal gradients of 700–1850 m a.s.l. especially for its metabolites to decode any environmental significance. Metabolites, namely, fatty acid (FA), amino acid (AA) and sugar, were quantified using LC-MS/MS technique. Fatty acid content showed maximum variation with abundance of unsaturated fatty acid, while amino acid profile is dominated by glutamic acid, proline and serine. Sugar content has maximum variation of lactose. Most of the common primary metabolites have physiological roles as osmoregulators and thermotolerants as well as deterrents quite evident

Vertika Shukla and Ankita Asati contributed equally to this work.

V. Shukla (✉)
School of Environmental Sciences, Babasaheb Bhimrao Ambedkar University, Lucknow, UP, India
e-mail: vertika_shukla@rediffmail.com

A. Asati • D.K. Patel (✉)
Analytical Chemistry Division, CSIR-Indian Institute of Toxicology Research, Lucknow, India
e-mail: dkpatel@iitr.res.in

M. Semwal
ICT Department, CSIR-Central Institute of Medicinal & Aromatic Plant, Lucknow, India

N. Kumar
Department of Environmental Science, School of Environmental Sciences, Babasaheb Bhimrao Ambedkar University, Lucknow, UP, India

D.K. Upreti
Lichenology Laboratory, CSIR-National Botanical Research Institute, Lucknow, India

V. Shukla et al. (eds.), *Plant Adaptation Strategies in Changing Environment*,
https://doi.org/10.1007/978-981-10-6744-0_9

by having high erucic acid content. In the changing temperature gradient, fatty acid profile appears to be a valuable biomarker as it attributes towards thermotolerance of the species. Depending on the environmental condition, *H. diademata* species displays variation in FA profile, which seems to primarily relate to adaptation. The ability to readily alter the metabolite profile especially fatty acid concentrations appears to be an important contributing factor determining the ubiquity of the species at different altitudes of Garhwal Himalayas.

Keywords
LC-MS/MS techniques · Primary metabolite · Stress · Temperature · Biomarkers · Osmoregulator

9.1 Introduction

Abiotic and biotic stresses, such as high irradiance, drought and salinity, are known to disrupt or alter the metabolic process in the organisms. Plants have evolved various strategies to adapt to stress conditions. Shifts in species distribution patterns have been documented at broad biogeographic scales where environmental conditions vary over small distances in terrestrial and marine ecosystems and correlate mainly with changes in one or more abiotic factors, particularly temperature (Somero 2012; Deutsch et al. 2015). As warmer temperature is anticipated with the changing global climate, it is expected to affect the biogeographic distribution of species along the different latitudes and altitude. The speciation changes with increasing temperature will probably result in spread of thermotolerant species and extinction of thermosensitive species as temperature is a primary factor that affects the rate of plant development (Hatfield and Prueger 2015).

Lichens are considered as sensitive receptors of microclimatic changes which affect the lichen diversity of tolerant and sensitive species. In the last century, increased urbanisation and industrialisation have resulted in increased light exposure due to land use changes. The rising temperature favours in increase in abundance of several thermophilous epiphytes. Certain ecological clads of lichens including terricolous, saxicolous lichen species and boulder fields as well as epiphytes of high-altitude mountain forests have been observed to decline which may be attributed to change the ambient temperature resulting in destruction of suitable habitat for lichenisation of sensitive species. Increase in thermophilous lichens has been recorded in high montane lichen diversity in Europe (Hauck 2009). In India the members lichen family Physciaceae such as *Phaeophyscia hispidula*, *Pyxine cocoes*, *Pyxine subcinerea* and *Heterodermia diademata* are more frequent and abundant at various latitudes and altitudes (Shukla et al. 2013).

A general stress response in organisms includes increased amounts of metabolites which regulate normal metabolism and are compatible solutes like sugars, sugar alcohols, fatty acids and amino acids (Falcone et al. 2004; Bohnert and Sheveleva 1998). Under high irradiance, plants synthesise antioxidant enzyme and low molecular weight antioxidants (Szymanska et al. 2014). Lichens too are known to biosynthesise wide array of metabolites including metabolites and unique class of secondary metabolites. Ubiquity of lichens in different climatic condition is mainly attributed to the occurrence of distinctive class of secondary metabolites, but as primary metabolites are known to play an important role in plant physiology and adaptation, therefore their protective role in lichens cannot be denied.

Metabolomics is a rapid and sophisticated technique based on the analysis of metabolites in complex biological fluids or extracts using hyphenated spectroscopic and spectrometric analysis on quantitative basis (e.g. GC/MS, LC/MS-MS and 1H-NMR). The analysis provides specific metabolic sets and/or single metabolic markers that are correlated with the genetic backgrounds and/or environmental factors (Mittermeier et al. 2015).

Therefore in the present study, a common foliose lichen *Heterodermia diademata*, having wide geographical distribution, known to be growing up to the altitude of 3200 m has been profiled for its metabolites. Fatty acid (FA), amino acid (AA) and sugar were analysed using LC-MS/MS techniques which afforded qualitative and quantitative characterisation of the individual compounds of different classes based on the standard procured from Sigma-Aldrich (USA).

9.2 Material and Methods

9.2.1 Study Area

Lansdowne is located at 29°50′N 78°41′E 29.83°N 78.68°E. It has an average elevation of 1850 m a.sl. (Fig. 9.1). The area is under protected area and has thick oak and blue pine forests. Lansdowne is an unexplored area of Garhwal Himalayas, Uttarakhand (India). Population of Lansdowne as per 2001 India census is 7902. The present study was carried out targeting forest localities of Lansdowne in Garhwal Himalayas; some local urban centres were also sampled to observe local contribution. The area has thick canopy forest with minimum anthropogenic activity.

9.2.2 Collection of Lichen Samples

Lichen samples were collected during 1st–6th April 2015, from Lansdowne region of Garhwal Himalayas between altitudinal ranges of 700 and 1850 m. Among epiphytic lichens, lichen *Heterodermia diademata* was selected for quantification of the metabolites, based on its wider distribution at all the locations and abundance. Its abundance presence at various environmental inches indicated that the species may be used for providing 'clue to ubiquitous distribution' at various altitudinal levels. Until analysis samples were kept at −4 °C.

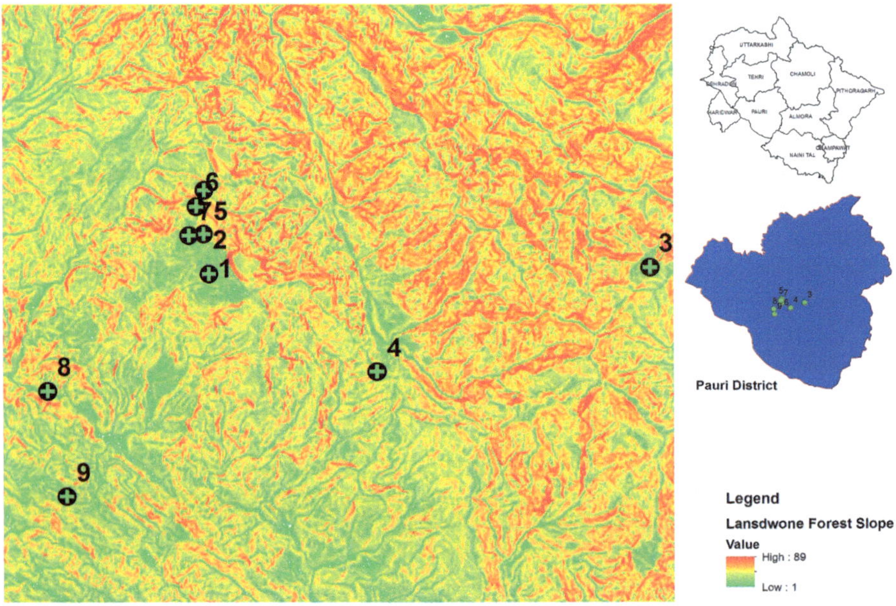

Fig. 9.1 Map of Lansdowne forest area in Uttarakhand surveyed for collection of lichen specimen

9.2.3 Sample Preparation for Fatty Acid (FA) Profiling

Extraction

0.1 gm homogenised lichen sample was Soxhlet extracted in $CHCl_3$: methanol (2:1) for 15 h. The extract was then dried in Buchi evaporator. The dried extract was redissolved in $CHCl_3$: methanol (1:2) (mass grade). The extract was filtered by 0.2 μ cellulose syringe filter and 10 μL injected to LC-MS/MS system for fatty acid analysis.

Fatty Acid Separation and Quantification

LC-MS/MS analysis of FAs was performed on Acquity UPLC (Waters) and coupled with electrospray ionisation (ESI) mass spectrometry (API 4000 Triple Quadrupole) (AB Sciex). The separation of FA compounds was achieved using UPLC®BEH C18 (1.7 μm × 2.1 × 50 mm) column with a gradient elution of solvent system (Table 9.1a). Triple quadrupole mass spectrometer (TQMS) equipped with ESI source has been used for the identification of 15 FA compounds in the separation of nine lichen species without derivatisation. The mass spectrometer was operated in negative ion mode. Optimisation of mass source parameters was set in Table 9.1b to facilitate separation of $[M-H]^-$ ion which was subsequently analysed (m/z) by the Q1-MS analyser. The daughter ion fragmentation ionisation of the deprotonated FA molecules into the collision cell (Q2-MS) was later analysed by the product ion

Table 9.1a Gradient flow rate for LC separation of fatty acids

Time (min)	Flow rate (mL/min)	A (%)	B (%)
Initial	0.3	95.0	5.0
0.2	0.3	95.0	5.0
0.7	0.4	96.0	4.0
1.2	0.4	96.0	4.0
1.7	0.5	97.0	3.0
2.2	0.5	97.0	3.0
2.5	0.6	98.0	2.0
3.0	0.6	98.0	2.0
3.5	0.3	95.0	5.0
4.0	0.3	95.0	5.0

A = Acetonitrile (LC-MS grade), B = 0.1% formic acid

Table 9.1b Optimisation of mass parameters for detection of fatty acids

Mass parameters	Fatty acids (−ve)
CAD (Collision-activated dissociation gas)	12 psi
CUR (Curtain gas)	20 psi
GS1 (Nebuliser gas)	18 psi
GS2 (Turbo gas)	15 psi
IS (Ion spray voltage)	4500v
Temperature	350 °C

mass analyser (MS2) Table 9.1c. The whole instrumental processing of ESI-MS/MS has been controlled using Analyst software (Version 1.6).

9.2.4 Sample Preparation for Amino Acid (AA) Profiling

Extraction

0.1 gm lichen was homogenised to a fine powder by use of a pestle. 2 ml aliquot of 0.1NHCl was added to homogenised powder. The homogenate was transferred to an Eppendorf tube and vortex for 2 min and heated followed by sonication for 1 min. The extract was then centrifuged for 3 min at 6000 rpm; the supernatant was pipette off and make up to 5 ml by adding MilliQ water. The supernant was filtered by 0.2 μ cellulose syringe filter and injected to LC-MS/MS system for amino acid analysis.

Amino Acid Separation and Quantification

Liquid chromatographic analysis of amino acid was performed Acquity UPLC (Waters) and coupled with ESI mass spectrometry equipped with a binary pump, a vacuum degasser system and a thermostated column oven. The separation of amino acid compounds was achieved using an Acquity UPLC BEH C18

Table 9.1c Optimisation of mass parameters for characterisation and quantification of individual fatty acid

	Unsaturated fatty acids	Parent ion [M-H]⁻ (m/z)	Quantifier ion (m/z)	DP (eV)	EP (eV)	CE (eV)	CXP (eV)
1	Linolenic acid (18:3) ω- 3	277.2	277.2	−20	−8	−25	−15
2	Docosahexaenoic acid (22:6) ω-3	327.2	283.2	−36	−6	−24	−8
3	Archidonic acid (20:4) ω-6	303.1	259.1	−34	−6	−14	−8
4	Palmitoleic acid (16:1) ω-7	253.2	114.9	−36	−7	−24	−10
5	Linoleic acid (18:2) ω-6	279.2	261.0	−60	−10	−25	−9
6	Elaidic acid (trans 18:1) ω-9	281.1	263.1	−92	−11	−28	−16
7	Oleic acid (cis 18:1) ω-9	281.1	143.0	−50	−7	−20	−15
8	Erucic acid (22:1) ω-9	337.3	97.0	−80	−9	−35	−10

	Saturated fatty acid	Parent ion [M-H]⁻ (m/z)	Quantifier ion (m/z)	DP (eV)	EP (eV)	CE (eV)	CXP (eV)
1	Hexanoic acid (6:0)	115.0	96.7	−62	−7	−20	−15
2	Decanoic acid (10:0)	171.1	153.1	−72	−11	−20	−11
3	Myristic acid (14:0)	227	209	−87	−11	−30	−11
4	Palmitic acid (16:0)	255.1	237.1	−95	−12	−32	−15
5	Eicosanoic acid (20:0)	311.2	183.0	−108	−8	−50	−8
6	Docosanoic acid (22:0)	339.0	183.0	−80	−6	−25	−9
7	Octadecanoic acid (18:0)	283.2	265.2	−65	−12	−33	−16

DP declustering potential, *EP* entrance potential, *CE* collision energy, *CXP* cell exit potential

(1.7 μm×2.1×50 mm) column at 40 °C temperature with an isocratic elution of solvent A, 15 mmol ammonium acetate (2%), and solvent B, MilliQ (98%) at constant flow rate of 0.3 mLmin^{-1}.

Mass has been used for the identification of AA compounds in the LC-based fractional extracts of lichen species. The decluster potential (DP), entrance potential (EP), collision-activated dissociation (CAD) gas, collision energy (CE) and cell exit potential (CXP) were thoroughly optimised and maintained (Table 9.2a, 9.2b) correspondingly to further assist the fragmentation ionisation of the protonated AA molecules into the collision cell (Q2-MS) which was later analysed by the product ion mass analyser (MS2).

9.2.5 Sample Preparation for Carbohydrate Profiling

Extraction

Lichen sample weighing 0.1 gm was extracted in 2 ml MilliQ/ACN/Propanol (5:2:3) and vortex for 2 min heated followed by sonication up to 2 min, centrifuged 6000 rpm for 4 min. The supernant was filtered by 0.2 μ cellulose syringe filter and injected to LC-MS/MS system for carbohydrates analysis.

Table 9.2a Optimisation of mass parameters for detection of amino acids

Mass parameters	Amino acids (+ve)
CAD (Collision-activated dissociation gas)	10 psi
CUR (Curtain gas)	10 psi
GS1 (Nebuliser gas)	60 psi
GS2 (Turbo gas)	40 psi
IS (Ion spray voltage)	5500
Temperature	200 °C

Carbohydrate Separation and Quantification

Analysis of carbohydrates has been performed on Acquity UPLC (Waters) and coupled with ESI mass spectrometry. Samples were separated on C-18 column at 40 °C temperature with conditions optimised as binary mobile phase A methanol (99%) and mobile phase B 0.1% formic acid (1%) with flow rate of 0.5 mLmin^{-1} up to 2 min in isocratic mode. LC-MS/MS was fully equipped with auto sampler and Analyst 1.6 software. The details of mass characterisation are in Table 9.3a and 9.3b.

9.2.6 Statistical Analysis

Data obtained (individual fatty acid, amino acid and carbohydrates) was subjected to principal component analysis (PCA) using statistical software program, Statistica (12.7) (USA).

9.3 Environmental Implication of Variation in Metabolite Profile

The samples collected from different niches of Garhwal Himalayas exhibit significant variation in the metabolite profile (Tables 9.4, 9.5 and 9.6; Fig. 9.2). FAs show highest variation (Fig. 9.3), and unsaturated fatty acid ranges between 171.54 and 558.67 μgg^{-1}, while saturated fatty acid varied from 78.10 to 230 μgg^{-1}. Unsaturated fatty acid profile is high mainly dominated by erucic acid (22:1) and linoleic acid (18:2), while saturated fatty acid is dominated by docosanoic acid (22:0), eicosanoic acid (20:0) and palmitic acid (16:0).

FAs, main source of metabolic energy, are the important molecules that regulate metabolism. Variation in FA composition of the cell membrane under different environmental conditions is known to play a key role in adaptation to environmental changes (Shukry 2007). Straight-chain saturated FAs, such as C14:0 or C16:0, are common in microorganisms including lichens (Dembitsky et al. 1991).

Variation in the lipid profile in poikilothermic organisms with the temperature of growth is well known (Falcone et al. 2004). If the ambient temperature is low, homeothermic organisms are also known to have a greater proportion of unsaturated

Table 9.2b Optimization of mass parameters for characterisation and quantification of individual amino acids

	Amino acids	Parent ion $[M\text{-}H]^+$ (m/z)	Quantifier ion (m/z)	DP (eV)	EP (eV)	CE (eV)	CXP (eV)
1	Glycine	76	30	20	10	40	18
2	Alanine	90	44	70	10	18	26
3	Aminobutyric acid	104.4	58	26	10	10	10
4	Serine	106.0	60	40	10	20	18
5	Proline	116	70	20	10	15	26
6	Threonine	120	74	67	10	10	8
7	Isoleucine	132	86	7	10	18	15
8	Norleucine	132	69	70	10	18	15
9	Leucine	132	30	29	10	18	15
10	Aspartic acid	134	74	50	10	20	18
11	Glutamic acid	148	84	50	10	21	10
12	Methionine	150	104	77	10	13	9
13	Histidine	156	110	70	10	18	15
14	Phenylalanine	166	120	50	10	11	11
15	Ornithine	170.7	111.2	25	10	7	7
16	Tyrosine	182	165.1	48	10	12	8
17	Trptophan	205	188	43	10	15	18
18	DOPA	198.2	181	48	10	20	15
19	Arginine	211.2	129	40	10	20	25

DP declustering potential, *EP* entrance potential, *CE* collision energy, *CXP* cell exit potential

Table 9.3a Optimisation of mass parameters for detection of carbohydrates

Mass parameters	Amino acids (+ve)
CAD (Collision-activated dissociation gas)	12 psi
CUR (Curtain gas)	10 psi
GS1 (Nebuliser gas)	13 psi
GS2 (Turbo gas)	0.0 psi
IS (Ion spray voltage)	5500
Temperature	300 °C

Table 9.3b Optimisation of mass parameters for characterisation and quantification of individual carbohydrates

Carbohydrates	Parent ion $[M\text{-}H]^+$ (m/z)	Quantifier ion (m/z)	DP (eV)	EP (eV)	CE (eV)	CXP (eV)
Dextrose	198.0	163.01	43	8	14	7
Mantose	198.1	145	36	8	17	13
Galactose	198.2	162.9	30	8	12	10
Fructose	198.3	163	43	8	12	7
Lactose	342.7	325	70	8	9	10
Maltose	360.1	325	66	8	14	7
Sugar	360.2	163.2	62	8	19	7

DP declustering potential, *EP* entrance potential, *CE* collision energy, *CXP* cell exit potential

fatty acids in the form of surface lipids (Marr and Ingraham 1962). Saturated and monounsaturated fatty acid contents are known to increase under high temperature, while trienoic unsaturated fatty acid and linolenic acid levels are higher at lower temperature. It has been observed that plants subjected to high temperature treatment produced seed with lowest linolenic acid contents and the highest levels of saturated and monounsaturated fatty acids (Fayyaz-ul-Hassan et al. 2005). Therefore occurrence of monounsaturated fatty acids along with saturated fatty acid in the present study may be attributed to high temperature conditions prevailing in the area as the study area is one of the dry regions of Garhwal Himalayas.

It has been observed that environmental factors which result in altering osmotic potential lead to changes in physiological functioning as it affects photosynthesis, sets off chain of reaction initiated by change in internal carbon dioxide concentration leading to disruption in photorespiratory cycle and affects carbon and nitrogen distribution, and now reactions that lead to the consumption of reducing power become favoured, which finally affects the development and growth of the organism (Bohnert and Sheveleva 1998).

Among different environmental *factors*, clear alterations are known to occur in the fatty acid profile. Variation in trienoic and dienoic fatty acid is used as valuable biomarker of temperature variation. A decline in trienoic fatty (16:3) correlates with the overall decrease in polyunsaturated fatty acid concentration with the increase in

Table 9.4 Fatty acid content in lichen, *Heterodermia diademata* sampled from different localities in and around Lansdowne, Garhwal Himalayas

	Fatty acid (FA)	1	2	3	4	5	6	7	8	9
Unsaturated (UFA)	Monoenoic FA									
	Palmitoleic acid (16:1)	0.15	0.13	0.27	0.19	0.16	0.09	0.13	0.11	0.08
	Erucic acid (22:1)	259.20	299.70	294.19	507.74	170.02	171.87	133.75	213.81	211.47
	Oleic acid (18:1)	0.30	0.40	0.29	0.40	0.26	0.30	0.23	0.30	0.27
	Elaidic acid (18:1)	3.97	1.48	3.13	1.92	1.76	1.86	1.24	1.49	1.98
	Dienoic FA									
	Linoleic acid (18:2)	34.03	27.14	17.62	32.83	61.10	25.42	26.47	27.64	23.09
	Trienoic FA									
	Linolenic acid (18:3)	20.00	12.28	8.91	14.68	9.75	12.36	9.04	9.74	4.94
	Polyenoic FA									
	Arachidonic acid (20:4)	1.74	0.75	0.23	0.66	0.58	0.41	0.56	0.11	0.07
	Docosahexaenoic acid (22:6)	0.23	0.25	0.27	0.24	0.24	0.20	0.11	0.10	0.12
	Σ UFA	319.63	342.14	324.92	558.67	243.88	212.52	171.54	253.30	242.02
Saturated (SFA)	Hexanoic acid	0.00	0.00	0.00	0.63	0.29	0.00	0.00	0.00	0.00
	Decanoic acid	9.65	9.47	8.03	17.37	10.66	5.83	6.15	4.76	6.14
	Myristic acid	8.10	4.06	13.97	5.04	2.46	5.18	4.75	3.85	4.12
	Palmitic acid	20.21	14.86	13.79	19.23	11.18	18.73	15.19	17.14	17.55
	Eicosanoic acid	59.86	32.59	35.25	85.23	30.49	43.95	31.98	38.61	37.44
	Docosanoic acid	57.78	47.52	44.42	98.14	20.44	80.78	41.54	87.22	66.99
	Octadecanoic acid	4.89	3.03	4.37	4.80	2.59	4.31	3.83	3.80	4.42
	Σ SFA	160.49	111.52	119.84	230.44	78.10	158.78	103.44	155.38	136.67

Table 9.5 Amino acid content in lichen, *Heterodermia diademata* sampled from different localities in and around Lansdowne, Garhwal Himalayas

Amino acid	1	2	3	4	5	6	7	8	9
Alanine	7.20	14.34	5.35	8.56	17.76	8.39	16.99	9.48	7.74
Glycine	1.59	3.53	0.77	1.29	1.56	1.55	3.48	2.02	3.73
Aminobutyric acid	0.43	0.50	0.23	0.33	0.37	0.35	0.43	0.39	0.57
Serine	16.76	20.23	12.71	18.94	19.64	12.42	25.96	26.28	26.54
Proline	17.30	14.86	10.28	10.15	17.77	5.18	4.05	6.81	4.27
Theronine	4.08	8.03	1.79	3.51	5.75	2.89	4.03	3.43	6.48
Isoleucine	3.48	6.61	2.37	4.44	6.60	3.79	6.08	3.71	7.26
Norleucine	3.41	7.52	0.50	3.38	5.84	3.03	4.81	4.39	6.47
Leucine	0.17	0.22	0.58	0.52	0.35	0.80	0.31	0.29	0.30
Aspartic acid	3.70	5.65	4.98	5.04	5.21	7.14	6.70	8.17	4.55
Glutamic acid	102.65	108.58	63.66	84.41	125.89	111.19	110.27	104.95	72.91
Methionine	0.67	0.89	1.30	1.33	1.05	1.94	0.77	0.79	0.85
Histidine	0.57	2.04	1.64	0.07	2.22	0.82	1.11	1.37	1.57
Phenylalanine	1.46	5.57	1.42	2.59	2.97	2.59	3.05	1.50	3.72
Ornithine	0.77	1.41	0.57	1.28	2.55	1.33	0.60	1.01	0.73
Tyrosine	7.49	16.60	6.76	6.08	7.34	7.31	16.37	9.51	17.56
Dopa	2.02	0.21	0.68	1.28	2.06	2.42	0.51	1.96	1.98
Tryptophan	0.39	1.70	0.41	0.74	0.86	0.47	0.99	0.72	1.98
Arginine	1.88	1.90	2.32	3.36	1.92	3.00	2.43	2.26	2.15
Σ amino acid	176.11	220.50	118.39	157.39	227.79	176.71	209.04	189.14	171.45

Table 9.6 Carbohydrate content in lichen, *Heterodermia diademata* sampled from different localities in and around Lansdowne, Garhwal Himalayas

Sugars	1	2	3	4	5	6	7	8	9
Dextrose	0.166	0.083	0.196	0.176	0.162	0.034	0.123	0.197	0.093
Mantose	0.091	0.051	0.173	0.185	0.128	0.059	0.075	0.113	0.050
Galactose	0.082	0.054	0.201	0.065	0.038	0.034	0.040	0.062	0.033
Fructose	0.060	0.041	0.180	0.090	0.071	0.022	0.062	0.061	0.043
Lactose	2.983	0.000	2.204	1.212	3.091	0.000	0.643	0.000	1.583
Maltose	0.110	0.037	0.136	0.102	0.235	0.125	0.470	0.062	0.331
Sugar	0.147	0.074	0.196	0.067	0.351	0.102	0.716	0.111	0.043
Σ sugar	4.64	2.34	6.29	5.89	9.08	6.38	9.13	8.61	11.18

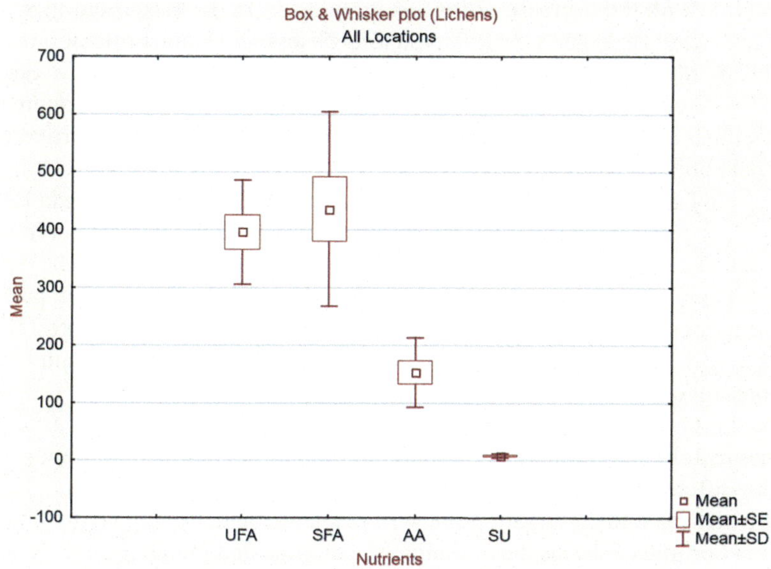

Fig. 9.2 Nutrient concentration represented as mean standard error and standard deviation

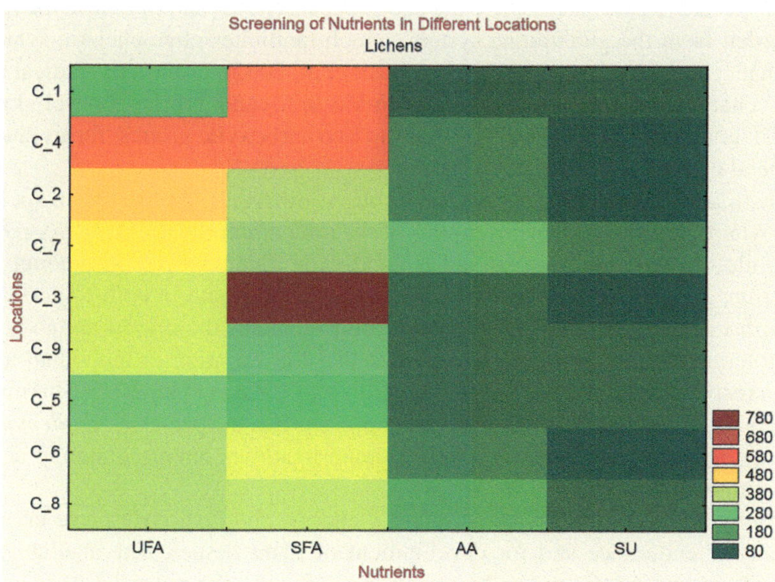

Fig. 9.3 Variation in nutrient profile at each location showing highest variation in fatty acids

temperature. With the decrease in trienoic fatty acid (18:3), concurrent increase in the linoleic acid (18:2) and (16:0) fatty acid is observed. Distinct role of individual fatty acid is also evident by high erucic acid content in all the lichen samples. Reason for detection of high erucic acid may be attributed to high temperature tolerance as well as characteristic feature of erucic acid to be an effective deterrent which lowers palatability (Vermorel et al. 1986).

Fatty acids are the main constituents of the cell membranes. Variation in FA profile has been studied extensively in microorganisms and their involvement in a variety of adaptations highlights its predominant role in survival of the species (Diomandé et al. 2015). Variation in fatty acid composition results in ecotype differences in temperature adaptation, as more warm-adapting fatty acids are reported from sites with greater solar exposure (Conner et al. 2010). Thus the result is based on the present observations and high abundance of *H. diademata* in wide geographical area of Garhwal Himalayas. *H. diademata* may be considered as a high-temperature tolerant species, which may be mainly attributed to variation in FA profile in different environmental condition.

In the present study, glutamic acid, serine, alanine and asparagine dominate the amino acid profile of the species. Alanine, asparagine and glutamic acid are known to increase under the influence of abiotic factors. Accumulation of N-metabolites, i.e. amino acid, displays wide array of functions ranging from metal complexation, radical scavenging, chaperone and signalling. Proline is known to quench reactive oxygen species (ROS) and reactive nitrogen species (RNS) and to relieve the oxidative burden from the glutathione system, which facilitates phytochelatin synthesis and enhances metal tolerance (Siripornadulsil et al. 2002). Osmolytic, radical scavenging characteristic of proteinogenic amino acid and proline is well known (Matysik et al. 2002). Glutamic acid, having two carboxylic groups, forms chelates with metal cations (Dutta et al. 2013).

Muhsin and Zwain (1989) studied the role of different amino acids in six halophytes where proline accounts over 60% of the total amino acid in *Suaeda vermiculata*, while tyrosine, methionine and leucine were detected in high amount in *Z. coccineum.* Arginine is the most abundant amino acid found in both *Cyperus conglomeratus* and *Matthiola oxyceras.* The result indicates the role of amino acid in stress tolerant. It has been further observed that the variation in individual amino acids in response to stress is a species-specific trait (Bhatia et al. 2005; Sharma and Dietz 2006), which seems to be true in the present lichen species *H. diademata*, as amino acid profile is consistent at all the nine localities having glutamic acid as major amino acid.

Carbohydrates are the primary source of energy, and carbohydrate reserves are essential for endurance and for development of plant tissues. The availability of carbohydrate levels is found to be useful in evaluating the potential of plants for preparing to rejuvenate under favourable condition. In the present study, sugars have been quantified from all the lichen samples from nine localities ranging between 2.34 and 11.18 $\mu g g^{-1}$.

Conditions of drought, salinity and temperature are known to affect osmotic potential in the organism. A complex combination of physiological, anatomical, morphological and genetical adaptations enables plants to maintain osmotic potential suitable for survival under adverse environmental conditions (Riederer and Schreiber 2001). During the past few years, research on changes in biochemical pathways that change during stress has gained attention, though detailed mechanistic explanation and interpretation of the data obtained are complex as it involves interaction among climatic and ecological factors, as well as direct environmental effect on the physiology of the organism in natural conditions (Somero 2012). More recently, it has been demonstrated that mycobiont biosynthesises fatty acids if it is unable to produce polyketides (Molina et al. 2003). Axenic cultures of *Physconia distorta* in nutrient-rich media produce mainly fatty acids and its triglyceride derivatives, which deposited on the surface of the mycelia as fat drops (Stocker-Wörgötter 2008). Hydrocarbons, monoacylglycerides and triacylglycerides are preferentially biosynthesised in aposymbiotically grown mycobiont. Metabolic switching has been reported in filamentous fungi, *Aspergillus nidulans* (Stocker-Wörgötter 2008). FASII is known to play significant role in major cell metabolic pathways, and FA synthesis seems to be explicitly regulated depending on the cellular requirements of the microorganisms (Diomandé et al. 2015). These experiments clearly indicate that FAS (fatty acid synthase) is switched on and activated, whereas PKS apparently inhibited under changing ambient conditions and cellular requirements (Stocker-Wörgötter 2008). But the evidence on metabolic switching in natural condition is rare in lichens; therefore detailed molecular and phenological research should be carried out extensively to provide mechanism of responses of organisms to external environmental changes (Somero 2012).

9.4 Conclusion

The present study shows that stress tolerance involves integrated complex mechanism which triggers or suppresses certain metabolic pathways to create inner environment fit to sustain external perturbations. Depending on the environmental condition, *Heterodermia* species displays variation in FA profile which favours biosynthesis of saturated and monosaturated fatty acids in response to varying temperatures along the altitudinal gradient, while variations in AA profile and sugar levels indicate species ability to sustain various abiotic and biotic stresses. The results clearly underline the environmental significance of variations in the metabolic profile of *H. diademata* which mainly seems to relate to adaptation and contribute towards the ubiquity of the species.

Acknowledgements Director, CSIR-National Botanical Research Institute, Lucknow is thankfully acknowledged for facilities and encouragements. VS is grateful to the Department of Science and Technology (DST-SERB), New Delhi, for award of Young Scientist fellowship (SR/FTP/ES-39/2013).

References

Bhatia NP, Walsh KB, AJM B (2005) Detection and quantification of ligands involved in nickel detoxification in the herbaceous Ni hyperaccumulator Stackhousia tryonii Bailey. J Exp Bot 56:1343–1349

Bohnert HJ, Sheveleva E (1998) Plant stress adaptations – making metabolism move. Curr Opin Plant Biol 1:267–274

Connor N, Sikorski J, Rooney AP, Kopac S, Koeppel AF, Burger A et al (2010) Ecology of speciation in the genus *Bacillus*. Appl Environ Microbiol 75(5):1349–1358

Dembitsky VM, Bychek IA, Shustov MV, Rozentsvet (1991) Phospholipid and fatty acid composition of some lichen species. Phytochemistry 30(3):837–839

Deutsch C, Ferrel A, Seibel B, Pörtner H-O, Huey RB (2015) Climate change tightens a metabolic constraint on marine habitats. Science 348:1132. https://doi.org/10.1126/science.aaa1605

Diomandé SE, Nguyen-The C, Guinebretière M-H, Broussolle V, Brillard J (2015) Role of fatty acids in Bacillus environmental adaptation. Front Microbiol 6:813. https://doi.org/10.3389/fmicb.2015.00813

Dutta S, Ray S, Nagarajan K (2013) Glutamic acid as anticancer agent: an overview. Saudi Pharm J 21:337–343

Falcone DL, Ogas JP, Somerville CR (2004) Regulation of membrane fatty acid composition by temperature in mutants of Arabidopsis with alterations in membrane lipid composition. BMC Plant Biol 4:17

Fayyaz-ul-Hassan AH, Cheema MA, Manaf A (2005) Effect of environmental variation on oil content and Fatty acid composition of Canola cultivars. J Res Sci., 2005 16(2):65–72

Hatfield JL, Prueger JH (2015) Temperature extremes: effect on plant growth and development. Weather Clim Extremes 10:4–10

Hauck M (2009) Global warming and alternative causes of decline in arctic-alpine and boreal-montane lichens in North-Western Central Europe. Glob Chang Biol 15:2653–2661. https://doi.org/10.1111/j.1365-2486.2009.01968.x

Marr AG, Ingraham JL (1962) Effect of temperature on the composition of fatty acid in Escherichia coli. J Bacteriol 84:1260–1267

Matysik J, Alia BB, Mohanty P (2002) Molecular mechanisms of quenching of reactive oxygen species by proline under stress in plants. Curr Sci 82:525–532

Mittermeier VK, Schmitt N, Volk LPM, Suárez JP, Beck A, Eisenreich W (2015) Metabolic profiling of alpine and Ecuadorian lichens. Molecules 20:18047–18065. https://doi.org/10.3390/molecules201018047

Molina MC, Crespo A, Vicente C, Elix JA (2003) Differences in the composition of phenolics and fatty acids of cultured mycobiont and thallus of *Physconia distorta*. Plant Physiol Biochem 41:175–180

Muhsin TM, Zwain KH (1989) Correlation between fungal populations and amino acid levels of salt desert plants. Sydowia 41:209–218

Riederer M, Schreiber L (2001) Protecting against water loss: analysis of the barrier properties of plant cuticles. J Exp Bot 52(363):2023–2032

Sharma SS, Dietz K-J (2006) The significance of amino acids and amino acid-derived molecules in plant responses and adaptation to heavy metal stress. J Exp Bot 57(4):711–726

Shukla V, Upreti DK, Bajpai R (2013) Selection of biomonitoring species. In: Lichens to biomonitor the environment. Springer, New Delhi, pp 97–170

Shukry WM (2007) Fatty acids and stress in three oil-producing plants: canola (*Brassica napus* L.), safflower (*Carthamus tinctorius* L.) and sunflower (*Helianthus annus* L.) Afr J Plant Sci Biotechnol 1(1):43–56

Siripornadulsil S, Traina S, Verma DPS, Sayre RT (2002) Molecular mechanisms of proline-mediated tolerance to toxic heavy metals in transgenic microalgae. Plant Cell 14:2837–2847

Somero GN (2012) The physiology of global change: linking patterns to mechanisms. Annu Rev Mar Sci 4:39–61

Stocker-Wörgötter E (2008) Metabolic diversity of lichen-forming ascomycetous fungi: culturing, polyketide and shikimate metabolite production, and PKS genes. Nat Prod Rep 25:188–200. https://doi.org/10.1039/b606983p

Szymanska R, Latowski D, Nowicka B, Strzalka (2014) Lipophilic molecules as a part of antioxidant system in plants. In: Ahmad P (ed) Oxidative damage to plants. Elsevier, pp 321–344. 10.1016/B978-0-12-799963-0.00010-1

Vermorel M, Heaney RK, Fenwick GR (1986) Nutritive value of rapeseed: Effects of individual glucosinolates. J Sci Food Agric 37:1197–1202

Heavy Metal Tolerance in Crop Plants: Physiological and Biochemical Aspects

10

Reshu Chauhan, Surabhi Awasthi, Amit Pal Singh, Sudhakar Srivastava, Veena Pande, Rudra Deo Tripathi, and Amit Kumar

Abstract

Plants are immobile and they have to adapt against adverse conditions of the environment for their survival. Heavy metal (HM) toxicity is posing a serious concern for the plant life and seriously hampering the food grain productivity. Heavy metals include the transition metals essential for plant nutrition as well as the nonessential elements. All these elements become toxic to crop plants when they are present at high tissue concentrations. Elevated concentration of HMs in soil may be due to natural as well as anthropogenic activities. Plants growing in HM-contaminated regions may accumulate a significant amount of these HMs.

Reshu Chauhan and Surabhi Awasthi contributed equally to this work.

R. Chauhan
CSIR, National Botanical Research Institute, Lucknow, India

Department of Biotechnology, Kumaun University, Bhimtal, Nainital, Uttarakhand, India

S. Awasthi
CSIR, National Botanical Research Institute, Lucknow, India

Institute of Environment & Sustainable Development, Banaras Hindu University, Varanasi, India

A.P. Singh • R.D. Tripathi
CSIR, National Botanical Research Institute, Lucknow, India

S. Srivastava
Institute of Environment & Sustainable Development, Banaras Hindu University, Varanasi, India

V. Pande
Department of Biotechnology, Kumaun University, Bhimtal, Nainital, Uttarakhand, India

A. Kumar (✉)
Cytogenetic and Plant Breeding Lab, Department of Botany, University of Lucknow, Lucknow, India
e-mail: amit.metagene@gmail.com

V. Shukla et al. (eds.), *Plant Adaptation Strategies in Changing Environment*, https://doi.org/10.1007/978-981-10-6744-0_10

This paves a way for HMs to enter into food chain posing serious health concerns for animal and human life. A number of morphological, biochemical, and physiological alterations occur during HM toxicity including alteration in uptake mechanism and transportation of water, root and shoot growth, oxidative stress, and changes in HM complexing ligands for sequestration of these HMs into vacuole to reduce the HM concentration in cytoplasm. The present chapter deals with the physiological and biochemical responses of plants to HM toxicity along with shedding light on uptake and transport mechanisms of HMs in brief.

Keywords
Antioxidants · Growth · Heavy metals · Oxidative stress · Thiols

Abbreviations

APS	5′-Adenylylsulfate reductase activity
As	Arsenic
Ca	Calcium
Cd	Cadmium
CS	Cysteine synthase activity
EC	Electrical conductivity
Fe	Iron
GST	Glutathione-S-transferase
H_2O_2	Hydrogen peroxide
Hg	Mercury
HMs	Heavy metals
MDA	Malondialdehyde
Mn	Manganese
OsLCT1	Low affinity cation transporter
P	Phosphorous
Pb	Lead
PCS	Phytochelatin synthase activity
S	Sulfur
SAT	Serine acetyl transferase activity
Se	Selenium
Si	Silica
Zn	Zinc

10.1 Introduction

According to the famous statistician Prof TR Malthus, "Population increases in geometrical progression while resources increase in arithmetic progression." His saying became well clear in the last century when geometrical increase in world population was observed while resources continued to become scarce. This was attributable to our substantially increased ability to grow sufficient food with the aid of a number of measures ranging from insecticides, pesticides, herbicides, and fertilizers to high-yielding varieties (green revolution). However, this led to increase in number and quantity of thousands of pollutants in our surroundings and also caused qualitative loss of our irreplaceable resources. Among these pollutants, heavy metals (HMs) found widespread applications in agriculture and industrial sectors and are presently regarded as contaminants throughout the world. Heavy metals also pose risk to growing sufficient food due to their detrimental impacts on plant growth, and these negative impacts would increase owing to growing demands in the coming decades.

Metals with a specific density of >5 g/cm^3 are generally considered as HMs (Järup 2003). Most of the metals have ubiquitous distribution in the earth crust, such as lead (Pb) and cadmium (Cd), but may be present at significantly high concentrations in particular geographical regions like arsenic (As) in Bangladesh and Gangetic plains in India (Srivastava et al. 2012). Heavy metals exert both common and specific effects on plants. On exposure of HMs, plants exhibit some common toxic symptoms that include reduced plant growth, disturbed photosynthetic machinery and distribution of essential nutrients, altered water balance, chlorosis, and senescence that finally results in plant death. Plants growing in HM-contaminated sites generally accumulate higher amounts of HMs, and, thus, contamination of food chain occurs (Zhao et al. 2010). Food chain contamination is the main cause/route of HM toxicity in human and other animals, making them prone to several diseases that include skin diseases and different types of cancers (Siddiqui 2015). The mechanisms by which HMs exert their toxicity in plants are (1) competition for their absorption with similar essential nutrients cations, for example, As competes with P and Cd with Zn, (2) HMs inactivate the proteins because of direct interaction with sulfhydryl group (−SH) of functional proteins, (3) collapse of function of enzymes by displacement of essential cations from their specific binding sites, and (4) ROS production damages the macromolecules (Sharma and Dietz 2009; DalCorso et al. 2013). Metal(loids), viz., Cd, Pb, mercury (Hg), and zinc (Zn), because of their high densities are considered as HMs (Oves et al. 2016), while As is listed in the same list because of similar properties (Chen et al. 1999). Manipulation with metal transporters is an effective strategy to improve metal tolerance and accumulation in different plant species. For developing new crops for metal tolerance, transmembrane transporter proteins are key targets.

10.2 Heavy Metals and Their Associated Toxicity

Heavy metals are well known for their toxic effects on plants and human including other animals. Consumption of toxic HM-contaminated crops poses several ill effects on human health. HMs can damage different cellular structures and a variety of tissues and organs. HMs exert toxicity by interacting with sulfhydryl groups and interference with the uptake of essential elements.

Arsenic is a non-threshold class-1 carcinogen. Several people worldwide suffer from As toxicity due to the intake of As-contaminated drinking water and food; thus, to reduce As concentration in drinking water is of critical importance. Phytoremediation of As-contaminated soil and its reduced accumulation in plants depends upon understanding of As uptake and transport in plants. Arsenate [As(V)] and arsenite [As(III)] are the predominant inorganic phyto-available forms of As in soil solution. Inorganic forms of As are highly toxic as As(V), that is structural analogue of phosphate, interferes with phosphate metabolism (such as phosphorylation and ATP synthesis), and As(III) binds to vicinal sulfhydryl groups of proteins affecting their structures or catalytic functions (Tripathi et al. 2007; Zhao et al. 2010). Arsenic contamination hampers many physiochemical processes of soils and leads to loss of soil fertility and crop yield.

Cadmium is the seventh most abundant toxic HM in the earth crust as per Agency for Toxic Substances and Disease Registry (ATSDR). Cadmium severely hampers the enzymatic activities, induces the oxidative stress, and causes the nutritional deficiency in plants (Irfan et al. 2013). Cadmium uptake in agrarian crops ultimately introduces it to human food chain. Cadmium shows significantly high affinity with cysteine, glutamate, histidine, and aspartate ligands and may lead to the deficiency of iron (Castagnetto et al. 2002). Cd^{2+} is the predominant form of Cd in the soil solution (Tudoreanu and Phillips 2004). Cd is known to interfere with calcium (Ca) metabolism in mammals, causing itai-itai disease, an osteoporotic disease upon severe exposure.

Lead, a toxic metal, is a serious concern globally. Use of Pb on commercial scale causes serious health problems and environmental pollution. Sources of Pb contamination include fossil fuel burning, industrial processes, drinking water, food, smoking, and domestic sources. Lead contamination of plant shows a severe negative effect on plant growth. Najeeb et al. (2014) reported that high Pb concentration in plant suppresses the overall plant growth by production of excessive ROS which causes damage of lipid membrane, chlorophyll, and photosynthetic processes. Exposure of Pb even at its low concentration hampers the ion uptake mechanism of plant which disturbs the plant metabolism and ultimately inhibits plant growth (Lamhamdi et al. 2013). Lead toxicity also severely affects biological processes, viz., enzyme regulation, cell adhesion, ionic transportation, intra- and intercellular signaling, protein folding, maturation, apoptosis, and release of neurotransmitters.

Mercury is a naturally occurring metal which is very toxic and remarkably bioaccumulative. Mercury in its different forms exerts their toxicity in different ways. In the environment, Hg is present in mainly three forms: metallic elements, inorganic salts, and organic compounds. Among all methylmercury is more toxic and more bioavailable in comparison to other inorganic Hg. The main source of human intake of methylmercury is via seafood; it readily bioaccumulates through aquatic food chains (EFSA 2012; Mergler et al. 2007). Mercuric ions by reacting with sulfhydryl groups of biomolecules induce the Fenton reaction which results in oxidative stress, lipid peroxidation (Cargnelutti et al. 2006; Niu et al. 2014), and cell damage and disrupt signaling in plants (Chen et al. 2012). Toxicity of Pb causes malfunctioning of nerves, kidneys, and muscles though the main target is the brain. Lead interrupts with intracellular calcium homeostasis and causes Minamata disease (Vallee and Ulmer 1972).

10.3 Trafficking of Heavy Metals in Plants

Plants use transporters for the uptake of beneficial and essential elements from soil (Table 10.1), but as their selectivity is imperfect, they can also take up nonessential elements. In case of As, which exists as different species, viz., As(V) and As(III), different mechanisms operate. Phosphate transporters, viz.,Pht1;1 and Pht1;4 (Shin et al. 2004) and AtPht1;9, AtPht1;7, and AtPht1;8 (Remy et al. 2012; LeBlanc et al. 2013), are involved in As(V) uptake and transport. OsPht1;8 is also reported as rice As(V) transporter (Wu et al. 2011). In contrast, As(III) is transported via aquaglyceroporin channels more predominantly through nodulin 26-like intrinsic proteins (NIPs) (Zhao et al. 2010; Mosa et al. 2012) that include OsNIP2;1 (Lsi1) (Ma et al.

Table 10.1 Transporters involved in the transport of heavy metals in plants

Heavy metals	Site of transporters	Transporters	References
As	Root cell membrane	Lsi1, Lsi2, OsPht1;8, OsNIP3;2, OsNIP3;3, OsPIP2;4, OsPIP2;6, and OsPIP2;7, NRAMP1	Li et al. (2015)
	Vacuolar sequestration efflux transporter for xylem loading	OsABCC1, OsABCC2, Lsi2	Song et al. (2014) Ma et al. (2008)
Cd	Root cell membranes	OsNRAMP5, OsIRT1/OsIRT2	Clemens et al. (2013)
	Vacuolar import through the tonoplast in root cells	OsHMA3	Ueno et al. (2010)
	Xylem loading at root pericycle cells	OsHMA2	Takahashi et al. (2012) Clemens et al. (2013)
Pb	Root cell membrane	PDR12, OsHMA9, NRAMPs	Lee et al. (2005), Takahashi et al. (2012), and Cailliatte et al. (2010)
	Vacuolar sequestration	ABCC1, ABCC2	Song et al. (2010) and Park et al. (2012)
Hg	Vacuolar sequestration	ABCC YCF1	Park et al. (2012)

Table 10.2 Replacement of essential metals through heavy metals by using their transporters

Heavy metal	Essential element	Deficiency symptoms	Reference
As	P, Si, S, Se	Less erect and stunted growth, loss of energy, imbalanced storage of carbohydrates	Ahsan et al. (2008), Ma et al. (2008), Srivastava et al. (2016), and Dixit et al. (2016), Chauhan et al. (2017)
Cd	Zn, Mn, Fe, Ca	Reduced chlorophyll results in reduced photosynthesis, interveinal chlorosis, slow plant growth, premature leaf fall, severe stunting of leaves	Clemens et al. (2013), Sasaki et al. (2012), Uraguchi and Fujiwara (2012), and Kim et al. (2002)
Pb	Zn, Ca	Interveinal chlorosis, hindered flowering, yellow brown spots on leaves surrounded by sharp brown outlined edges	Clemens et al. (2013) and Kim et al. (2002)
Hg	Zn, Ca	Necrotic leaf margins on young leaves, death of terminal buds and root tips, stunted plant growth	Meagher and Heaton (2005)

2008), OsNIP3;2 (Bienert et al. 2008), and OsNIP3;3 (Katsuhara et al. 2014). Another class of aquaporins, plasma membrane intrinsic proteins (PIPs), is also involved in As(III) uptake and transport; it includes OsPIP2;4, OsPIP2;6, and OsPIP2;7 (Mosa et al. 2012). Among other classes of transporters, NRAMP1 (natural resistance-associated macrophage protein 1) (Tiwari et al. 2014) is also suggested for As(III) uptake and transport (Table 10.2).

For Cd, OsNramp1, OsNramp5, OsIRT1 (Iron Transporter 1), and OsIRT2 have been reported as main uptake transporters in rice (Nakanishi et al. 2006; Takahashi et al. 2011; Sasaki et al. 2012). While in *Arabidopsis thaliana*, AtNRAMP1, AtNRAMP3, and AtNRAMP4 are responsible for Cd uptake and transport (Babula et al. 2012). PDR8 (pleiotropic drug resistance), MRP6 (multidrug resistance protein), and MRP7 have also been reported for the transportation of Cd in *Arabidopsis* (Kim et al. 2007; Gaillard et al. 2008; Wojas et al. 2009). In addition OsHMA3 (heavy metal ATPase) is also shown to be involved in regulation of xylem loading of Cd in rice by mediating its sequestration in root cell (Miyadate et al. 2011; Ueno et al. 2010). OsHMA9 is also identified as Cd transporter (Lee et al. 2007). Uraguchi et al. (2011) have reported the role of OsLCT1 in regulating transportation of Cd into rice grains. Lead also exploits existing Ca channels of plant cells for its entry into plants (Sunkar et al. 2000). NRAMPs and P-type ATPase also transport Pb in different plant species (Simões et al. 2012; Argüello et al. 2007). Similar to Cd, Hg is also a toxic analogue of Zn and uses Zn transporters for its uptake into plants. Mercury finds the way for its uptake in plants through calcium channels (Zhao et al. 2010), aquaporins, and water channels. Johansson et al. (2000) have reported the blockage of aquaporins and water channels in white lupin because of high Hg uptake.

10.4 Physiological Changes

Physiological reactions like water uptake and transport in plants, transpiration, photosynthesis, and respiration are the crucial processes for sustaining normal plant growth (Table 10.3). However, in the presence of HMs, several of these processes are known to be affected. Several researchers reported that exposure of HMs reduced the water status of plants (Rucińska-Sobkowiak 2016; Srivastava et al. 2013), altered photosynthetic pigment level and photosynthetic efficiency (Bai et al. 2015), brought changes in respiratory processes, and also affected transpiration (Garg and Bhandari 2016; Kumar et al. 2016; Singh et al. 2017). When concentration of metals reaches beyond their optimal level in plants, it adversely affects the plant growth by

Table 10.3 Physiological and biochemical changes affected in different plant species under heavy metal stress

Heavy metals	Plant species	Remarks	References
As	*Arabidopsis*, *H. verticillata*, *C. demersum*, *B. juncea*, *O. sativa*, *A. thaliana*, tobacco	Physiological parameters (root-shoot length, biomass, and chlorophyll) decrease	Li et al. (2004), Gasic and Korban (2007), Abercrombie et al. (2008), Mishra et al. (2008), Srivastava et al. (2007), Sung et al. (2009), Wojas et al. (2010), Kumar et al. (2014), and Chauhan et al. (2017)
		APS, SAT, CS, GST, PCS activity and transcripts were found to be elevated	
		As exposure increased the level of PCs	
Cd	*Ricinus communis*	Reduction in dry biomass, root, and shoot length	Iannelli et al. (2002), Boominathan and Doran (2003); Metwally et al. (2003), Cuypers et al. (2009), Bauddh et al. (2016), and Ahmad et al. (2017)
	Brassica juncea	Increase level of GR, GST, GPX activities and AsA, GSH, and proline level	
	Phragmites australis	H_2O_2 and MDA level increased	
	Pisum sativum		
	Thlaspi caerulescens		
Pb	*Phaseolus vulgaris* L., *Najas indica*, *H. vulgare*, *Avena sativa*	Decreases in growth and photosynthetic parameters	Kaznina et al. (2005), Cuypers et al. (2009), Singh et al. 2010, Gupta et al. (2010), and Aldoobie and Beltagi (2013)
	Sedum alfredii, *Nicotiana tabacum*, *O. sativa*	Increased MDA, EC, and H_2O_2 content	Ashraf and Tang (2017)
		High GSH level	
Hg	*Lupinus albus*, *Sesbania drummondii*, *Z. mays*, *Medicago sativa*, *Phaseolus vulgaris*, *Lemna minor*	Decrease water content	Israr et al. (2006), Rellán-Álvarez et al. (2006), Ortega-Villasante et al. (2007), Esteban et al. (2008), Sharma and Shrivastava (2014), and Zhang et al. (2017)
		Induced oxidative stress by modulating nonenzymatic antioxidants (GSH and NPSH) GSH content and GSH/GSSG ratio increased H_2O_2	

inhibiting activity of cytoplasmic enzymes and through damage of cellular structures because of oxidative stress (Jadia and Fulekar 2008) or via replacement of essential cation ions from their specific sites in plants (Taiz and Zeiger 2002). Essential enzyme activities for plant metabolism may also be hampered due to heavy metal interference. These toxic effects hamper the growth of plants, which sometimes results in plant death (Schaller and Diez 1991). The molecular mechanisms of these physiological effects have been delineated in certain cases. Water status of plants has been found to be altered due to changes in expression of aquaporin channels upon exposure to HMs like As (Srivastava et al. 2013). Aquaporin channels are a major route for water uptake in plants, and it was found in case of As that plants decreased their expression so as to prevent As(III) entry in plants. However, this tolerance mechanism negatively affected water homeostasis and in turn led to reduced growth of plants (Srivastava et al. 2013). In case of Cd also negative changes in expression of aquaporin channels are reported (Yamaguchi et al. 2009). Photosynthetic, transpiration, and respiratory processes are directly or indirectly linked to stomatal opening. It is well known that water stress of plants in turn leads to closure of stomata so as to save water. This also affects, however, gaseous exchange and subsequently photosynthesis and respiration. The negative effects of HMs on gaseous exchange are reported (Azevedo et al. 2005; Srivastava et al. 2013). There is a need to understand detailed molecular features of physiological changes in plants under HM toxicity in future.

10.5 Biochemical Changes

10.5.1 Oxidative Stress

The exposure of HMs to plants stimulates the ROS production and generation of free radicals in plants, thus leading to oxidative stress (Meharg and Hartley-Whitaker 2002). The negative effects of photooxidative stress are marked as production of ROS by light-dependent processes as the reactions linked with photosynthesis and photorespiration are the main source of ROS within plant cells (Foyer and Noctor 2003). Protein oxidation and generation of ROS in mitochondria are considered as a causative factor to the "oxidative stress" syndrome in plants (Sweetlove et al. 2002; Kristensen et al. 2004). Both As species (AsIII and AsV) induce ROS production above optimal level, which is controlled by plant through the involvement of various enzymatic and nonenzymatic antioxidants (Srivastava et al. 2007). Chauhan et al. (2017) also reported the induced formation of ROS in As-treated root and shoot, and it was also confirmed by histochemical detection of ROS via nitroblue tetrazolium (NBT) and diaminobenzidine (DAB) staining for superoxide and H_2O_2, respectively, in both root and shoot. The sensitivity of thiol-modulated enzymes of the Benson–Calvin cycle to oxygen and ROS is well known since the 1970s (Kaiser 1979). Much more recently, sensitivity to lipid peroxides has been described for plant mitochondrial proteins. However, when ROS synthesis increases at a high rate under HM toxicity, this leads to oxidative stress development as scavenging mechanisms can no longer operate at the same rate (Srivastava et al. 2007). Hence, there is an optimum tolerance limit to each HM, viz., As, Cu, and Cd, in plants up to which they regulate ROS effectively, and beyond that oxidative stress ensues (Srivastava et al. 2007; Chauhan et al. 2017). Oxidative stress becomes evident through peroxidation of membrane lipids leading to increased production of malondialdehyde (MDA) and other lipid peroxides. Further, proteins are broken down due to oxidative stress which is experimentally seen through increased concentration of carbonyls (Srivastava et al. 2012).

10.5.2 Thiolic Compounds

Plants require the HMs in trace, but when the concentration of these elements reaches beyond the certain tolerance level, then plants cope up with this enhanced level of HMs by various mechanisms. It includes complexation of HMs and vacuolar sequestration through chelating ligands. Among which the phytochelatins (PCs) and metallothioneins (MTs) are the best metal-chelating ligands in plant cells. Both MTs and PCs are the polypeptides which are rich in cysteine; however, MTs are encoded by a family of genes and PCs are enzymatically synthesized. Grill et al. (1987) first demonstrated PCs to be specific plant-derived polypeptides functioning specifically for complexation of HMs. The synthesis of PCs occurs from binding of glutathione moieties together with the help of enzyme phytochelatin synthase leading to increasing repetitions of the γ-GluCys dipeptide followed by a terminal Gly, (γ-GluCys)n-Gly, where n is generally in the range of two to five. In response to HMs, synthesis of PCs has been reported in a variety of plants including some microorganisms (Grill et al. 1989). PCs complex with various HMs, viz.,Pb, Ag, As, Cd, and Hg, and sequester them into vacuole, thus reducing the concentration of free HM ion in the cytoplasm. This has been referred to as primary detoxification mechanisms in plants as this avoids HM interaction with cellular components and consequently any oxidative stress too (Srivastava et al. 2007). However, the capacity of primary detoxification mechanism through increased synthesis of PCs depends on the activity of enzymes of thiol metabolism (both anabolic and catabolic) and sulfur availability (Srivastava et al. 2016). PCs have been found to play a major role against toxic metals like Cd, As, and Hg (Carrasco-Gil et al. 2011; Song et al. 2010, Kim et al. 2017), while MTs are predominantly active for the homeostasis of essential metals like Cu and Zn. Nonetheless, both PCs and MTs play some role in HM homeostasis and complexation of all metals (Zimeri et al. 2005). Manipulating the expression of PCs and MTs is a possible important mechanism to enhance plant tolerance against HMs stress.

10.6 Conclusions

The last century witnessed the human population burst that enhanced the pressure on natural resources. To support this population burst, excessive use of HM-containing pesticides caused the enhanced level of HMs in the atmosphere. These HMs are toxic for humans as well as plants with a varied degree of toxicity. Heavy metals find its route to enter into human food chain through their accumulation in plants where they find place through uptake and transport via transporters of essential element according to their structural analogy. Heavy metal uptake severely hampers the plant growth and development from germination to their grain yield. Oxidative stress is a common symptom of almost all of HM toxicity. To withstand the HMs toxicity, plants are equipped with various physiological processes. One of most important processes is the vacuole sequestration HMs effected by the increased synthesis of MTs and PCs. With the modern omics tools, more and more knowledge is being gathered with the hope of delineation of detailed molecular features of physiological and biochemical processes in the near future.

Acknowledgments The authors are thankful to the director, CSIR-NBRI, Lucknow and Botany Department, University of Lucknow, Lucknow, for the facilities and financial support. Reshu Chauhan is thankful to the Department of Science and Technology (DST), New Delhi, for the award of DST-Inspire Junior/Senior Research Fellowship. Surabhi Awasthi is thankful to CSIR, New Delhi, for the award of CSIR-Senior Research Fellowship. APS is thankful to CSIR, New Delhi, for the award of RAship in Emeritus Scientist project. Amit Kumar is thankful to DSKPDF Cell, Pune, India, and University Grant Commission, New Delhi, India, for the award of the D.S. Kothari Postdoctoral Fellowship.

References

Abercrombie JM, Halfhill MD, Ranjan P, Rao MR, Saxton AM, Yuan JS, Stewart CN Jr (2008) Transcriptional responses of *Arabidopsis thaliana* plants to As(V) stress. BMC Plant Biol 8:87–96

Ahmad P, Alyemeni MN, Wijaya L, Alam P, Ahanger MA, Alamri SA (2017) Jasmonic acid alleviates negative impacts of cadmium stress by modifying osmolytes and antioxidants in faba bean (*Vicia faba L.*). Arch Agron Soil Sci 1–11

Ahsan N, Lee DG, Alam I, Kim PJ, Lee JJ, Ahn YO, Kwak SS, Lee IJ, Bahk JD, Kang KY, Renaut J (2008) Comparative proteomic study of arsenic-induced differentially expressed proteins in rice roots reveals glutathione plays a central role during as stress. Proteomics 8(17):3561–3576

Aldoobie NF, Beltagi MS (2013) Physiological, biochemical and molecular responses of common bean (Phaseolus vulgaris L.) plants to heavy metals stress. Afr J Biotechnol 12(29):4614–4622

Argüello JM, Eren E, González-Guerrero M (2007) The structure and function of heavy metal transport P1B-ATPases. Biometals 20(3–4):233

Ashraf U, Tang X (2017) Yield and quality responses, plant metabolism and metal distribution pattern in aromatic rice under lead (Pb) toxicity. Chemosphere 176:141–155

Azevedo H, Glória Pinto CG, Fernandes J, Loureiro S, Santos C (2005) Cadmium effects on sunflower growth and photosynthesis. J Plant Nutr 28(12):2211–2220

Babula P, Masarik M, Adam V, Eckschlager T, Stiborova M, Trnkova L, Skutkova H, Provaznik I, Hubalek J, Kizek R (2012) Mammalian metallothioneins: properties and functions. Metallomics 4:739–750

Bai XY, Dong YJ, Wang QH, Xu LL, Kong J, Liu S (2015) Effects of lead and nitric oxide on photosynthesis, antioxidative ability, and mineral element content of perennial ryegrass. Biol Plant 59(1):163–170

Bauddh K, Kumar A, Srivastava S, Singh RP, Tripathi RD (2016) A study on the effect of cadmium on the antioxidative defense system and alteration in different functional groups in castor bean and Indian mustard. Arch Agron Soil Sci 62(6):877–891

Bienert GP, Thorsen M, Schüssler MD, Nilsson HR, Wagner A, Tamás MJ, Jahn TP (2008) A subgroup of plant aquaporins facilitate the bi-directional diffusion of As (OH) 3 and Sb (OH) 3 across membranes. BMC Biol 6(1):26

Boominathan R, Doran PM (2003) Cadmium tolerance and antioxidative defenses in hairy roots of the cadmium hyperaccumulator, Thlaspi caerulescens. Biotechnol Bioeng 83(2):158–167

Cailliatte R, Schikora A, Briat JF, Mari S, Curie C (2010) High-affinity manganese uptake by the metal transporter NRAMP1 is essential for Arabidopsis growth in low manganese conditions. Plant Cell 22(3):904–917

Cargnelutti D, Tabaldi LA, Spanevello RM, de Oliveira Jucoski G, Battisti V, Redin M, Linares CEB, Dressler VL, de Moraes Flores EM, Nicoloso FT, Morsch VM (2006) Mercury toxicity induces oxidative stress in growing cucumber seedlings. Chemosphere 65(6):999–1006

Carrasco-Gil SANDRA, Álvarez-Fernández ANA, Sobrino-Plata JUAN, Millan R, Carpena-Ruiz RO, Leduc DL, Andrews JC, Abadia J, Hernandez LE (2011) Complexation of Hg with phytochelatins is important for plant Hg tolerance. Plant Cell Environ 34(5):778–791

Castagnetto JM, Hennessy SW, Roberts VA, Getzoff ED, Tainer JA, Pique ME (2002) MDB: the metalloprotein database and browser at the Scripps Research Institute. Nucleic Acids Res 30(1):379–382

Chauhan R, Awasthi S, Tripathi P, Mishra S, Dwivedi S, Niranjan A, Mallick S, Tripathi P, Pande V, Tripathi RD (2017) Selenite modulates the level of phenolics and nutrient element to alleviate the toxicity of arsenite in rice (Oryzasativa L.) Ecotoxicol Environ Saf 138:47–55

Chen H, Zheng C, Tu, C. Z., YG (1999) Heavy metal pollution in soils in China: status and countermeasures. Ambio 28:130–131

Chen J, Yang ZM (2012) Mercury toxicity, molecular response and tolerance in higher plants. Biometals 25(5):847–857

Clemens S, Aarts MG, Thomine S, Verbruggen N (2013) Plant science: the key to preventing slow cadmium poisoning. Trends Plant Sci 18:92–99

Cuypers A, Smeets K, Vangronsveld J (2009) Heavy metal stress in plants. Plant stress biology: from genomics to systems biology. Wiley-VCH Verlag, Weinheim, pp 161–178

DalCorso G, Manara A, Furini A (2013) An overview of heavy metal challenge in plants: from roots to shoots. Metallomics 5(9):1117–1132

Dixit G, Singh AP, Kumar A, Mishra S, Dwivedi S, Kumar S, Trivedi PK, Pandey V, Tripathi RD (2016) Reduced arsenic accumulation in rice (*Oryza sativa* L.) shoot involves sulfur mediated improved thiol metabolism, antioxidant system and altered arsenic transporters. Plant Physiol Biochem 99:86–96

Esteban E, Moreno E, Peñalosa J, Cabrero JI, Millán R, Zornoza P (2008) Short and long-term uptake of Hg in white lupin plants: kinetics and stress indicators. Environ Exp Bot 62:316–322

European Food Safety System (EFSA) (2012) ISBN: 978-92-9199-543-1: 53

Foyer CH, Noctor G (2003) Redox sensing and signalling associated with reactive oxygen in chloroplasts, peroxisomes and mitochondria. Physiol Plant 119(3):355–364

Gaillard S, Jacquet H, Vavasseur A, Leonhardt N, Forestier C (2008) AtMRP6/AtABCC6, an ATP-binding cassette transporter gene expressed during early steps of seedling development and up-regulated by cadmium in Arabidopsis thaliana. BMC Plant Biol 8(1):22

Garg N, Bhandari P (2016) Interactive effects of silicon and arbuscular mycorrhiza in modulating ascorbate-glutathione cycle and antioxidant scavenging capacity in differentially salt-tolerant Cicer arietinum L. genotypes subjected to long-term salinity. Protoplasma 253(5):1325–1345

Gasic K, Korban SS (2007) Transgenic Indian mustard (*Brassica juncea*) plants expressing an *Arabidopsis* phytochelatin synthase (*AtPCS1*) exhibit enhanced As and Cd tolerance. Plant Mol Biol 64:361–369

Grill E, Winnacker EL, Zenk MH (1987) Phytochelatins, a class of heavy-metal-binding peptides from plants, are functionally analogous to metallothioneins. Proc Natl Acad Sci 84(2):439–443

Grill E, Löffler S, Winnacker EL, Zenk MH (1989) Phytochelatins, the heavy-metal-binding peptides of plants, are synthesized from glutathione by a specific γ-glutamylcysteine dipeptidyl transpeptidase (phytochelatin synthase). Proc Natl Acad Sci 86(18):6838–6842

Gupta DK, Huang HG, Yang XE, Razafindrabe BHN, Inouhe M (2010) The detoxification of lead in Sedum alfredii H. is not related to phytochelatins but the glutathione. J Hazard Mater 177:1–3. 437–444

Iannelli MA, Pietrini F, Fiore L, Petrilli L, Massacci A (2002) Antioxidant response to cadmium in Phragmites australis plants. Plant Physiol Biochem 40:977–982

Irfan M, Hayat S, Ahmad A, Alyemeni MN (2013) Soil cadmium enrichment: allocation and plant physiological manifestations. Saudi J Biol Sci 20(1):1–10

Israr M, Sahi S, Datta R, Sarkar D (2006) Bioaccumulation and physiological effects of mercury in Sesbania drummondii. Chemosphere 65(4):591–598

Jadia CD, Fulekar MH (2008) Phytotoxicity and remediation of heavy metals by fibrous root grass (sorghum). J Appl Biosci 10:491–499

Järup L (2003) Hazards of heavy metal contamination. Br Med Bull 68(1):167–182

Johansson I, Karlsson M, Johanson U, Larsson C, Kjellbom P (2000) The role of aquaporins in cellular and whole plant water balance. Biochim Biophys Acta 1465:324–342

Kaiser WM (1979) Reversible inhibition of the Calvin cycle and activation of oxidative pentose phosphate cycle in isolated intact chloroplasts by hydrogen peroxide. Planta 145(4):377–382

Katsuhara M, Sasano S, Horie T, Matsumoto T, Rhee J, Shibasaka M (2014) Functional and molecular characteristics of rice and barley NIP aquaporins transporting water, hydrogen peroxide and arsenite. Plant Biotechnol 31(3):213–219

Kaznina NM, Laidinen GF, Titov AF, Talanov AV (2005) Effect of lead on the photosynthetic apparatus of annual grasses. Biol Bull 32:147–150

Kim YY, Yang YY, Lee Y (2002) Pb and Cd uptake in rice roots. Physiol Plant 116(3):368–372

Kim DY, Bovet L, Maeshima M, Martinoia E, Lee Y (2007) The ABC transporter AtPDR8 is a cadmium extrusion pump conferring heavy metal resistance. Plant J 50(2):207–218

Kim YO, Bae HJ, Cho E, Kang H (2017) Exogenous glutathione enhances mercury tolerance by inhibiting mercury entry into plant cells. Front Plant Sci 8:683

Kristensen BK, Askerlund P, Bykova NV, Egsgaard H, Møller IM (2004) Identification of oxidised proteins in the matrix of rice leaf mitochondria by immunoprecipitation and two-dimensional liquid chromatography-tandem mass spectrometry. Phytochemistry 65(12):839–1851

Kumar A, Singh RP, Singh PK, Awasthi S, Chakrabarty D, Trivedi PK, Tripathi RD (2014) Selenium ameliorates arsenic induced oxidative stress through modulation of antioxidant enzymes and thiols in rice (*Oryza sativa* L.) Ecotoxicology 23:1153–1163

Kumar A, Dixit G, Singh AP, Srivastava S, Mishra K, Tripathi RD (2016) Selenate mitigates arsenite toxicity in rice (*Oryza sativa* L.) by reducing arsenic uptake and ameliorates amino acid content and thiol metabolism. Ecotox Environ Safety 133:350–359

Lamhamdi M, El Galiou O, Bakrim A, Nóvoa-Muñoz JC, Arias-Estévez M, Aarab A, Lafont R (2013) Effect of lead stress on mineral content and growth of wheat (Triticum aestivum) and spinach (Spinacia oleracea) seedlings. Saudi J Biol Sci 20(1):29–36

LeBlanc MS, McKinney EC, Meagher RB, Smith AP (2013) Hijacking membrane transporters for arsenic phytoextraction. J Biotechnol 163(1):1–9

Lee M, Lee K, Lee J, Noh EW, Lee Y (2005) AtPDR12 contributes to lead resistance in Arabidopsis. Plant Physiol 138:827–836

Lee S, Kim YY, Lee Y, An G (2007) Rice P1B-type heavy-metal ATPase, OsHMA9, is a metal efflux protein. Plant Physiol 145(3):831–842

Li Y, Dhankher O, Carreira L, Lee D, Chen A, Schroeder J, Balish R, Meagher R (2004) Overexpression of phytochelatin synthase in *Arabidopsis* leads to enhanced arsenic tolerance and cadmium sensitivity. Plant Cell Physiol 45:1787–1797

Li N, Wang J, Song WY (2015) Arsenic uptake and translocation in plants. Plant Cell Physiol 57(1):4–13

Ma JF, Yamaji N, Mitani N, Xu XY, Su YH, McGrath SP, Zhao FJ (2008) Transporters of arsenite in rice and their role in arsenic accumulation in rice grain. Proc Natl Acad Sci 105(29):9931–9935

Meagher RB, Heaton AC (2005) Strategies for the engineered phytoremediation of toxic element pollution: mercury and arsenic. J Ind Microbiol Biotechnol 32:502–513

Meharg AA, Hartley-Whitaker J (2002) Arsenic uptake and metabolism in arsenic resistant and nonresistant plant species. New Phytol 154(1):29–43

Mergler D, Anderson HA, Chan LHM, Mahaffey KR, Murray M, Sakamoto M, Stern AH (2007) Methylmercury exposure and health effects in humans: a worldwide concern. AMBIO J Hum Environ 36(1):3–11

Metwally A, Finkemeier I, Georgi M, Dietz KJ (2003) Salicylic acid alleviates the cadmium toxicity in barley seedlings. Plant Physiol 132:272–281

Mishra S, Srivastava S, Tripathi RD, Dwivedi S, Trivedi PK (2008) Thiol metabolism and antioxidant systems complement each other during arsenate detoxification in *Ceratophyllum demersum* L. Aquat Toxicol 86:205–215

Miyadate H, Adachi S, Hiraizumi A, Tezuka K, Nakazawa N, Kawamoto T, Katou K, Kodama I, Sakurai K, Takahashi H, Satoh-Nagasawa N (2011) OsHMA3, a P1B-type of ATPase affects root-to-shoot cadmium translocation in rice by mediating efflux into vacuoles. New Phytol 189:190–199

Mosa KA, Kumar K, Chhikara S, Mcdermott J, Liu Z, Musante C, White JC, Dhankher OP (2012) Members of rice plasma membrane intrinsic proteins subfamily are involved in arsenite permeability and tolerance in plants. Transgenic Res 21(6):1265–1277

Najeeb U, Ahmad W, Zia MH, Zaffar M, Zhou W (2014) Enhancing the lead phytostabilization in wetland plant Juncuseffusus L. through somaclonal manipulation and EDTA enrichment. Arab J Chem

Nakanishi H, Ogawa I, Ishimaru Y, Mori S, Nishizawa NK (2006) Iron deficiency enhances cadmium uptake and translocation mediated by the Fe2+ transporters OsIRT1 and OsIRT2 in rice. Soil Sci Plant Nutr 52(4):464–469

Niu Z, Zhang X, Wang S, Zeng M, Wang Z, Zhang Y, Ci Z (2014) Field controlled experiments on the physiological responses of maize (Zea mays L.) leaves to low-level air and soil mercury exposures. Environ Sci Pollut Res 21(2):1541–1547

Ortega-Villasante C, Hernández LE, Rellán-Álvarez R, Del Campo FF, Carpena-Ruiz RO (2007) Rapid alteration of cellular redox homeostasis upon exposure to cadmium and mercury in alfalfa seedlings. New Phytol 176:96–107

Oves M, Saghir Khan M, Huda Qari A, NadeenFelemban M, Almeelbi T (2016) Heavy metals: biological importance and detoxification strategies. J Bioremed Biodegr 7(334):2

Park J, Song WY, Ko D, Eom Y, Hansen TH, Schiller M, Lee TG, Martinoia E, Lee Y (2012) The phytochelatin transporters AtABCC1 and AtABCC2 mediate tolerance to cadmium and mercury. Plant J 69(2):278–288

Rellán-Álvarez R, Ortega-Villasante C, Álvarez-Fernández A, del Campo FF, Hernández LE (2006) Stress responses of Zea mays to cadmium and mercury. Plant Soil 79:41–50

Remy E, Cabrito TR, Batista RA, Teixeira MC, Sá-Correia I, Duque P (2012) The Pht1; 9 and Pht1; 8 transporters mediate inorganic phosphate acquisition by the Arabidopsis thaliana root during phosphorus starvation. New Phytol 195(2):356–371

Rucińska-Sobkowiak R (2016) Water relations in plants subjected to heavy metal stresses. Acta Physiol Plant 38(11):257

Sasaki A, Yamaji N, Yokosho K, Ma JF (2012) Nramp5 is a major transporter responsible for manganese and cadmium uptake in rice. Plant Cell 24(5):2155–2167

Schaller A, Diez T (1991) Plant specific aspects of heavy metal uptake and comparison with quality standards for food and forage crops. Der Einfluß von festenAbfällen auf Böden, Pflanzen 92–125

Sharma SS, Dietz KJ (2009) The relationship between metal toxicity and cellular redox imbalance. Trends Plant Sci 14(1):43–50

Sharma J, Shrivastava S (2014) Physiological and morphological responses of Phaseolus vulgaris caused by mercury stress under lab conditions. Recent Adv Biol Med 1:7

Shin H, Shin HS, Dewbre GR, Harrison MJ (2004) Phosphate transport in Arabidopsis: Pht1; 1 and Pht1; 4 play a major role in phosphate acquisition from both low-and high-phosphate environments. Plant J 39(4):629–642

Siddiqui S (2015) DNA damage in Cicer plant grown on soil polluted with heavy metal. J King Saud Univ Sci 27:217–223

Simões CC, Melo JO, Magalhaes JV, Guimarães CT (2012) Genetic and molecular mechanisms of aluminum tolerance in plants. Genet Mol Res 11(3):1949–1957

Singh AP, Dixit G, Kumar A, Mishra S, Kumar N, Dixit S, Singh PK, Dwivedi S et al (2017) A protective role for nitric oxide and salicylic acid for arsenite phytotoxicity in rice (*Oryzasativa* L.) Plant Physiol Biochem 115:163–173

Singh R, Tripathi RD, Dwivedi S, Kumar A, Trivedi PK, Chakrabarty D (2010) Lead bioaccumulation potential of an aquatic macrophyte Najas indica are related to antioxidant system. Bioresour Technol 101(9):3025–3032

Song WY, Park J, Mendoza-Cózatl DG, Suter-Grotemeyer M, Shim D, Hörtensteiner S, Geisler M et al (2010) Arsenic tolerance in Arabidopsis is mediated by two ABCC-type phytochelatin transporters. Proc Natl Acad Sci U S A 107:21187–21192

Song WY, Yamaki T, Yamaji N, Ko D, Jung KH, Fujii-Kashino M et al (2014) A rice ABC transporter, OsABCC1, reduces arsenic accumulation in the grain. Proc Natl Acad Sci U S A 111:15699–15704

Srivastava S, Mishra S, Tripathi RD, Dwivedi S, Trivedi PK, Tandon PK (2007) Phytochelatins and antioxidant systems respond differentially during arsenite and arsenate stress in Hydrillaverticillata (Lf) Royle. Environ Sci Technol 41(8):2930–2936

Srivastava S, Suprasanna P, D'souza SF (2012) Mechanisms of arsenic tolerance and detoxification in plants and their application in transgenic technology: a critical appraisal. Int J Phytoremediat 14(5):506–517

Srivastava S, Akkarakaran JJ, Sounderajan S, Shrivastava M, Suprasanna P (2016) Arsenic toxicity in rice (Oryza sativa L.) is influenced by sulphur supply: impact on the expression of transporters and thiol metabolism. Geoderma 270:33–42

Srivastava S, Srivastava AK, Suprasanna P, D'souza SF (2013) Identification and profiling of arsenic stress-induced microRNAs in Brassica juncea. J Exp Bot 64:303–315

Sung DY, Kim TH, Komives EA, Mendoza-Cózatl DG, Schroeder JI (2009) ARS5 is a component of the 26S proteasome complex, and negatively regulates thiol biosynthesis and arsenic tolerance in *Arabidopsis*. Plant J 59:802–812

Sunkar R, Kaplan B, Bouche N, Arazi T, Dolev D, Talke IN, Maathuis FJ, Sanders D, Bouchez D, Fromm H (2000) Expression of a truncated tobacco NtCBP4 channel in transgenic plants and disruption of the homologous Arabidopsis CNGC1 gene confer Pb2+ tolerance. Plant J 24(4):533–542

Sweetlove LJ, Heazlewood JL, Herald V, Holtzapffel R, Day DA, Leaver CJ, Millar AH (2002) The impact of oxidative stress on Arabidopsis mitochondria. Plant J 32(6):891–904

Taiz L, Zeiger E (2002) Plant physiology. Sinauer Associates, Sunderland

Takahashi R, Ishimaru Y, Senoura T, Shimo H, Ishikawa S, Arao T, Nakanishi H, Nishizawa NK (2011) The OsNRAMP1 iron transporter is involved in Cd accumulation in rice. J Exp Bot 62(14):4843–4850

Takahashi R, Bashir K, Ishimaru Y, Nishizawa NK, Nakanishi H (2012) The role of heavy-metal ATPases, HMAs, in zinc and cadmium transport in rice. Plant Signal Behav 7(12):1605–1607

Tiwari M, Sharma D, Dwivedi S, Singh M, Tripathi RD, Trivedi PK (2014) Expression in Arabidopsis and cellular localization reveal involvement of rice NRAMP, OsNRAMP1, in arsenic transport and tolerance. Plant Cell Environ 37(1):140–152

Tripathi RD, Srivastava S, Mishra S, Singh N, Tuli R, Gupta DK et al (2007) Arsenic hazards: strategies for tolerance and remediation by plants. Trends Biotechnol 25:158–165

Tudoreanu L, Phillips CJC (2004) Modeling cadmium uptake and accumulation in plants. Advances în. Agronomy 84:121–157

Ueno D, Yamaji N, Kono I, Huang CF, Ando T, Yano M, Ma JF (2010) Gene limiting cadmium accumulation in rice. Proc Natl Acad Sci 107(38):16500–16505

Uraguchi S, Fujiwara T (2012) Cadmium transport and tolerance in rice: perspectives for reducing grain cadmium accumulation. Rice 5(1):5

Uraguchi S, Kamiya T, Sakamoto T, Kasai K, Sato Y, Nagamura Y, Yoshida A, Kyozuka J, Ishikawa S, Fujiwara T (2011) Low-affinity cation transporter (OsLCT1) regulates cadmium transport into rice grains. Proc Natl Acad Sci 108(52):20959–20964

Vallee BL, Ulmer DD (1972) Biochemical effects of mercury, cadmium, and lead. Annu Rev Biochem 41(1):91–128

Wojas S, Hennig J, Plaza S, Geisler M, Siemianowski O, Skłodowska A, Ruszczyńska A, Bulska E, Antosiewicz DM (2009) Ectopic expression of Arabidopsis ABC transporter MRP7 modifies cadmium root-to-shoot transport and accumulation. Environ Pollut 157(10):2781–2789

Wojas S, Clemens S, Sklodowska A, Antosiewicz DA (2010) Arsenic response of AtPCS1-and CePCS-expressing plants–effects of external As(V) concentration on As-accumulation pattern and NPT metabolism. J Plant Physiol 167:169–175

Wu C, Ye Z, Shu W, Zhu Y, Wong M (2011) Arsenic accumulation and speciation in rice are affected by root aeration and variation of genotypes. J Exp Bot 62(8):2889–2898

Yamaguchi H, Fukuoka H, Arao T, Ohyama A, Nunome T, Miyatake K, Negoro S (2009) Gene expression analysis in cadmium-stressed roots of a low cadmium-accumulating solanaceous plant, Solanum torvum. J Exp Bot 61(2):423–437

Zhang T, Lu Q, Su C, Yang Y, Hu D, Xu Q (2017) Mercury induced oxidative stress, DNA damage, and activation of antioxidative system and Hsp70 induction in duckweed (*Lemna minor*). Ecotoxicol Environ Saf 143:46–56

Zhao FJ, McGrath SP, Meharg AA (2010) Arsenic as a food chain contaminant: mechanisms of plant uptake and metabolism and mitigation strategies. Annu Rev Plant Biol 61:535–559

Zimeri AM, Dhankher OP, McCaig B, Meagher RB (2005) The plant MT1 metallothioneins are stabilized by binding cadmiums and are required for cadmium tolerance and accumulation. Plant mol Boil 58(6):839–855

Plant Adaptation to Recalcitrant Chemicals

11

Hina Khatoon, Apourv Pant, and J.P.N. Rai

Abstract

The massive growth of urbanization, agriculture, and other anthropogenic activities has created serious environmental pollution with the use of numerous recalcitrant chemicals. Various polycyclic aromatic hydrocarbons, aromatic hydrocarbons, nitroaromatic compounds, polychlorinated biphenyls, dioxins, and their derivatives are greatly toxic, mutagenic, and carcinogenic to natural microflora as well as to humans. In the biodegradation of toxic organic compounds, xenobiotics, and heavy metals, hyperaccumulating plants and plant-associated bacteria (including endophytic bacteria and rhizospheric bacteria) have enormous potential to improve phytoremediation. In particular, endophytic and rhizospheric bacteria may break down recalcitrant chemicals in polluted soil. Approaches from toxicology, stress physiology, and ecotoxicology will be helpful for the characterization of plant responses to recalcitrant chemicals. Therefore, it is necessary to understand plants' adaptation strategies against various harmful recalcitrant chemicals.

Keywords

Xenobiotics · Phytoremediation · Endophytic bacteria · Ecotoxicology · Recalcitrant chemicals

H. Khatoon (✉) • A. Pant • J.P.N. Rai
Department of Environmental Science, G.B. Pant University of Agriculture and Technology, Pantnagar, Uttarakhand, India
e-mail: hinakatoon29@gmail.com

V. Shukla et al. (eds.), *Plant Adaptation Strategies in Changing Environment*,
https://doi.org/10.1007/978-981-10-6744-0_11

11.1 Introduction

Synthetic chemicals are found in large quantities in the environment. Anthropogenic chemicals are foreign compounds made by man. The chemicals are distributed in various parts of the environment and can be transported long or short distances. These chemicals generally can undergo various chemical reactions and transformations. A hyperaccumulating plant is an important part of the terrestrial food web and thus is central to both agricultural and natural ecosystems. Similar attributes have been reported for precipitation and the concentration of adulterated resources; these also predispose plants to the accumulation of some anthropogenic substances, which is called bioconcentration (Bernes 1998). Conditions such as hydrophobicity, water solubility, and vapor pressure are functions of the chemical and physical properties of each pollutant; by these pathways, organic pollutants can penetrate the vegetation. Some environmental characteristics, such as temperature, organic content of the soil, and plant species, also affect uptake. Most toxic pollutants are hydrophobic; thus, the large, lipid-covered surface of leaves is an ideal location for the accumulation of these chemicals.

Numerous mechanisms are present in plants to protect them from the uptake and transport of anthropogenic chemicals. Because of the nature of these chemicals, hydrophobic organic pollutants are not readily transported in the phloem. Active enzymes secreted by plants are responsible for the rapid degradation of recalcitrant chemicals (Trapp and McFarlane 1995). In the degradation of anthropogenic chemicals in plants and soil, water is also considered to be a solvent, transport medium, and reactant for uptake and translocation (Hopkins 1999). The area outside the endodermis, which has an extensive surface area, may be involved in the movement and adsorption of pollutants (Trapp and McFarlane 1995).

In pollution studies, the characteristics of recalcitrant chemicals are often investigated, including their structural complexity, slow biodegradability, and biomagnification potential. Two aspects of plant functioning are of concern in the remediation process: favorable conditions for microbial degradation and assessing contaminants in the soil via the roots.

11.2 Recalcitrant Chemicals

Artificial compounds that are somewhat biodegradable or nonbiodegradable are known as *recalcitrant compounds*. These compounds range from simple halogenated hydrocarbons to complex polymers.

11.2.1 Heavy Metals

Generally, higher-density heavy metals are considered to be toxic if their concentrations are higher than the permissible limit. Many heavy metals are important for the growth of living cells, including iron, manganese, zinc, copper, cobalt, and molybdenum. Other heavy metals show toxicity toward these living cells, such as uranium, lead, cadmium, thallium, chromium, silver, and mercury. The heavy metals present in plants have a definite threshold value of toxicity (Ernst WHO 1982). Plants that have been are particularly modified by heavy metals in soil are called metallophytes. The concentrations of heavy metals vary in different soils: for example, zinc-rich soils (0.1–10.0% Zn) may contain a high content of Pb but not Cd, whereas higher concentrations of other heavy metals (e.g., Zn, Pb, Co, Ni, Cd) may be found in soils with a high copper content (0.1–3.2% Cu) (Ernst WHO 1974).

Many heavy metal soils have been widely exploited since early times by human activity. However, heavy metal soils are still widespread and support metallophyte vegetation (Becker and Dierschke 2008; Lucassen et al. 2010). Heavy metals are nonbiodegradable and persistent in nature; thus, they are the cause of several serious health problems. Because they are stored in the kidneys, bones, and liver, they may lead to the malfunctioning of these essential organs (Duruibe et al. 2007). Various heavy metals (e.g. Zn, Cu, Pb, Mn, Ni, Cr, Cd, As) are dispersed in the environment via wastewater and cause serious hazards to different levels of the food chain (Singh et al. 2010).

Heavy metals cause dissimilar problems in humans and plants. At concentrations below the acceptable limit, some heavy metals are necessary for plant growth and development (Li et al. 2013). The soil residence times of heavy metals are in the thousands of years, which leads to health dangers for higher organisms (Vinodhini and Narayanan 2009). Plant growth and productivity are inhibited by heavy metal contamination; via the biomagnification process, heavy metals also cause impact animal and human health (Chhonkar et al. 2007; Sun et al. 2008; Zhang et al. 2010; Garbisu and Alkorta 2001). Heavy metals have a negative impact on soil microflora and are also known to reduce plant growth (Duruibe et al. 2007). A promising solution is bioremediation—the removal, attenuation, or conversion of pollutants through the use of biological processes with microorganisms to degrade organic pollutants (Yadav et al. 2014).

11.2.2 Xenobiotics

Xenobiotics are unfamiliar chemical substances without natural production that are found within an organism in much higher concentrations than expected (i.e. artificial substances). Drugs such as antibiotics are examples of xenobiotics because the human body does not manufacture them. Natural compounds are said to be xenobiotics when they are utilized by another organism (Mansuy 2013).

Xenobiotics did not exist in nature before their manufacturing by humans. They are often considered to be pollutants (e.g. polychlorinated biphenyls, dioxins) because of their effects on the biota. The term *xenobiotic* comes from the Greek words *xenos* (foreigner, stranger) and *bios* (life), plus the Greek suffix *-tic* for adjectives. Xenobiotics are classified as environmental pollutants, carcinogens, drugs, hydrocarbons, food additives, and pesticides.

11.2.3 Organic Pollutants

Organic compounds that do not undergo environmental degradation through chemical, biological, and photolytic processes are known as organic pollutants (Ritter 2007). These toxic chemicals bioaccumulate and unfavorably affect human health and the environment due to their persistent nature. They can be distributed by wind and water and accumulate in various species through the food chain.

11.3 Phytoremediation: Detoxification of Recalcitrant Chemicals

Phytoremediation is a promising technology that uses plants to remove, contain, sequester, or degrade organic and inorganic contaminants in soils, sediments, surface water, and groundwater (Raskin and Ensley 2000). The term *phytoremediation* comes from the Latin words *phyto* (plant) and *remedium* (restoring balance). It refers to the use of living plants called hyperaccumulators to bioaccumulate, degrade, and remediate soil, air, and water contaminated with harmful chemicals (Reichenauer and Germida 2008). Phytoremediation is a plant-based approach to remediation that concentrates contaminants and compounds from the environment. It is a cost-effective technique used to metabolize various molecules in their tissues. In recent years, various mechanisms of phytoremediation became known, such as physiological and molecular mechanisms coupled with biological and engineering strategies that are intended to optimize and improve phytoremediation. Several field trials established the viability of using plants for environmental cleanup, with their main targets being toxic heavy metals and organic pollutants (Salt 1998). Studies on phytoremediation for polycyclic hydrocarbons contaminants in soil are presented in Table 11.1.

The use of microbial communities in the rhizosphere is a potential biological approach for the remediation and degradation of chemically organic substances from the soil. Plant roots have the ability to modify their environments. For bioremediation, the rhizosphere can be used to enhance the remediation of contaminated soils with bioaccumulation. The uptake and metabolism of the herbicide atrazine in poplar trees (*Populus*) grown in a bioreactor was studied by Burken and Schnoor (1997). The adsorption of naphthalene by the roots of two plant species (tall fescue, *Festuca arundinacea* Schreber and alfa-alfa, *Medicago sativa* L.) with different lipid content was quantified in a greenhouse experiment; the study showed that lipid content is a controlling factor in the adsorption of naphthalene into plant roots (Schwab et al. 1998). Studies on bioremediation using rhizospheric bacteria (Liu 2008) are presented in Table 11.2.

Phytoremediation is a green technology that utilizes plants for the environmental cleanup of a wide array of contaminants in soil, sediment, and surface water (Pradhan et al. 1998). It is a cost-effective, ecofriendly method to remove contaminants such as petroleum hydrocarbons, chlorinated solvents, pesticides, and metals (Anderson et al. 1993; Cunningham and Lee 1995). This technology is appropriate

Table 11.1 Phytoremediation for polycyclic hydrocarbon contaminants in soil

Common name of plant	Scientific name of plant	Research findings	References
Cocks foot	*Dactylis glomerata*	Naphthalene decreased to about 20% and other PAHs decreased with an increase in molecular weight, except for pyrene, the only PAH not to show any significant decrease.	Smith et al. (2006)
Tall fescue	*Festuca arundinacea*		
Red fescue	*Festuca rubra*		
Rye grass	*Lolium perenne*		
Birdsfoot-trefoil	*Lotus carniculatus*		
Red clover	*Trifolium pretense*		
White clover	*Trifolium repens*		
Cat claw	*Mimosa monancistra*	Dissipation of benzo alpha pyrene was significantly faster in vegetated soil.	Bernal et al. (2007)
Baby bear	*Cucurbita pepo* ssp	As soil moisture, the plant density increased as rates of contaminant accumulation by plant roots increased. Concentrations of the compound in plant roots were inversely related to plant age, whereas no change in the bioavailability of the compound was observed in successive generations of plants grown in the same contaminated soil.	Kelsey et al. (2006)
Annual ryegrass	*Lolium multiflorum*	Maturity of plant root contributes to reduction in the bioavailability of target. PAHs	Parrish et al. (2005)
Tall Fescue	*Festuca arundinacea*		
Yellow Sweet	*Schreb*		
Clover	*Melilotus officinalis* Lam.		
Sunflower	*Helianthus annuus* L.	Vegetation increases total numbers of beneficial fungi and bacteria in contaminated soil.	Olson and Fletcher (2000)
Bermuda grass	*Cynodon dactylon* L.		
Southern	*Digitaria ciliaris*		
Crabgrass	*(Retz.)Koeler*		
Maize	*Zea mays* L.	Increase in hydrocarbon bioavailability stimulates bacterial population.	Radwan et al. (1995)
			Chaineau et al. (2000)

(continued)

Table 11.1 (continued)

Common name of plant	Scientific name of plant	Research findings	References
Switch grass	*Panicum virgatum*	Total PAH concentration was reduced after 6 months of treatment.	Pradhan et al. (1998)
Little bluestem	*Schizachyrium scoparium*		Wiltse et al. (1998)
grass Alfalfa	*Medicago sativa* L.		
Rice	*Oryza sativa L.cv.*	TPH concentration significantly decreased under vegetated conditions.	Kaimi et al. (2007)
Naked Spinach	*Spinacia oleracea*		
Devil's	*L.cv.Ohrai.*		
beggartick	*Pueraria lobata* (Willd.)		
Slender oat	*Avena barbata*	A large phenanthrene degrader population in the rhizosphere is related to root debris and soil exudates.	Miya and Firestone (2001)
Alfalfa	*Medicago sativa* L.	Stimulation of bioremediation around plant roots due to higher number of organic chemical degraders shows potential.	Nichols et al. (1997)
Alpine blue grass	*Poa alpine* L.		
Tall fescue	*Festuca arundinacea*	Greater total bacterial numbers and PAH-degrading bacteria were found in rhizosphere soil.	Ho and Banks (2006)
Alfalfa	*Medicago sativa* L.	Rhizosphere microflora of alfalfa were less inhibited by hydrocarbon contamination with higher degradative potential compared to reed.	Muratova et al. (2005)
Reed	*Phragmitis australis*		

PAH polycyclic aromatic hydrocarbons

for various regions and climates because plants preserve the natural structure and texture of soil using solar energy.

The phytoremediation process includes the following: (1) uptake of organic compounds from soil and water; (2) accumulation of these chemicals via lignification, metabolization, volatilization, and mineralization; (3) degradation of complex organic molecules into simpler molecules (eventually carbon dioxide and water) facilitated by enzymes; and (4) increased carbon and oxygen content in the soil around roots and increased microbial/fungal activity.

11.4 Phytoremediation Mechanisms

Various mechanisms are involved in the cleanup and remediation of contaminants via phytoremediation. The uptake of contaminants in plants occurs by the root system, which provides a huge surface area for the absorption and accumulation of

Table 11.2 Bioremediation using rhizospheric bacteria (Liu 2008)

Compound	Plants used	Microbes used	References
Polychlorinated biphenyls s	Alfalfa (*Medicago sativa*)	*Pseudomonas fluorescens*	Brazil et al. (1995)
	Sugar beet (*Beta vulgaris* L.)	*Pseudomonas putida* Flav1-1	Narasimhan et al. (2003)
	Rockcress (*Arabidopsis*)	*Pseudomonas putida* PML2	
	Switchgrass	Indigenous degraders	Chekol et al. (2004)
	(*Panicum virogatum* L.)	*Pseudomonas fluorescens*	Villacieros et al. (2005)
	Alfalfa (*Medicago sativa*)		
	Sugar beet (*Beta vulgaris* L.)		
Pesticides	Barley (*Hordeum sativum* L.)	*Burkholderia cepacia*	Jacobsen et al. (1997)
2,4-D	Red clover (*Trifolium pratense*) ryegrass (*Lolium perenne* L.)	Indigenous degraders	Shaw and Burns (2004)
PCP (Phencyclidine)	Ryegrass (*Lolium perenne* L.)	Indigenous degraders	He et al. (2005)
			Yee et al. (1998)
			Liste and Prutz (2006)
			Shaw and Burns (2004)
			Huang et al. (2004)
			Kuiper et al. (2001)
Phenanthracene	Barley (*Hordeum sativum* L.)	Degrading rhizosphere colonizing *Pseudomonas*	Ankohina et al. (2004)
Chrysene	White clover (*Trifolium repens* L.)		Johnson et al. (2004)
		PAH tolerant *Rhizobium leguminosarum*	

water and nutrients that are necessary for growth, along with other contaminants (Ouyang 2002). Organic and inorganic exudates are released in the rhizosphere and affect the number and activity of microorganisms, such as the aggregation and stability of soil particles around the root, the availability of contaminants through changes in soil characteristics, the discharge of organic substances, and changes in chemical composition.

The type of contaminant, bioavailability, and soil properties affect the mechanisms and efficiency of phytoremediation (Cunningham and Ow 1996). In the bioaccumulation of extremely high concentrations of metals, some accumulator species have evolved precise mechanisms for detoxifying the high metal levels that accumulate in the cells. Another group of plants, called *indicators*, have poor control over metal uptake and transport processes. In these plants, the extent of metal accumulation reflects the metal concentration in the rhizospheric soil (Raskin and Ensley 2000).

Membrane proteins with transport functions (generically known as *transporters*) are required for the transportation of metal ions across the cellular membranes. These transporters have an extracellular binding domain to which the ions attach just before transport, as well as a transmembrane structure that connects extracellular and intracellular media. The transfer of bound ions from the extracellular space through the hydrophobic environment of the membrane into the cell is facilitated by the transmembrane structure. The translocated metals and ions are subsequently inhibited in the shoot and also become complexed and sequestered in cellular structures (Lasat et al. 1998). Specific mechanisms to limit metal uptake into the roots occur in some plants, which are known as *excluders* (Peterson 1983). The translocation of metal-containing sap occurs from the root to the shoot in a process that is mainly controlled by root pressure and leaf transpiration. By this process, metals are reabsorbed from the sap into leaf cells.

11.5 Phytoremediation Processes

Phytoremediation has several benefits: it is in situ, effective, inexpensive, and green (Schnoor et al. 1995). Trees, grasses, herbs, and associated fungi and microorganisms are widely used for the remediation of pollutants. The various processes that are responsible for the remediation of pollutants are presented in Fig. 11.1 and are described in the following sections.

11.5.1 Phytoextraction

Phytoextraction (also known as phytoaccumulation) is a technique for the uptake, translocation, and accumulation of recalcitrant chemicals in the soil by plant roots and other parts of a plant. It is mainly used for the treatment of heavy metals in contaminated soils (US EPA 2000). This approach uses plants to absorb, concentrate, and precipitate toxic metals from contaminated soils into the aboveground biomass (shoots, leaves, etc.). Hyperaccumulator plant species have the potential to eliminate metals from contaminated soils (Raskin and Ensley 2000), with the ability to accumulate 100 times more metal than a common nonaccumulator plant (UNEP, Undated). In phytoextraction, metals such as nickel, zinc, and copper are preferred by a majority of plants (approximately 400) that uptake and absorb unusually large amounts of metals and remove them from contaminated sites. The uptake of metals into root cells—the point of entry into living tissues—is the main step in the process of phytoextraction.

Phytoextraction have several advantages: it is cost effective, fairly inexpensive compared to other conventional methods, and able to permanently remove contaminants from the soil. The amount of waste material that must be disposed of is considerably decreased (up to 95%); in some cases, the contaminant can be recycled

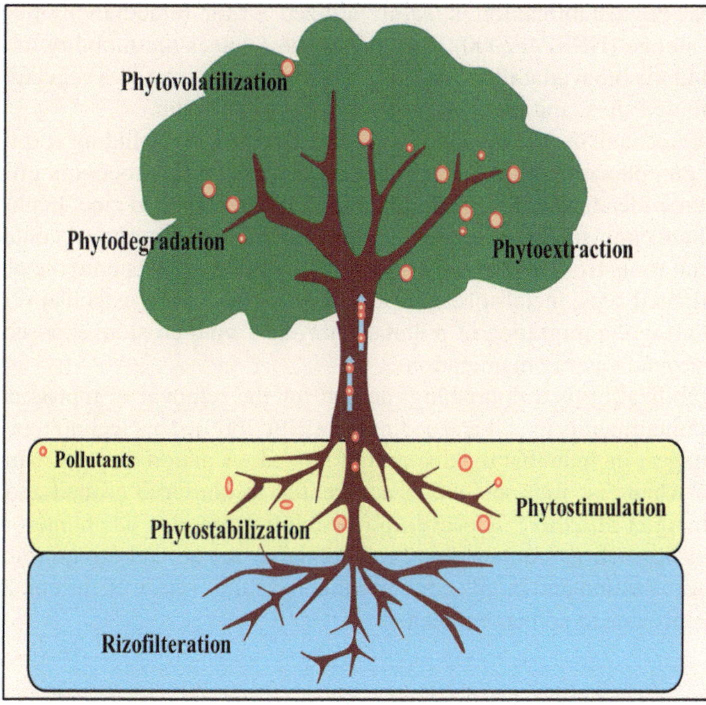

Fig. 11.1 Schematic representation of various processes responsible for phytoremediation of pollutants

from the contaminated plant biomass (US EPA 2000). Slow growth, a shallow root system, and low biomass production are the major limitations for the use of hyperaccumulator species.

After the remediation process, the plant biomass must be harvested and disposed of correctly, complying with standards (Raskin and Ensley 2000). The phytoextraction method is also usually limited to metals and other inorganic compounds in soil or sediment (EPA 2000). For this clean-up method to be practicable, the plants must (1) translocate the heavy metal into the surface biomass, (2) remove large concentrations of heavy metals into their roots, and (3) produce a large amount of plant biomass. To detoxify and/or accept high metal concentrations accumulated in their shoots, remediative plants must have some compatible mechanisms (Brennan and Shelley 1999).

11.5.2 Phytostabilization

Phytostabilization is the application of a plant species to immobilize pollutants in the soil and groundwater by absorption and accumulation in the roots. Pollutants are converted from a soluble form to a nonsoluble form in a process known as in-place inactivation. Phytostabilization is mainly utilized for the remediation of soil, sediment, and sludge (USEPA 2000). This technique reduces the mobility of the contaminant and the bioavailability of metal in the food chain, restores vegetation cover at contaminated sites, and prevents migration to groundwater.

Various mechanisms are involved in phytostabilization, including sorption, precipitation, complexation, and metal valence reduction. The process is effective in the treatment of lead, arsenic, cadmium, chromium, copper, and zinc. In phytostabilization, changes in soil chemistry may encourage the adsorption of contaminants into the plant roots from soil or cause the precipitation of metals onto the plant root. At contaminated sites, metal-tolerant species can be used to re-establish vegetation, thereby limiting the migration of pollutants through wind erosion, exposed surface soils, and groundwater contamination.

Phytostabilization is a flourishing method for the removal of metals and other inorganic contaminants in soil and sediments (EPA 2000). This technology has several advantages, include that the disposal of hazardous material is not required and its success when fast immobilization is essential to conserve ground and surface waters (Zhang et al. 2009). However, phytostabilization also has numerous major disadvantages, such as contaminants remaining in the soil, the requirements for extensive fertilization and monitoring, and that the stabilization of the contaminants may be mainly due to soil amendments.

11.5.3 Phytotransformation

The breakdown of complex organic molecules to simple molecules or the assimilation of these molecules into plant tissues is called phytotransformation (Trap et al. 2005). Like phytoextraction and phytovolatilization, plant uptake normally occurs when the contaminants' solubility and hydrophobicity fall into a definite tolerable range; the contaminants are broken down after they have been taken up by the plant. Phytotransformation has been used to remediate some organic contaminants, including chlorinated solvents, herbicides, and other recalcitrant chemicals in soil, sediment, and groundwater (EPA 2000).

11.5.4 Phytovolatilization

In phytovolatilization, plants are used for the extraction and successive outgassing of compounds from soil. This process is used mainly for the removal of *m*-xylene, chlorobenzene, trichloroethene, organically bound mercury, and other volatile compounds as well as inorganic chemicals that have volatile forms, such as selenium and arsenic, present in soil, sediment, or water (EPA 2000). In phytovolatilization, contaminants are converted into volatile forms by transpiring them into the atmosphere (USEPA 2000). Contaminants may be contained in the body of the plant, but volatile degradation products are transpired with water vapor from the leaves (EPA 2000).

Phytovolatilization transmits contaminants from the stems or other plant parts before the leaves (Raskin and Ensley 2000). Certain organic pollutants that are recalcitrant in the subsurface environment react quickly in the atmosphere with hydroxyl radicals, an oxidant produced in the photochemical cycle. Very few contaminants are adequately water soluble, nontoxic to plants, and volatile enough to reach atmospheric concentrations that would be of concern by evapotranspiration (Davis et al. 1996). The process has been studied for contaminants such as trichloroethylene, polychlorinated biphenyls, and total petroleum hydrocarbons (Sandermann 1994; Schwab and Banks 1994; Davis et al. 1996; Macek et al. 2000). For in situ degradation, it is not desirable to transfer the contaminants from the soil or groundwater to the atmosphere; however, it may be preferable for long-lasting exposure in the soil environment and when there is a threat of groundwater contamination.

11.5.5 Phytostimulation

Phytostimulation is the combination of phytoremediation and bioaugmentation. It is a useful rhizoremediation technology in which plant roots release compounds or root exudates enhance microbial activity in the rhizosphere. This low-cost approach to remove organic pollutants from the soil is also known as rhizosphere phytoremediation. Phytostimulation is significant in rhizoremediation because the rhizosphere provides a precise niche for soil microorganisms to survive. Furthermore, the root exudates have several benefits for microorganism growth and activity. The niche is constantly increasing as roots grow and penetrate new soil zones, and the rhizosphere serves as an exclusive soil habitat, Root exudates may serve as energy sources for microorganisms, mimic man-made pollutants, and function as inducers and co-metabolites for microbial genes involved in the degradation of pollutants; hence, they eventually participate in phytostimulation process. Another characteristic feature of phytostimulation is the promotion of new microorganisms in the soil by inoculating specific microrganisms on plant seeds to introduce them to the growing rhizosphere. Genetically engineered or naturally occurring strains are used as inoculated organisms that execute a precise function in the interest of remediation.

11.5.6 Phytotransformation

It occurs through metabolic processes within the plant and by the effect of compounds, such as enzymes, formed by the plant. The organic contaminants are broken down into simpler compounds, which are incorporated with plant tissue and promote plant growth. Phytotransformation depends on the direct uptake of pollutants from the media and accumulation in the vegetation.

11.5.7 Rhizofiltration

Rhizofiltration is the adsorption, precipitation, or absorption of contaminants in the solution surrounding the root zone or in plant roots. This method is primarily used to clean up extracted groundwater, surface water, and wastewater for the elimination of metals, low-concentration contaminants, or other inorganic compounds (EPA 2000). Lead, cadmium, copper, nickel, zinc, and chromium, which are mainly retained within the roots, can be remediated by rhizofiltration (US EPA 2000).

Rhizofiltration is similar to phytoextraction. However, it is mainly used for contaminated groundwater rather than soil, often in greenhouses with the plant roots in water rather than soil. Once a large root system has been established to regulate the plants, polluted water is gathered from a waste site and then used as the plants' water source. The plants are then planted in the polluted area, where the roots capture contaminants and water along with them. As the roots are saturated by the contaminants, they are harvested. The ability of plants to remove lead from water have been studied for sunflowers, Indian mustard, tobacco, rye, spinach, and corn (Raskin and Ensley 2000).

11.5.8 Hydraulic Control

In hydraulic control (also known as rhizodegradation), the water table and the soil field capacity are controlled by plant canopies. Phreatophytic trees and plants that have the ability to transpire large volumes of water (and thereby affect the existing water balance at the site) are generally used for hydraulic control. In groundwater plumes, the increased transpiration reduces the infiltration of precipitation or increases the transpiration of groundwater; thus, migration from the contaminated site decreases. Hydraulic control is also a potential mechanism to manage groundwater contamination and can remediate a wide range of contaminants in soil, sediment, and groundwater (EPA 2000).

In this technique, the degradation of contaminants occurs within the plant root zone, or rhizosphere. It is carried out by bacteria or other microorganisms whose numbers naturally prosper in the rhizosphere (USEPA 2000). Plant exudes of sugars, amino acids, enzymes, and other compounds can stimulate bacterial growth and enhance the occurrence of microorganisms in the rhizosphere. The roots also provide a pathway for oxygen transfer from the environment and additional surface area for microbes to grow on.

A wide variety of mostly organic chemicals—including petroleum hydrocarbons, polycyclic aromatic hydrocarbons, chlorinated solvents, pesticides, polychlorinated biphenyls, benzene, toluene, ethylbenzene, and xylenes—can be cleaned up from contaminated sites through the rhizodegradation process (EPA 2000). Hydraulic control also acts as plant-assisted bioremediation, with microbial inspiration and fungal degradation by exonerates of root exudates into rhizosphere (Zhang et al. 2005).

11.6 Applications

Phytoremediation is a cost-effective process for the remediation of contaminated sites that uses plants to filter and metabolize foreign substances. Plants have successfully remediated areas contaminated with organic chemical waste, including petroleum products, solvents, wood preservatives, pesticides, and metals such as lead, cadmium, zinc, and radionuclides. Phytoremediation is an efficient alternative to conventional, engineered remedial plans, which usually require costly activities such as excavation and treatment of storm water, industrial wastewater, and sewage in ecofriendly manner.

11.7 Factors Affecting Plant Adaptation to Recalcitrant Chemicals

A number of biotic and abiotic factors affect the uptake of metals in plants, such as temperature, soil pH, soil aeration, moisture, plant type, plant size, root system, competition among plants, and the accessibility of elements in the soil (Nagajyoti et al. 2010; Yamamoto and Kozlowski 1987). The availability of heavy metals in the environment may be influenced by several anthropogenic activities. Several factors limit the extent of metal phytoextraction, including the following:

- Metal bioavailability within the rhizosphere
- Rate of metal uptake by roots
- Quantity of metal "fixed" within the roots
- Rate of xylem loading/translocation to shoots
- Cellular tolerance to toxic metals

A schematic summary of the factors affecting the heavy metal tolerance of plants is provided in Fig. 11.2.

11.8 Contaminant Metabolism During Phytoremediation in Plants

Organic compounds can be translocated to other plant tissues and undergo partial or complete degradation. Depending on the chemical properties of the contaminant and the specific relationship of a pollutant with the soil, the physiological properties of the introduced plant species and the contaminated medium may differ (Salt et al. 1998). Whether a contaminant should be subjected to phytoextraction, phytodegradation, phytovolatilization, or rhizodegradation is determined by the phytoremediation process, beginning with the contaminant being transported to the plant. Some amount of transformation occurs in plant cells before they are sequestered in vacuoles or bound to insoluble cellular structures, such as lignin. In wetland plant

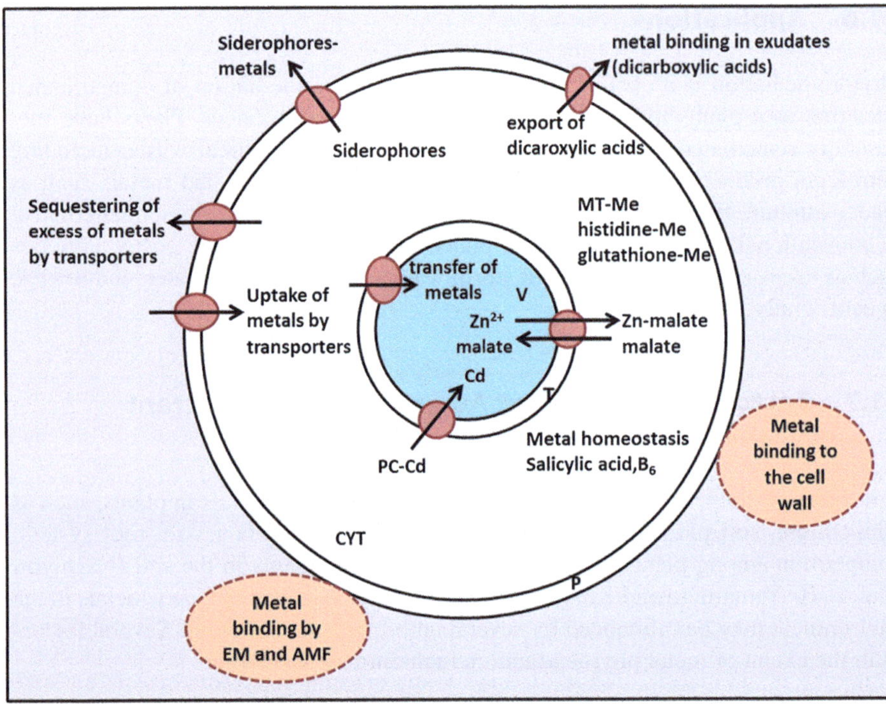

Fig. 11.2 Schematic summary of the factors involved in heavy metal tolerance of plants. *Abbreviations*: *P* plasma membrane (cytoplasmic membrane), *CYT* cytoplasm, cytosol, *T* tonoplast, *V* vacuole, *Me* metal, *PC* phytochelatin, *MT* metallothionein

species, contaminants can penetrate through the roots and separation can occur from the water column directly into plant tissues.

11.9 Biotic Interactions of Hyperaccumulators

A plant is considered to be a hyperaccumulator if it can concentrate pollutants to a small percentage, which varies according to the pollutant involved (e.g. more than 1000 mg/kg of dry weight for nickel, copper, cobalt, chromium, or lead; more than 10,000 mg/kg for zinc or manganese) (Hannink 2001). This potential for accumulation is due to hypertolerance or phytotolerance—the outcome of a plant's adaptive development to aggressive environments through many generations. Various interactions may be affected by metal hyperaccumulation, including protection, interference with neighboring plants of different species, mutualism, and commensalism.

In the rhizosphere of Ni-hyperaccumulators, bacteria are able to tolerate high concentrations of Ni and also acquire nickel uptake potential. Nickel bioremediation may use these Ni-hyperaccumulators in combination with the Ni-resistant bacteria as an ideal tool. The high concentration of nickel in the rhizosphere of

Ni-hyperaccumulators facilitates a niche for nickel-resistant microflora. The principal Ni-resistant bacteria in the rhizosphere of the Ni-hyperaccumulator *Alyssum bertolonii* are *Pseudomonas* (Mengoni et al. 2001), whereas the principal Ni-resistant bacteria in the rhizosphere of *Thlaspi goesingense* are *Methylobacterium*, *Rhodococcus* and *Okibacterium* (Idris et al. 2004). The rhizosphere of *Rinorea bengalensis* and *Dichapetalum gelonioides* ssp. *andamanicum*—the endemic Ni-hyperaccumulators from the serpentines of Andaman—harbor Ni-resistant bacteria that are capable of accumulating and tolerating >8 mM of nickel; feasible cells were able to accumulate Ni from an aqueous solution. This may be credited to the occurrence of Ni-binding sites on the cell surface of metallophiles (Beveridge and Doyle 1989).

For bioremediation, plant-associated endophytes are more advantageous than plant-associated rhizospheric bacteria for several reasons: (1) the use of endophytes that inhabit the host plant reduces competition between bacterial strains and may eradicate the requirement for reinoculation; (2) toxic organic contaminants can stay in the plant xylem for up to 2 days, facilitating their degradation by endophytes; and (3) endophytes can be isolated from host plants of interest and genetically enhanced with genes encoding degradation enzymes of interest before reinoculation for bioremediation.

When developing bioremediation systems using plant-associated bacteria, it is important to choose wild-type bacteria or bacteria enhanced by natural gene transfer. This avoids the complications of national and international legislation restricting and monitoring the use of genetically modified materials. However, bioremediation can provide an efficient, economic, and sustainable green remediation technology for our modern world, considering the global political shift towards sustainability and the use of plant-associated bacteria to break down toxic synthetic organic compounds in environmental soil. Studies on bioremediation using endophytic bacteria are presented in Table 11.3 (Ryan et al. 2008).

Bioremediation is an environmentally-friendly management technique for polluted soil because it uses solar-driven biological processes for the treatment of pollutants. The success of a phytoextraction technique is largely dependent on the continuous availability of the metal of interest to the phytoextracting plants. In contrast to most other remediation technologies, it is not invasive and, in principle, delivers intact and biologically active soil. However, there is a need to improve research efforts on this emerging technology. Research should focus on identifying remediating plants that have adapted to the limited climate and soil conditions. Because phytoremediation is a slow process, both biotechnological and classical hybridization techniques should be used to develop more proficient metal-hyperaccumulating plant species with increasing pollutant tolerance, root and shoot biomass, root architecture and morphology, pollutant uptake properties, and degradation capabilities for organic pollutants. The management of microbial consortia, the selection and engineering of microorganisms with capabilities for pollutant degradation, advantageous effects on phytoremediation crops, and modification of the effects on pollutant bioavailability are other considerations. The selection and engineering of plants and microbial strains that can modify the solubility and transport

Table 11.3 Bioremediation using endophytic bacteria (Ryan et al. 2008)

Compound	Plants used	Microbes used	References
Polychlorinated biphenyls, TCP	Wheat (*Triticum* spp.)	*Herbaspirillum* sp. K1	Mannisto et al. (2001)
Chlorobenzoic acids	Wild rye (*Elymus dauricus*)	*Pseudomonas aeruginosa* R75	Siciliano et al. (1998)
		Pseudomonas savastanoi CB35	
Pesticide	Pea (*Pisum sativum*)	*Pseudomonas putida* VM1450	Germaine et al. (2006)
2,4-D			
Volatile organic compounds and toluene	Yellow lupine (*Lupinus luteus* L.)	*Burkholderia cepacia G4*	Barac et al. (2004)
Toluene		*Burkholderia cepacia* Bu61	Taghavi et al. (2005)
MTBE (Methyl tert-butyl ether), BTEX (Benzene, toluene, ethylbenzene and xylene), trichloroethylene	Poplar (*Populus*)	(pTOM-Bu61)	Germaine et al. (2004)
	Poplar (*Populus* cv. *Hazendans* and cv. *Hoogvorst*)	*Pseudomonas* sp.	Porteus-Moore et al. (2006)
Hydrocarbons	Pea (*Pisum sativum*)	*Pseudomonas putida* VM1441 (pNAH7)	Germaine et al. (2006)
Naphthalene			
Explosives	Poplar tissues	*Methylobacterium populi* BJ001	Van Aken et al. (2004a, b)
TNT (Trinitrotoluene), RDX (Research Department Explosive), HMX (Octogen)	(*Populus deltoidesnigra* DN34)		

of organic pollutants through the exudation of biosurfactants holds great promise (Ankohina et al. 2004).

Additional strategies should include appropriate management of the soil, such as via fertilization or chelant addition, to increase pollutant bioavailability and the use of bioresources for remediation of soil pollution remediation crops, such as via optimization of coppicing, harvest cycles, and the development of mixed cropping systems. In long-contaminated soil, the enhancement of bioavailability appears to be the key to successful biodegradation. Several benefits are associated with the use of endophytic bacteria versus unaccompanied plants in phytoremediation: (1) the efficiency of the remediation process can be enhanced through the quantitative gene expression of bacterial pollutant catabolic genes; (2) the genetic engineering of a bacterial catabolic pathway is better than the plant catabolic pathway; and (3) the toxic effects of pollutants in environmental soil are diminished by endophytic degraders when toxic pollutants are taken up by the plants.

Several disadvantages should also be noted for plant-associated bioremediation of pollutant-rich soil with the use of bacteria: (1) the technology is limited for the cleanup of shallow contaminants in environmental soil; (2) it is slower than traditional remediation technologies; (3) some plants are only seasonally effective; (4) there can be phytotoxic effects from the contaminants; (5) the environmental contaminants or their metabolites can enter the food chain when consumed by local fauna if the contaminants are not absolutely detoxified (Ryan et al. 2008). Weyens et al. (2009) reviewed the advantages of plant-associated endophytes in bioremediation. The authors emphasized that although it has been effectively functional in a number of laboratory-scale experiments, several issues may restrict the large-scale field application of this technology, including the levels of contaminants tolerated by plants, the restricted bioavailability of organic pollutants, and the undesirable levels of volatile organic compounds that are evaportranspired into the atmosphere.

There are some disadvantages to the degradation of volatile organic compounds in environmental soil by using plant-associated endophytic bacteria, but there is potential for these bacteria to make a considerable contribution to sustainable bioremediation. As Doty (2008) stated, "The benefit of using endophytic bacteria over rhizospheric bacteria in phytoremediation is that while a rhizospheric bacterial population is difficult to control, and competition between rhizospheric bacterial strains often reduces the number of the desired strains, the use of endophytes that naturally occupy the internal tissues of plants reduces the problem of competition between bacterial strains." Rhizoremediation is a precise form of phytoremediation using plants and their related rhizospheric microorganisms (bacteria and fungi) that occur in nature or can be inoculated in soil with microorganisms to break down environmental contaminants.

11.10 Transgenic Plants for Phytoremediation of Recalcitrant Chemicals

Genetic engineering is an important technique to increase the potential of natural phytoremediation by introducing new properties into plants. Microorganisms have some desirable genes for phytoremediation; these genes are transferred from one plant to another variety to improve its ability in the cleanup of environmental contaminants. For example, genes from a bacterium were inserted into tobacco to encode a nitroreductase enzyme and represent an earlier exclusion of TNT to provide resistance against toxic chemicals (Beveridge and Doyle 1989). Researchers have also discovered pollutant-resistant plants in their efforts to decrease pollutant concentrations in soil. Some natural, biodegradable compounds, such as exogenous polyamines, allow plants to tolerate and absorb pollutants at concentrations 500 times greater than unprocessed plants.

Unfavorable environmental conditions and the hyperaccumulation of pollutants in soil may prevent the endurance of competent plants in accepted ecosystems; thus, the efficiency of phytoremediation may be lower than that under laboratory conditions. The hyperaccumulation of various contaminants via phytoremediation occurs in the presence of engineered plants. Various biotechnological techniques are able to develop an efficient hyperaccumulator plant to remediate contaminated environments, as well as also achieve the idea of a "magic tree" for efficient environmental cleanup.

Molecular biology approaches have already been used for phytoremediation and the elimination of toxicity from contaminated sites (Rugh et al. 1998; Bizily et al. 1999; Kramer and Chardonnens 2001). This methodology is used to formulate transgenes to shorten the juvenile phase, modify lignin biosynthesis, and increase cellulase accumulation in forest trees (Pena and Seguin 2001). A few advances have come to light in the expansion of a transgenic approach for metal phytoremediation, which are efficient in plants with the potential for environmental application (Rugh et al. 1998; Bizily et al. 1999).

11.11 Conclusion and Future Perspectives

A number of plants can grow in contaminated soils, and their different biological characteristics provide interesting research perspectives. Metallophytes—either alone or in combination with microorganisms—may be used to alleviate the harmful effects of recalcitrant chemicals. Broad-scale applications are feasible, but the cost–benefit outcome has not yet been properly assessed. Plant-associated endophytic and rhizospheric bacteria may be used to degrade a wide range of toxic organic compounds in contaminated soil, which is essential for a cleaner environment. Specially selected or engineered plants can be used in the processes of phytoremediation, which is an environmentally friendly option for polluted soil because it uses solar-driven biological processes to remove pollutants.

However, research efforts on this emerging and environment-friendly technology need to be improved. Future studies should focus on identifying phytoremediators that are modified to the local climate and soil conditions. Although phytoremediation is a new field, it holds immense potential if it can be investigated in a multidisciplinary manner that includes plant biology, agricultural engineering, agronomy, soil science, microbiology, and genetic engineering, among others. Furthermore, the applicability and wider acceptability of these bioapproaches should be fully established in these fields for the treatment of contaminated soils.

References

Anderson TA, Guthrie EA, Walton BT (1993) Bioremediation in the rhizosphere. Plant roots and associated microbes clean contaminated soil. Environ Sci Technol 27(13):2630–2636

Ankohina T, Kochetkov V, Zelenkova N, Balakshina V, Boronin A (2004) Biodegradation of phenanthrene by *Pseudomonas* bacteria bearing rhizospheric plasmids in model plant-microbial associations. Appl Biochem Microbiol 40:568–572

Barac T, Taghavi S, Borremans B, Provoost A, Oeyen L, Colpaert JV, Vangronsveld J, van der Lelie D (2004) Engineered endophytic bacteria improve phytoremediation of water-soluble, volatile, organic pollutants. Nat Biotechnol 22:583–588

Becker T, Dierschke T (2008) Vegetation response to high concentrations of heavy metals in the Harz Mountains. Phytocoenologia 38:255–265

Bernal D, Contreras-Ramos S, Marsch R, Dendooven L (2007) Influence of Catclaw mimosa monancistra on the dissipation of soil PAHs. Int J Phytoremed 9:79–90

Bernes C (1998) Persistent organic pollutants. Swedish Environmental Protection Agency, Stockholm

Beveridge TJ, Doyle R (1989) Metal ions and bacteria. Wiley, New York

Bizily S, Rugh CL, Summers AO, Meagher RB (1999) Phytoremediation of methylmercury pollution: Mer B expression in Arabidopsis thaliana confers resistance to organomercurials. Proc Natl Acad Sci 96:6808–6813

Brazil GM, Kenefick L, Callanan M, Haro A, De Lorenzo V, Dowling DN, O'Gara F (1995) Construction of a rhizosphere pseudomonad with potential to degrade polychlorinated biphenyls and detection of *bph* gene expression in the rhizosphere. Appl Environ Microbiol 61:1946–1952

Brennan MA, Shelley ML (1999) A model of the uptake, translocation, and accumulation of lead (Pb) by maize for the purpose of phytoextraction. Ecol Eng 12:271–297

Burken JG, Schnoor JL (1997) Uptake and metabolism of atrazine by poplar trees. Environ Sci Technol 31:1399–1406

Chaineau CH, Morel JL, Oudot J (2000) Biodegradation of fuel oil hydrocarbons in the rhizosphere of maize. J Environ Qual 29:569–578

Chekol T, Vough LR, Chaney RL (2004) Phytoremediation of polychlorinated biphenyl-contaminated soils: the rhizosphere effect. Environ Int 30:799–804

Chhonkar PK, Bhadraray S, Purakayastha TJ (2007) Phytoremediation of heavy metal contaminated soils, Monograph. Division of Soil Science and Agricultural Chemistry, Indian Agricultural Research Institute, New Delhi

Cunningham SD, Lee CR (1995) Phytoremediation: Plant-based remediation of contaminated soils and sediments, Soil Science Society of America, Madison. Bioremediation: Science and Applications. SSSA Special Publication 43

Cunningham SD, Ow DW (1996) Promises and prospect of phytoremediation. Plant Physiol 110:715–719

Davis LC, Banks ML, Schwab AP, Muralidharan N, Erickson LE, Tracy JC (1996) Plant based bioremediation. In: Sikdar I (ed) Bioremediation. Technomics, Basel

Doty S (2008) Enhancing phytoremediation through the use of transgenics and endophytes. New Phytol 179:318–333

Duruibe JO, Ogwuegbu MOC, Egwurugwu JN (2007) Heavy metal pollution and human biotoxic effects. Int J Phys Sci 5:112–118

EPA (2000) A citizen's guide to phytoremediation. EPA 542-F-98-011. United States Environmental Protection Agency, p 6. Available at: http//www.bugsatwork.com/XYCLONYX/EPA_GUIDES/PHYTO.PDF

Ernst WHO (1974) Schwermetallvegetation der Erde. Gustav Fischer, Stuttgart

Ernst WHO (1982) Metal tolerance in plants. Acta Bot Neerl 41:229–248

Garbisu C, Alkorta I (2001) Phytoextraction: a costeffective plant-based technology for the removal of metals from the environment. Bioresour Technol 77(3):229–236

Germaine K, Keogh E, Garcia-Cabellos G, Borremans B, van der Lelie D, Barac T, Oeyen L, Vangronsveld J, Moore FP, Moore ERB, Campbell CB, Ryan D, Dowling DN (2004) Colonisation of poplar trees by *gfp* expressing bacterial endophytes. FEMS Microbiol Ecol 48:109–118

Germaine KJ, Liu X, Cabellos GG, Hogan JP, Ryan D, Dowling DN (2006) Bacterial endophyte-enhanced phytoremediation of the organochlorine herbicide 2,4-dichlorophenoxyacetic acid. FEMS Microbiol Ecol 57:302–310

Hannink N, Rosser SJ, French CE, Basran A, Murray JA, Nicklin S, Bruce NC (2001) Phytodetoxification of TNT by transgenic plants expressing a bacterial nitroreductase. Nat Biotechnol 19(12):1168–1172. https://doi.org/10.1038/nbt1201-1168. PMID 11731787

He Y, Xu JM, Tang CX, Wu YP (2005) Facilitation of pentachlorophenol degradation in the rhizosphere of ryegrass (*Lolium perenne* L.) Soil Biol Biochem 37:2017–2024

Ho C, Banks MK (2006) Degradation of polycyclic aromatic hydrocarbons in the rhizosphere of *Festuca arundinacea* and associated microbial community changes. Biorem J 10(3):93–104

Hopkins WG (1999) Introduction to plant physiology. Lewis, New York

Huang XD, El-Alawi Y, Penrose DM, Glick BR, Greenberg BM (2004) A multi-process phytoremediation system for removal of polyaromatic hydrocarbons from contaminated soils. Environ Pollut 130:465–476

Idris R, Trifonova R, Puschenreiter M, Wenzel WW, Sessitsch A (2004) Bacterial communities associated with flowering plants of Ni Hyperaccumulator *Thlaspi Goesingense*. Appl Environ Microbiol 70(5):2667–2677

Jacobsen CS (1997) Plant protection and rhizosphere colonization of barley by seed inoculated herbicide degrading *Burkholderia (Pseudomonas) cepacia* DB01 (pR0101) in 2,4-D contaminated soil. Plant Soil 189:139–144

Johnson DL, Maguire KL, Anderson DR, McGrath SP (2004) Enhanced dissipation of chrysene in planted soil; the impact of a rhizobial inoculum. Soil Biol Biochem 36:33–38

Kaimi E, Mukaidani T, Miyoshi S, Tamaki M (2007) Screening of twelve plant species for phytoremediation of petroleum hydrocarbon contaminated soil. Plant Prod Sci 10(2):211–218

Kelsey JW, Colino A, Koberle M, White JC (2006) Growth conditions impact 2,2-bis (p-chlorophenyl)-1,1- dichloroethylene (p,p′-DDE) accumulation by Cucurbita pepo. Int J Phytoremediation 8:261–271

Kramer U, Chardonnens AN (2001) The use of transgenic plants in the bioremediation of soils contaminated with trace elements. Appl Microbiol Biotechnol 55:661–672

Lasat MM, Fuhrman M, Ebbs SD, Cornish JE, Kochian LV (1998) Phytoremediation of Radiocesium-contaminated soil: evaluation of Cesium-137 bioaccumulation in the shoots of three plant species. J Environ Qual 27(1):165–169

Li X, Yang Y, Jia L, Chen H, Wei X (2013) Zinc-induced oxidative damage, antioxidant enzyme response and proline metabolism in roots and leaves of wheat plants. Ecotoxicol Environ Saf 89:150–157

Liste HH, Prutz I (2006) Plant performance, dioxygenase-expressing rhizosphere bacteria, and biodegradation of weathered hydrocarbons in contaminated soil. Chemosphere 62:1411–1420

Liu X (2008) Development and application of biosensor technologies for the biodegradation of environmental pollutants. Ph.D. thesis, Higher Education and Training Awards Council (HETAC): Dublin, Ireland

Lucassen E, Van Kempen MML, Roelofs JGM, Van der Velde G (2010) Decline in metallophytes in tertiary polluted floodplain grassland in the Netherlands, experimental evidence for metal and nutritional changes in soil as driver factors. Chem Ecol 26:273–287

Macek T, Mackova M, Kas J (2000) Exploitation of plants for the removal of organics in environmental remediation. Biotechnol Adv 18:23–34

Mannisto MK, Tiirola MA, Puhakka JA (2001) Degradation of 2,3,4,6-tetraclorophenol at low temperature and low dioxygen concentrations by phylogenetically different groundwater and bioreactor bacteria. Biodegradation 12:291–301

Mansuy D (2013) Metabolism of xenobiotics: beneficial and adverse effects. Biol Aujourdhui 207(1):33–37. https://doi.org/10.1051/jbio/2013003. PMID 23694723

Mengoni A, Barzanti R, Gonnelli C, Gabbrielli R, Bazzicalupo M (2001) Characterization of nickel-resistant bacteria isolated from serpentine soil. Environ Microbiol 3:691–708

Miya RK, Firestone MK (2001) Enhanced phenanthrene biodegradation in soil by slender oat root exudates and root debris. J Environ Qual 30:1911–1918

Muratova AY, Turkovskaya OV, Antonyuk LP, Makarov OE, Pozdnyakova LI, Ignatov VV (2005) Oil-oxidizing potential of associative rhizobacteria of the genus *Azospirillum*. Microbiology 74:210–215

Nagajyoti PC, Lee KD, Sreekanth TVM (2010) Heavy metals, occurrence and toxicity for plants: a review. Environ Chem Lett 8:199–216

Narasimhan K, Basheer C, Bajic VB, Swarup S (2003) Enhancement of plant-microbe interactions using a rhizosphere metabolomics-driven approach and its application in the removal of polychlorinated biphenyls. Plant Physiol 132:146–153

Nichols TD, Wolf DC, Rogers HB, Beyrouty CA, Reynolds CM (1997) Rhizosphere microbial populations in contaminated soils. Water Air Soil Pollut 95:165–178

Olson PE, Fletcher JS (2000) Ecological recovery of vegetation at a former industrial sludge basin and its implications to phytoremediation. Environ Sci Pollut Res 7:1–10
Ouyang Y (2002) Phytoremediation of modeling plant uptake and contaminant transport in the soil plant atmosphere continuum. J Hydrol 266(1–2):66–82
Parrish ZD, Banks MK, Schwab AP (2005) Effect of root death and decay on dissipation of polycyclic aromatic hydrocarbons in the rhizosphere of yellow sweet clover and tall fescue. J Environ Qual 34:207–216
Pena L, Seguin A (2001) Recent advances in the genetic transformation of trees. Trends Biotechnol 19:500–506
Peterson PJ (1983) Adaptation to toxic metals. In: Robb DA, Pierpoint WS (eds) Metals and micronutrients: uptake and utilization by plants. Academic, London, pp 51–69
Porteus-Moore F, Barac T, Borremans B, Oeyen L, Vangronsveld J, van der Lelie D, Campbell D, Moore ERB (2006) Endophytic bacterial diversity in poplar trees growing on a BTEX-contaminated site: the characterization of isolates with potential to enhance phytoremediation. Syst Appl Microbiol 29:539–556
Pradhan SP, Conrad JR, Paterek JR, Srivastava VJ (1998) Potential of phytoremediation for treatment of PAHs in soil at MGP sites. J Soil Contam 7:467–480
Radwan SS, Sorkhoh NA, Fardoun F, Al-Hasan RH (1995) Soil management enhancing hydrocarbon biodegradation in the polluted Kuwaiti desert. Appl Microbiol Biotechnol 44(1–2):265–270
Raskin I, Ensley BD (2000) Phytoremediation of toxic metals: using plants to clean-up the environment. Wiley, New York
Reichenauer TG, Germida JJ (2008) Phytoremediation of organic contaminants in soil and groundwater. Chemsuschem 1(8–9):708–717. https://doi.org/10.1002/cssc.200800125
Ritter L, Solomon KR, Forget J, Stemeroff M, O'Leary C (2007) Persistent organic pollutants *(PDF)*. United Nations Environment Programme. Retrieved 2007-09-16
Rugh CL, Senecoff JF, Meagher RB, Merkle SA (1998) Development of transgenic yellow poplar for mercury phytoremediation. Nat Biotechnol 16:925–928
Ryan RP, Germaine K, Franks A, Ryan DJ, Dowling DN (2008) Bacterial endophytes; recent developments and applications. FEMS Microbiol Ecol 278:1–9
Salt DE, Smith RD, Raskin I (1998) Phytoremediation. Annu Rev Plant Physiol Plant Mol Biol 49:643–668
Sandermann H (1994) Higher plant metabolism of xenobiotics: the green liver concept. Pharmacogenetics 4:225–241
Schwab AP, Al-Assi AA, Banks, MK (1998) Adsorption of Naphthalene onto Plant Roots. J Environ Qual 27(1):220–224
Schnoor JL, Licht LA, McCutcheon SC, Wolfe NL, Carreira LH (1995) Phytoremediation of organic and nutrient contaminants. Environ Sci Technol 29:318–323
Schwab AP, Banks MK (1994) Biologically mediated dissipation of polyaromatic hydrocarbons in the root zone. In: Anderson TA, Coats JR (eds) Bioremediation through rhizosphere technology. American Chemical Society, Washington, DC
Shaw LJ, Burns RG (2004) Enhanced mineralization of [U14-C] 2,4-dichlorophenoxyacetic acid in the soil from the rhizosphere of *trifolium pratense*. Appl Environ Microbiol 70:4766–4774
Siciliano SD, Goldie H, Germida JJ (1998) Enzymatic activity in root exudates of dohurian wild rye (Elymus dauricus) that degrades 2-chlorobenzoic acid. J Agric Food Chem 46:5–7
Singh A, Sharma RK, Agrawal M, Marshall FM (2010) Health risk assessment of heavy metals via dietary intake of foodstuffs from the wastewater irrigated site of a dry tropical area of India. Food Chem Toxicol 48:611–619
Smith MJ, Flowers TH, Duncan HJ, Alder J (2006) Effects of polycyclic aromatic hydrocarbons on germination and subsequent growth of grasses and legumes in freshly contaminated soil and soil with aged PAHs residues. Environ Pollut 141:519–525
Sun YB, Zhou QX, Diao CY (2008) Effects of cadmium and arsenic on growth and metal accumulation of Cd-hyperaccumulator *Solanum nigrum* L. Bioresour Technol 99(5):1103–1110

Taghavi S, Barac T, Greenberg B, Borremans B, Vangronsveld J, van der Lelie D (2005) Horizontal gene transfer to endogenous endophytic bacteria from poplar improves phytoremediation of toluene. Appl Environ Microbiol 71:8500–8505

Trap S, Kohler A, Larsen LC, Zambrano KC, Karlson U (2005) Phytotoxicity of fresh and weathered diesel and gasoline to willow and poplar trees. J Soil Sediments 1:71–76

Trapp M, McFarlane C (1995) Plant contamination. Lewis Publisher, Boca Raton

United States Environmental Protection Agency (USEPA) (2000) Introduction to phytoremediation, EPA 600/R-99/107, U.S. Environmental Protection Agency, Office of Research and Development, Cincinnati

Van Aken B, Peres C, Doty S, Yoon J, Schnoor J (2004a) *Methylobacterium populi* sp. nov., a novel aerobic, pink-pigmented, facultative methylotrophic, methane-utilising bacterium isolated from poplar trees (*Populus deltoids x nigra* DN34). Evol Microbiol 54:1191–1196

Van Aken B, Yoon J, Schnoor J (2004b) Biodegradation of nitro-substituted explosives 2,4,6-trinitrotoluene, hexahydro-1,3,5-trinitro-1,3,5-triazine inside poplar tissues (*Populus deltoids x nigra* DN34). Appl Environ Microbiol 70:508–517

Villacieros M, Whelan C, Mackova M, Molgaard J, Sanchez-Contreras M, Lloret J, Aguirre de Carcer D, Oruezabal RI, Bolanos L, Macek T, Karlson U, Dowling DN, Martin M, Rivilla R (2005) Polychlorinated biphenyl rhizoremediation by *Pseudomonas fluorescens* F113 derivatives, using a *Sinorhizobium meliloti* nod system to drive *bph* gene expression. Appl Environ Microbiol 71:2687–2694

Vinodhini R, Narayanan M (2009) Heavy metal induced histopathological alterations in selected organs of the Cyprinus carpio L (Common Carp). Int J Environ Res 1:95–100

Weyens N, van der Lelie D, Taghavi S, Vangronsveld J (2009) Phytoremediation: plant-endophyte partnerships take the challenge. Curr Opin Biotechnol 20:248–254

Wiltse CC, Rooney WL, Chen Z, Schwab AP, Banks, MK (1998) Greenhouse evaluation of agronomic and crude oil-phytoremediation potential among alfalfa genotypes. J Environ Qual 27(1):169–173

Yadav G, Srivastava PK, Singh VP, Prasad SM (2014) Light intensity alters the extent of arsenic toxicity in Helianthus annuus L. seedlings. Biol Trace Elem Res 158:410–421

Yamamoto F, Kozlowski TT (1987) Effect of flooding, tilting of stem, and ethrel application on growth, stem anatomy, and ethylene production of *Acer platanoides* seedlings. Scand J For Res 2:141–156

Yee DC, Maynard JA, Wood TK (1998) Rhizoremediation of trichloroethylene by a recombinant, root-colonising *Pseudomonas fluorescens* stain expressing toluene *ortho*-monooxygenase constituitively. Appl Environ Microbiol 64:112–118

Zhang P, Ye ZH, Lan CY, Xie ZW, Hsu WS (2005) Chemically assisted phytoextraction of heavy metal contaminated soils using three plant species. Plant Soil S276:153–162

Zhang H, Zheng LC, Yi XY (2009) Remediation of soil co-contaminated with pyrene and cadmium by growing maize (*Zea mays* L.) Int J Environ Sci Tech 6:249–258

Zhang X, Xia H, Li Z, Zhang P, Gao B (2010) Potential of four forage grasses in remediation of Cd and Zn contaminated soils. Bioresour Technol 101:2063–2066

Plant Growth Under Stress Conditions: Boon or Bane

12

Poonam, Shamshad Ahmad, Narendra Kumar, Paromita Chakraborty, and Richa Kothari

Abstract

Plants are very prone to various stress conditions due to excessive use of chemical-based products and impacts of various living organisms (bacteria, virus, fungi, etc.) with their natural environment. These stresses, if continued for a long period, may affect the growth of plant and productivity to a great extent. Among the various types of stress conditions associated with nutrient, water, soil, and climate-specific stresses can be categorized for better understanding the concept of adaptation in the changing environment. Different types of stresses affect the physiological or metabolic pathways in plants. Water stresses adversely impact the plant physiology especially photosynthetic capacity, drought affects water potential and reduces growth of the plants, excess of nitrogen may cause dehydration of plants, etc. Hence, with these stress conditions arising due to the

Poonam • S. Ahmad
Department of Environmental Sciences,
Babasaheb Bhimrao Ambedkar University, Lucknow, UP, India

N. Kumar
Department of Environmental Science, School of Environmental Sciences, Babasaheb Bhimrao Ambedkar University, Lucknow, UP, India

P. Chakraborty
Department of Civil Engineering, SRM Research Institute, SRM University, Kattankulathur, Tamil Nadu, India

Water Science Laboratory, University of Nebraska-Lincoln, Lincoln, NE, USA

R. Kothari (✉)
Department of Environmental Sciences, Babasaheb Bhimrao Ambedkar University, Lucknow, UP, India

Robert B. Daugherty Water for Food Institute, University of Nebraska-Lincoln, Lincoln, NE, USA
e-mail: kothariricha21@gmail.com

V. Shukla et al. (eds.), *Plant Adaptation Strategies in Changing Environment*,
https://doi.org/10.1007/978-981-10-6744-0_12

changes in surrounding environment, plants adapt to different types of physiological, morphological, and genetic changes to ensure their existence on this planet. In the present chapter, different kinds of stresses and their tolerance and adaption mechanism by various plant groups have been discussed in general, but the main focus is given to the C^3 and C^4 plants in particular. Similarly, impacts of anthropogenic activities responsible for climatic changes or stresses will also be investigated with growth of plants in different habitats and their responses toward these stress conditions.

Keywords
Plant growth · Stress · Mechanisms · Productivity

12.1 Introduction

Plants have the ability to adapt themselves toward the change in their environment by controlling different mechanisms, e.g., by transpiration, photorespiration, photosynthesis, etc. For example, xerophytic plants have converted their leaves in the form of thorns to lower the transpiration rate to minimize water stress. Altered climatic conditions have increased the possibility of different stresses to occur (Ramegowda and Senthil-Kumar 2015). Other abiotic and biotic stresses (temperature, drought, salinity, scarcity of nutrients, increase in CO_2 concentration, etc.) also affect the growth and productivity of plants in their natural environment (Mahajan and Tuteja 2005). Environmental stresses have varying effects on different organs and tissues within a plant, and all the molecular, cellular, and morphological responses to these stresses vary among tissues and organs of plants throughout its lifetime (Gray and Brady 2016). To cope up with these stress conditions, plants adapt to different types of physiological, morphological, and genetic changes. These stresses not only affect the growth and production rate but also the survival of plants in particular stress. As per the recent census records, population of the world is also increasing at an alarming rate and expected to be 6 billion at the end of 2050. To fulfill the nutritional demand of the population, the use of chemical fertilizers, herbicides, and pesticides is increased for the crop production in the last few years. But these practices have retarded the natural properties of soil. Soil has become contaminated by a number of pollutants like heavy metals (As, Cr, Cd, Ni, Pb, Hg, etc.), organic pollutants (pesticides, persistent organic pollutants, volatile organic carbon, polycyclic aromatic hydrocarbons, etc.), and inorganic pollutants (nitrate, phosphate, salinity, sodicity, etc.). These anthropogenic pollutants have also attenuated natural stresses on plants.

12.2 Stresses

When different environmental or anthropogenic factors interfere with the expresses of genotype of plants, the condition is termed as stress. Stresses are broadly categorized into abiotic and biotic stress.

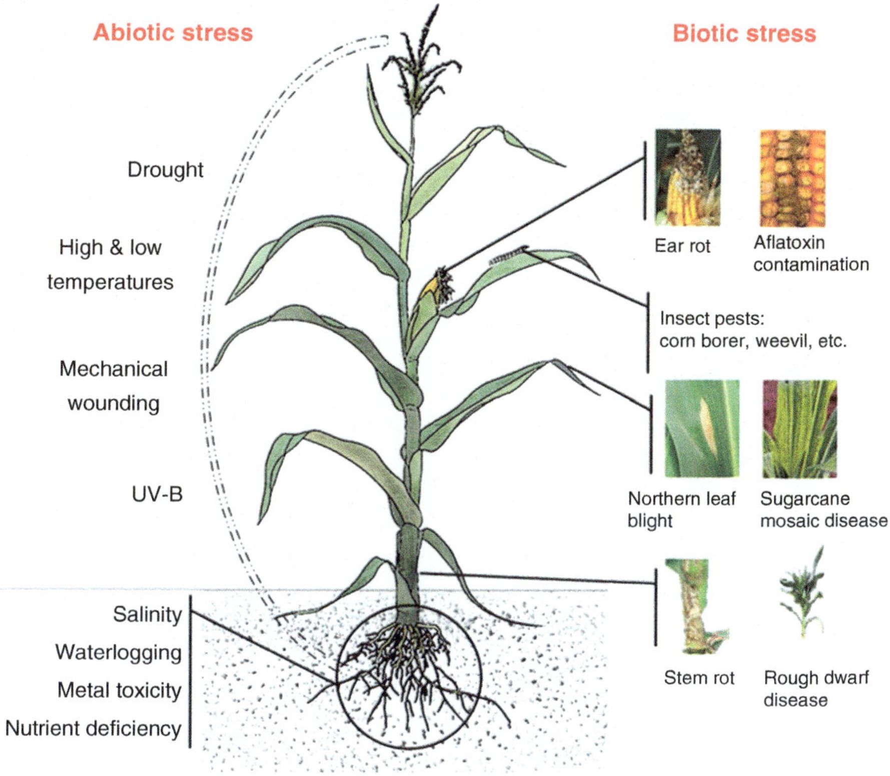

Fig. 12.1 Abiotic and biotic stresses for plants (Fangping et al. 2014)

12.2.1 Abiotic Stresses

Abiotic stresses are caused by the physical/chemical factors to such a level that may affect the plant growth and productivity. Some of the important physicochemical parameters are temperature, salinity, sodicity, radiations, drought, flood, CO_2 concentration, etc. These stresses are found to be unfavorable in most of the cases for plant growth and development, as illustrated in Fig. 12.1.

12.2.2 Biotic Stresses

Involvement of living organism may be beneficial or injurious for the growth and development of plants. Plant growth-promoting bacteria, viruses, fungi, insects, etc. may cause severe biotic stresses affecting metabolic process of the plants.

12.2.3 On the Basis of Environmental Components

12.2.3.1 Salinity Stress

Salinity is a main environmental problem and is a significant limiting factor for plant production. It is caused by an increased level of Na^+, K^+, H^+, and Ca^{2+} ions which are also responsible for maintaining homeostasis in the cell. Higher concentration of salinity in soil causes both hyperionic and hyperosmotic stress perturbing the development and diversity of higher plants and organisms (microorganism, protozoa, and nematodes) existing in soil. It is caused by an excessive amount of evaporation leading to an increase in the salt concentration. Plants, which are highly sensitive toward salinity, are rice, maize, soybean, etc., and stress limits the expansion of the leaves. It affects transpiration, cell division, premature aging of leaves, photosynthesis, and growth of plants. High salt concentration can alter the soil texture and decrease porosity soil aeration and water conductance (Ohran 2016; Mahajan and Tuteja 2005).

12.2.3.2 CO_2 Concentration

Industrial pollution, GHGs, and rise in temperature have increased the level of CO_2 in the atmosphere. CO_2 directly affects photosynthesis, gas exchange, and downstream developmental processes in plants (Ainsworth and Long 2005). Higher concentration of CO_2 in the atmosphere is also a kind of stress for growth and development of plants. CO_2 is an essential element for photosynthesis, and elevated concentration of CO_2 may stimulate higher CO_2 assimilation in C^3 plants in comparison to C^4 plants. The photorespiration in C^4 plants is inhibited by increasing the concentration of CO_2 at the site of RUBISCO by inhibiting the activity of oxygenase enzyme. But, increase in CO_2 concentration also increases the atmospheric temperature which will decrease the amount of water in the atmosphere (Gray and Brady 2016). Further, plants supplemented with CO_2 will need extra water both to maintain their healthy growth and also to compensate the huge losses by evaporation under heat stress. This again works as a stress. Therefore, CO_2 alone will not be able to increase plant growth, but certain other factors like water availability to plants, nutrients, and optimum temperature are also required for plant growth.

12.2.3.3 Temperature

Global warming has caused temperature of the Earth to increase by 0.2 °C per decade for the past 30 years and is expected to increase the global average surface temperature by 1.0–3.7 °C by the end of the century (IPCC 2014). Growth rate of plant is dependent upon the surrounding temperature, and each plant species has specific temperature range for their growth and development. Late flowering, affected pollination, sterility, and low production rate are often increased with increasing temperatures. Exposure of plants to elevated temperatures limits the ability of the plant to produce fruit due to disturbance in pollination (Hatfield and Prueger 2015). Higher temperature can affect rate of photosynthesis by affecting the intonation of photosynthetic enzymes, electron transport chain, and transpiration rates. The optimum ambient temperature for plant growth varies between species depending upon the climate where they grow. The root, shoot, and leaf development is also adversely affected by elevated temperatures (Sage et al. 2008). Even abnormal low temperature is responsible for chilling stress or freezing stress. It directly affects the rate of uptake of water and nutrients, leading to cell desiccation and starvation and leading to death of plants.

12.2.3.4 Availability of Water

Water stress may be because of excess of water due to flooding or due to scarcity of water due to less rain/drought. Surplus water reduces supply of oxygen to the roots which limits nutrient uptake of plants as well as also interfere with respiration of plants.

Drought Condition

Drought stress disrupts cell membrane making the membrane very porous when desiccated. Furthermore, the drought stresses displacement of membrane proteins contributing the loss of integrity of membrane, selectivity, and cellular disruption, compartmentalization, and loss of activity of enzymes. Dehydration may also cause increase in the concentration of electrolytes resulting in disruption of cellular metabolism and osmotic disturbances (Alam et al. 2013). Further, this kind of stress reduces vegetative growth, slower cell division, and reduced leaf expansion reducing transpiration which thus affects agricultural production (Hasanuzzaman et al. 2012).

Flood

Excess of water is also a kind of stress for the development and growth of plants. Excess of water in soil deplete the oxygen concentration which harms the health of roots. If the life-threatening with life-sustaining functions such as respiration, water uptake, root growth, etc. are inhibited considerably, then such conditions might result in the death of plants. Further, toxic compounds such as ethanol and hydrogen sulfide can also be formed which are harmful for plants during anaerobic fermentation. Excessive moisture also favors the growth of different harmful soil microbes like *Fusarium* spp., *Phytophora* spp., and *Rhizoctonia solani*, which can infect the plants with different diseases.

12.2.3.5 Climate

Climate is specific weather condition of a particular place. Optimal temperature required for the proper growth of plant may vary. Plants of colder regions have abilities to resist the bitter temperature during late-night spring or premature autumn frost. But plants, native to warm habitats, exhibit symptoms like lesions or abrasions when exposed to low temperatures (Mahajan and Tuteja 2005), e.g., maize, soybean, cotton, tomato, and banana. These plants are very sensitive to psychrophilic range of temperatures below 10–15 °C and exhibit signs of injuries. This suppresses their full growth and development. Other phenotypic symptoms which may appear are wilting of leaves, leaf expansion, chlorosis, and necrosis (Jiang et al. 2002) shown in Fig. 12.2. Ice formation is the actual cause behind freeze-induced injury. According to specific climatic conditions, plants are classified under the following categories:

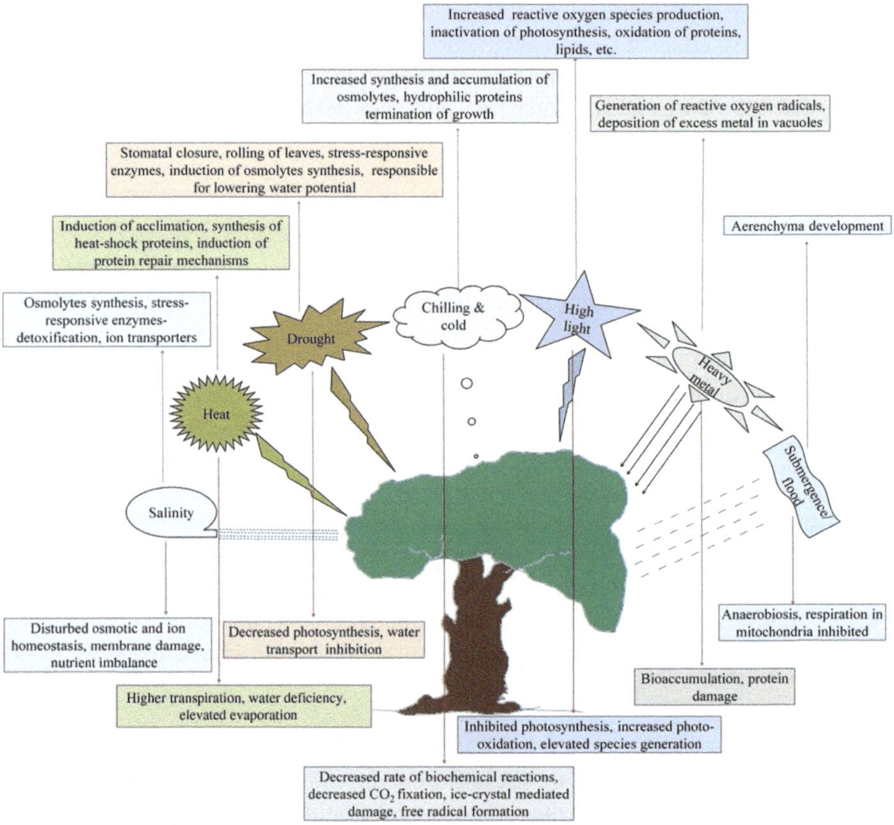

Fig. 12.2 Different kinds of stresses acting upon plants (Kamlesh et al. 2017)

Temperate

Plants found in cold region are characterized by four different seasons, i.e., winter, spring, summer, and fall. Temperature is neither too hot nor too cold and rain is optimum throughout the year. The plants found in this region have stress in the form of cold and humidity stress. These plants have adaptations in such conditions as they go to dormancy by shedding off their leaves during winters, for example, apple, cherry plums, oaks, aspens, walnut trees, lime trees, chestnut trees, almond, etc.

Tropical Climate

These are the plants which cannot tolerate severe cold and grow in warm temperatures of about 100 °F. Tropical regions have only two seasons, i.e., wet and dry, and are frost-free with very small change in solar angles. These plants experience stress conditions in the form of heat, water scarcity, and high radiation. Plants have adapted themselves with such climate by having very large leaves and aerial and buttress root systems. Tropical plants require humidity, sufficient sunshine, warmth, and mild winters. They cannot withstand frost, e.g., papaya, cotton, rice, banana, cocoa shrubs, pineapple, etc.

Subtropical Climate

Subtropical climates have summers from warm to hot and winters ranging from cool to mild weather with rare frost. The subtropical climate is of two types – humid subtropical in which rainfall is often concentrated in the warmest months and dry summer (or Mediterranean) where rainfall occurs in the colder months. Plants grown in subtropical climate have stress in the form of water deficiency or flooding, e.g., *Pinus*, orange, litchi, fig, mango, and cashew nut. These plants have broad leaves and well-developed root system to cope up with the climatic conditions with optimized limits.

12.3 Plant Adaptation Toward Stress Conditions

Adaptation is a trait that allows any organism to survive in a particular environmental condition and facilitate it to evolve accordingly. Climate change is mainly caused by increasing the temperature of the earth because of emission of different greenhouse gases (GHGs). This is changing the weather conditions, rainfall pattern, change in time or duration in fruiting and flowering of plants, and properties of soil which affect the growth of plants. Plants respond to these stresses by altering time of developmental events, initiating of different kinds of mechanisms, or changing developmental processes, e.g., the presence of supplementary leaf nodules with larger leaves size in response to elevated CO_2 (Dermody et al. 2006). To overcome such stresses, plants have developed different kinds of mechanism, which are discussed below:

12.3.1 Role of Plant Growth-Promoting Regulators

Plant growth-promoting regulators (PGPR) are a chemical ingredient that stimulates the growth and cell division of plant, tissues, and organs and also perform the work of chemical messengers for intercellular communications (Rao Srinivasa et al. 2016). These chemicals have the ability to amend nearly all characteristics of plant growth including the stress tolerance and help the plant to adapt to stressful conditions. These chemicals are:

12.3.1.1 Auxin

Auxin (indoleacetic acid, IAA) is a plant growth-promoting hormone which regulates growth of roots, shoot, buds formation, and stress control (Zhao 2010). The level of IAA was found to decrease during stress caused by salinity in root system of wheat plants (Sakhabutdinova et al. 2003). Further, seed germination was improved with increasing salinity by IAA treatments (Akbari et al. 2007). Auxin levels were found to decrease in lower temperatures hindering the root gravity response (Shibasaki et al. 2009). Development of adventitious roots shows undesirable effects of flooding stress on plant growth (Kinoshita et al. 2012). These mechanisms suggest that IAA plays a significant role during different stress conditions.

12.3.1.2 Gibberellins

Gibberellins are responsible for various growth and development like elongation, germination, dormancy, vernalization, flowering, sex appearance, and enzyme activity. Ashraf et al. (2002) reported that it increases salt stress tolerance by increasing the uptake of nutrients, height of the plant, area of the leaf, and wheat production. Further it also reduces the growth of the plant to overcome the stress conditions. Wen et al. (2010) reported that NaCl-induced stress was lowered by gibberellins through growth inhibition of rice plants.

12.3.1.3 Abscisic Acid (ABA)

It facilitates stomatal functioning, root hydraulic conductivity, photosynthesis, plant-water relations, osmotic pressure, and stress-responsive protein synthesis and confers stress tolerance gene (Kim et al. 2010). It helps plant to tolerate different stress conditions specially drought and salinity (Keskin et al. 2010). Drought and salinity stress induced the accumulation of ABA in several plant species (Satisha et al. 2005). Higher levels of ABA stimulate closure of stomata during transpiration which reduce water and an increased water status in plant with increased root hydraulic conductivity (Thompson et al. 2007). This also indicates that ABA works as chemical signaling agent between roots and aerial parts and helps to transport the stressed root condition (drought and salinity) to the plant (Zhang et al. 2006). Further, stomatal sensitivity to ABA changes with different species of the plants and depends upon climatic features. Thus as climate change causes stress, ABA is secreted by plants to adopt the stress conditions.

12.3.1.4 Cytokinins

It helps the plant during seed germination, fruit and leaf abscission, and stomatal functioning with root-to-shoot communication in stress-experiencing plants (Rao Srinivasa et al. 2016). The secretion of this plant hormone is lowered down during stress conditions, and after release of stress, it comes back to normal levels. Its levels were found to alter during drought conditions in different plants (Vankova et al. 2011; Zhou et al. 2004; Upreti and Murti 2004).

12.3.1.5 Ethylene

It is a gaseous hormone which regulates seed germination, abscission, fruit ripening, root formation, etc. (Rao Srinivasa et al. 2016). During stress conditions especially drought, the levels of ethylene are increased. It assists the plants to adapt stress conditions by minimizing water loss during fruits/leaves senescence and reduced growth. Ethylene regulates salt, cold, heat, and drought stress effectively and helps the plants to adapt with such stressful conditions (Zhao et al. 2014; Jiang et al. 2013; Wang et al. 2009).

12.4 Genes Responsible for Adaption of Plants for Different Stresses

12.4.1 Cold Stress

Cold stress produces species of reactive oxygen (RO) in response to damages in the cell membrane. Further, it results in integrity of membrane failure which leads to leakage of solute. The phenomenon developed by plants to withstand chilling stress by increasing their freezing tolerance is termed as "cold acclimatization." This process preserves the cell membranes from injury and increases the sugar level intracellular space as well as accumulation of other cryoprotein which increases the fraction of unsaturated fatty acids (Mahajan and Tuteja 2005).

12.4.2 Salinity Stress

Salt overly sensitive (SOS) was identified by Liu and Zhu (1998) in which excess Na^+ ions were removed out of the cell via plasma membrane by using Na^+/H^+ antiporter and helping in reestablishing cell homeostasis. Zhu (2002) described the interpretation signaling of Ca^{2+} pathway in response to a salt stress made possible through SOS discovery.

12.4.3 Drought Stress

Drought stress brings about change in expression of late embryogenesis abundant (LEA) dehydrin-type genes. It helps in protection of partner protein from degradation and removes damaged or denatured proteins (Mahajan and Tuteja 2005). The response of these genes including pectin lyases and expansions is dependent upon the stages of the leaves as these are upregulated in young leaves in comparison to older ones. These genes enable plants to grow continually at low rates at the time of drought stress (Gray and Brady 2016). Further, ABA and ROR enzymes also play significant roles to overcome drought conditions.

12.4.4 Temperature Stress

Excess of heat causes heat stress which affects the cellular and metabolic responses of plants. Alterations at molecular levels were observed after exposure at high temperature that modifies the expression of gene with transcripts accumulation, thereby synthesizing stress-related proteins as a stress tolerance strategy. This includes modifications in normal proteins synthesis and heat-shock proteins (HSPs) and production of ABA and antioxidants (Bita and Gerats 2013; Bray et al.2000). Heat-shock elements (HDEs) interact with heat-shock factor (HSF) proteins to increase HSP genes which are responsible for tolerating the heat stress in plants improving the photosynthesis, assimilation of food, partitioning in tissues, water and nutrient efficiency, and stability of membrane(Wahid et al. 2007; Ahn and Zimmerman 2000).

12.5 Adaption Through Different Metabolic Processes

Plants have adapted themselves by different biochemical and metabolic processes to endure different stresses. These mechanisms include transpiration, photosynthesis, and photorespiration.

12.5.1 Transpiration

Different stress conditions like drought, temperature increase due to GHGs, salinity, humidity, etc. are tolerated successfully by regulating transpiration rate, and stomatal opening controlled the metabolic process. Under such conditions, plants close their stomata to slower down water loss due to transpiration. This may occur due to the direct evaporation of water from the guard cells without any metabolic activity. Metabolically, stomatal closure involves processes of reverting ion refluxes that result in stomatal opening which is regulated by ABA. It assists reflux of K^+ ions from guard cells, resulting in the loss of turgor pressure (Wilkinson and Davies 2002).

12.5.2 Photosynthesis

It is the process which converts radiant energy into chemical energy by means of green plants (Tkemaladze and Makhashvili 2016). The plant productivity greatly depends on the speed and efficiency of photosynthesis. The best temperature range for photosynthesis in almost all plants is 10–35 °C. Increase in temperature due to various stress factors like increasing temperature, drought, and higher concentration of CO_2 results in checking the process of photosynthesis due to deterioration of activity of RUBISCO enzyme (Mahajan and Tuteja 2005). On the other side, GHGs increase CO_2 concentration in the atmosphere resulting in increase in the rate of photosynthesis. But, on the other hand, water scarcity and increasing temperature may lower down the rate of photosynthesis. Calvin-Benson cycle is used for the fixation of CO_2 in which RUBISCO acts as a catalyst in the first step to produce a three-carbon (C-C-C) phosphoglycerate. Therefore, its known as C_3 Cycle and plants which follow this mechanism are known as C_3 plants (Fig. 12.3), whereas C_4 cycle is an adaptation to C_3 cycle with exception of photorespiration, which improves photosynthesis efficiency and water loss in hot and dry environment. Thus, plants following this cycle are known as C_4 plants and are found in tropics and temperate zones. These plants have adapted the stress of higher temperature, salinity, and water deficiency and exhibit higher photosynthetic and growth rates.

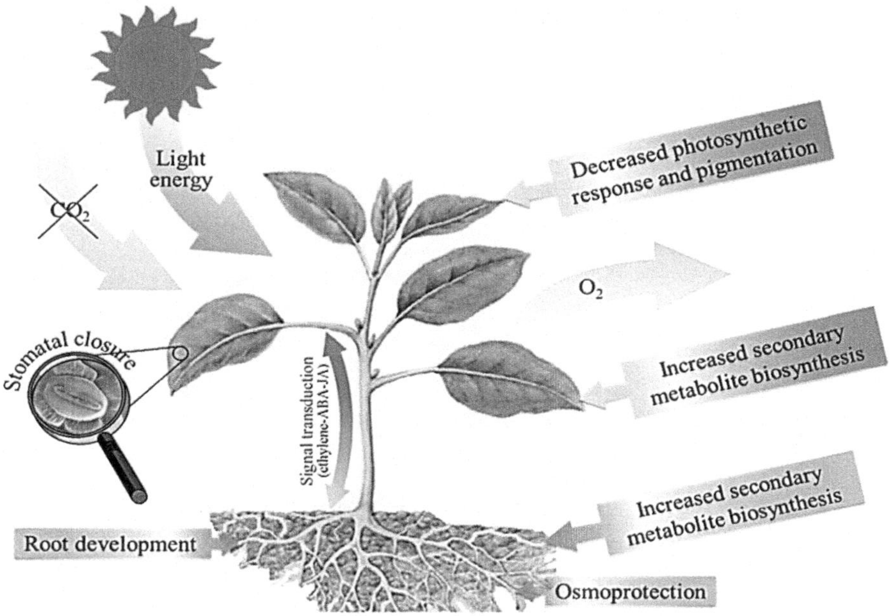

Fig. 12.3 Plant's responses toward different stresses (Sonia et al. 2013)

12.5.3 Photorespiration

Photorespiration is the biochemical process which occurs in C_4 plants found in humid and dry environments. When the level of CO_2 inside the leaf decreases, plants are forced to close stomata to prevent excess water loss. It results in the uptake of O_2 from the environment and release of CO_2. It takes place concurrently with photosynthesis and is wastage of energy. Most of the times, plants keep their stomata open, so that photorespiration is low. When temperature is high and there is water scarcity, stomata are closed by the plants to conserve water loss by transpiration. This restricts normal gas exchange and level of CO_2 slowly rises. So, to cope up with this situation, photorespiration rate increases and efficiency of photosynthesis decreases. Photorespiration thus can reduce the plant growth. C_4 and crassulacean acid metabolism (CAM) plants have adapted such situation by developing a special kind of anatomy of leaves known as "Kranz anatomy" in many tropical grasses (corn and sugarcane) and succulent plants (pineapples, cacti, and stonecrops).

12.6 Effect of Different Stress Conditions on Plant Growth

Several stresses cause a no. of morphological as well as physiological changes in the plants. The effect of different stresses on plants and their adoptive responses have been discussed in Table.12.1.

Table 12.1 Effect of different stresses on plants ant their adoptive responses

Stress	Plant	Effect on growth and development parameters	Plant's adaptation/ responses	References
Elevated CO_2	*Deschampsia flexuosa*	Leaf area index (LAi) reduced due to leaf wilting	Net photosynthesis increased No reduction in stomatal conductance	Albert et al. (2011)
Drought	*Deschampsia flexuosa*	Aboveground biomass was very low	Reduced photosynthetic capacity Stomatal conductance was very low	Albert et al. (2011)
Global warming	*Deschampsia flexuosa*	Increase in leaf respiration	Improved net photosynthesis but did not affect net photosynthetic capacity of grass	Albert et al. (2011)
Drought and heat stress	Barley (*Hordeum vulgare*)	Drought causes intense reductions of biomass, plant height, and yield. In combination, these two stresses caused stronger senescence of lower leaves and lesser growth of the plant	Drought did not change photosynthetic performance and the proteome protein Heat stress affected relative water content, photosynthetic performance, and floret fertility	Rollins et al. (2013)
Salt stress	Castor oil (*Ricinus communis*)	Retarded seed germination and elongation of stem reduced expansion of leaves Dry weight and water content of plants decreased with increasing salinity Reduced the content of N, P, K^+, and Ca^{2+} in plant tissues	Addition of Ca^{2+} restored N, P, K^+ and Ca^{2+} content in plant tissues	Joshi et al. (2012)
Salt stress	*Atriplex halimus* and *Atriplex nummularia*	Initially shoot and leave growth for both were promoted till 300 mM of NaCl After that, reduction in growth increasing conc. of NaCl and KCl concentration were observed	High amount of proline were identified under NaCl treatments Inhibitory effects may be contributed to Cl^- toxicity	Belkheiri and Mulas (2013)

(continued)

Table 12.1 (continued)

Stress	Plant	Effect on growth and development parameters	Plant's adaptation/responses	References
Salt stress	Sugarcane clones, CP-4333 (tolerant) and HSF-240 (sensitive)	Rate of photosynthesis decreased the leaf area, length, and dry weight of shoots	Both of the clones had the tendency to accumulate Na^+, Cl, and little K^+	Wahid and Ghazanfar (2006)
			Secondary metabolites (soluble phenolics, anthocyanins, and flavones) level was higher in sensitive clone than in tolerant clone which may be because of tolerance levels of the plant	
Heat stress	Mung bean	Reduction in plant growth due to unavailability of nutrient and amino acid content	Application of triacontanol reduced deleterious effects of heat stress by increasing ABA and JA levels	Waqas et al. (2016)
		Decreased levels of abscisic acid (ABA) and jasmonic acid (JA)	Its application increased plant growth and enhanced nutrient and amino acid in the plants	

12.6.1 Drought

Climate change has also influenced the precipitation pattern on the earth surface leading to drought due to water scarcity in many regions. Stress hampers the growth of plants with loss of production yields (Verelst et al. 2010). It causes plants to invest nutrient and other organic content in root tissues at the price of shoot which is proved by root-to-shoot ratio (Poorter and Nagel 2000). Root grows by elongation of primary roots as well as initiation in lateral roots; enhanced roots are shifted deep in the soil in the search of water to fulfill the requirement of water. Thus, root develops well in comparison to shoot and leaves. This enhances the elongation of root but ceases the growth of shoots (Gray and Brady 2016). Further, leaf tissues are also less developed as compared to root which reduces the loss of water through transpiration. Two major processes for the development of plant, viz., division and expansion of cells, are badly inhibited by drought. Drought causes cell turgor to reduce due to osmotic stress and affects plant-water relationship which also affects the developmental processes.

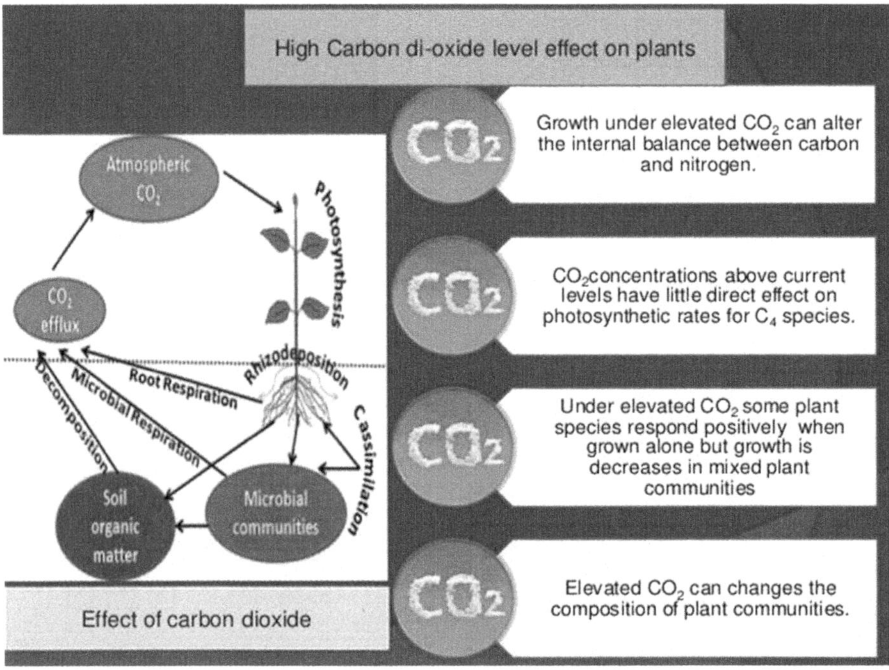

Fig. 12.4 Adaptation of plants toward higher concentrations of CO_2 (https://www.slideshare.net/SwatiSagar1/environmental-stress-and-microorganisms)

12.6.2 Carbon Dioxide (CO_2)

Higher concentrations of CO_2 stimulate the aboveground biomass, increase no. of leaf nodes, lead expansion, and enhance root branching and root elongation. But all these do not merely depend upon elevated CO_2 but also upon availability of water and nutrient uptake by plants (Bishop et al. 2014; Reich et al. 2014; Madhu and Hatfield 2013; Dermody et al. 2006). All these positive responses of plants toward higher concentration of CO_2 are related to increased cell division and cell production and/or cell expansion (Gray and Brady 2016). Further, elevated concentration of CO_2 concentration increases photosynthesis in C3 plants, thereby resulting in increase in aboveground biomass (Fig. 12.4). This process is not merely dependent upon higher CO_2 concentrations; water and nutrient availability is also important (Reich et al. 2014). Root biomass is also found to increase seed yield significantly in many plants, e.g., wheat, soybean, rice, peanut, etc. (Madhu and Hatfield 2013; Hatfield et al. 2011). Plants may increase the number of leaf node, root length, and pod number toward response to elevated CO_2 (Bishop et al. 2014).

12.6.3 Temperature

The temperature of the earth has been found to increase about 8.5 °C from 1880 to 2013 (Hartmann et al. 2013). Amplifying frequency, strength, and period of heat waves has put heat stress over the plant and affected the plant community (IPCC 2014). It affects the plant growth and development by varying rate of photosynthetic carbon assimilation. Optimum temperature for full growth and development of plants that vary among various species depends upon the climate and environment (Sage and Kubien 2007). Rate of root and shoot elongation, leaf initiation, and their expansion increases with increasing temperature till threshold limits are achieved. After that it works as a stress brings about change in photosynthesis as well as respiration leading to a reduced life cycle and reduces the productivity of plant. A number of physical injuries may be caused by elevated temperature of leaves, leaf burns and wilt, abscission and senescence, and inhibition of root-shoot growth (Bita and Gerats 2013). Higher temperature affects shoot net assimilation rates, and thus, total dry weight is reduced, thereby inhibiting the plant growth (Wahid et al. 2007). Increasing temperatures, most of the times, are associated with elevated concentration of CO_2 vapor pressure deficit (VPD) and drought. It affects photosynthetic metabolism directly by influencing the activity of RUBISCO processes through the Calvin cycle (Lloyd and Farquhar 2008). For growth and development, plants uptake nutrients from decaying organic matters present in the soil in the form of leaf litters and other dead organic substances by undergoing various physiochemical processes. These processes are affected by rise in temperature, soil moisture content, decomposition of organic matters, and uptake of nutrients from soil, thus affecting plant growth and development. Higher temperature also causes increased photorespiration and affects rate of photosynthesis. Further, at low temperature, the reaction processes become much slowed, thereby affecting plant morphology, growth, and root turnover. Thus, both high and low temperatures are harmful for growth and development of plants in Fig. 12.5.

12.6.4 Salinity

Salts are naturally present in the soil, but higher concentration works as stress affecting the growth, development, and productivity of plants. Saline soils have higher osmotic pressure which unable plants for water uptake and nutrients from soil (Hogarth 2015). Now, plants will need more energy to extract water from the soil as an added stress on plants and may cause wilting of the plant tissues. Leaves may be small in size, marginal necrosis is also possible, and thus plants' health is affected badly. Further, salinity is often caused by higher levels of water table which may be accompanied by water logging. It restricts growth of the plant, and also the ability of the roots to eliminate salt is decreased. This increases the salt uptake and its accumulation in shoots (Fig. 12.6).

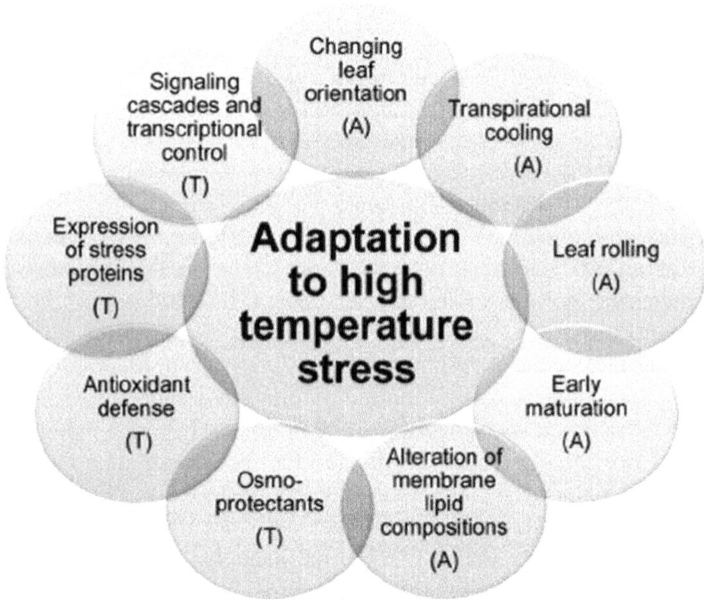

Fig. 12.5 Temperature adaptations in plant (Mirza et al. 2013)

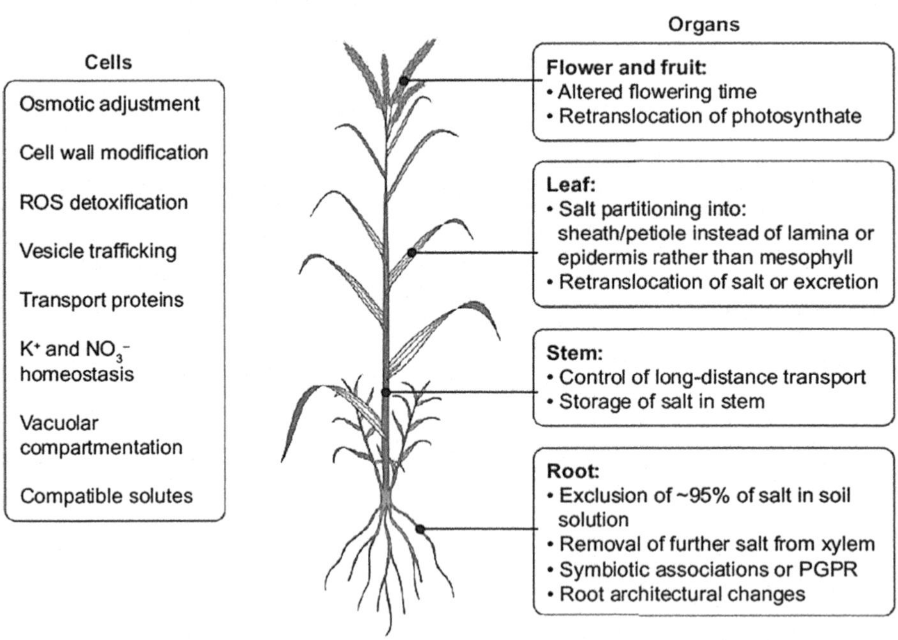

Fig. 12.6 Plants' adaptation toward high salt concentrations (http://www.plantransig.com/research/gilliham-lab/salinity/)

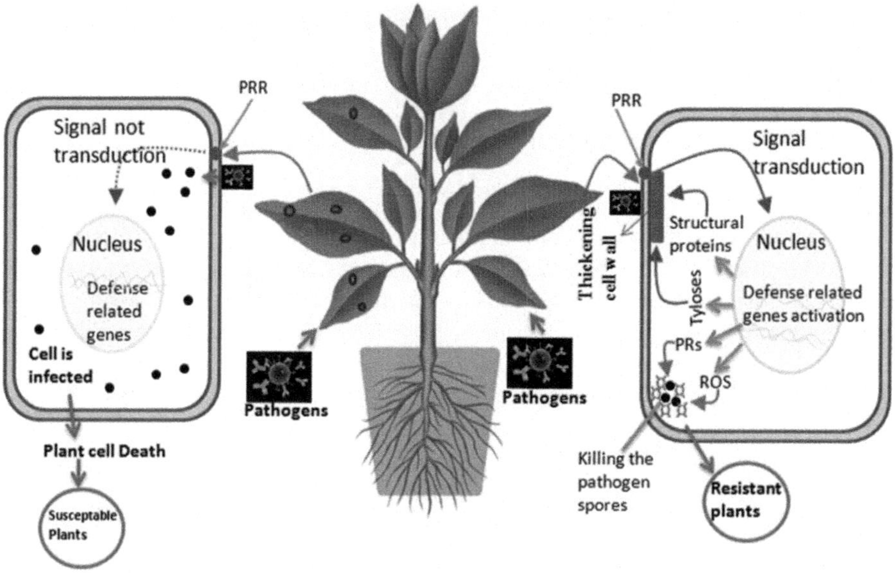

Fig. 12.7 Resistance mechanism of plant by virus (Md Abdul et al. 2016)

12.7 Plant Resistance Mechanism Toward Different Stress

In the past several decades, researchers mainly focused and concentrated on sympathetic responses of biotic and abiotic stresses on plant (Stotz et al. 2014), although the environmental responses produce synchronized stresses to much more complex scenario. Higher plants are able to accomplish synchronizing the acquaintance in the abiotic and biotic stress and produce link between responses (Ramirez et al. 2009). Usually, both the stress (abiotic and biotic) can reassure the environmental pressure of resistance on plants. However, few plants antagonized the susceptibility of stress compared with synchronized exposure to two different stresses (Suzuki et al. 2014). So, there is cross-tolerance between both abiotic and biotic factors thereby boosting the resistance in plants and has significant effect on agricultural production. Amusingly, the defense mechanism at the site of pathogen infection is regulated by abiotic stress in different parts of plant, certifying and increasing the plant's innate immunity system. Similarly, hydrogen ion and water stress are inducing resistance against powdery mildew in barley, and resistance depends on the cellulose-containing papilla which is capable of blocking growth of fungus (Fig. 12.7) and is analogue to the chemically induced resistance by isonicotinic acid (Achuo et al. 2006) Consequently, in order to put the plant for battle for the activation of signaling pathways, regulator hormones, various detoxifying enzymes, and expression of gene are obligatory (Koornneef and Pieterse 2014). The plant defense response exposed for different stressors is anticipated with different interconnection using signaling pathways that help in the regulation of numerous metabolic networks (Nakashima et al. 2009).

12.7.1 Tolerance Mechanisms in Plants

Specific trait is closely associated with the higher species of plant, and even different tissue and organ of the similar plant may vary and thriving under different tolerance level. They include long-term evolutionary acclimatization short term with avoidance. Several tolerance mechanisms, including transporter ion, osmoprotectants, late embryogenesis abundant proteins (LEAP), antioxidant defense mechanism, and aspects convoluted in cascades signaling with transcriptional control, are essentially significant and counteract the effect of stress (Mirza et al. 2013). In case of stress, with short-term response, i.e., orientation of leaf, transpiration mechanism and variations in lipid composition of membrane play a significant role for the survival of plant (Rodríguez et al. 2005). Different kinds of tissues and organs illustrate dissimilarities in development, acquaintance, and responses toward the applied stress types. By an initial signaling, stress-responsive mechanism of initial signaling maintains the ionic and osmotic effect of changing fluidity. This helps to restoration of the homeostasis with protection of damaged proteins.

12.7.2 Avoidance Mechanism in Plants

Angiospermic plants show innumerable mechanisms for the survival, which comprise long-term phenological as well as morphological adaptations with temporary acclimation or avoidance mechanisms. During stress, plant shows some mechanisms, e.g. Orientation in leaf, transpirational behavior, and changing the lipid composition. Loss of water reduced by closure mechanism of stomata, trichomatous densities increases around the stomata and size of xylam increase, common feature appear during the heat stress (Mirza et al. 2014). Plants growing in a warm and humid atmosphere evaded the heat stress by decreasing the solar flux with modification in internal as well as external parts of a plant like the presence of small hair, thick coat of cuticle, and protective covering of wax on the surface of the leaf. Many species of plant having advanced life cycle permit growth round of the year. Such anatomical, physiological, and genetical adaptations are generally favoring in the adaptation in biochemical process photosynthesis in C^4 and CAM, but C^3 vegetation is very common in high-temperature zone (Fitter and Hay 2002). High temperature plays a significant role in the rolling of leaf which maintains the adaptation by increasing water metabolism efficiency (Sarieva et al. 2010). Only in the summer, few favored terrestrial plant species increase the resistivity toward heat, whereas others exhibit the highest level of tolerance during dormancy period. Upon reaching a developmental stage, dormant plant species grow which are induced by climatic factors other than high environmental temperature. Perceptible changes in heat tolerance are not observed in many land plant species. By the crop management practices, avoiding the high temperature stress, such as cultivars, method of irrigation, choosing proper sowing technique, selecting date of sowing, etc., for example, lettuce sown in the late summer, may show appearance due to high temperature of soil and incomplete germination. During the day, sowing the seed of lettuce into dry beds has the incomplete germination problem. In contrast, tropical crops may face inadequate development in plant which can regulate the efficiency of numerous annual crops due to high temperature of soil.

12.8 Stress: Boon or Bane

12.8.1 Boon

The increasing temperature has a lot of adverse effects on the plant diversity, if plants fail to cope up with or adapt with the increasing temperature. But if they adapt with this temperature increases, it would be beneficial for the balance in ecosystem and from food security point of view. Seasonal plants can adapt quickly, at shifting climate conditions by their emerging response when the weather shifts. If plants are adapting, we could readily get the important food crops and cash crops, i.e., from food security point of view, and this plant adaptation will prove to be a boon. Plants are required for the ecological balance of the Earth. Due to the effect of the changing climatic conditions, those plants that could not cope up might be lost. But if they can tolerate the extreme temperature and get adapted, it would be beneficial from the ecological point of view. Though all plants are important, some plants are also important from the economic point of view as some of the plants are listed under cash crops like coffee, tea, cotton, sugarcane, etc. If these cannot tolerate and adapt as per the change of climatic conditions, it would prove to be a great loss to the economy of a country, but if they get adapted to the climatic conditions, it would prove to be a great boon to the economy of a country. Several plants are flourishing in the areas where their growth was limited before, just because of change in temperature, e.g., American flowering plants like *columbines* and wild *geraniums* are flourishing earlier than before. These events show that the plant diversity is getting adapted to the changing environmental conditions.

12.8.2 Bane

The changing climate has always put a serious threat to the surviving organisms on the earth. This climate change mostly proves to be a trouble for the healthy plant survival. Plant adaptation is a boon with respect to climate change only if those plants are food crops or cash crops, but what if those are the notorious weeds? Many weed crops are extremely adaptive at changing environment, such as high temperatures, flood, and drought, which may help them for critical crops. But due to the maladaptation of the different food crops and cash crop, it would affect the disturbance in the dynamics of interspecific competition for favoring weedy invaders, which could ultimately lead if inhabitants failed to adapt and compete.

12.9 Conclusion

Stress condition will alter the development of plant as well as significant effects on the function of natural ecosystems of plant. Predictions of the plants growing pattern in the world affected by the raising temperature, amplified the number of incidence of life-threatening events e.g. heat waves, drought, humidity, radiation and composition of atmospheric gases in the atmosphere. Flexibility found in the development of plant in response to climate change will be critical in maintaining ecosystem function and agricultural productivity in the future.

References

Abdul K, Hoy-T, Ujjal KN, Jong-In P, Kang HK, Yong-GC, Ill-Sup N (2016) Research on biotic and abiotic stress related genes exploration and prediction in Brassica rapa and B. oleracea: a review. Plant Breed Biotech 4(2):135–144

Achuo EA, Prinsen E, Hofle M (2006) Influence of drought; salt stress and abscisic acid on the resistance of tomato to *Botrytis cinerea* and *Oidium neolycopersici*. Plant Pathol 55:178–186

Ahn YJ, Zimmerman JL (2000) Introduction of the carrot HSP17.7 into potato (*Solanum tuberosum* L.) enhances cellular membrane stability and tuberization in vitro. Plant Cell Environ 29:95–104

Ainsworth EA, Long SP (2005) Tansley review: what have we learned from 15 years of free-air CO_2 enrichment (FACE) a meta-analytic review of the responses of photosynthesis, canopy properties and plant production to rising CO_2. New Phytol 165:351–371

Akbari G, Sanavy S, Yousefzadeh S (2007) Effect of auxin and salt stress (NaCl) on seed germination of wheat cultivars (*Triticum aestivum* L.) Pak J Biol Sci 10:2557–2561

Alam MM, Hasanuzzaman M, Nahar K, Fujita M (2013) Exogenous salicylic acid ameliorates short-term drought stress in mustard (*Brassica juncea* L.) seedlings by up-regulating the antioxidant defense and glyoxalase system. Aust J Crop Sci 7:1053–1063

Albert KR, Ro-Poulsen H, Mikkelsen TN, Michelsen A, Van D, Linden L, Beier C (2011) Interactive effects of elevated CO_2, warming, and drought on photosynthesis of *Deschampsia flexuosa* in a temperate heath ecosystem. J Exp Bot 62:4253–4266

Ashraf M, Karim F, Rasul E (2002) Interactive effects of gibberellic acid (GA) and salt stress on growth, ion accumulation and photosynthetic capacity of two spring wheat (*Triticum aestivum* L.) cultivars differing in salt tolerance. Plant Growth Regul 36:49–593

Belkheiri O, Mulas M (2013) Effect of water stress on growth, water use efficiency and gas exchange as related to osmotic adjustment of two halophytes *Atriplex* spp. Funct Plant Bio 40:466–474

Bishop KA, Betzelberger AM, Long SP, Ainsworth EA (2014) Is there potential to adapt soybean (Glycine max Merr.) to future [CO2]? An analysis of the yield response of 18 genotypes in free-air CO2 enrichment. Plant Cell Environ 38:1765–1774

Bita CE, Gerats T (2013) Plant tolerance to high temperature in a changing environment: scientific fundamentals and production of heat stress-tolerant crops. Front Plant Sci 4:273–291

Bray EA, Bailey-Serres J, Weretilnyk E (2000) Responses to abiotic stresses. In: Buchanan BB, Gruissem W, Jones RL (eds) Biochemistry and molecular biology of plants. American Society of Plant Physiologists, Rockville, pp 1158–1203

Dermody O, Long SP, DeLucia EH (2006) How does elevated CO_2 or ozone affect the leaf-area index of soybean when applied independently. New Phytol 169:145–155
Fangping G, Le Y, Fuju T, Xiuli H, Wei W (2014) "Omics" of maize stress response for sustainable food production: opportunities and challenges. OMICS A J Integr Bio 18(12):714–732
Fitter AH, Hay RKM (2002) Environmental physiology of plants, 3rd edn. Academic, London
Gray SB, Brady SM (2016) Plant developmental responses to climate change. Dev Biol 419:64–77
Hartmann DL, Klein Tank AMG, Rusticucci M, Alexander LV, Brönnimann S, Charabi Y, Dentener FJ, Dlugokencky EJ, Easterling DR, Kaplan A, Soden BJ, Thorne PW, Wild M, Zhai PM (2013) Observations: atmosphere and surface. In: Stocker TF, Qin D, Plattner G-K, Tignor M, Allen SK, Boschung J, Nauels A, Xia Y, Bex V, Midgley PM (eds) Climate climate change (2013): the physical science basis. Contribution of working group I to the fifth as- sessment report of the intergovernmental panel on climate change. Cambridge University Press, Cambridge/New York
Hasanuzzaman M, Hossain MA, Teixeira da Silva JA, Fujita M (2012) Plant response and tolerance to abiotic oxidative stress: antioxidant defense is a key factor. In: Venkateswarlu B, Shanker SC, Maheswari M (eds) Crop stress and its management: perspectives and strategies. Springer, New York, pp 261–315
Hatfield JL, Prueger JH (2015) Temperature extremes: effect on plant growth and development. Weather Clim Extremes 10:4–10
Hatfield JL, Boote KJ, Kimball BA, Ziska LH, Izaurralde RC, Ort D, Thomson AM, Wolfe D (2011) Climate impacts on agriculture: implications for crop production. Agron J 103:351–370
Hogarth PJ (2015) The biology of mangroves and seagrasses. Oxford University Press, Oxford. ISBN:978-0-19-871657-9
IPCC (2014) In: Core Writing Team, Pachauriand RK, Meyer LA (eds) Climate change 2014: synthesis report. contribution of Working Groups I, II and III to the fifth assessment report of the Intergovernmental Panel on Climate Change. IPCC, Geneva. 151pp
IPCC (Intergovernmental Panel Climate Change) Climate Change (2007) Impacts, adaptation and vulnerability: contribution of Working Group II to the fourth assessment report of the Intergovernmental Panel on Climate Change. Cambridge University Press, Cambridge/New York
IPCC, (Intergovernmental Panel Climate Change) Climate Change (2014) In: Core Writing Team, Pachauri RK, Meyer LA (eds) Synthesis report. Contribution of Working Groups I, II and III to the fifth assessment report of the Intergovernmental Panel on Climate Change. IPCC, Geneva, p 151
Jiang QW, Kiyoharu O, Ryozo I (2002) Two novel mitogen-activated protein signaling components, OsMEK1 and OsMAP1, are involved in a moderate low-temperature signaling pathway in Rice1. Plant Physiol 129:1880–1891
Jiang C, Belfield EJ, Cao Y, Smith JA, Harberd NP (2013) An Arabidopsis soil-salinity-tolerance mutation confers ethylene-mediated enhancement of sodium/potassium homeostasis. Plant Cell 25:3535–3552
Joshi M, Mishra A, Jha B (2012) NaCl plays a key role for in vitro micropropagation of Salicornia brachiata, an extreme halophyte. Ind Crop Prod 35:313–316
Kamlesh KM, Ajay MS, Utkarsh MB, Khushboo C, Priyanka G, Ashwani P, Dhananjaya PS, Ratna PR, Pramod KS, Vijai KG, Harikesh BS, Kishor KK, Paramjit SM (2017) Abiotic stress responses and microbe-mediated mitigation in plants: the omics strategies. Front Plant Sci 8:172
Keskin BC, Saikaya AT, Yuksel B, Memoh AR (2010) Abscisic acid regulated gene expression in bread heat. Aust J Crop Sci 4:617–625
Kim TH, Bohmer M, Nishilmura N, Schorder JI (2010) Guard cell signal transduction network: advances in understanding abscisic acid, CO_2 and Ca^{+2}signalling. Ann Rev Plant Biol 61:561–569
Kinoshita N, Wang H, Kasahara H, Liu J, MacPherson C, Machida Y, Kamiya Y, Matthew A, Hannah MA, Chua N (2012) IAA-Ala Resistant3, an evolutionarily conserved target of

miR167, mediates arabidopsis root architecture changes during high osmotic stress. Plant Cell 24:3590–3602

Koornneef A, Pieterse CMJ (2014) Cross talk in defense signaling. Plant Physiol 146:839–844

Liu J, Zhu JK (1998) A calcium sensor homolog required for plant salt tolerance. Science 280:1943–1945

Lloyd J, Farquhar G (2008) Effects of rising temperatures and [CO_2] on the physiology of tropical forest trees. Phil Trans R Soc 363:1811–1817

Madhu M, Hatfield JL (2013) Dynamics of plant root growth under increased atmospheric carbon dioxide. Agron J 105:657–669

Mahajan S, Tuteja N (2005) Cold, salinity and drought stresses: an overview. Arch Biochem Biophys 444:139–158

Mirza H, Kamrun N, Mahabub A, Rajib R, Masayuki F (2013) Physiological, biochemical, and molecular mechanisms of heat stress tolerance in plants. Int J Mol Sci 14:9643–9684

Mirza N, Mahmood Q, Maroof SM, Pervez A, Sultan S (2014) Plants as useful vectors to reduce environmental toxic arsenic content. Scient World J 2014:921581

Nakashima K, Ito Y, Yamaguchi-Shinozaki K (2009) Transcriptional regulatory networks in response to abiotic stresses in arabidopsis and grasses. Plant Physiol 149:88–95

Ohran F (2016) Alleviation of salt stress by halotolerant and halophilic plant growth-promoting bacteria in wheat (*Triticum aestivum*). Braz J Microbiol 47:621–627

Poorter HA, Nagel O (2000) The role of biomass allocation in the growth response of plants to different levels of light, CO_2, nutrients and water: a quantitative review. IMF Occas Pap 27:595–607

Qu Y, Wege S, Qiu J, Henderson S, Wu Y, Byrt C, Xu B, Sohami H, Jiaen Qiuplant Transport and Signaling Group (n.d.) http://www.plantransig.com/research/gilliham-lab/salinity/

Ramegowda V, Senthil-Kumar M (2015) The interactive effects of simultaneous biotic and abiotic stresses on plants: mechanistic understanding from drought and pathogen combination. J Plant Physiol 176:47–54

Ramirez V, Coego A, Lopez A, Agorio A, Flors V, Vera P (2009) Drought tolerance in Arabidopsis is controlled by the *OCP3* disease resistance regulator. Plant J 58:578–591

Rao Srinivasa NK, Shivashankara KS, Laxman RH (2016) Abiotic stress physiology of horticultural crops. Springer India, New Delhi. ISBN:978-81-322-2725-02

Reich PB, Hobbie SE, Lee TD (2014) Plant growth enhancement by elevated CO_2. Eliminated by joint water and nitrogen limitation. Nat Geosci 7:920–924

Rodríguez M, Canales E, Borrás-Hidalgo O (2005) Molecular aspects of abiotic stress in plants. Biotechnology 22:1–10

Rollins JA, Habte E, Templer SE, Colby T, Schmidt J, von Korff M (2013) Leaf proteome alterations in the context of physiological and morphological responses to drought and heat stress in barley (*Hordeum vulgare* L.) J Exp Bot 64(11):3201–3212

Sagar S (2017)https://www.slideshare.net/SwatiSagar1/environmental-stress-and microorganisms; Can microorganism crop plants in combating climate change associated variabilities?

Sage RF, Kubien DS (2007) The temperature response of C3a nd C4 photosynthesis. Plant Cell Environ 30:1086–1106

Sage RF, Way DA, Kubien DS (2008) Rubisco, Rubisco activase, and global climate change. J Exp Bot 59(7):1581–1595

Sakhabutdinova AR, Fatkhutdinova DR, Bezrukova MV, Shakirova FM (2003) Salicylic acid prevents the damaging action of stress factors on wheat plants. Bulg J Plant Physiol 29. (Special issue 2013:314–319

Sarieva GE, Kenzhebaeva SS, Lichtenthaler HK (2010) Adaptation potential of photosynthesis in wheat cultivars with a capability of leaf rolling under high temperature conditions. Russ J Plant Physiol 57:28–36

Satisha J, Prakash GS, MurtiM GSR, Upreti KK (2005) Response of grape genotypes to water deficit: root, shoot growth and endogenous hormones. Indian J Plant Physiol 10:225–230

Shibasaki K, Uemura M, Tsurumi S, Rahman A (2009) Auxin response in Arabidopsis under cold stress: underlying molecular mechanisms. Plant Cell 21:3823–3838

Sonia MZ, Marielle CI, Lívia de MP, Tiago AP, Suzelei de CF (2013) Water stress and agriculture. edited book Agricultural and biological sciences "responses of organisms to water stress", book edited by Sener Akıncı, ISBN 978–953–51-0933-4
Stotz HU, Mitrousia GK, De Wit PJGM, Fitt BDL (2014) Effector-triggered defence against apoplastic fungal pathogens. Trends Plant Sci 19:491–500
Suzuki N, Rivero RM, Shulaev V, Blumwald E, Mittler R (2014) Abiotic and biotic stress combinations. New Phytol 203:32–43
Thompson AJ, Barry JA, Mulholland J, McKee JMT, Howard W, Hilton HW, Horridge JS, Farquhar GD, Smeeton RC, Smillie IRA, Black CR, Taylor IB (2007) Overproduction of abscisic acid (ABA) in tomato increases transpiration efficiency and root hydraulic conductivity and influences leaf expansion. Plant Physiol 143:1905–1917
Tkemaladze GS, Makhashvili KA (2016) Climate changes and photosynthesis. Ann Agrar Sci 14:119–126
Upreti KK, Murti GSR (2004) Changes in electrolyte leakage, chlorophyll concentration, endogenous hormones and bulb weight in onion in response to water stress. Trop Agric 81:127–132
Vankova R, Gaudinova A, Dobrev P, Malbeck J, Haisel D, Motyka V (2011) Comparison of salinity and drought stress effects on abscisic acid metabolites activity of cytokinin oxidase/dehydrogenase and chlorophyll levels in radish and tabacco. Eco Quest 14:99–100
Verelst W, Skirycz A, Inze D (2010) Abscisic acid, ethylene and gibberellic acid act at different developmental stages to instruct the adaptation of young leaves to stress. Plant Signal Behav 5:473–475
Wahid A, Ghazanfar A (2006) Possible involvement of some secondary metabolites in salt tolerance of sugarcane. J Plant Physiol 163:723–730
Wahid A, Gelani S, Ashraf M, Foolad MR (2007) Heat tolerance in plants: an overview. Environ Exp Bot 61:199–223
Wang Y, Bao ZL, Zhu Y, Hua J (2009) Analysis of temperature modulation of plant defense against biotrophic microbes. Mol Plant-Microbe Interact 22:498–506
Waqas M, Raheem S, Abdul LK, Sajjad A, Yoon-Ha K, Sang-Mo K, Saqib B, Muhammad H, In-Jung L (2016) Salvaging effect of triacontanol on plant growth, thermo-tolerance, macronutrient content, amino acid concentration and modulation of defense hormonal levels under heat stress. Plant Physiol Biochem 99:118–125
Wen F, Zhang Z, Bai T, Xu Q, Pan Y (2010) Proteomics reveals the effects of gibberellic acid (GA) on salts tressed rice (*Oryza sativa* L.) shoots. Plant Sci 178:170–175
Wilkinson S, Davies WJ (2002) ABA-based chemical signaling: the coordination of responses to stress in plants. Plant Cell Environ 25(2002):195–210
Zhang J, Jia W, Yang J, Ismail AM (2006) Role of ABA in integrating plant responses to drought and salt stresses. Field Crop Res 97:111–119
Zhao CZ, Xia H, Frazier TP, Yao YY, Bi YP, Li AQ, Li MJ, Li CS, Zhang BH, Wang XJ (2010) Deep sequencing identifies novel and conserved microRNAs in peanuts (Arachis hypogaea L.). BMC Plant Biol 10:3
Zhao M, Wenjing L, Xiuzhi X, Tianzuo W, Wen-Hao Z (2014) Cold acclimation-induced freezing tolerance of *Medicago truncatula* seedlings is negatively regulated by ethylene. Physiol Plant 152:115–129
Zhou R, Cutler AJ, Ambrose SJ, Galka MM, Nelson KM, Squires TM, Loewen MK, Ashok S, Jadhav AS, Ross ARS, Taylor DC, Suzanne R, Abrams SR (2004) A new abscisic acid catabolic pathway. Plant Physiol 134:361–369
Zhu JK (2002) Salt and drought stress signal transduction in plants. Annu Rev Plant Physiol Plant Mol Biol 53:247–273

Adaptation Strategies of Plants Against Common Inorganic Pollutants and Metals

13

Dhananjay Kumar, Sanjeev Kumar, Vertika Shukla, and Narendra Kumar

Abstract

One of the most distress environmental issues of the recent decades is the burden of pollution in soil and water. Till date, the quality of water and soil has been degraded as a result of excess inputs of toxic metals and several other inorganic pollutants through many anthropogenic activities like agricultural runoff, sewage, and industrial waste disposals. There are so many problems associated with inorganic contamination in soil which directly or indirectly affects plant health. When the consequences of inorganic contaminants are lethal, plants may die out, while the sublethal effects are more precarious, as the surviving plants and animals can accumulate some of the pollutants in their body parts giving rise to biomagnification. In order to cope with the toxicity caused by inorganic pollutants like metal and metalloids, plant has developed a number of adaptation strategies. Adaptive strategies developed by plants against the inorganic contaminants comprise antioxidant defense system, metal complexation at cell wall plasma membrane, sequestration within vacuoles, induction of stress proteins, chelation and sequestration through specific ligands, etc.

D. Kumar
Department of Environmental Science, Babasaheb Bhimrao Ambedkar (Central) University, Lucknow, UP, India

S. Kumar • V. Shukla
School of Environmental Sciences, Babasaheb Bhimrao Ambedkar University, Lucknow, UP, India

N. Kumar (✉)
Department of Environmental Science, School of Environmental Sciences, Babasaheb Bhimrao Ambedkar University, Lucknow, UP, India
e-mail: narendrakumar_lko@yahoo.co.in

V. Shukla et al. (eds.), *Plant Adaptation Strategies in Changing Environment*, https://doi.org/10.1007/978-981-10-6744-0_13

Keywords
Antioxidant · Chelation · Metal · Pollution · Toxicity · Vacuole compartmentalization

13.1 Introduction

Over the last decades, accelerated population growth, urbanization, and industrialization have liberated massive amount of waste and are accepted as the most important cause of major environmental changes. Environmental alterations are natural progression, but the anthropogenic activities like inadequate waste management practices, landfill operations, mining, application of chemical fertilizers, sewage sludge, etc. have accelerated the rate, level, and varieties of contamination to ecosystem. Every industry generates its own specific wastes that pollute ecosystem all around. Contamination in water being an obvious health hazard has been acceptably well looked into, but contamination in agricultural water and soil still needs much necessary attention. Routinely tons of industrial effluents containing various amounts of toxic contaminants such as metals and various organic chemicals are released into all components of environment (Kumar et al. 2013; Nagajyoti et al. 2010; Rascio and Navari-Izzo 2011; Ali et al. 2013a, b; Abolghassem et al. 2015). After entering to environment, these wastes can contaminate agricultural field and surface and groundwater. When this contaminated water and agricultural field are used, various toxic chemicals particularly heavy metals become a part of the vegetation. Further, some of these heavy metals are highly toxic to the plants and animals even at low concentration (Singh and Prasad 2014). However, the repercussions of various pollutants are specific for specific life form and also influenced by the physicochemical properties of the recipient environment (Krämer and Clemens 2005). Therefore the proper assessment of irrigation water and agricultural land required specific attention according to the purpose of use. It is a well-established fact that dumped solid waste and waste effluents degrade the quality of soil and water directly. The polluted water contaminates both good soil and unpolluted water. Further, there are so many problems which are associated with industrial waste discharge ranging from the heavy metal pollution to complex organic compounds affecting plant life in several ways. When the effect of such pollution is lethal, some plants may completely die out, while the sublethal effects are more precarious, as the surviving flora and fauna can accumulate some of the pollutants in their body giving rise to biomagnifications, and if they enter into the food chain, directly or indirectly, human beings and other animals are endangered (Abolghassem et al. 2015; Kumar et al. 2013).

13.2 Sources of Common Inorganic Pollutants and Metals

The several fractions or ionic forms of heavy metals in soil can considerably influence their fate in terms of bioavailability or soil profile migration. Heavy metals present in soil are generally articulated in terms of total concentration, and the distribution among their chemical forms remains unknown. The quantity and type of pollutants present in the effluent depend on the manufacturing processes involved and raw materials used in the industry. Several industries like tannery, metal, paints, textile, electroplating, battery manufacturing, etc. generate hazardous wastes (McIntyre 2003; Singh et al. 2010a, b; Kumar et al. 2013; Chen et al. 2012). Various workers have reported the presence of heavy metals, chemicals, salts, and suspended inorganic and organic solids in the industrial effluents (Munzuroglu and Geckil 2002). Further, polluted water not only changes the soil properties of the industrial and agricultural fields, but also they affect the riverbeds by generating secondary pollutants (Barman et al. 2000; Kisku et al. 2000). Some of the macro- and micronutrients also present in the industrial effluent which are of immense importance for plant growth and yield can become toxic if they exceed the tolerance limit (Chandra et al. 2002). Various industries have been uninterruptedly adding a lot of their wastewater containing high level of inorganics, trace metals, and hazardous xenobiotics to the agricultural land. This activity has been detrimental to the quality of soil and water and causes plant toxicity (Chandra et al. 2002).

Sharma et al. (1999) review EIA of textile printing industries in Jaipur, Rajasthan, and found high levels of Na^+ and Cl^-, high BOD and COD, high metal concentration, dissolved solids, conductivity, Ca, Na, and nitrates. Avasn and Rao (2000) and Kumar and Gopal (2001) had assessed the physicochemical parameters of sugar mill and distillery effluent, respectively, and their impact on local fish, *Channa punctatus* L. They reported high levels of BOD which revealed the presence of high concentration of biodegradable organic matter in the effluent (Avasn and Rao 2000; Kumar and Gopal 2001). In a study on the effect of paper mill effluent on soils and crops, Rajanan and Oblisami (1979) found that the effluent was alkaline containing large amounts of suspended and dissolved solids and has low BOD, COD, carbonate, and phytoplankton. The discharge of industrial effluent on cultivated land causes increases in the pH of soil; retarded germination of seed rice, black gram, and tomato; and decrease in the growth of seedlings.

13.3 Effects of Common Inorganic Pollutants and Metals on Plant Physiology and Biochemistry

Impact of contamination of irrigation water mixed with industrial effluents on crops has displayed many adverse effects. However, different varieties of crops have shown different degrees of sensitivity. It is also observed that not only the entire crop is affected by the pollutants; even the grain filling and ripening time are also adversely affected (Barman et al. 2000; Sahu et al. 2008). Oilseeds like mustard are adversely affected by contaminated water ranging from germination to yield. Plants growing in soil polluted with carbon black factory wastes exhibit reduced height, leaf dimensions, and all other yield components (Inoni et al. 2006; Agbogidi 2009; Agbogidi and Enujeke 2012). Sewage sludge generated from wastewater treatment plants is frequently applied to soil to make available nutrients for crop production. However, this exercise has potential of creating pollution problems because sludge has high content of toxic heavy metals which get accumulated in soil. Further, on repeated application of sludge, the increased concentration may become toxic to microorganisms, soil organisms, and/or plants. Soils are principally important in the diminution of heavy metals in the environment because they comprise surface active minerals and humic acid which play an important role in metal retention. The surface active minerals comprise of both constant negative charge sites associated with oxides (oxy), hydroxides, and humic material. Both inner- and outer-sphere complexes can be formed between the reactive sites and metallic ions (Evans 1995). Thus, information of total amount of heavy metals in soil without considering their speciation is not adequate to assess the environmental impact of contamination. The fate of metal contamination in the soil is controlled by their chemical forms in which the metals exist in the soil, and it is important to have information about the metal distribution in each soil zone. In soil, metals are associated with a number of fractions, viz.:

- In solution form metal exists as free ion and soluble metal complexes.
- Adsorbed to inorganic soil parts at ion exchange sites.
- Bound to soil organic matter.
- Precipitated as oxides, hydroxides, and carbonates.
- Immobilized with the silicate minerals.

13.3.1 Toxicity to Plant

Generally, the term "phytotoxicity" has been related with all phenomena whereby a potentially harmful element gets accumulated in plant tissues to a level affecting optimum plant growth and development (Li et al. 2013; Singh and Prasad 2014). Phytotoxicity may result into reduced yield or death of plants by the toxic contaminants present in soil. But such a definition is not suitable for developing the phytotoxicity criteria because plants that experience varying degree of phytotoxicity display several numbers of symptoms during their course of growth. However, reduction of plant growth may not be limited to accumulation of toxic substances since several other environmental factors are also associated with plant growth such as nutrient deficiencies, root disease, and water and salt stress. Further, several organic chemical exposures may also produce similar visual symptoms (Pilon et al. 2009). Confirmation about the occurrence of phytotoxicity due to metal toxicity requires certain symptoms, viz.:

- Plants have persistent injuries.
- A potentially phytotoxic metal has accumulated in the plant roots/shoots.
- Observed anomalies are not due to other disorders of plant growth.
- Biochemical mechanisms that cause metal toxicity in plants are observed during the progression of growth and development (Chang et al. 1992).

The effects of phytotoxicants on plants are due to the result of combination of several factors of heavy metals and their compounds resulting in deformity in growth and cellular structure and decreased metabolic rate. Plants are extremely important for screening the sensitivity and susceptibility (especially of common food crop species) against the toxic substances. Wide variations in Cu, Pb, and Cd concentrations were observed in soil and plants during vegetation in all studied areas (Pilon et al. 2009; Beesley et al. 2014). The root system of the plants acts as the first barrier limiting the penetration of heavy metals. Despite different metal mobility in plants, the root system accumulates them more intensively than aboveground parts. Therefore mostly roots are affected by metal toxicity (Yamamoto and Kozlowski 1987; Nagajyoti et al. 2010). Heavy metals affect a number of physiological functions due to the distributed functioning of the root system as roots accumulate heavy metal from soil and subsequently translocate them to other parts of the plant like stem, leaf, and fruits. Consequently, analysis of plant response to the impact of heavy metals is quite complicated (Hall 2002; McIntyre 2003; Li et al. 2013).

Heavy metal accumulation by plants grown in metal-contaminated soils has been studied to extensive extent. Elevated levels of heavy metals in soils may lead to uptake by plants, but the enrichment of metal in plant tissues is not generally found, because it depends on various factors such as plant growth, metal distribution to plant tissue, and soil metal bioavailability (Logan et al. 1997; Kidd et al. 2007; Mench et al. 2009). Therefore, metal mobility and its bioavailability are very important factors while assessing the effect of soil contamination, metal uptake, and related phytotoxicity. Interaction among the metals occurs at the root surface and within the plant, which can affect metal uptake as well as translocation and toxicity (Mench et al. 2010). High concentration of heavy metals in the soil disrupts essential physiological functions in plants, causes nutrient imbalance, and has many adverse effects on the biosynthesis and functioning of several biologically important compounds such as enzymes, vitamins, and hormones (Reeves and Baker 2000).

Heavy metals have been reported for the reduction in biosynthesis of chlorophyll by interacting with sulfhydryl requiring enzymes like α-aminolevulinic acid (ALA) dehydratase and porphobilinogen deaminase leading to the accumulation of intermediates of chlorophyll synthesis like ALA and porphyrins (Lewis et al. 2001; Somashekaraiah et al. 1992; Smirnoval et al. 2006; Vikram et al. 2011). However, higher metal concentrations lead to physiological, biochemical, and cellular injuries that result in K^+ ion leakage, chlorosis, and growth retardation of tissue. Heavy metals generally inhibit the physiological processes, e.g., photosynthesis, phloem translocation, and transpiration; however, respiration is less sensitive (Clysters and Van-Assche 1985; Somashekaraiah et al. 1992; Smirnoval et al. 2006). Heavy metals can bind to functional main domains of biomolecules, thereby inactivating them and resulting in inhibition of enzymatic reaction and disturbances in metabolic process. Furthermore, heavy metals also encourage formation of free radicals (FR) and reactive oxygen species (ROS), either by direct involvement in electron transfer or as a consequence of metal-assisted inhibition of metabolic processes (Elstner 1982; Clysters and Van-Assche 1985). Plant cells cannot cope with the increased rate of FR and ROS formation by increased activity of the antioxidative mechanism. Uncontrolled oxidation and chain reactions producing free radicals will result in oxidative stress. Therefore, the degree of cell damage due to metal stress depends on the rate of formation of FR and ROS as well as on the efficiency and ability of their detoxification and reparation mechanism. Oxy radicals are formed when an oxygen molecule receives electrons from other molecules and/or by reduction of oxygen to superoxide (O^{2-}) or hydrogen peroxide (H_2O_2) by several intracellular reactions. Moreover, oxy radicals are not very reactive while they get converted to high reactive hydroxyl radical ($^{\bullet}OH$), which are probably responsible for most of oxidative damage in biological systems (Elstner 1982; Schützendübel and Polle 2002; Mithöfer et al. 2004; Gill and Tuteja 2010). Reduction of one electron of molecular oxygen to superoxide radical is thermodynamically unfavorable but takes place by interaction with other paramagnetic center. Generally, heavy metals like iron and copper (M) have unpaired electrons in their outer shell and, therefore, act as very good catalysts for reductions of oxygen by the following reaction:

$$\mathrm{Mn} + \mathrm{O}_2 \rightarrow \mathrm{Mn}^{+1} + \mathrm{O}^{2-}$$

In solutions having neutral pH, superoxide (O^{2-}) can form H_2O_2, which subsequently decompose to produce hydroxyl radicals by the Haber-Weiss reaction. Copper or iron (M) acts as a catalyst in this reaction:

$$\mathrm{Mn}^{+1} + \mathrm{O}^{2-} \rightarrow \mathrm{Mn} + \mathrm{O}_2$$

$$\mathrm{Mn} + \mathrm{H}_2\mathrm{O}_2 \rightarrow \mathrm{Mn}^{+} + \mathrm{OH}^{-} + \mathrm{OH}$$

The above reactions are generally summarized as

$$\mathrm{O}^{2-} + \mathrm{H}_2\mathrm{O}_2 \rightarrow \mathrm{O}_2 + \mathrm{OH}^{-} + \mathrm{OH}$$

Hydroxyl radicals ($^{\bullet}OH$) formed during the Haber-Weiss reaction can oxidize biological molecules, leading to major cell damages and finally cell deaths.

13.4 Adaptation Strategies Adopted by Plants Against Common Inorganic Pollutants and Metals

Plant has developed a number of adaptation strategies to cope with the metal stress condition. Adaptive strategies developed by plants against the metal stress include antioxidant defense system, cellular exclusion, metal complexation at cell wall plasma membrane, sequestration within vacuoles, induction of stress proteins and chelation, sequestration through specific ligands, etc. (Cobbett 2000; Clemens 2006; Dalcorso et al. 2008; Hossain et al. 2009; Sharma and Dietz 2009; Hossain and Fujita 2009; Hossain et al. 2012).

13.4.1 Antioxidant Defense System

To avoid oxidation and radical damage, the plants have evolved a complex antioxidation defense system, which can be regulated according to the environmental conditions. Under the metal stress environment, increased production of reactive oxygen species (ROS) such as superoxide ($O_2^{\cdot-}$) and hydroxyl ($OH^{\cdot}$) radicals and hydrogen peroxide (H_2O_2) is the primary plant cell response (Vranova et al. 2002;

Moura et al. 2012). At lower level this prevents oxidative damage while at higher level causes cell damages (Abolghassem et al. 2015). At moderate level ROS helps to acclimatize stress by initiation of antioxidant responses. A complex antioxidant defense system at molecular and cellular levels decreases the oxidative damage and increased tolerance to heavy metals (Chen et al. 2009).The most prominent antioxidative enzyme is the ROS-detoxifying superoxide dismutase (SOD), Cu–Zn SOD, Mn–SOD, and Fe–SOD, catalase (CAT), and a large number of H_2O_2-reducing peroxidases (POD) (Kumar et al. 2016; Halliwell and Gutteridge 1999; Das and Roychoudhury 2014).

Besides, enzymatic antioxidants, glutathione, ascorbate, carotenoids, flavonoids, tocopherol, uric acid, lipoic acid, various amino acids, polyamine, and a variety of phenolic compounds and thiols are important nonenzymatic antioxidants (Dalvi and Bhalerao 2013), some of which are regenerated by antioxidant enzymes. Various intracellular compartments and apoplast have been reported to contain millimolar concentration of one or several of these low molecular weight antioxidants (Prasad 1999; Salt and Rauser 1995; Sharma and Dietz 2009).

13.4.2 Vacuole Compartmentalization

In plant cells, vacuole is generally considered as the main storage site for heavy metals. Vacuole compartmentalization is quite effective in controlling the distribution and concentration of metal ions. Over the years, several researchers have focused on mechanisms involved in metal accumulation and tolerance. Vacuole compartmentalization has been reported as an essential metal detoxification mechanism (Sharma and Dietz 2009). Accumulation of metals in plant cell wall and vacuole is also another detoxification tactic, which disallowed the metal from entering to sensitive metabolic sites. To compartmentalize, vacuole arrests nonessential metal and confines and squeezes them in a limited site. Thus other sensitive parts of the cell have no admittance of such nonessential metal ions and cell safety from metal toxicity ensured. Salt and Rauser (1995) had demonstrated the vacuole compartmentalization involved in metal tolerance mechanism (Salt and Rauser 1995). They reported that the Cd induces the synthesis of phytochelatins (PCs) and formation of Cd-PC molecule, which is then transferred into the vacuole by a Cd/H antiport and an ATP-dependent PC transporter. Furthermore, the metal tolerance in *T. goesingense* was significantly improved by confining the most of the intracellular Ni into vacuole, which endorses the role of vacuole in metal tolerance.

Some studies related to cadmium accumulation shows that trichomes of *B. juncea* (Salt and Rauser 1995) and *N. tabacum* (Choi et al. 2001) can play a promising role in the metal detoxification. It is reported that in some metal-hyperaccumulating plant, viz., *T. caerulescens* and *A. halleri*, the leaf mesophyll is the main accumulating compartment, but Kupper et al. (2000) have found high Cd content in the leaf trichomes of *A. halleri* (Kupper et al. 2000; Ma et al. 2005). In contrast to these, Wojcik et al. (2005) have found that the epidermis of *T. caerulescens* was the Cd richest tissue. Furthermore, the localization of Cd was also found in cortex parenchyma cells, endodermis, parenchyma cells, and xylem vessels which show that compartmentalization of metals is a tissue/plant-part specific (Wojcik et al. 2005). Several researches have suggested the metal compartmentalization in apoplastic tissues, e.g., trichome and cell walls, and metal chelation with specific ligands, followed by the sequestration of the metal–ligand complex into the vacuole via tonoplast-facilitated metal transporters (MacDiarmid et al. 2000; Lyons et al. 2000). There are several research reports which suggest that the phytochelatin–metal complexes are driven into the vacuole in fission yeast and in plants (Ortiz 1995; Salt and Rauser 1995).

13.4.3 Chelation

Chelation of metal by means of high-affinity cellular ligands has been reported as one of the predominant adoptive mechanisms for metal tolerance and its detoxification. Mechanism involved in metal detoxification processes is achieved through complexation with cellular ligands such as organic acids, cysteine phytochelatins, glutathione, etc. The role of commonly found two classes of metal-binding ligands, viz., phytochelatins (PCs) and metallothioneins and phytochelatins, in metal detoxification and tolerance has been extensively studied in plants by several authors (Zenk 1996; Cobbett 2000; Schat et al. 2002; Kupper et al. 2004; Mishra et al. 2006). Phytochelatins have been reported in fungi and other organisms (Grill et al. 1987; Gekeler et al. 1988; Piechalak et al. 2002). Stimulation of metal-binding proteins associated with metallothioneins (MTs) and/or phytochelatins (PCs) (γ-glutmylcysteinyl-isopeptides) upsurges the level of tolerance on exposure of excess of metal ions. These metal-binding proteins are cysteine-rich polypeptides which sequestrate heavy metals efficiently from sensitive location and contribute to detoxification and accumulation in plant cells. PCs are synthesized enzymatically from GSH or related peptides. PCs are low molecular weight proteins or peptides having high cysteine content and specific metal-binding capacity. The general structure formula of PCs is (g-Glu-Cys)n-Gly where $n = 2–11$. PCs are enzymatically synthesized from tripeptide glutathione (Glu-Cys-Gly) GSH and catalyzed by PC synthase (PCS). Further, the GSH is synthesized by gamma-glutamylcysteine synthetase (g-ECS) and glutathione synthetase (GS). GSH have a high affinity to conjugate with several metals and metalloids and executes as a crucial metabolite in cellular redox balance. Cumulative GSH synthesis is also considered as systematic approach to enhance the cellular defense against oxidative stress. Further, on exposure of heavy metal, tripeptide GSH also executes as an important product in defying oxidative stress and conjugated to various xenobiotics as an essential for their vacuolar sequestration (Grill et al. 1987; Cobbett 2000; Guo et al. 2008).

Metallothioneins (MTs), small, low molecular weight, cysteine-rich polypeptides, bind metals via mercaptide bonds. For the first time, in 1957, MT was identified from horse kidneys as cadmium-binding proteins (Margoshes and Vallee 1957). Today, a number of MT genes have been reported from a widespread variety of bacteria, fungi, and all eukaryotic plant and animal species (Robinson et al. 1993; Robinson and Winge 2010). MTs have been reported for their significant role in metal tolerance and detoxification. MT–metal complexes seem to reside in the cytosolic compartment. The spatial structures of MTs resemble as a dumbbell-like structure having two distinct domains, i.e., α and β. These two domains resided in core clusters are an association of numerous tetrahedral metal–Cys units. The reactivity and affinity of two domains toward the different metal prompt diverse functional roles. *N*-terminal of β-domain involves in the homeostasis of essential metal ions (Willner et al. 1987; Kagi and Schaffer 1988), and C-terminal of α-domain takes part in tight binding sequestration of nonessential/excess and/or toxic metal ions (Cherian et al. 1994; Wright et al. 1987). Further, the spacer region linking α- and β-domains, contributing stability or subcellular localization of MT proteins (Domenech et al. 2005), is essential for the role of metal detoxification (Domenech et al. 2005; Zhou et al. 2006).

13.4.4 Role of Genes in Metal Uptake and Their Transportation

Over the past years, noteworthy researches have been made in explaining the role of genetic makeup of plant in metal accumulation and tolerance. A number of genes have been reported for their expression in activating the specific enzymes to overcome metal stress condition (Clemens 2006; Plaza et al. 2007; Willems et al. 2007; Nagata et al. 2010; Wojas et al. 2010). Nagata et al. (2010) had reported that the genetically modified tobacco callus displayed high resistance to methyl mercury (CH_3Hg^+) than the wild variety of tobacco callus. They reported that the merB gene is responsible for the synthesis of MerB enzyme that converted the Hg^+ to Hg^{2+} which is less toxic. Further, the mercury ion gets accumulated in the form of Hg–polyP complex in cells (Nagata et al. 2010). Furthermore, it has been reported that the overexpression of gene like AtPCS1 and CePCS under As and Cd stress condition and increased level of PCs increase the potential of detoxification in tobacco plants (Nagata et al. 2010; Wojas et al. 2010). Siripornadulsil et al. (2002) have reported that the transgenic *C. reinhardtii* produces 80% higher proline (important role in countering water stress, saline stress, metal stress) than the wild type and shows rapid growth rate under higher Cd concentrations in growing environment. They also reported that transgenic *C. reinhardtii* show the expressing gene encoding moth bean D1-pyrroline-5-carboxylate synthetase (P5CS) (catalyzes the initial step of proline synthesis) which initiates the proline synthesis and bring together into the nuclear genome of *C. reinhardtii* responsible for enhanced tolerance capacity of transgenic *C. reinhardtii*. In order to detail the genetic profiles, particularly related to know how a genetic code translates to phenotypic variation and the resultant adaptive physiological traits, a method has been developed to profile and compare the transcriptome of two metal-accumulating plants. Furthermore, acquisition of knowledge about transcript-profiling platforms, protein and metabolite profiling, tools depending on RNA interference (RNAi), and collections of insertion line mutants and analysis of genetic profile of metal-accumulating plant helps to find out the ideal phytoremediaters and its closest relatives that display the highest level (Rigola et al. 2006; Van de Mortel 2006; Freeman and Salt 2007; Milner and Kochian 2008; Roosens et al. 2008).

13.5 Conclusions

Soil contamination with inorganic pollutants causes a number of toxic effects on plants like growth and cellular deformities, disturbed metabolism, etc. Plant has developed a number of adaptation strategies to render the effects of metal stress. The strategies adapted by plant against the metal stress include antioxidant defense system, cellular exclusion, vacuole sequestration and compartmentalization, stimulation of stress proteins, chelation and sequestration through specific ligands, etc. During the normal metabolic processes, ROS are formed as by-products and play an important role in regulation of several physiological processes but at higher levels are also responsible for many complications. Over the time, on increasing abiotic contamination in soil, this situation gets worsened. Further, overgeneration of ROS subsequently causes damage to essential macromolecules such as lipids, proteins, and nucleic acids and, thus, constrains to plant physiology and productivity. Despite the availability of ample literature, the understanding of relationship between metal stress and proline interaction at the physiological, biochemical, as well as molecular level is still an area of scientific thrust.

References

Abolghassem E, Ding Y, Mokhberdoran F, Xie Y (2015) Heavy metal stress and some mechanisms of plant defense response. Sci World J. https://doi.org/10.1155/2015/756120

Agbogidi OM (2009) Effects of soil contaminated with spent lubricating oil on the germination of Gmelina arborea (Roxb.) seeds. Afr J Nat Sci 12:11–14

Agbogidi OM, Enujeke EC (2012) Effects of spent motor oil on soil physico- chemical properties and growth of Arachis hypogaea L. Glob J Biosci Biotechnol 1(1):71–74

Ali S, Farooq MA, Yasmeen T, Hussain S, Arif MS, Abbas F, Zhang G (2013a) The influence of silicon on barley growth, photosynthesis and ultrastructure under chromium stress. Ecotoxicol Environ Saf 89:66–72

Ali H, Khan E, Sajad MA (2013b) Phytoremediation of heavy metals – concepts and applications. Chemosphere 91(7):869–881

Avasn YM, Rao SR (2000) Effect of sugar mill effluent on organic reserves of fish. Pollut Res 19:391–393

Barman SC, Sahu RK, Bhargava SK, Chaterjee C (2000) Distribution of heavy metals in wheat, mustard, and weed grown in field irrigated with industrial effluents. Bull Environ Contam Toxicol 64:489–496

Beesley L, Inneh OS, Norton GJ, Moreno-Jimenez E, Pardo T (2014) Assessing the influence of compost and biochar amendments on the mobility and toxicity of metals and arsenic in a naturally contaminated mine soil. Environ Pollut 186:195–202

Chandra S, Joshi HC, Pathak H, Jain MC, Kalra N (2002) Effect of potassium salts and distillery effluent on carbon mineralization in soil. Bioresour Technol 83:255–257

Chang AC, Granato TC, Page AL (1992) A methodology for establishing phytotoxicity criteria for chromium, copper, nickel and zinc in agricultural land application of municipal sewage sludges. J Environ Qual 21:521–536

Chen S, Olbrich A, Langenfeld-Heyser R, Fritz E, Polle A (2009) Quantitative X-ray microanalysis of hydrogen peroxide within plant cells. Microsc Res Tech 72(1):49–60

Chen CW, Chen CF, Dong CD (2012) Distribution and accumulation of mercury in sediments of Kaohsiung River Mouth, Taiwan. APCBEE Procedia 1:153–158

Cherian MG, Howell SB, Imura N, Klaassen CD, Koropatnick J, Lazo JS, Waalkes MP (1994) Role of metallothionein in carcinogenesis. Toxicol Appl Pharmacol 126:15

Choi Y, Harada E, Wada M, Tsuboi H, Morita Y, Kusano SHT (2001) Detoxification of cadmium in tobacco plants: formation and active excretion of crystals containing cadmium and calcium through trichomes. Planta 213:45–50

Clemens S (2006) Toxic metal accumulation, responses to exposure and mechanisms of tolerance in plants. Biochimie 88(11):1707–1719

Clysters H, Van-Assche F (1985) Inhibition of photosynthesis by metals. Photosynth Res 7:31–40

Cobbett CS (2000) Phytochelatins and their roles in heavy metal detoxification. Plant Physiol 123(3):825–832

DalCorso G, Farinati S, Maistri S, Furini A (2008) How plants cope with cadmium: staking all on metabolism and gene expression. J Integr Plant Biol 50(10):1268–1280

Dalvi AA, Bhalerao SA (2013) Response of plants towards heavy metal toxicity: an overview of avoidance, tolerance and uptake mechanism. Ann Plant Sci 2(9):362–368

Das K, Roychoudhury A (2014) Reactive oxygen species (ROS) and response of antioxidants as ROS scavengers during environmental stress in plants. Front Environ Sci 2(53):1–13

Domenech J, Mir G, Huguet G, Capdevila M, Molinas M, Atrian S (2005) Plant metallothionein domains: functional insight into physiological metal binding and protein folding. Biochimie 88:583–589

Elstner EF (1982) Oxygen activation and oxygen toxicity. Annu Rev Plant Physiol 33:73–96

Evans LJ (1995) Chemical aspects of heavy metal solubility with reference to sewage sludge amended soils. Int J Environ Anal Chem 59:291–302

Freeman JL, Salt DE (2007) The metal tolerance profile of *Thlaspi goesingense* is mimicked in Arabdopsis thaliana heterologously expressing serine acetyl-transferase. BMC Plant Biol 7(63):1–10

Gekeler W, Grill E, Winnacker EL, Zenk MH (1988) Algae sequester heavy metals via synthesis of phytochelatin complexes. Arch Microbiol 150:197–202

Gill SS, Tuteja N (2010) Reactive oxygen species and antioxidant machinery in abiotic stress tolerance in crop plants. Plant Physiol Biochem 48:909–930

Grill E, Winnacker E-L, Zenk MH (1987) Phytochelatins, a class of heavy-metal-binding peptides from plants are functionally analogous to metallothioneins. Proc Natl Acad Sci U S A 84:439–443

Guo J, Dai X, Xu W, Ma M (2008) Overexpressing GSH1 and AsPCS1 simultaneously increase the tolerance and accumulation of cadmium and arsenic in Arabidopsis thaliana. Chemosphere 72:1020–1026

Hall JL (2002) Cellular mechanism for heavy metal detoxification and tolerance. J Exp Bot 53:1–11

Halliwell B, Gutteridge JMC (1999) Free radicals in biology and medicine, 3rd edn. Oxford University Press, Oxford

Hossain MA, Fujita M (2009) Purification of glyoxalase I from onion bulbs and molecular cloning of its cDNA. Biosci Biotechnol Biochem 73(9):2007–2013

Hossain MA, Hossain MZ, Fujita M (2009) Stress-induced changes of methylglyoxal level and glyoxalase I activity in pumpkin seedlings and cDNA cloning of glyoxalase I gene. Aust J Crop Sci 3(2):53–64

Hossain MA, Piyatida P, Teixeira da Silva JA, Fujita M (2012) Molecular mechanism of heavy metal toxicity and tolerance in plants: central role of glutathione in detoxification of reactive oxygen species and methylglyoxal and in heavy metal chelation. J Bot. https://doi.org/10.1155/2012/872875

Inoni OE, Omotor DG, Adun FN (2006) The effect of oil spillage on crop yield and farm income in Delta state, Nigeria. J Cent Eur Agric 7(1):41–49

Kagi JH, Schaffer A (1988) Biochemistry of metallothionein. Biochemistry 27:8509–8515

Kidd PS, Dominguez-Rodriguez MJ, Diez J, Monterroso C (2007) Bioavailability and plant accumulation of heavy metals and phosphorus in agricultural soils amended by long-term application of sewage sludge. Chemosphere 66:1458–1467

Kisku GC, Barman SC, Singh DC, Bhargawa SC (2000) Contamination of soil and plants with potentially toxic elements irrigated with mixed industrial effluent and its impact on the environment. Water Air Soil Pollut 120:121–137

Krämer U, Clemens S (2005) Molecular biology of metal homeostasis and detoxification. In: Martinoia E (ed) Topics in current genetics M. Tamäs. Springer, New York, pp 216–271

Kumar S, Gopal K (2001) Impact of distillery effluent on physiological consequences in the fresh water teleost, Channa punctatus. Bull Environ Contam Toxicol 66:617–622

Kumar N, Bauddha K, Kumar S, Dwivedi N, Singh DC, Barman SC (2013) Accumulation of metals in weed species grown on the soil contaminated with industrial waste and their phytoremediation potential. Ecol Eng 61:491–495

Kumar D, Singh DP, Barman SC, Kumar N (2016) Heavy metal and their regulation in plant system: an overview. Plant responses to xenobiotics. Springer, New York, pp 19–38

Kupper H, Lombi E, Zhao FJ, McGrath SP (2000) Cellular compartmentation of cadmium and zinc in relation to other elements in the hyperaccumulator *Arabidopsis halleri*. Planta 212:75–84

Kupper H, Mijovilovich A, Meyer-Klaucke W, Kroneck PHM (2004) Tissue- and age-dependent differences in the complexation of cadmium and zinc in the cadmium/zinc hyperaccumulator *Thlaspi caerulescens* (Ganges ecotype) revealed by X-ray absorption spectroscopy. Plant Physiol 134:748–757

Lewis S, Donkin ME, Depledge MH (2001) Hsp70 expression in Enteromorpha intestinalis (Chlorophyta) exposed to environmental stressors. Aquat Toxicol 51:277–291

Li X, Yang Y, Jia L, Chen H, Wei X (2013) Zinc-induced oxidative damage, antioxidant enzyme response and proline metabolism in roots and leaves of wheat plants. Ecotoxicol Environ Saf 89:150–157

Logan TJ, Linday BJ, Goins LE, Ryan JA (1997) Field assessment of sludge metal bioavailability to crops: sludge rate response. J Environ Qual 26:534–550

Lyons TJ, Gasch A, Gaither LA, Botstein D, Brown PO, Eide D (2000) Genome-wide characterization of the Zap1p zinc- responsive regulon in yeast. Proc Natl Acad Sci U S A 97:7957–7962

Ma JF, Ueno D, Zhao FJ, McGrath SP (2005) Subcellular localisation of Cd and Zn in the leaves of a Cd-hyperaccumulating ecotype of Thlaspi caerulescens. Planta 220:731–736

MacDiarmid CW, Gaithe LA, Eide D (2000) Zinc transporters that regulate vacuolar zinc storage in *Saccharomyces cerevisiae*. EMBO J 19:2845–2855

Margoshes M, Vallee BL (1957) A cadmium protein from equine kidney cortex. J Am Chem Soc 79(17):4813–4814

McIntyre T (2003) Phytoremediation of heavy metals from soils. In: Springer T (ed) Advances in biochemical engineering/biotechnology, vol 78. Springer, Heidelberg, pp 97–123

Mench M, Schwitzguébel JP, Schröder P, Bert V, Gawronski S, Gupta S (2009) Assessment of successful experiments and limitations of phytotechnologies: contaminant uptake, detoxification and sequestration, and consequences for food safety. Environ Sci Pollut Res 16:876–900

Mench M, Lepp N, Bert V, Schwitzguébel JP, Gawronski SW, Schröder P, Vangronsveld J (2010) Successes and limitations of phytotechnologies at field scale: outcomes, assessment and outlook from COST Action 859. J Soils Sediments 10:1039–1070

Milner MJ, Kochian LV (2008) Investigating heavy-metal hyperaccumulation using Thlaspi caerulescens as a model system. Ann Bot 102:3–13

Mishra S, Srivastava S, Tripathi RD, Kumar R, Seth CS, Gupta DK (2006) Lead detoxification by coontail (*Ceratophyllum demersum* L.) involves induction of phytochelatins and antioxidant system in response to its accumulation. Chemosphere 65:1027–1039

Mithöfer A, Schulze B, Boland W (2004) Biotic and heavy metal stress response in plants: evidence for common signals. FEBS Lett 566:1–5

Moura DJ, Peres VF, Jacques RA, Saffi J (2012) Heavy metal toxicity oxidative stress parameters and DNA repair. In: Gupta DK, Sandalio LM (eds) Metal toxicity in plants: perception, signaling and remediation. Springer, Berlin/Heidelberg, pp 187–205

Munzuroglu O, Geckil H (2002) Effects of metals on seed germination, root elongation, and coleoptiles and hypocotyls growth in *Triticum aestivum* and *Cucumis sativus*. Arch. Environ Contam Toxicol 43:203–213

Nagajyoti PC, Lee KD, Sreekanth TVM (2010) Heavy metals, occurrence and toxicity for plants: a review. Environ Chem Lett 8:199–216

Nagata T, Morita H, Akizawa T, Pan-Hou H (2010) Development of a transgenic tobacco plant for phytoremediation of methyl mercury pollution. Appl Microbiol Biotechnol 87(2):781–786

Ortiz DW (1995) Transport of metal-binding peptides by HMT1, a fission yeast ABC-type vacuolar membrane protein. J Biol Chem 270:4721–4728

Piechalak A, Tomaszewska B, Baralkiewicz D, Malecka A (2002) Accumulation and detoxification of lead ions in legumes. Phytochemistry 60:153–162

Pilon M, Cohu CM, Ravet K, Abdel-Ghany SE, Gaymard F (2009) Essential transition metal homeostasis in plants. Curr Opin Plant Biol 12:347–357

Plaza S, Tearall KL, Zhao FJ, Buchner P, McGrath SP, Hawkesford MJ (2007) Expression and functional analysis of metal transporter genes in two contrasting ecotypes of the hyperaccumulator *Thlaspi caerulescens*. J Exp Bot 58:1717–1728

Prasad MNV (1999) Metallothioneins and metal binding complexes in plants; heavy metal stress in plants. Springer, Berlin/Heidelberg, pp 51–72

Rajanan G, Oblisami G (1979) Effects of paper factory wastewaters on soil and crop plants. Indian J Environ Health 21:120–130

Rascio N, Navari-Izzo F (2011) Heavy metal hyperaccumulating plants: how and why do they do it? And what makes them so interesting? Plant Sci 180(2):169–181

Reeves RD, Baker AJM (2000) Metal-accumulating plants. In: Raskin I, Ensley BD (eds) Phytoremediation of toxic metals: using plants to clean up the environment. Wiley, New York, pp 193–229

Rigola D, Fiers M, Vurro E, Aarts MGM (2006) The heavy metal hyperaccumulator *Thlaspi caerulescens* expresses many species specific genes, as identified by comparative expressed sequence tag analysis. New Phytol 170:753–766

Robinson NJ, Winge DR (2010) Copper metallochaperones. Annu Rev Biochem 79:537–562

Robinson NJ, Tommey AM, Kuske C, Jackson PJ (1993) Plant metallothioneins. Biochem J 295:1–10

Roosens NHCJ, Willems G, Saumitou-Laprade P (2008) Using Arabidopsis to explore zinc tolerance and hyperaccumulation. Trends Plant Sci 13:208–215

Sahu RK, Katiyar S, Yadav AK, Kumar N, Srivastava J (2008) Toxicity assessment of industrial effluent by bioassays. Clean 5–6:517–520

Salt DE, Rauser WE (1995) MgATP-dependent transport of phytochelatins across the tonoplast of oat roots. Plant Physiol 107:1293–1301

Schat H, Llugany M, Vooijs R, Hartley-Whitaker J, Bleeker PM (2002) The role of phytochelatins in constitutive and adaptive heavy metal tolerance in hyperaccumulator and non-hyperaccumulator metallophytes. J Exp Bot 53:2381–2392

Schützendübel A, Polle A (2002) Plant responses to abiotic stresses: heavy metal-induced oxidative stress and protection by mycorrhization. J Exp Bot 53:1351–1365

Sharma SS, Dietz KJ (2009) The relationship between metal toxicity and cellular redox imbalance. Trends Plant Sci 14(1):43–50

Sharma KP, Sharma K, Bhardwaj SM, Chaturvedi RK, Sharma S (1999) Environmental impact assessment of textile printing industries in Sanganer, Jaipur: a case study. J Ind Bot Soc 78:71–85

Singh S, Prasad SM (2014) Growth photosynthesis and oxidative responses of *Solanum melongena* L. seedlings to cadmium stress: mechanism of toxicity amelioration by kinetin. Sci Hortic 176:1–10

Singh R, Singh DP, Kumar N, Bhargava SK, Barman SC (2010a) Accumulation and translocation of heavy metals in soil and plants from fly ash contamination. Environ Biol 3:421–430
Singh A, Sharma RK, Agrawal M, Marshall FM (2010b) Health risk assessment of heavy metals via dietary intake of foodstuffs from the wastewater irrigated site of a dry tropical area of India. Food Chem Toxicol 48:611–619
Siripornadulsil S, Traina S, Verma DP, Sayre RT (2002) Molecular mechanisms of proline-mediated tolerance to toxic heavy metals in transgenic microalgae. Plant Cell 14:2837–2847
Smirnoval TA, Kolomiitseval YG, Prusoval AN, Vanyushin BF (2006) Zinc and copper content in developing and aging coleoptiles of wheat seedling. Russ J Plant Physiol 53:535–540
Somashekaraiah BV, Padmaja K, Prasad ARK (1992) Phytotoxicity of cadmium ions on germinating seedlings of mung bean (*Phaseolus vulgaris*): involvement of lipid peroxides in chlorophyll degradation. Physiol Plant. https://doi.org/10.1111/j.1399-3054.1992.tb05267.x
Van de Mortel JE (2006) Large expression differences in genes for iron and zinc homeostasis, stress response, and lignin biosynthesis distinguish roots of Arabidopsis thaliana and the related metal hyperaccumulator Thlaspi caerulescens. Plant Physiol 142:1127–1147
Vikram A, Johri T, Tandon PK (2011) Effect of chromium (IV) on growth and metabolism of *Spinacia oleracea* (Spinach) plants. Res Environ Life Sci 4(3):119–124
Vranova E, Inze D, Van Breusegem F (2002) Signal transduction during oxidative stress. J Exp Bot 53(372):1227–1236
Willems G, Dräger DB, Courbot M (2007) The genetic basis of zinc tolerance in the metallophyte *Arabidopsis halleri* ssp. *halleri* (Brassicaceae) an analysis of quantitative trait loci. Genetics 176:659–674
Willner H, Vasak M, Kagi JH (1987) Cadmium thiolate clusters in metallothionein: spectrophotometric and spectropolarimetric features. Biochemistry 26:6287–6292
Wojas S, Clemens S, Skłodowska A, Maria Antosiewicz D (2010) Arsenic response of AtPCS1- and CePCS-expressing plants – effects of external As(V) concentration on As-accumulation pattern and NPT metabolism. J Plant Physiol 167:169–175
Wojcik M, Vangronsveld J, Tukiendorf DHJA (2005) Cadmium tolerance in Thlaspi caerulescens-II. Localization of cadmium in *Thlaspi caerulescens*. Environ Exp Bot 53:163–171
Wright CF, Mckenney K, Hamer DH, Byrd J, Winge DR (1987) Structural and functional studies of the amino terminus of yeast metallothionein. J Biol Chem 262:219–258
Yamamoto F, Kozlowski TT (1987) Effect of flooding, tilting of stem, and ethrel application on growth, stem anatomy, and ethylene production of Acer platanoides seedlings. Scand J For Res 2:141–156
Zenk MH (1996) Heavy metal detoxification in higher plants: a review. Gene 179:21–30
Zhou G, Xu Y, Li J, Yang L, Liu JL (2006) Molecular analyses of the metallothionein gene family in rice (*Oryza sativa* L.) J Biochem Mol Biol 39(5):595–606

Impacts of Climate Change on Agriculture: Adaptation, Mitigation, and Environmental Policy

14

Ravindra Prasad, Sanjeev Kumar, Anuj Kumar Yadav, Shailender Kumar, Mahesh Kumar, Ritu Singh, and Narendra Kumar

Abstract

Agriculture sector which is already facing challenges due to rising food demand of growing population is getting severely affected by climate change. Climate change-induced effects like temperature increase, variation in rainfall pattern, extreme weather events, etc. are exacerbating pressure on the agriculture ecosystem globally. In fact, agriculture itself is contributing a significant share in greenhouse gas (GHG) emissions that are causing climate change, ~ 17% directly through agricultural activities and an additional 7–14% through changes in land use. In the Indian subcontinent, agriculture is the major part of the GDP, and 58% of population is employed in the agriculture sector. The Government of India is

R. Prasad • S. Kumar
Department of Environmental Studies, University of Delhi, New Delhi, Delhi, India

S. Kumar (✉)
School of Environmental Sciences, Babasaheb Bhimrao Ambedkar University, Lucknow, UP, India
e-mail: sanjivk845@gmail.com

A.K. Yadav
Department of Zoology & Environmental Sciences, Gurukula Kangri Vishwavidyalaya, Haridwar, Uttrakhand, India

M. Kumar
Department of Environmental Science, Babasaheb Bhimrao Ambedkar University, Lucknow, UP, India

R. Singh (✉)
Department of Environmental Science, School of Earth Sciences, Central University of Rajasthan, Ajmer, Rajasthan, India
e-mail: ritu2735@gmail.com

N. Kumar
Department of Environmental Science, School of Environmental Sciences, Babasaheb Bhimrao Ambedkar University, Lucknow, UP, India

V. Shukla et al. (eds.), *Plant Adaptation Strategies in Changing Environment*,
https://doi.org/10.1007/978-981-10-6744-0_14

making efforts to counteract the effects of climate change through shifting current agricultural practices and strategies toward climate-smart agriculture diversified adaptation and mitigation policies. The present chapter provides insight over adaptation and mitigation strategies which are available for combating negative effects of climate change on agriculture sector. Various policies on which Indian government is working for supporting sustainable productivity were also discussed.

Keywords
Climate change · Agriculture · Soil fertility · Greenhouse gases · Mitigation strategies

14.1 Introduction

Climate change is a long-term change in the weather pattern at a global or regional scale over extended period of time which is brought about by natural activities or human interventions. The alteration in the global atmospheric composition is natural climate variability observed over comparable time periods (IPCC 2007a, 2014). IPCC 2014 studied that "each of the last three decades has been successively warmer at the earth's surface than any past decade since 1850." 97% published literature claims that the human/anthropogenic activities are responsible for the global climate change (mainly the increasing level of carbon dioxide (CO_2) (Cook et al. 2016; Chalise et al. 2017). Climate change is unavoidable, and it is predicted to exert negative impacts on human, wildlife, and agriculture ecosystem (Chalise et al. 2017; Salamanca et al. 2017). The impacts of climate change are noticeable and mandatory actions are needed (Klein 2011; Chalise et al. 2017; Salamanca et al. 2017).

Increasing levels of CO_2 is gradually altering the weather patterns which in turn is adversely affecting the crop productivity. Although an increase in crop yield of C3 plants have been observed due to increased levels of CO_2, but CO_2 rise is also associated with higher temperature which is causing declination in overall yield of plants due to excessive evaporative loss of water. Elevated levels of CO_2 could adversely affect agriculture via (1) limiting the growth rate of crops, (2) decelerating agricultural productivity because of increasing global temperature and frequency of extreme weather events, and (3) elevating levels of global temperature responsible for rise in sea level which directly or indirectly affects the agricultural by losing the farmland areas, increased levels of salinity, etc.

INCCA (Indian Network for Climate Change Assessment, 2016) reported that there is indication of loss in wheat crop productivity with every 1 °C rise in the atmospheric temperature in the current land use system (Saseendran et al. 2000; Mall and Aggarwal 2002; Aggarwal 2003). The Indo-Gangetic plains of India and Asian countries are important regions for agricultural production and grain productivity contributing to ~50% of the total food grain production and supporting food security of about 40% population of Indo-Gangetic plains (Pal et al. 2009; Kumar et al. 2014a, b, 2016; Singh et al. 2017). In most of the countries, applications of chemical fertilizers, i.e., urea, DAP, ammonium nitrate, etc., pesticides, energy-based tools and equipments, and high water consumption for irrigation to obtain high yields in agriculture have raised cost of production, on one hand, and have degraded soil, water, and biosphere, on the other (Kumar et al. 2013a, b, 2014a, b; Trost et al. 2016; Wang et al. 2017).

The aim of this chapter is to display the impact of climate change on agriculture sector including production productivity, soil enrichment, and plant-microbial interaction.

14.2 Impact of Climate Change on Agriculture

14.2.1 Effect on Crop Production

The total contribution of agriculture in gross domestic product (GDP) of India as given by The Economic Survey (2016–2017) is 17.1% which is very significant. ICAR report made predictions about crop production in the medium-term climate change scenarios and projected that the crop yield is likely to be reduced from 4.5% to 9% by 2039. According to this report, it has been estimated that food grain requirement till 2020 will increase by 30–35% (Paroda and Kumar 2000). The projected temperature increase by the end of this era is predictable to be in the range 1.8–4.0 °C (IPCC 2007b). It is projected to be in the range 0.5–1.2 °C rise in temperature by 2020, 0.88–3.16 °C by 2050, and 1.56–5.44 °C by 2080 in Indian/South Asian region (IPCC 2007b, c). These temperature changes are likely to impact the Asian continental agriculture. Agriculture affects everything from climate change, i.e., greenhouse gas emissions, to biological diversity, water quality, soil erosion, pollination services, carbon sequestration, human health, livelihoods, and food security (Tscharntke et al. 2012; DeLonge et al. 2016; Wang et al. 2017). Agricultural practices are contributing to the degradation of ecological processes that underpin life on Earth, driving climate change, loss of biosphere integrity, destructive land system changes, and the eutrophication of oceans from phosphorus and nitrogen fertilizers.

Greenhouse gases (GHGs) are responsible for climate change in India and worldwide. GHGs are those that absorb infrared radiation and then heat the surface of the Earth. The three GHGs associated with agriculture are methane (CH_4), nitrous oxide (N_2O), and CO_2. Among them, the most important GHG is N_2O, coming mainly from soil, and N from fertilizers to crop and soil systems. So, agriculture is considered to be the major source of N_2O, and it is linked to soil management and excessive use of fertilizer in agriculture. N_2O emission results from two processes: nitrification and denitrification. Both processes are complex because the amount of nitrous oxide produced depends on the oxygen concentration in the soil.

Elevated CO_2 and temperature directly affect the photosynthesis of plants, growth, and yield (Rozema 1993; Wang et al. 2013). Some examples show that excess availability of CO_2 results in the decrease of N concentration in vegetative parts as well as seeds and grains. Many researchers have reported interactive effect of atmospheric CO_2 and temperature on crop biomass and nitrogen use efficiency (Mitchell et al. 1995; Batts et al. 1997; Gavito et al. 2001; Tian et al. 2014; Han et al. 2015). In Indian subcontinent, similar kind of declining trends in agricultural crop productivity due to climate change has been reported by several groups (Saseendran et al. 2000; Mall and Aggarwal 2002; Aggarwal 2003; Table 14.1).

Atmospheric concentrations of CO_2 have increased from approximately 280 ppm in preindustrial times to 400 ppm (IPCC 2013) and will continue to rise due to

Table 14.1 Projected impacts for crops in the Indian subcontinents under future scenario

Crop	Region	Estimated yield (%) change	Period	References
Maize	India	−31.0. −28.2, −26.3, −22.9, −19.8, −18.6, −16.9, −14.6	2050/2069	Deryng et al. (2011)
Soya bean	India	−32.9, −27.8, −24.6, −24.5, −21.8, −20.0, −17.4, −15.5	2050/2069	Deryng et al. (2011)
Millet	Africa and India	−90.5,–44.3, −41.0, −25.8, −25.1, −24.6, −23.1, −23.0, −22.5, −22.5, −22.0, −21.5, −20.5, −20.0, −18.4, −18.0, −17.8, −17.4, −17.2, −16.9, −15.3, −14.6, −14.1, −13.6,–12.6, −12.5, −12.4, −11.2, −11.1, −11.0, −10.8, −10.2, −9.2, −8.2, −8.0, −5.7, −5.6, −4.8, −3.8, −3.6, −3.2, 7.9, 18.9, 23.0, 45.8, 48.6, 56.4, 62.2	2030/2049	Berg et al. (2013)
Rice	Central plains of India, Southern India, Sri Lanka	3.0,18.0	2010/2039	Lal (2012)
Wheat	–	23.0, 25.0	2010/2039	–
Rice	–	−6.0, 1.0	2050/2069	–
Wheat	–	−3.0, 9.0	2070/2089	–
Groundnut	South Asia	1.2	2010/2029	Lobell et al. (2008)
Millet	–	−2.1	–	–
Wheat	–	−2.9	–	–
Maize	–	−4.8	–	–
Rapeseed	–	−6.5	–	–
Rice	–	−3.3	–	–
Soya bean	–	3.9	–	–
Sugarcane	–	0.0	–	–
Sorghum	–	0.1	–	–

(−) value shows the percent reduction of crop yield

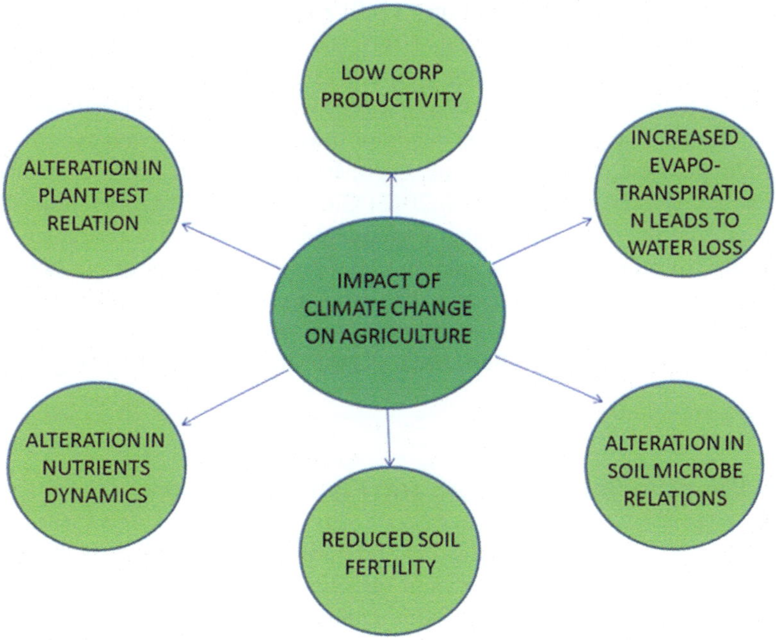

Fig. 14.1 Schematic representation of impact of climate change on agriculture

burning of fossil fuels and changing land use patterns. It is expected that in the middle of this century, CO_2 would surpass 550 ppm (Meehl et al. 2006). Dieleman et al. (2012) and Zinta et al. (2014) studied the responses of crops to high CO_2 concentrations at various levels ranging from the ecosystem to the genomic level. Indian agriculture is facing challenges due to climate variability and associated factors. The schematic representation of overall impacts of climate change is depicted in Fig. 14.1.

14.2.2 Effect on Soil Fertility

Climate changes have strongly affected the agricultural stability; ecological, agricultural, and ecosystem services; etc. Soils play a vital role in the multifaceted system that creates climate change. The response of soil for climate change is dependent on the nature and the structure of soil (Ciric et al. 2017). The increased level of CO_2 increases the soil organic matter, while increased level of temperature increases the rate of evapotranspiration which in turn leads to the reduced soil moisture content. Soil-climate models predicted that the increased level of atmospheric CO_2 directly affects the soil function such as poor soil structure, stability, topsoil water holding capacity, nutrient availability, soil erosion, etc. (Karmakar et al. 2016).

Ciric et al. (2017) reported that the increase in the mean annual soil temperature up to 3.5 °C in the surface layer, i.e., 0–10 cm, and up to 1.8 °C in average, i.e., 0–200 cm depth, decreases the mean soil moisture up to 0.039 kg kg^{-1} in the subsurface layer (40–100 cm) and up to 0.019 kg kg^{-1} in average i.e. 0–200 cm depth, which corresponds to decrease of 16% and 7%, respectively. Many researchers revealed that the increase in soil temperature can create disbalance in the soil respiration, i.e., release of CO_2, microbial processes (nitrification and denitrification), soil organic carbon and matter decomposition, disrupt nitrogen use efficiency, crop production, etc. (Davidson et al. 1998; Rankinen et al. 2004; Conant et al. 2008; Paul et al. 2004; Gremer et al. 2015; Ciric et al. 2017). Duran et al. (2016) determined that the projected changes in climate will affect a wide range of soil biogeochemical properties in organic and mineral soil horizons of northern hardwood forests. Many researchers revealed that the climate change will affect the amounts of soil inorganic N, N transformation rates in soil, and N availability in soil (Brooks et al. 2011; Evans and Burke 2012; Duran et al. 2013, 2016; Campbell et al. 2014).

14.2.3 Effect on Soil Microflora and Plant Interaction

Microflora is an essential part of the soil and plays a significant role in the nutrient cycling, incorporation of nutrients, soil structure, and porosity. Climate change adversely affects the soil microflora. Although the soil microflora has a wide range of optimum temperature, the increased level of CO_2 concentration affects the soil microflora in so many ways, for example, increased level of CO_2 increased the plant growth and disturbed the belowground allocation of nutrient distribution or microbial activity (Classen et al. 2015). Migration of soil microflora and fauna due to temperature change also adversely affects the plant growth. Zak et al. (1996) Although elevated concentration of CO_2 enhances plant litter production and biomass of soil microorganisms but it also negatively impacts microbial rhizospheric community in natural soil system (Zak et al. 1996). Elhottova et al. (1997) observed that the elevated CO_2 affects the rhizospheric microflora with short-term changes in C partitioning in plants and also affected availability of C source in microbial processes. Increase in atmospheric CO_2 not only has an impact on plant growth through accelerated photosynthesis and altered carbon allocation but also causes ultrastructural modifications (Hao et al. 2013). High CO_2 is also found to mitigate the negative impacts of various abiotic stresses (Naudts et al. 2014; Zinta et al. 2014). Therefore, stress-mitigating effects of high CO_2 concentration are likely to become increasingly relevant for plant growth in the face of global changes.

14.2.4 Effect on Pest and Predator Relation

The 1943 Great Bengal Famine in India which was caused by a simple fungus led to the deaths of around three million people. Even today, when farmers are better equipped to control pests and pathogens, still 10–16% of crop production is lost to biological threats each year. It is very difficult to predict the pest-predator relationship in the changing environmental condition because the insect pest has ability to respond to small changes in the environmental conditions. Increased levels of temperature may affect the pest and predator relation either in a positive way or in a negative way; both possibilities are there. Roos et al. (2011) reported that warmer winter and autumns have more risk from viruses to infect winter crops (wheat, barley, oilseed rape, etc.). Climate change can influence productivity and yield of stable crop as well as can induce changes in crop pest distributions worldwide (Yan et al. 2017). Rao et al. (2015) studied that temperature is very relevant in the life cycle of insect pests; if global temperature increases, then the life cycle of pests will become relatively shorter, resulting in faster growth pest population. The issue of invasive species is also a challenge for sustainable agriculture as invasive species can survive and grow well in local environment along with native species, fastly complete life cycle and directly affect the functioning of agricultural ecosystem (Bradley 2009; Yan et al. 2017). Global warming also facilitates dispersal of these invasive species (Ziska et al. 2011).

14.2.5 Effect on Nutrient Dynamics

Global climate change influences the global carbon and nitrogen cycles which are the most important components of soil organic matter. The alteration in the soil organic matter is responsible for the soil properties, including soil structure, water holding capacity, and ion exchange capacity, and for the supply of nutrients to the soil ecosystem (Hillel and Rosenzweig 2011; Brevik 2009, 2013; Ciric et al. 2017). Niklaus and Korner (2004) reported that elevated levels of CO_2 lead to the limitation of nitrogen and phosphate within 2 years of periods. Although in some studies it has been reported that the CO_2 fertilization increases the growth of plants while the increased levels of temperature together with increased levels of CO_2 lead to reduced crop yield (Hattenschwiler et al. 2002; Jarvis et al. 2010; Ciric et al. 2017). The CO_2 fertilization effect is not as effective as it was originally thought. It has been reported elsewhere that nitrogen limitation negatively affects the plant growth. Ciric et al. 2017 reported that change in weather and higher temperatures will bring definite changes in soil processes and evolution as well as in growing vegetation in upper layers.

14.3 Climate Change Adaptation Strategies

14.3.1 Screening and Evaluation of Climate-Resilient Crops

In order to satisfy the food demand of growing population, agricultural productivity needs to be increased to significant levels. The impact of climate change on the agricultural crop productivity is likely to be a major factor. To feed the growing population, it is necessary to develop climate-resilient crop variety and improve agriculture system focusing over climate-smart agriculture. Researchers across the globe are engaged in the development of climate-resilient crop variety, but it is not an easy task. Henry (2014) suggested that genomic characterization strategies can be employed for the development of climate-resilient crop varieties. Assessment of the impact of climate change simultaneously with formulation of adaptive strategies is the prime approach under strategic research across all sectors of agriculture. Development of climate-resilient agriculture system in the changing environment and development of technology are the major focused area. In order to improve the agriculture productivity, four modules were taken into account, i.e., natural resource management, improving soil health, crop production, and livestock. India has started a number of initiatives in collaboration with national institute as well as international institute to build climate-resilient agriculture system.

14.3.2 Soil Quality and Fertility Management

Soil is an essential component in ecosystem and support system. The main processes which attribute to soil degradation include erosion, salinization, nutrient and carbon depletion, drought, and soil structure changes. Climate changes deteriorate soil quality and fertility to higher extent. Increase in temperature leads to rise in decomposition rates of soil carbon cycling and water availability. The lower rainfall and uncovered lands cause wind erosion hazards. The higher rate of soil degradation alters the efflux of GHGs from soil to environment. The regions with high population load and climate change may have severe risk to soil erosion (Lal 2012). The rise in population needs more food and land for settlements. The rapid growth of population has led to land clearance and expansion of cultivated land. One of the most significant causes of soil degradation include anthropogenic activities which are triggered by demographic pressure, human, and industrial needs. Climatic model studies revealed that rise in global warming may decrease soil moisture in semiarid regions (Sivakumar 2007). Practicing agroforestry could be a solution for improving the soil quality and fertility. Organic farming also improves soil balance and increases crop yield. Organic fertilizers, modern agronomic techniques, and hybrid seeds can significantly improve crop production and soil ecosystem.

14.3.3 Irrigation Management

In India, agriculture mainly depends on irrigation for crop production. Worldwide, surface water irrigation covers about 60% and groundwater irrigation about 40% of the net area irrigated (Siebert et al. 2010). Intensive use of groundwater can adversely affect aquifers and causes groundwater depletion. In India, surface water irrigates 24% and 25% of rice and wheat production, and groundwater irrigates 49% and 72%, respectively (Smilovic et al. 2015). Drip irrigation is an efficient irrigation technique which allows uniform distribution of water over the field and improves crop yield (Benouniche et al. 2014). Pivot techniques are also advantageous which consist of distributing water in the form of rains on ground (Kendouci et al. 2013). Drier climate will decrease water availability. Farmer should be trained for modern irrigation techniques. Rainwater harvesting projects are also having a significant role in groundwater recharge and irrigation.

Olayide et al. (2016) suggested minimization of the impact of climate-induced production risks through climate-smart agriculture (CSA). Under CSA, irrigation would also enhance complementary agricultural water management with sustainable food production. The resilience of agriculture to climate change, i.e., management of irrigation techniques, could enhance sustainable agricultural production (Nwanze and Fan 2016; Olayide et al. 2016).

14.3.4 Disease and Pest Control

It was estimated that worldwide 40% crop loss occurs due to pest (Oerke 2006). Global climate change affects the distribution of pest like insects and fungal and bacterial pathogens. Pest can regulate their body temperature according to the environment along a wide range. Insect physiology is mainly triggered by temperature exposure. Higher temperature or elevated CO_2 will enhance life cycle of pathogenic fungi. The microvariation in climate is significant for pest due to their small structure. Plant height and canopy structure are important variable which are affected by microclimate and ultimately cause barrier to pathogen dispersal (Ando and Grumet 2006). The metabolism and translocation of pesticide are affected by morphological and physiological changes in plant canopy (Pangga et al. 2013).

14.4 Mitigation Strategies for Climate Change

14.4.1 Greenhouse Gas Emission Reduction

Several natural phenomena and anthropogenic activities have caused continuous increase in the emission of greenhouse gases like CO_2, CH_4, and N_2O which are largely affecting the agricultural productivity, soil fertility, and human health. Agricultural crop field is also an important source of GHGs. For instance, CH_4 is emitted largely from rice fields and N_2O emits from wheat fields. GHG emission from agricultural fields is sharing approximately 13% of global emissions and one-fifth of the annual increase in radioactive forcing (IPCC 2007a). Since the preindustrial era, the concentrations of CO_2, N_2O, and CH_4 have globally increased by 36%, 18%, and 148%, respectively (IPCC 2007a; Kumar et al. 2017). The need of the hour is to develop sustainable agricultural practices to protect the environment. Agricultural activities are source of greenhouse gases (GHGs) and ammonia (NH_3) which can cause acidification and eutrophication in the environment (Novak and Fiorelli 2010). Rice cultivation majorly releases two GHGs, one is methane (CH_4) and other is nitrous oxide (N_2O). C_4 crops can be favorable in temperate areas due to heat tolerance capacity (Smith and Almaraz 2004). Increased CO_2 levels will increase the photosynthesis rate and water use efficiency of C_3 plants leading to higher crop production. O'Mara (2012) revealed that the pasturelands could contribute in reducing GHG emission and enhancing carbon sinks in grazing lands.

Nutrient mismanagement is responsible for about 80% of the economic losses from environmental damage caused by traditional and modern agriculture practices in the form of greenhouse gas emission, volatalization, and groundwater pollution (Subash et al. 2015). Nutrient management is among the most cost-effective ways of achieving sustainable development, mitigating climate change, and improving farm-scale economics and resource use at national and international level. Integrated nutrient management (INM) refers to combining traditional and modern methods of plant nutrient management into ecologically sound and economically optimal farming system that uses the benefits from all possible sources in an efficient and integrated manner (Kumar et al. 2014a, b; Bauddh et al. 2015; Kumar et al. 2016; Singh et al. 2017). INM enhances nitrogen use efficiency of plants on one hand and on other minimizes volatilization, gasesous emission, inorganic pollutants leaching and immobilization (Zhang et al. 2012; Wu and Ma 2015; Singh et al. 2017). INM also controls nitrogen losses and its consequent effects while maintaining high crop productivity (Gruhn et al. 2000; Ashok et al. 2014; Kumar et al. 2016; Singh et al. 2017).

14.4.2 Use of Alternate Energy Resources

The use of alternate energy resources has been given attention in the IPCC special report known as Special Report on Renewable Energy Sources and Climate Change Mitigation (SRREN). The agriculture is the third highest energy consuming sector after transportation and industrial sector. The major portion of energy gets utilized in mechanical (farm, machine, human labor, animal draft), chemical (fertilizer, pesticide, herbicide), and electrical works (irrigation) (Chaudhary and Kumar 2013). The use of alternate or renewable energy in the agriculture system could be a better option for the climate change mitigation. Although the government has provided subsidy on the installation of renewable energy based equipment, still utilization of renewable energy resources is a costly affair for farmers (Fig. 14.2). Agricultural practices can contribute to climate change mitigation through reducing fossils fuel combustions, increasing carbon sequestration in soil and plant biomass, and decreasing the soil-related emissions. The use of alternate energy or green energy will reduce the power consumption in agricultural machinery with less fuel consumption, lesser working time, emission reduction, and slower depreciation rates of equipment (Sapkota et al. 2015). Listed below are given few practices which could help in mitigating the effect of climate change on agriculture.

- Improving nitrogen fertilizer management, i.e., use of slow-release fertilizers, organic matric entrapped fertilizers, neem-coated urea, customized fertilizers, and controlled release fertilizers

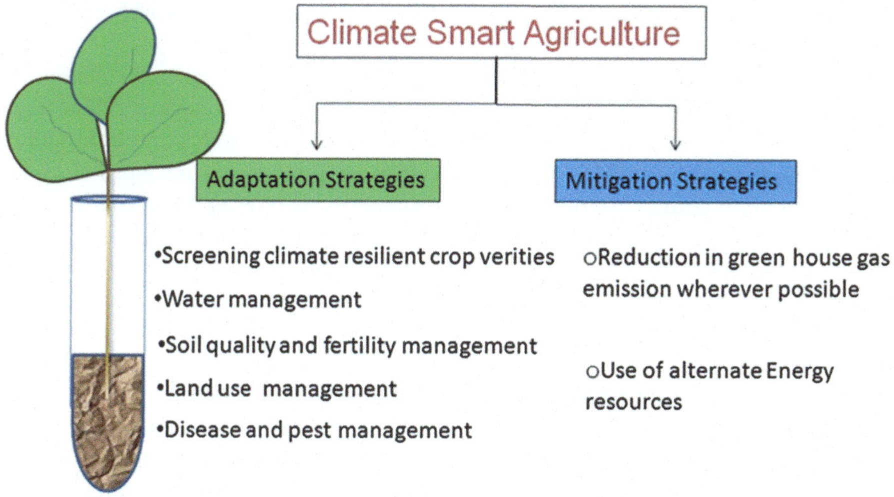

Fig. 14.2 Schematic depiction of climate-smart agriculture in the changing climate scenarios

- Improving preharvest and postharvesting techniques to reduce losses
- Livestock management
- Improved management of soil fertility, water losses, field pest, crop disease, weeding, etc.
- Use of ecological agriculture and agroforestry
- Reducing CH_4 and N_2O gas emission from rice and wheat field particularly in developing countries
- Sustainable intensification, i.e., used of less emission products
- Carbon sequestration techniques on large scale
- Use of social networking for farmers' awareness to climate change

14.5 Environmental Policies for Climate Change in India

14.5.1 National Action Plan on Climate Change (NAPCC)

NAPCC is a comprehensive action plan which outlines climate change-related adaptation and mitigation strategies. There are eight core missions under NAPCC which include (1) the Jawaharlal Nehru National Solar Mission (JNNSM), (2) National Mission for Enhanced Energy Efficiency (NMEEE), (3) National Mission on Sustainable Habitat, (4) National Water Mission, (5) National Mission for Sustainable Himalayan Ecosystem (NMSHE), (6) National mission for "Green India," (7) National Mission for Sustainable Agriculture (NMSA), and (8) National Mission on Strategic Knowledge for Climate Change (NMSKCC). All these plans have their own specific goals to combat climate change-induced effects on agriculture.

14.5.2 National Mission for Sustainable Agriculture (NMSA)

NMSA is focused to improve the quality and quantity of food crops in the changing environmental scenarios. Researchers are looking for climate change-resilient new varieties of crops (e.g., thermally resistant crops) and improving the alternative cropping system under NMSA program. The major focus of the mission is to improve the agriculture system through the adaptation and mitigation strategies; to achieve this goal, different initiatives have been started, for example, (1) Rainfed Area Development (RAD), (2) On-Farm Water Management (OFWM), (3) Soil Health Management, etc. (Rockström et al. 2010).

14.5.3 Implication and Challenges in Agricultural Policy

Agriculture continues to be the predominant sector of the Indian economy even though its share in national output has declined from more than 50% in the initial years after Independence from colonial rule in 1947 to less than 25% in recent years. In terms of employment and livelihood, still, around 58% of the workforce is engaged in agriculture as their principal occupation. The major difficulties in the implications are their own infrastructure and funding budgets.

Combined with the output price support, India's border measures have led to higher domestic food prices. Food subsidies were instituted to minimize the impact of higher food prices on consumers. Input subsidies have contributed to the excessive use of inputs and resulted in a number of agro-environmental problems, such as soil salinity and groundwater depletion. The growing cost of input and food subsidies has also contributed to fiscal deficits in many states. Expenditures on subsidies could have been invested in research, education, and infrastructure to improve productivity and competitiveness of the sector. India's institutional, policy, and regulatory systems have favored the production of food staples and hindered vertical coordination in the value chain as well as the development of its horticultural industry. Under India's Constitution, most agriculture-related responsibilities, including water and electricity, are state jurisdiction. In addition, India has a highly decentralized system of governance. Until recently, this decentralized system of governance has proven an obstacle to undertake fundamental reform in the sector. Existing policies have benefited special interests which oppose significant reforms.

14.6 Future Prospects

In the present scenario of climate change, one of the biggest challenges in front of the government and scientific community around the globe is to fulfill the feed demands of growing population. The Indian economy is largely dependent on the agriculture and agricultural productivity depends on the monsoon rains. Around 58% of total employment comes from agriculture sector, so the adverse impact of climate change on the Indian economy is substantial. Although Government of

India has initiated vast programs and policies to combat climate change, still we need to give more attention toward building "climate-smart agriculture system" through effective climate change mitigation and adoption policy measures. Further research is also needed to support innovations in sustainable climate-friendly and climate-proof agriculture productivity.

References

Aggarwal PK (2003) Impact of climate change on Indian agriculture. J Plant Biol 30:189–198

Ando K, Grumet R (2006) Evaluation of altered cucumber plant architecture as a means to reduce *Phytophthora capsici* disease on cucumber fruit. J Am Soc Hortic Sci 131:491–498

Ashok V, Kumar S, Singh RP (2014) Response of organic matrix entrapped biofertilizers on growth, yield and soil properties of rice (*Oryza sativa* L.) Asian J Agric Food Sci 2(3):211–221

Batts GR, Morison JIL, Ellis EH, Hadley P, Wheeler TR (1997) Effects of CO_2 and temperature on growth and yield of crops of winter wheat over several seasons. Eur J Agron 7:43–52

Bauddh K, Sainger M, Kumar S, Sainger PA, Jaiwal PA, Singh RP (2015) Biotechnological approaches to mitigate adverse effects of extreme climatic factors on plant productivity. In: Jaiwal PK et al (eds) Genetic manipulation in plants for mitigation of climate change. Springer India Publishing House, New Delhi, pp 187–203

Benouniche M, Kuper M, Hamani A, Boesveld H (2014) Making the user visible: analysis irrigation practices and farmer logic to explain actual drip irrigation performance. Irrig Sci 32:405–420

Berg A, de Noblet-Ducoudre M, Sultan B, Langaigne M, Guimberteau M (2013) Projections of climate change impacts on potential C_4 crop productivity over tropical regions. Agric For Meteorol 170:89–102

Bradley BA (2009) Regional analysis of the impacts of climate change on cheatgrass invasion shows potential risk and opportunity. Glob Chang Biol 15:196–208

Brevik EC (2009) Soil health and productivity. In: Verheye W (ed) Soils, plant growth and crop production, Encyclopedia of Life Support Systems (EOLSS), Developed under the Auspices of the UNESCO. EOLSS Publishers, Oxford

Brevik EC (2013) An introduction to soil science basics. In: Brevik EC, Burgess LC (eds) Soils and human health. CRC Press, Boca Raton, pp 3–28

Brooks PD, Grogan P, Templer PH, Groffman P, Oquist MG, Schime J (2011) Carbon and nitrogen cycling in snow-covered environments. Geog Comp 5:682–699

Campbell JL, Socci AM, Templer PH (2014) Increased nitrogen leaching following soil freezing is due to decreased root uptake in a northern hardwood forest. Glob Chang Biol 20:2663–2673

Chalise S, Naranpanawa A, Bandara JS, Sarker T (2017) A general equilibrium assessment of climate change-induced loss of agricultural productivity in Nepal. Econ Model 62:43–50

Chaudhary VK, Kumar PS (2013) Crop and water productivity, profitability and energy consumption pattern of a maize-based crop sequence in the North Eastern Himalayan Region, India. Arch Agron Soil Sci 59(5):653–669

Ciric VI, Dreskovic N, Mihailovic DT, Mimic G, Arsenic I, Durdevic V (2017) Which is the response of soils in the Vojvodina Region (Serbia) to climate change using regional climate simulations under the SRES-A1B. Catena 158:171–183

Classen AT, Sundqvist M, Henning JA, Newman GS, Moore JAM, Cregger M et al (2015) Direct and indirect effects of climate change on soil microbial and soil microbial-plant interactions: what lies ahead. Ecosphere 6:130

Conant RT, Steinweg JM, Haddix ML, Paul EA, Plante AF, Six J (2008) Experimental warming shows that decomposition temperature sensitivity increases with soil organic matter recalcitrance. Ecology 89:2384–2391

Cook J, Oreskes N, Doran PT, Anderegg WRL, Barterheggen, Maibach EDW, Carlton JS, Lewandowsky S, Skuce AG, Green A, Nuccitelli D, Jacobs P, Richardson M, Winkler B,

Painting R, Rice K (2016) Consensus on consensus: a synthesis of consensus estimates on human-caused global warming. Environ Res Lett 11(04):8002

Davidson E, Belk E, Boone RD (1998) Soil water content and temperature as independent or confounded factors controlling soil respiration in a temperate mixed hardwood forest. Glob Chang Biol 4:217–227

DeLonge MS, Miles A, Carlisle L (2016) Investing in the transition to sustainable agriculture. Environ Sci Pol 55:266–273

Deryng D, Sacks WJ, Barford CC, Ramankutty N (2011) Simulating the effects of climate and agricultural management practices on global crop yield. Glob Biogeochem Cycles 25(2):1–18

Dieleman WIJ, Vicca S, Dijkstra FA, Hagedorn F, Hovenden MJ, Larsen KS (2012) Simple additive effects are rare: a quantitative review of plant biomass and soil process responses to combined manipulations of CO_2 and temperature. Glob Chang Biol 18:2681–2693

Duran J, Rodríguez A, Morse JL, Groffman PM (2013) Winter climate change effects on soil C and N cycles in urban grasslands. Glob Chang Biol 19:2826–2837

Duran J, Morse JL, Rodríguez A, Campbell JL, Christenson LM, Driscoll CT, Fahey TJ, Fisk MC, Mitchell, Templer PH, Groffman PM (2016) Differential sensitivity to climate change of C and N cycling processes across soil horizons in a northern hardwood forest. Soil Biol Biochem 107:77–84

Elhottova D, Triska JN, Santrokova H, Kveton J, Santrucek J, Simkova M (1997) Rhizosphere microflora of winter wheat plants cultivated under elevated CO_2. Plant Soil 197:251–259

Evans SE, Burke IC (2012) Carbon and nitrogen decoupling under an 11-year drought in the short-grass steppe. Ecosystems 16:20–33

Gavito ME, Curtis PS, Mikkelsen TN, Jakobsen I (2001) Interactive effects of soil temperature, atmospheric carbon di-oxide and soil N on root development, biomass and nutrient uptake of winter wheat during vegetative growth. J Exp Bot 52:1913–1923

Gremer JR, Bradford JB, Munson SM, Duniway MC (2015) Desert grassland responses to climate and soil moisture suggest divergent vulnerabilities across the southwestern United States. Glob Chang Biol 21:4049–4062

Gruhn P, Goletti F, Yudelman M (2000) Integrated nutrient management, soil fertility and sustainable agriculture: current issues and future challenges. International Food Policy Research Institute, Washington, DC, p 2

Han X, Hao X, Lam SK, Wang H, Li Y, Wheeler T, Hui J, Lin E (2015) Yield and nitrogen accumulation and partitioning in winter wheat under elevated CO_2: a 3-year free-air CO_2 enrichment experiment. Agric Ecosyst Environ 209:132–137

Hao X, Li P, Feng Y, Han X, Gao J, Lin E (2013) Effects of fully open-air (CO_2) elevation on leaf photosynthesis and ultrastructure of Isatis indigotica. Fort. PLoS One 8:e74600. https://doi.org/10.1371/journal.pone.0074600

Hattenschwiler S, Handa IT, Egli L, Asshoff R, Ammann W, Körner C (2002) Atmospheric CO_2 enrichment of alpine treeline conifers. New Phytol 156:363–375

Henry RJ (2014) Genomics strategies for germplasm characterization and the development of climate resilient crops. Front Plant Sci 5:68

Hillel D, Rosenzweig C (2011) The role of soils in climate change. In: Hillel D, Rosenzweig C (eds) Handbook of climate change and agroecosystems. Imperial College Press, London

IPCC (2007a) Climate change 2007: the physical science basis. Inter-governmental Panel for Climate Change, Cambridge University Press, Cambridge

IPCC (2007b) Summary for policymakers. In: Solomon S, Qin D, Manning M, Chen Z, Marquis M, Averyt AB, Tignor M, Miller HL (eds) Climate change 2007: the physical science basis. Contribution of working group I to the fourth assessment report of the intergovernmental panel on climate change. Cambridge University Press, Cambridge/New York

IPCC (2007c) Summary for policymakers. In: Parry ML, Canziani OF, Palutikof JP, Van der Linden PJ (eds) Climate change, impacts, adaptation and vulnerability. Contribution of Working Group II to the fourth assessment report of the Intergovernmental Panel on Climate Change. Cambridge University Press, Cambridge, p 976

IPCC (2013) Summary for policymakers. In: Stocker TF, Qin D, Plattner GK, Tignor M, Allen SK, Boschung J et al (eds) Climate change 2013: the physical science basis. Contribution of Working Group I to the fifth assessment report of the Intergovernmental Panel on Climate Change. Cambridge University Press, Cambridge/New York, pp 1–28

IPCC (2014) Annex II: glossary, Agard J, Schipper ELF, Birkmann J, Campos M, Dubeux C, Nojiri Y, Olsson L, Osman-Elasha B, Pelling M, Prather MJ, RiveraFerre MG, Ruppel OC, Sallenger A, Smith KR, St Clair AL, Mach KJ, Mastrandrea MD, Bilir TE, Climate change 2014: impacts, adaptation, and vulnerability. Part B: Regional aspects. WG II 5AR IPCC. Cambridge University Press, Cambridge/New York 1757–1776

Jarvis A, Ramirez J, Anderson B, Leibing C, Aggarwal P (2010) Scenarios of climate change within the context of agriculture. In: Reynolds MP (ed) Climate change and crop production. CPI Antony Rowe, Chippenham, pp 9–37

Karmakar R, Das I, Rakshit A (2016) Potential effects of climate change on soil properties: a review. Sci Int 4:51–73

Kendouci MA, Bendida A, Khelfaouri R, Kharroubi B (2013) The impact of traditional irrigation (Foggara) and modern (drip, pivot) on the resources, non-renewable groundwater in the Algerian Sahara. Energy Procedia 36:154–162

Klein RJT (2011) Adaptation to climate change. More than technology. In: Linkov I, Bridges TS (eds) Climate: global change and local adaptation, NATO Science for Peace and Security Series. Springer, Dordrecht, pp 157–168

Kumar S, Bauddh K, Barman SC, Singh RP (2013a) Evaluation of conventional and organic matrix entrapped urea and di-ammonium phosphate for growth and productivity of *Triticum aestivum* L. and mobilization of NO_3^-, NO_2^+, NH_4^+ and PO_4^{-3} from soil to plant leaves. Int J Agro Plant Prod 4(6):1357–1368

Kumar M, Bauddh K, Kumar S, Sainger M, Sainger PA, Singh RP (2013b) Increase in growth, productivity and nutritional status of wheat (*Triticum aestivum* L. cv. Wh-711) and enrichment in soil fertility applied with organic matrix entrapped urea. J Environ Biol 34:1–9

Kumar S, Bauddh K, Barman SC, Singh RP (2014a) Organic matrix entrapped bio-fertilizers increase growth, productivity and yield of *Triticum aestivum* L. and mobilization of NO_3^-, NO_2^-, NH_4^+ and PO_4^{-3} from soil to plant leaves. J Agric Sci Technol 16(2):315–329

Kumar S, Bauddh K, Barman SC, Singh RP (2014b) Amendments of microbial biofertilizers and organic substances reduces requirement of urea and DAP with enhanced nutrient availability and productivity of wheat (*Triticum aestivum* L.) Ecol Eng 71:432–437

Kumar S, Bauddh K, Singh R, Rani A (2016) Nitrogen use efficiency of *Triticum aestivum* (L.) under condition of elevated carbon dioxide and temperature. In: Sharma D, Sharma KC (eds) Climate change and environment. New India Publishing Agency, pp 71–86

Kumar V, Sachdev S, Kumar S (2017) A study on methane emissions and its mitigation strategies in present scenario. Samriddhi JPSET 8(2):87–92

Lal R (2012) Climate change and soil degradation mitigation by sustainable management of soil and other natural resources. Agron Res 1(3):199–212

Lobell DB, Burke MB, Tebaldi C, Mastrandrea MD, Falcon WP, Naylor RL (2008) Prioritizing climate change adaptation needs for food security in 2030. Science 319:607–610

Mall RK, Aggarwal PK (2002) Climate change and rice yields in diverse agro-environments of India. I. Evaluation of impact assessment models. Climate Change 52:315–330

Meehl GA, Washington WM, Santer BD, Collins WD, Arblaster JM, Hu A (2006) Climate change projections for the twenty-first century and climate change commitment in the CCSM3. J Clim 19:2597–2616

Mitchell RAC, Lawlor DW, Mitchell UJ, Gibbard CL, White EM, Porter JR (1995) Effects of elevated CO_2 concentration and temperature on growth and yield of winter wheat: test of ARCWHEAT1 simulation model. Plant Cell Environ 18:736–748

Naudts K, Van den Berge J, Farfan E, Rose P, AbdElgawad H, Ceulemans R (2014) Future climate alleviates stress impact on grassland productivity through altered antioxidant capacity. Environ Exp Bot 99:150–158

Niklaus PA, Korner C (2004) Synthesis of a six-year study of calcareous grassland responses to in situ CO_2 enrichment. Ecol Monogr 74:491–511
Novak SM, Fiorelli JL (2010) Greenhouse gases and ammonia emissions from organic mixed crop-dairy systems: a critical review of mitigation options. Agron Sustain Dev 30:215–236
Nwanze KF, Fan S (2016) Climate change and agriculture: strengthening the role of smallholders. In: Global food policy report. International Food Policy Research Institute, Washington, DC, pp 13–21
O'Mara FP (2012) The role of grasslands in food security and climate change. Ann Bot 110:1263–1270
Oerke EC (2006) Crop losses to pests. J Agric Sci 144:31–43
Olayide OE, Tetteh IK, Popoola L (2016) Differential impacts of rainfall and irrigation on agricultural production in Nigeria: any lessons for climate-smart agriculture. Agric Water Manag 178:30–36
Pal DK, Mandal DK, Bhattacharyya T, Mandal C, Sarkar D (2009) Revisiting the agro-ecological zones for map evaluation. Indian J Genet Pl Br 69:315–318
Pangga IB, Hanan J, Chakraborty S (2013) Climate change impacts in plant canopy architecture implications for pest and pathogen management. Eur J Plant Pathol 135:595–610
Paroda RS, Kumar P (2000) Food production and demand in South Asia. Agric Econ Res Rev 13(1):1–24
Paul KI, Polglase PJ, Smethurst PJ, O'Connell AM, Carlyle CJ, Khanna PK (2004) Soil temperature under forests: a simple model for predicting soil temperature under a range of forest types. Agric F Meteorol 121:167–182
Rankinen K, Karvonen T, Butterfield D (2004) A simple model for predicting soil temperature in snow-covered and seasonally frozen soil: model description and testing. Hydrol Earth Syst Sci Discuss 8:706–716
Rao MS, Manimanjari D, Rao CAR, Maheswari M (2015) Prediction of pest scenarios of Spodoptera litura fab in peanut growing areas of India during future climate change. Natl Acad Sci Lett 38(6):465–468
Rockström J, Karlberg L, Wani SP, Barron J, Hatibu N, Oweis T, Bruggeman A, Farahani J, Qiang Z (2010) Managing water in rainfed agriculture – the need for a paradigm shift. Agric Water Manag Compr Assess Water Manag Agric 97(4):543–550
Roos J, Hopkins R, Kavarnheden A, Dixelius C (2011) The impact of global warming in plant diseases and insect vector in Sweden. Eur J Plant Pathol 129:9–19
Rozema J (1993) Plant response to atmospheric carbon di oxide enrichment: interactions with some soil and atmospheric conditions. Vegetation 104(105):173–190
Salamanca AE, Sierra RD, Aranda RMM, Santos MJ (2017) Environmental impacts of climate change adaptation. Environ Impact Assess Rev 64:87–96
Sapkota TB, Jat ML, Aryal JP, Jat JK, Cheetri AK (2015) Climate change adaptation, greenhouse gas mitigation and economic profitability of conservation agriculture: some examples from cereals systems of Indo-Gangetic Plans. J Integr Agric 14(8):1524–1533
Saseendran SA, Singh KK, Rathore LS, Singh SV, Sinha SK (2000) Effects of climate change on rice production in the tropical humid climate of Kerala, India. Climate Change 44:495–414
Siebert S, Burke J, Faures JM, Frenken K, Hoogeveen J, Doll P et al (2010) Groundwater use for irrigation-a global inventory. Hydro Earth Sci 14:1863–1886
Singh RP, Kumar S, Sainger M, Sainger PA, Barnawal D (2017) Eco-friendly nitrogen fertilizers for sustainable agriculture. In: Rakshit A, Abhilash PC, Singh HB, Ghosh S (eds) Adaptive soil management: from theory to practices. Springer Publication, pp 227–246
Sivakumar MVK (2007) Interactions between climate and desertification. Agric Meteorol 142:143–155
Smilovic M, Gleeson T, Siebert S (2015) The limits of increasing food production with irrigation in India. Food Sec 7:835–856
Smith DL, Almaraz JJ (2004) Climate change and crop production: contributors, impact and applications. Can J Pathol 26:253–266

Subash N, Gangwar B, Singh R, Sikka RA (2015) Identification of climate-resilient integrated nutrient management practices for rice-rice cropping system-an empirical approach to uphold food security. Int J Biometeorol 59:65–78
The Economic Survey (2016–17) Agricultural and Processed Food Products Export Development Authority (APEDA), Department of Commerce and Industry, Union Budget 2017–18, Press Information Bureau, Ministry of Statistics and Programme Implementation, Press Releases, Media Reports
Tian Y, Zheng C, Chen J, Chen C, Deng A, Song Z, Zhang B, Zhang W (2014) Climatic warming increases winter wheat yield but reduces grain nitrogen concentration in east china. PLoS One 9(4):e95108. https://doi.org/10.1371/journal.pone.0095108
Trost B, Prochnowa A, Meyer-Auricha A, Drastiga K, Baumecker M, Ellmer F (2016) Effects of irrigation and nitrogen fertilization on the greenhouse gas emissions of a cropping system on a sandy soil in northeast Germany. Eur J Agron 81:117–128
Tscharntke T, Clough Y, Wanger TC, Jackson L, Motzke I, Perfecto I, Whitbread A (2012) Global food security, biodiversity conservation and the future of agricultural intensification. Biol Conserv 151(1):53–59
Wang W, Liao Y, Guo Q (2013) Seasonal and annual variations of CO_2 fluxes in rain-fed winter wheat agro-ecosystem of loess plateau, China. J Integr Agric 12(1):147–158
Wang Z, Chen J, Mao S, Han Y, Chen F, Zhang L, Li Y, Li C (2017) Comparison of greenhouse gas emissions of chemical fertilizer types in China's crop production. J Clean Prod 141:1267–1274
Wu W, Ma B (2015) Integrated nutrient management (INM) for sustaining crop productivity and reducing environmental impact: a review. Sci Total Environ 512–513:415–427
Yan Y, Wang C, Feng C, Wan P, Chang KT (2017) Potential distributional changes of invasive crop pest species associated with global climate change. Appl Geochem 82:83–92
Zak DR, Ringelberg DB, Randlett DL, White DC, Curtis PS (1996) Soil microbial communities beneath *Populus grandidentata* grown under elevated atmospheric CO_2. Ecol Appl 6:257–262
Zhang F, Cui Z, Chen X, Ju X, Shen J, Chen Q, Liu X, Zhang W, Mi G, Fan M, Jiang R (2012) Integrated nutrient management for food security and environmental quality in China. Adv Agron 116:1–40
Zinta G, AbElgawad H, Domagalska MA, Vergauwen L, Knapen D, Nijs I (2014) Physiological, biochemical and genome-wide transcriptional analysis reveals that elevated CO_2 mitigates the impact of combined heat wave and drought stress in Arabidopsis thaliana at multiple organizational levels. Glob Chang Biol 20:3670–3685
Ziska LH, Blumenthal DM, Runion GB, Hunt ER, Diaz-Soltero H (2011) Invasive species and climate change: an agronomic perspective. Clim Chang 105:13–42

Nitric Oxide (NO) and Physio-biochemical Adaptation in Plants Against Stress

15

Arun Kumar Maurya and Anita Rani

Abstract

Nitric oxide synthase (NOS) enzyme is responsible for NO generation by catalyzing the oxidation of L-arginine to L-citrulline. Presence of this enzyme is reported from unicellular prokaryotes to multicellular higher plants and animals. Plants employ diverse enzymes to generate NO that can be grouped into animal-like NOS or alternative enzymes. Alternative enzymes that are structurally unlike animal NOS but that generate NO include NR, PM NiNOR, XOR, and polyamine oxidase. NO is also produced by nonenzymatic reactions in plants. Both in natural and agriculture conditions, plants are frequently exposed to stress. NO has been found to be involved in an array of plant physiological processes including pollen germination, seed germination, root development, stomatal movements, flowering, senescence, and programmed cell death along with abiotic and biotic stress conditions such as temperature, salinity, drought, heavy metal stress, and pathogen attack or herbivory. Some environmental factors such as temperature, cold or heat, can become stressful in just a few minutes; others may take days to weeks (soil water) or even months (mineral nutrients) to become stressful. All these conditions induce the generation of reactive oxygen species (ROS) and reactive nitrogen species (RNS). These species not only initiate several oxidatively destructive processes, but also trigger various signaling pathways that maintain appropriate cellular ROS levels. It has been found that NO generated during various stress conditions interacts with ROS in various ways and serves as an antioxidant molecule that helps plants adapt in adverse environmental condition.

A.K. Maurya (✉)
Department of Botany, Multanimal Modi PG College, Modinagar Ghaziabad 201204, UP, India
e-mail: botany25@gmail.com

A. Rani
Department of Botany, Dyal Singh College, Delhi University, Delhi, India

V. Shukla et al. (eds.), *Plant Adaptation Strategies in Changing Environment*,
https://doi.org/10.1007/978-981-10-6744-0_15

Keywords
Nitric oxide · Nitric oxide synthase · Nitrate reductase · Oxidative stress · PM-NiNOR · Protein S-nitrosylation · Polyamine oxidase · Reactive oxygen species · Regulation · XOR

15.1 Introduction

Nitric oxide (NO) is a diatomic inorganic gaseous signaling molecule. NO synthesis by nitric oxide synthase (NOS) enzyme, its signaling, and physiological role is well established in animal system (Katsuki et al. 1977; Ignarro et al. 1987; Palmer et al. 1987). NO emission from plants was firstly observed by Klepper in 1975 in soybean plants treated with herbicides (Klepper 1979), but the presence of NO and NOS-like activity in higher plants by using the method of conversion of radiolabeled arginine, the substrate of NOS, into radiolabeled citrulline was firstly shown by Cueto et al. (1996) and Ninnemann and Maier (1996). NO research gained momentum in plant when Durner et al. (1998) revealed its role in plant defense. Since then NO became active area in plant biology research reflected by publications of various groups of scientists across globe, which established the role of NO in plant physiological processes such as seed germination (Beligni and Lamattina 2000; Sirova et al. 2011), flowering (Guo et al. 2003), stomatal movements (Neill et al.2008; Fan and Liu 2012), seed dormancy (Gniazdowska et al. 2010; Arc et al. 2013), root organogenesis and development (Pagnussat et al. 2002; Correa-Aragunde et al. 2004), pollen tube growth (Wang et al. 2012), posttranslational modification (S-nitrosylation) that regulates gene transcription (Mengel et al. 2013), gene expression, and the mobilization of second messenger (Astier and Lindermayr 2012). Nitric oxide also plays a crucial role in drought, salinity, low and high temperature, UV and ozone exposure, heavy metal, mineral deficiency, pathogen-induced stress, herbivory, and cross talk with plant hormones that contributes in relevant gene expression and adaptation.

15.2 NO-Generating Enzymes

NO production is reported in species ranging from unicellular bacteria to highly evolved angiosperms and mammals. NO-generating enzyme is well characterized in mammals. The major obstacle in plants is that no animallike NOS enzyme has been isolated and purified till date. Only two reports (an alga (*O. tauri*) (Foresi et al. 2010) and slime mold (*Physarum polycephalum*) (Messner et al. 2009) are reported apart from the enzyme having NOS-like activity in bacteria *Bacillus subtilis* (Adak et al. 2002) *Staphylococcus* (Bird et al. 2002) and *Deinococcus* (Adak et al. 2002) and fungi *F. velutipes* (Song et al. 2000), *P. blankesleeanus* (Maier et al. 2001), and *Neurospora* (Ninnemann and Maier 1996).

Mammalian nitric oxide synthase (NOS) (EC 1.14.13.39) was first identified and described in 1989. NOS have three major isoforms which were cloned and purified between 1991 and 1996 (Stuehr 1996; Stuehr et al. 1991), and first X-ray crystallography of NOS domains was published in 1998 and 1999 in humans (Li et al. 1999; Raman et al. 1998). In 1998, the noble prize was awarded jointly to R. Furchgott, L. Iganarro, and F. Murad for the work that led to the discovery of nitric oxide as a biosignaling molecule produced by mammalian cells. Now it is well established that active NOS enzyme exists as dimer and contains relatively tightly bound four cofactors – tetrahydrobiopterin (BH_4), FAD, FMN, and iron protoporphyrin IX (heme). NOS catalyzes a reaction by using L-arginine, NADPH, and O_2 as a substrate to the free radical NO, citrulline, and $NADP^+$ via L-hydroxyarginine as an intermediate (Knowles and Moncada 1994). NADPH donates elections to reductase domain of the enzyme and moves via redox carrier (FAD, FMN) at the active site to catalyze the reaction. Electron flow from reductase domain to oxygenase domain requires the presence of bound calcium/calmodulin. Animal NOS exhibit bi-domain structure with *N*-terminal oxygenase domain containing binding site for heme, tetrahydrobiopterin, and L-arginine and linked by a CaM recognition site to a C-terminal reductase domain that contains binding site for FAD, FMN, and NADPH (Ghosh and Stuehr 1995). Three distinct genes for the human neuronal, inducible, and endothelial NOS with a single copy each were located on a different chromosome in the haploid genome (Hall et al. 1994). The NOS genes have a similar genomic structure but not identical suggesting a common origin, and human isoforms are 51–57% homologous. Each NOS has different localization, regulation, catalytic properties, and inhibitor sensitivity. The NOS types in human are type I (nNOS/NOS-I), type II (iNOS/NOS-II), and type III (eNOS/NOS-III). Type I was the first NOS to be identified, predominating in neuronal tissue and located on the 12th chromosome in humans. Type II is an inducible type and present in a wide range of cells and tissues and located on the 17th chromosome in humans. Type III is found predominately in vascular endothelial cells and located on the seventh chromosome in humans. These isoforms have also been grouped into the following two categories, depending upon their nature and calcium requirement. Group I includes eNOS and nNOS types, which are constitutive and calcium dependent, but group II contains iNOS type which is inducible and calcium independent. Above limited information reveals a systematic evolution that occurred and led to establish the diversity in NO-synthesizing enzyme.

Structurally diverse NO-generating enzymes with different mechanism are reported in plant species (Gupta et al. 2011). They have been grouped into homologous and analogous in comparison with mammalian NOS (Fig. 15.1).

15.2.1 Homologous NO-Generating Enzyme/Mammalian-Like NOS Enzyme

Three isoforms of NOS are responsible for the generation of NO in mammalian systems with an array of physiological roles. To investigate the presence of mammalian-like NOS in plants, initial detection was based on anti-NOS antibodies raised against mammalian NOS. Using mouse anti-NOS from brain (nNOS)

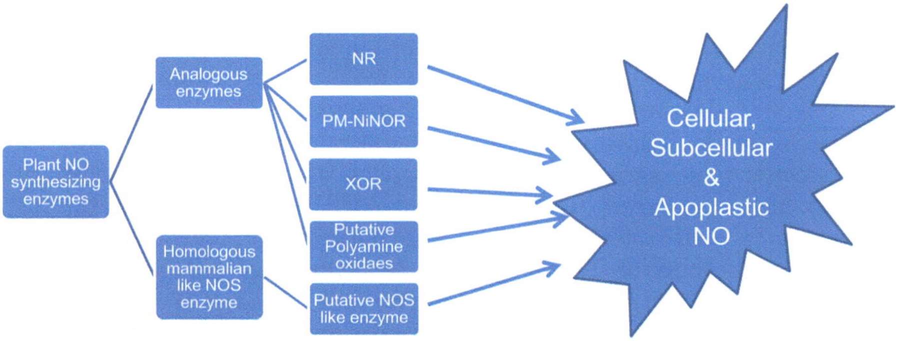

Fig. 15.1 Plant NO synthesizing enzyme classification

antibody, Western blot analysis showed the presence of positive immunoreactivity to NOS protein in yeast and wheat germ (Kuo et al. 1995). Similarly rabbit anti-NOS from the brain used in pea embryonic axis revealed the presence of a single band of 105.4 kDa and two bands of 57.5 and 89.7 kDa in wheat germ observed on blot (Sen and Chema 1995). Antibodies generated against mouse macrophage NOS and rabbit brain NOS helped in detection of a protein band about 166 kDa in soluble fraction of root tips and young leaves of maize seedling which was capable of converting [U-^{14}C] arginine to L-[U-^{14}C] citrulline. Presence of reactivity and bands on Western blot raised hope in scientists to find out the enzyme. Further evidence was generated by the use of mammalian nitric oxide synthase inhibitors and substrate labeled with radioactive element that confirmed the presence of putative NOS activity in higher plants (*Mucuna hassjoo*) in the green husk (pericarp) of seeds (Ninnemann and Maier 1996), roots, and nodules of *Lupinus albus* (Cueto et al. 1996). The use of antibody inhibitor again confirms the presence of enzyme, but their exact location could not be ascertained. Thus, to get more insight, immunotechnique, radiolabeling, and inhibitors were fused with fluorescent dye. Immunofluorescence study in maize revealed the presence of translocatory NOS in cytosol and nucleus depending on the phase of cell growth (Ribeiro et al. 1999). In soybean cell extract, NOS activity was found to be calcium dependent and present primarily in cytosolic fraction. It was further proved by using NOS inhibitors L-NNA and PBITU (Delledonne et al. 1998). NOS-like activity was reported in TMV-resistant tobacco, infected with TMV that was inhibited by NOS inhibitor (Durner et al. 1998). The presence of NOS in peroxisome and chloroplast from leaves of pea using activity assay in purified intact peroxisomes, a Ca^{2+}-dependent NOS activity, have been reported. The peroxisome NOS activity was clearly inhibited by well-characterized inhibitor of mammalian NOS, and the immunoblot analysis of peroxisome with a polyclonal antibody against the C-terminal region of murine iNOS revealed a protein of 130 kDa. It was further confirmed by electron microscopy immunogold labeling that shows the subcellular localization of NOS in the matrix of peroxisome as well as chloroplast (Barroso et al. 1999). Fluorometric analysis and EPR using Fe-MGD further substantiated the presence of NO in peroxisome (Corpas et al. 2002).

Existence of NOS activity by the use of mammalian NOS inhibitor was seen in dermal layer structure such as guard cell and epidermal cell of *Kalanchoe daigremontiana* and callus of gymnosperm *Taxus brevifolia* (Pedroso et al. 2000) and in *Taxus cuspidata* (Gong and Yuan 2006). The enzyme activity localization was seen in photosynthetically active cells or nearby cells like epidermal cytosol, guard cell chloroplast, and leaf parenchyma cell confirmed by the use of fluorescent dye DAF-2DA (Pedroso et al. 2000). NOS-like protein activity was also found in nucleolar region by DAF-2DA in *Nicotiana* (Foissner et al. 2000) and in cotyledon of soybean on the basis of citrulline assay and confirmed by mammalian NOS inhibitor (Modolo et al. 2002). Later on development of better NO detection technique like EPR spin trap technique fused with enzyme assay and use of inhibitors showed the presence of NOS-like activity in soybean chloroplast. But unlike other putative plant NOS, this was found calcium independent and heat labile (Jasid et al. 2006). Legume plant (*Vicia faba*) showed the NOS-like activity after wound stress and jasmonic acid treatment in guard cell and activity was Ca^{++} dependent and localized in the nucleus, cytoplasm, chloroplast, mitochondria, and cell wall of guard cell immunofluorescence technique (Xin et al. 2003, 2005) (Fig. 15.2).

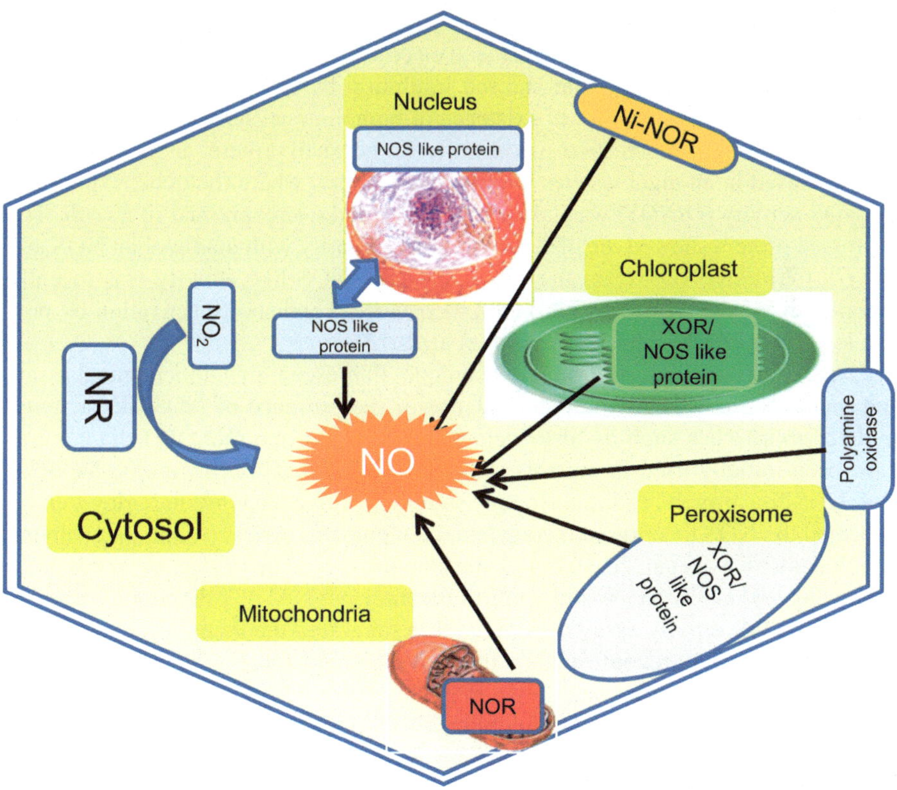

Fig. 15.2 Diverse locations of NO synthesis in plant cell

NO studies in *Arabidopsis* were initially carried out by the use of NOS inhibitors that inhibited NO production suggesting the presence of inducible form of NOS (Garces et al. 2001). But Guo et al. (2003) showed that an *Arabidopsis* gene *AtNOS1* encodes a protein involved in nitric oxide synthesis and confirmed by using activity assay, inhibitor, mutants, and fluorescent dye. Further experiment revealed that recombinant *AtNOS1* was failed to exhibit NOS activity in vitro and later identified as a plastid GTPase, involved in the posttranscriptional regulation of plastid-localized methylerythritol phosphate pathway enzyme (Gas et al. 2009; Moreau et al. 2008) and ribosome assembly (Flores-Pèrez et al. 2008); therefore it was renamed as *Arabidopsis thaliana* nitric oxide-associated protein 1 (AtNOA1), and now it is well established that it regulates physiological processes like salt stress tolerance (Zhao et al. 2007), chlorophyll biosynthesis, and Rubisco formation in rice (Yang et al. 2011).

Knowing the fact that NOS-like protein occurs in peroxisome (Barrosso et al. 1999), NOS-like activity was tracked in peroxisome of cotyledon and root of *Arabidopsis* seedling by using target signal 1 (PTS1) in vivo, and to get correct temporal and expressional location of NOS-like activity during various stages of *Arabidopsis*, green fluorescent protein (GFP) was used (Corpas et al. 2009). The presence of NOS-like protein in peroxisome matrix was also reported by Palma et al. (2009). Temperate area growing blue mustered (*Chorispora bungeana*) suspension culture cells after chilling and enhanced Ca^{++} treatment showed NO generation and NOS-like activity that showed inhibition by NOS inhibitors (Liu et al. 2010). Although many reports of existence of homologous NOS are available, protein purification is not achieved successfully. To get such protein, an alternate pathway was used in an algal species (*Ostreococcus tauri*) where the gene responsible for NOS activity (*OtNOS*) was taken from algae and overexpressed in *E. coli*. The expressed protein showed the 45% amino acid similarity with mammalian NOS and *OtNOS* fold-like eNOS (Foresi et al. 2010). The NOS-like activity was recently reported in Indian yellow mustard (*Brassica juncea*) and shows activation by protein kinase C (PKC) with Ca^{++} (Talwar et al. 2012).

Use of pharmacological, immunotechnique, fluorescent dye, biochemical assay, and mutant's analysis gave direct evidence of the presence of NOS-like activity. Indirect approaches such as histochemical marker enzyme (NADPH diaphorase) have been initially used in animals (Hope et al. 1991). This enzyme co-localizes with NOS-like protein. Indirect strategy showed positive response in barley root tip indicated the NOS location in metaxylem, metaphloem, pericycle, and parenchyma cell (Valentovicova et al. 2010) of phloem element of root tip in *Vicia faba* pericycle (Zou et al. 2012). To locate and confirm the enzyme, NADPH diaphorase activity assay and NOS activity assay were done that suggested the location of enzyme in root region and were reconfirmed by the use of mutants (Negi et al. 2010). But the co-localization of enzyme was shown in guard cell of leaf epidermal peel and was cross-checked by the use of NOS inhibitors that reduced the NO generation (Talwar et al. 2012). These results give augmenting evidence to support the previous report claiming the presence of NOS-like enzyme in leaf guard cells.

The presence of NOS-like enzymes is also reported in tomato seedlings (*Lycopersicon esculentum*) (Diao et al. 2016; Jin et al. 2011; Leonetti et al. 2011), *Arabidopsis* (Liu et al. 2017), white clover (Peng et al. 2016), rice (Cai et al. 2015), *Scutellaria baicalensis* (Zhang et al. 2014), soybean (Santa-cruz et al. 2014), maize (Tossi et al. 2009; Hao et al. 2008; An et al. 2005), *Camellia sinensis* (Zhang et al. 2012), *Chorispora bungeana* (Liu et al. 2010), and *Pleurotus eryngii* var. *tuoliensis* (Kong et al. 2012) by the use of NOS inhibitor or other pharmacological tools.

In addition to indication of existence of homologous NOS in plants, NO production was not inhibited by NOS inhibitors as in the green algae *Scenedesmus obliquus* (Mallick et al. 2000), *Arabidopsis thaliana* (Clarke et al. 2000), *Chlamydomonas reinhardtii* (Sakihama et al. 2002), and sunflower (Rockel et al. 2002) suggesting that some alternative pathway (He et al. 2007; Deng et al. 2016) or analogous forms of NOS might be operative in plants that evolved to work and adopt according to the need or conditions of plant species. Till date there is no concrete report of plant NOS enzyme from higher plant that has been reported except an alga *O. tauri* (Foresi et al. 2010; Hu et al. 2015).

15.2.2 Analogous NO-Generating Enzyme in Plants

There are structurally different enzymes from mammalian NOS enzyme existing for generation of NO in plant systems, namely, nitrate reductase (NR), plasma membrane-bound Ni-NOR (PM-NiNOR), xanthine oxidoreductase (XOR), polyamine oxidase, catalase, and hemeprotein or protein with molybdo-Fe-S cluster (Fröhlich and Durner 2011).

15.2.2.1 Nitrate Reductase (NR)

Mostly nitrate is absorbed by their roots from soil and assimilated in utilizable organic nitrogen form by plants. Nitrate assimilation is carried out by a two-step mechanism. The first step is the reduction of nitrate to nitrite in the cytosol via nitrate reductase using either NADH or NAD(P)H (Oaks 1994), and the second step is conversion of nitrite into ammonia in the chloroplast in green tissue and plastids in root by a ferredoxin-dependent nitrite reductase (NiR).

$$\mathrm{NO_3}^- + NADPH + 2e^- \rightarrow \mathrm{NO_2}^- + \mathrm{NAD(P)}^+ + \mathrm{H_2O}$$

Nitrate reductase (NR) converts nitrate to nitrite. It is a homodimer made up of two 100 KDa monomers, each having three prosthetic groups, i.e., FAD (flavin adenine dinucleotide), heme, and a molybdenum complex, which are present. (Campbell 1996). Molybdenum is bound to the enzyme via a complex with an organic molecule called pterin, which acts as a chelator of metal (Mendel and Stellmeyer 1995). NR mostly uses only NADH as an electron donor. Another form of NR, located mostly in non-green tissue such as roots, uses either NADH or NADPH (Warner and Kleinhofs 1992). NR has FAD, heme, and pterin that are also found in nitric oxide synthase (NOS). Nitrate reductase is the principal molybdenum-containing protein in vegetative tissue involved in nitrate assimilation (Lea 1999). Key symptom of molybdenum deficiency is the accumulation of nitrate that results from diminished nitrate reductase activity.

NR was started to be considered as alternate enzyme to produce NO (Yamasaki et al. 1999) and reported in spinach (Rockel et al. 2002), *Chlamydomonas reinhardtii* (Sakihama et al. 2002), and *Arabidopsis* (Wilkinson and Crawford 1993). NR under hypoxia reduces nitrite into NO (Planchet et al. 2005; Gupta et al. 2012), but reports about NOx-like product formation in vivo assay were reported three decades ago in soybean (Harper 1981). It was reconfirmed by the use of NR mutants comparing with wild-type soybean (Dean and Harper 1986). NR-based NO generation has been reported; NR activity is not seen by addition of tungstate (Deng et al. 1989), cyanide (Notton and Hewitt 1971), and azide (Yamasaki and Sakihama 2000) but not affected by NOS inhibitors like L-NAME. The NR activity was absent in soybean NR mutant (Dean and Harper 1986), *nia1* and *nia2* mutants of *Arabidopsis*, and cc-2929 strain of *C. reinhardtii.* NR also help in NO generation by supplying nitrite to other enzymes located on cellular PM (PM-NiNOR) (Stohr et al. 2001) or xanthine oxidase located in peroxisome.

15.2.2.2 Plasma Membrane-Bound Ni-NOR (PM-NiNOR)

PM-NiNOR is another enzyme reported from purified plasma membrane of tobacco responsible for the generation of nitric oxide as *Nicotiana tabacum* L. cv. Samsun roots that reduces NO_2 to NO at pH 6.0, and a reduced cytochrome c acts as an electron donor. Enzyme activity was absent in soluble protein fraction or in plasma membrane vesicle of leaves. Ni-NOR was insensitive to cyanide and anti-NR IgG. Such characteristic distinguishes Ni-NOR from PM-NR. Its molecular mass was 310 kDa in comparison to 200 kDa for PM-NR. It is also suggested that the PM-associated Ni-NOR may reduce apoplastic nitrite which is produced by PM-NR in vivo and play a key role in nitrate signaling via NO formation. Because the root tissue has limited access to oxygen, it might have evolved an additional system that is active at low-oxygen concentration because nitric oxide synthase requires oxygen for NO production (Stohr et al. 2001).

15.2.2.3 Xanthine Oxidoreductase (XOR)

NO generation occurs in plant peroxisome by xanthine oxidoreductase via reduction of nitrite into NO, during anaerobic condition, by using NADH or xanthine as reductant and has been reported in pea and white lupine (Gupta et al. 2011).

15.2.2.4 Polyamine Oxidase

Polyamine oxidase role has been linked with oxidation of polyamine like spermine or spermidine (Tun et al. 2006) and is dependent on availability of L-arginine. Polyamine-based NO generation is region specific and has higher NO level in elongation zone of root tip and in primary leaves, especially in the veins and trichomes, while in cotyledons it had little or no effect (Tun et al. 2006). These results are based on NO generation detection by polyamine application, and till date no such enzyme has been biochemically characterized (Gupta et al. 2011). Catalase and hemeprotein

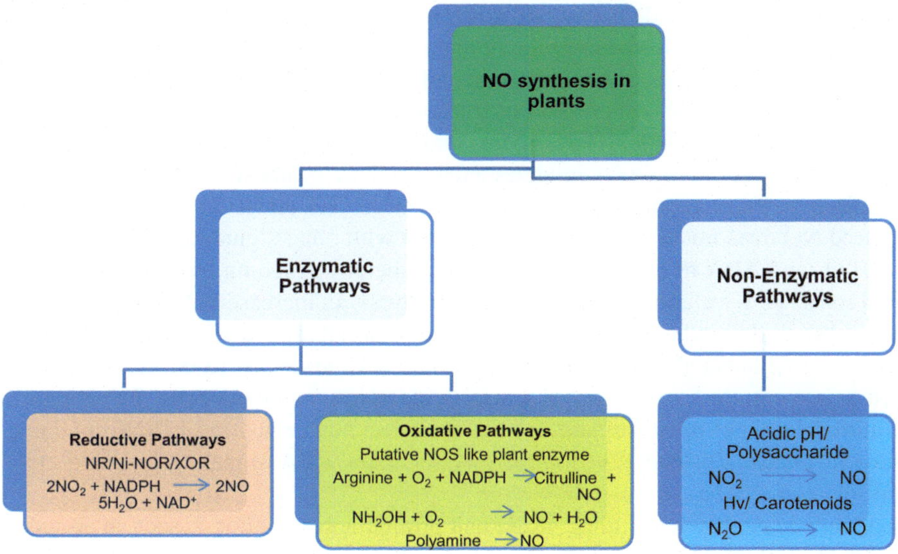

Fig. 15.3 Diverse biochemical pathways for NO synthesis in plants

or protein with molybdo-Fe-S cluster has also been proposed to participate in plant NO synthesis, but no characterization has been established to prove their role (Fig. 15.3).

15.3 NO and Abiotic Stress

Under both natural and agriculture conditions, plants are frequently exposed to stress. Some environmental factors (such as cold or heat) can become stressful in just a few minutes; others may take days to weeks (soil water) or even months (some mineral nutrients) to become stressful. There are various abiotic stresses such as water, drought, salinity, low and high temperature, heavy metal, mineral deficiency, UV, and ozone exposure. All these abiotic stress conditions induce the generation of reactive oxygen species (ROS) (Neill et al. 2002) as well as nitric oxide. ROS and NO initiate several oxidatively destructive processes and also trigger various signaling pathways. Thus, maintenance of appropriate levels in plant system provides adaptive response.

15.3.1 Water Stress/Drought

Day by day the requirement for water is increasing, and the water availability is decreasing, contributing to emergence of water stress. To cope up the physical or physiological water stress-like condition, plants have evolved protective mechanism. Nitric oxide is one of the key molecules playing an important role via various strategies such as closure of stomata.

Nitric oxide induces stomatal closure and enhances the adaptive plant responses against drought stress by preventing transpiration in wheat and other species by application of NO donors (Garcia-Mata and Lamattina 2001). ABA is synthesized following turgor loss and stimulates guard cell to synthesize NO. Leshem and Haramaty (1996) reported that wilting increased NO emission from pea plants. Indeed NO does not act alone but may interact with other signaling molecules such as H_2O_2 and ABA to effect stomatal closure, thereby reducing transportation level and conserving water. In response to drought stress, an increase in NOS-like activity was observed in wheat seedling, and ABA accumulation was inhibited by NOS inhibitors (Zhao et al. 2001) showing cross talk between ABA, ROS, and NO levels. ABA induces the production of nitric oxide (NO) in guard cells. NO through S-nitrosylation of open stomata 1 (OST1)/sucrose nonfermenting 1 (SNF1)-related protein kinase 2.6 (SnRK2.6) negatively regulates ABA signaling in guard cells (Wang et al. 2015a).

Abscisic acid-induced nitric oxide and proline accumulation is seen in an independent pathway under water-deficit stress during seedling establishment in *Medicago truncatula* (Planchet et al. 2014). *Arabidopsis* WD40-REPEAT 5a (WDR5a), which is a homolog of yeast TUP1, complemented H_2O_2-induced NO accumulation of a yeast mutant Δtup1, suggesting the conserved role of WDR5a in regulating NO accumulation and NOS-like activity and stomatal closure in drought stress tolerance (Liu et al. 2017). The nitric oxide made available through NO donor enhances the tolerance responses against moderate water deficit in leaves of *Phaseolus vulgaris* and *Vigna unguiculata* (Zimmer-prados et al. 2014). Expression of the tetrahydrofolate-dependent nitric oxide synthase from the green alga *Ostreococcus tauri* increases tolerance to abiotic stresses and influences stomatal development in *Arabidopsis* (Foresi et al. 2010). *Poncirus trifoliata* seedlings showed dehydration/drought tolerance through regulation of antioxidant systems and stomatal response with NO involvement (Fan and Liu 2012).

It has been found that spermidine (Spd) and NO can enhance drought tolerance via stress-activated NR and NOS pathways associated with NO release, which mediated antioxidant defense (Peng et al. 2016). Transgenic *Arabidopsis* plants containing nNOS displayed high levels of osmolytes and increased antioxidant enzyme activities. It is correlated with transcriptomic analysis which identified 601 or 510 genes that were differentially expressed as a consequence of drought stress or nNOS transformation, respectively. It suggests that NO leads to enhanced drought stress resistance and extensive transcriptional reprogramming (Shi et al. 2014).

S-Nitrosoglutathione (GSNO) acts as reservoir of NO and as an antioxidant. It is less degraded and that the drought-tolerant genotype has a higher natural reservoir of NO than the drought-sensitive one. Those differences in intracellular and extracellular NO contents and enzymatic activities were associated with higher leaf hydration in the drought-tolerant genotype as compared to the sensitive one under water deficit (Silveira et al. 2017).

NO signaling probably plays a key role in water stress-induced tanshinone production in *Salvia miltiorrhiza* hairy roots. NO donor SNP stimulated the MEP pathway to increase tanshinone accumulation (Du et al. 2015). Dehydration stress enhances both NOS activity as well as NO release substantially. Exogenous NO application increases the prevention of water loss as well as oxidative damage by inducing high SOD activity, less transpiration, more water retention, and less membrane damage to maize leaves (Hao et al. 2008) and in rice (Xiong et al. 2012). The rhizospheric application of melatonin (10 μmol L^{-1}) remarkably enhanced the drought tolerance of alfalfa plants by modulating nitro-oxidative homeostasis and proline metabolism acting as a promising priming agent for the enhancement of plant tolerance to drought conditions (Antoniou et al. 2017). Some rhizobacteria produce a volatile compound (2R,3R-butanediol) that induces drought tolerance by stomatal closure in *Arabidopsis thaliana*, by acting as an active player involving nitric oxide and hydrogen peroxide production. Similarly nitric oxide and hydrogen peroxide alleviate drought stress in marigold (*Tagetes erecta* L.) explants and promote its adventitious root development (Liao et al. 2012).

Wheat seedlings showed protection against drought stress by NO-mediated alternative pathway (AP) through dissipating excessive photosynthetic reducing equivalents. Moreover, NO-modulated AP under drought stress by increasing AOX1a expression and pyruvate content suggests that NO-mediated AP is involved in optimizing photosynthesis under drought stress by avoiding the over-reduction of photosynthetic electron heat chain, thus reducing reactive oxygen species production and oxidative damage in wheat leaves (Wang et al. 2016a).

15.3.2 Salt Stress

Abiotic stress is one of the main threats affecting crop growth and production (Fancy et al. 2017). Nitric oxide (NO) is a free radical molecule involved in an array of functions under various abiotic and biotic stress conditions (Asgher et al. 2017). Salt stress is one of the main abiotic stresses which affects reduction in biomass of sunflower (*Helianthus annuus* L.) seedling growth (Arora et al. 2016), injurious to germinating *P. sativum* L. (var. Shubhra IM-9101) reduction in growth and biomass yield, leaf relative water content (LRWC), and chlorophyll content of chickpea plants (Ahmad et al. 2016).

Physiologically high salinity increases electrolyte leakage, carotenoid content, and the levels of osmolytes (proline, glycine betaine, soluble proteins, and soluble sugars), hydrogen peroxide (H_2O_2), and malondialdehyde (MDA) along with antioxidant enzyme activities (superoxide dismutase (SOD), catalase (CAT), ascorbate peroxidase (APX), and glutathione reductase) in chickpea (*Cicer arietinum* L.) plants (Ahmad et al. 2016) and affects redox and NO homeostasis in tomato roots (Manai et al. 2014).

Ethylene promotes germination of *Arabidopsis* seed under salinity stress by decreasing reactive oxygen species (Lin et al. 2013) and increased NO via EIN3 protein which is an ethylene signaling transduction transcription factor. NO activates ethylene biosynthesis, possibly via posttranslational enzymatic modification (*S*-nitrosylation) involved in ethylene synthesis like ACC synthase and ACC oxidase. EIN3 transcript expression was enhanced by high salt concentrations. In nia1nia2 mutants, plants showed large attenuation in NO biogenesis as compared to wild types (Li et al. 2016b). It has also been observed that chemicals like 1-amin ocyclopropane-1-carboxylate (ACC, the immediate precursor of ethylene), nitrite, GA(4), and BA improved seed germination in the presence of salt in *Suaeda salsa* plant seeds (Li et al. 2005).

Salt stress also induced an increase in endogenous carbon monoxide (CO) production and the CO synthetic enzyme heme oxygenase (HO) in wheat seedling roots that improved salt tolerance by nitric oxide-mediated maintenance of ion homeostasis and upregulation of antioxidant defense (Xie et al. 2008). NO also affected the salt-induced changes in free amino acid levels and cadaverine levels that may be involved in salt stress regulating in maize (Simon-Sarkadi etal 2014). Exogenous application of NO (donor) has capability to mitigate the adverse effects of high salinity by improving LRWC, photosynthetic pigment biosyntheses, osmolyte accumulation, and antioxidative defense system in chickpea plants (Ahmad et al. 2016) and during seedling establishment by inducing an effective antioxidant system and limiting toxic ion and reactive oxygen species (ROS) accumulation in *J. curcas* (Gadelha et al. 2017) or by increasing the accumulation of proline content, total protein content, and total soluble sugar along with increasing antioxidant enzyme system in raspberry (*Rubus idaeus* var. Danehdrosht) (Ghadakchiasl et al. 2017). The cocktail of SA and NO donor (SNP) showed positive response against salinity stress in *Pisum sativum* L. (Yadu et al. 2017) and even atmospheric trace amounts of nitric oxide application enhances tolerance to salt stress and improves nutritional quality in spinach (*Spinacia oleracea* L.) (Du et al. 2015).

Salt stress and salicylic acid (SA)-induced cell death gets activated by many signaling pathways including ethylene (ET) signaling as seen in tomato plants (Poór et al. 2013). Endogenous NO, NaCl stress, and iron homeostasis modulates HO-1 activity at the early stage of sunflower seedling growth. The interactive nitrate reductase (NiA/NR) and NOA1-dependent NO production and HY1 expression showed that upon salinity stress, the majority of NO production was attributed to NIA/NR/NOA1 which was confirmed by the use of mutants in *Arabidopsis* (Xie et al. 2013).

Melatonin acts as an antioxidant through its interaction with nitric oxide (NO) and has a role in combating deleterious effects of ROS and reactive nitrogen species (RNS) by modulation of two SOD isoforms, i.e., Cu/Zn SOD and Mn SOD (Arora et al. 2016).

Unicellular algae (*Chlamydomonas reinhardtii*) use multiple strategies to adapt to salt stress, and GSNOR activity and protein S-nitrosylation levels play important roles (Chen et al. 2016), and also in pea mitochondrial proteins (Camejo et al. 2013) and protein carbonylation, nitration and *S*-nitrosylation are involved in acclimation to salinity stress in the roots and leaves of sour orange plants (*Citrus aurantium* L.) (Tanou et al. 2012). Caspase-like enzymatic activity and the ascorbate-glutathione cycle also participate in salt stress tolerance of maize conferred by exogenously applied nitric oxide (Keyster et al. 2012).

NO and salt stress are two important regulators of GST gene and enzyme expression in soybean (*Glycine max* L.), and both ABA-dependent and ABA-independent pathways are involved in regulation (Dinler et al. 2014).

Nitric oxide (NO) enhances salt tolerance of glycophytes and contributes to $K^{(+)}/Na^{(+)}$ balance in high salinity-treated *Kandelia obovata* roots, by activating AKT1-type $K^{(+)}$ channel and $Na^{(+)}/H^{(+)}$ antiporter (Chen et al. 2013) and serving as a signal molecule by increasing the K to Na ratio, which is dependent on the increased plasma membrane $H^{(+)}$-ATPase activity (Zhao et al. 2004). Salinity increases NO production selectively in mesophyll cells of sorghum leaves (Monreal et al. 2013). All these strategies involved by NO help in adaptation against the salt stress.

15.3.3 Heat Stress

Nitric oxide (NO) is a well-established signaling molecule in different physiological processes of animals as well as plants. NO involvement in response to heat stress is also observed in various plants. Possibly the antioxidant property of NO is via suppression of the high levels of ROS that accumulates following exposure to chilling or heat stress (Neill et al. 2002). High temperature (HT) and light had activating effect on the different indicators of reactive nitrogen species (RNS) metabolism in plants like pea, and results suggest that low and high temperature, continuous light, and high light intensity are abiotic stress conditions that can induce nitrosative stress. Protein tyrosine nitration is a potential marker of nitrosative stress (Corpas et al. 2008) and correlated by tyrosine nitration that provokes inhibition of sunflower carbonic anhydrase (β-CA) activity under high temperature stress (Chaki et al. 2013).

Heat Stress results in a marked decrease in membrane thermostability and cell viability, whereas content of lipid peroxide and activities of antioxidant enzymes, viz., superoxide dismutase (SOD), catalase, ascorbate peroxidase, guaiacol peroxidase, and glutathione reductase increased in both the cultivars. The exogenous application of NO donor (SNP) caused marked increase in antioxidant enzymes in both cultivars, and lipid peroxide content was reduced more in shoots of heat-tolerant cultivar suggesting that HT-induced oxidative damage is tolerated by NO (Bavita et al. 2012). Similarly in highbush blueberry, PSII photochemical activity was restored and antioxidant system elevated to provide protection under high temperature stress (Wei et al. 2010).

Heat stress inhibits NO accumulation by guard cell protoplasts (GCPs) and suppresses activation of the BA auxin-responsive effects such as expansion of protoplast 20- to 30-fold, cell wall regeneration, dedifferentiation, reentry in the cell cycle, and division. In tobacco plant, it reduces mitotic index of primary root meristems along with inhibition of the elongation of lateral root (Beard et al. 2012).

NO also interact with ROS in various ways and might serve an antioxidant function during various stress conditions (Beligni and Lamattina 1999). Nitric oxide and ethylene emission from pea foliage increases with duration of stress and/or senescence promoting conditions, and NO appears to exert a stress-coping or contrary inhibitory effect on leaf growth such as marked increase of chlorophyll fluorescence in red region suggestive of impairment of photosynthetic electron transport (Leshem and Haramaty 1996). In this respect, one of its effects may be biophysical by increasing photosynthetic membrane surface tension and a probable site of action being the C=C region of MGDG and decrease in LOX activity because these are very much prone to free radicals like NO and superoxides. Its decrease could involve lessened biosynthesis of jasmonic acid or methyl jasmonate-like compounds and add to the harmful effect induced by NO under stressful circumstances (Leshem et al. 1997). It has been shown that endogenously produced nitric oxide (NO) emission decreases with the progress of maturation, and senescence goes and ethylene upsurge takes place. This phenomenon is also being geared to ethylene increment under short-term heat stress (Leshem et al. 1998). It can be said that NO has not been found as key molecule in heat stress management, but it plays its positive role at certain extant.

15.3.4 Cold Effect

Chilling is another abiotic factor that induces rapid molecular responses that allows the maintenance of plant cell homeostasis and plant adaptation. Short-term abiotic stress caused an increase in NO production in alfalfa (Leshem 2001). NO donor mediates chilling resistance observed in tomato, wheat, and corn (Lamattina et al. 2001) and was found dose dependent in loquat leaves that could enhance the activity of antioxidant system and alleviated the cell injury (Wu et al. 2009). The reduction of oxidative stress and proline accumulation are the key factor (Wang et al. 2013). NO confers an increased tolerance to chilling in *C. bungeana* suspension cultures by chilling-induced NO accumulation. Here, NO might act as an antioxidant by directly scavenging ROS or operate as a signal activating antioxidant defense mechanism (Liu et al. 2010). Chilling stress as well as spermidine (Spd) and spermine (Spm) given to tomato seedlings [(*Lycopersicon esculentum* Mill.) cv. Moneymaker] (4 °C; 0, 12, and 24 h) led to elevated NO and H_2O_2 levels, enhanced nitrite reductase (NR), nitric oxide synthase (NOS)-like and polyamine oxidase activities, and upregulated LeNR relative expression (Diao et al. 2017). Spermidine (Spd) induced superoxide dismutase (SOD), peroxidase (POD), catalase (CAT), and ascorbate peroxidase (APX) in tomato leaves, and NO scavenger or inhibitor reversed such events in tomato (Diao et al. 2016) and wheat seedling (Esim et al. 2014).

Plant response to cold stress includes rapid nitric oxide (NO) formation that regulates cold-responsive gene expression. Nitric oxide (NO) causes polarized pollen tube growth in *Camellia sinensis* by alterations to gene expression under low temperature stress (Pan et al. 2016). Contrary to that NO production from NOS-like enzyme reaction decreased the cold-responsive pollen germination, inhibited tube growth, and reduced proline accumulation. It was mediated partly via cGMP signaling pathway in *C. sinensis* (Wang et al. 2012, 2016c). The effects of 24-epibrassinolide (EBR, a synthetic BR) affected the seed activity index and seedling growth by increasing the activities of some antioxidative enzymes including SOD, POD, CAT, and GR and the contents of nonenzymatic antioxidants, such as GSH and proline, and induced the accumulation of nitric oxide (NO). The expression of genes like P5CS1, CBF1, CBF3, and COR15a was induced by low temperature stress and further increased by EBR treatment in maize embryo. These expressions were suppressed by NO synthesis inhibitors. NO might be a downstream signal mediating 5-aminolevulinic acid (ALA) that induced chilling resistance in *E. nutans* as it enhanced antioxidant defense and activated plasma membrane (PM) H^+-ATPase and decreased the accumulation of ROS induced by chilling stress (Fu et al. 2015). It has been observed that CO acts as regulator to improve the tolerance of recalcitrant seeds to low temperatures through NO-mediated glutathione homeostasis (Bai et al. 2012).

NO production in shoot apoplast of *Brassica juncea* seedlings showed nonenzymatic nitrite reduction to NO in cold stress where thiol pool enrichment and *S*-nitrosylated proteins, including novel targets, provided the preview of RNS-mediated cold stress signaling in the apoplast (Sehrawat and Deswal 2014). Low temperature-induced S-nitrosylation is responsible for significant (~40%) inactivation of Rubisco (Abat and Deswal 2009). Cold, abscisic acid (ABA), H_2O_2, and nitric oxide (NO) cause induction of MfSAMS1 expression and cold acclimation in *Medicago sativa* subsp. falcata that caused upregulation of polyamine oxidation along with production of SAM, putrescine, spermidine and spermine levels, and ethylene (Guo et al. 2014).

Phytosphingosine phosphate (PHS-P) and a ceramide phosphate (Cer-P) are two phosphorylated sphingolipids which are specifically synthesized upon cold exposure as well as by cellular NO (Guillas et al. 2011). NO production was detected after 1–4 h of chilling. By contrast, the formation of phytosphingosine phosphate and ceramide phosphate, which are transiently synthesized upon chilling, was negatively regulated by NO, suggesting that NO acts as an intermediate in gene regulation and lipid-based signaling during cold transduction in *Arabidopsis thaliana* (Cantrel et al. 2011) and contributes in cold stress mitigation.

15.3.5 UV Stress

Ultraviolet B (UV-B) is an abiotic stress factor received by the earth from the sun. Its elevated level induces the accumulation of a variety of intracellular reactive oxygen species (ROS) from multiple sources, as well as NO through increased NOS activity. These molecules cause oxidative stress for plants and initiate parallel signaling pathways mediating responses of specific genes to UV-B radiation (Mackerness et al. 2001). Nitric oxide (NO) also acts as protective molecule against UV-B by inducing antioxidant system (Shi et al. 2005). Exogenous NO treatment on oxidative damages in wheat seedlings under elevated UV-B stress showed decline in ROS levels and membrane damage parameters, with an increment in antioxidant contents and antioxidant enzyme activity which was caused by a combination with He-Ne laser and exogenous NO treatment, which were greater than those of each individual treatment and had a similar effect on transcriptional activities of plant antioxidant enzymes (Li et al. 2016c). Pretreatments with sodium nitroprusside (SNP), a NO donor, prevented lipid peroxidation, ion leakage, and H_2O_2 and superoxide anion accumulation in leaves of UV-B-treated soybean plants. It was correlated with transcripts levels of superoxide dismutase (SOD), catalase (CAT), and two isoforms of ascorbate peroxidase (APX), which were significantly induced in SNP-treated plants compared to controls, respectively (Santa-Cruz et al. 2014). NO alleviates oxidative damage caused by UV-B stress by increasing the activities of antioxidant enzymes such as SOD, peroxidase, and CAT and enhances accumulation of GSH, by eliminating O^{-2} in a cyanobacterial (*S. platensis*-794) cells (Xue et al. 2007).

UV-B radiation-induced nitric oxide synthase (NOS) enzyme caused decrease of leaf biomass and an exo- or endo-beta-glucanase activity in maize (An et al. 2005) and stimulated the activity of plasma membrane H^+-ATPase, 5′-nucleotidase, peroxidase, ascorbate peroxidase, and glutathione reductase in ultraviolet B (UV-B)-irradiated *Chlorella pyrenoidosa* and negatively regulated the activity of nitrate reductase, nitrite reductase, and glutamine synthetase (Chen et al. 2003, 2010).

Flavonoids are UV-B-absorbing compounds whose concentration increases in plant cells stimulated by UV-B irradiation along with a rise in NO level. NO meditates the induction of chalcone synthase (Chs) by UV-B which appears along with isoflavone accumulation enhancement of antioxidant system in soybean sprouts (Jiao et al. 2016). Chalcone synthase (CHS) and chalcone isomerase (CHI) genes are involved in the flavonoid biosynthetic pathway, and their expression is systemically induced by UV-B in a NO-dependent pathway (Tossi et al. 2012). GSNO and SNAP increased in Chs transcript levels in the absence of UV-B, suggesting the importance of NO in regulating gene expression in response to UV-B radiation (Mackerness et al. 2001). NO also acts as a second messenger of ultraviolet B in inhibiting mesocotyl elongations in maize seedlings (Zhang et al. 2003). On the other hand, ultraviolet-B-induced flavonoid accumulation in *Betula pendula* leaves was found to be dependent upon nitrate reductase-mediated nitric oxide signaling by inducing upregulation of NIA1, but it does not affect NIA2 expression during UV-B-triggered NO generation (Zhang et al. 2011).

UV-B radiation inhibits pollen germination and tube growth partly via promoting NO production in pollen grains as well as pollen tube by a NOS-like enzyme of *Paulownia tomentosa* in vitro (He et al. 2007). UV-B irradiation significantly showed an increase in NOS activity, NO release, and baicalin biosynthesis in *S. baicalensis* cells that act as part of the defense response (Zhang et al. 2014).

An increase in ABA concentration was observed in UV-B-irradiated leaves of maize (*Zea mays*) which further activates pNOX from NOS-like dependent mechanism and H_2O_2 that maintain cell homeostasis and attenuate UV-B-derived cell damage (Tossi et al. 2009, 2012). UV-B perception increases NO concentration, which protects plant against UV-B by two mechanisms, firstly, by scavenging ROS and, secondly, by upregulating the expression of HY5, MYB12, and ZmP, resulting in the phenylpropanoid biosynthetic pathway (PPBP) activation in *Zea mays* and *Arabidopsis* (Tossi et al. 2011).

Heme oxygenase (HO) enzyme has antioxidant properties and gets upregulated by reactive oxygen species (ROS) in soybean plants and by nitric oxide synthase (NOS)-like activity in ultraviolet B irradiation (Santa Cruz et al. 2010). In the Krebs cycle, aconitases are the major NO targets in animals (Drapier 1997; Mott et al. 1997) and directly affect the aconitase functionality (Gardner et al. 1998). Aconitase is an iron-sulfur (4Fe-4S)-containing enzyme that catalyzes the reversible isomerization of citrate to isocitrate. It has two isoforms; one is located in the mitochondria and the other in the cytosol. Conversely, aconitase inactivation has been proposed to increase ROS generation due to the accumulation of reduced metabolite, a condition termed "reductive stress" (Yan et al. 1997). Aconitase is exquisitely sensitive to ROS and is vulnerable than other iron-sulfur- or heme-containing enzyme to inhibition by NO (Castro et al. 1994) and other ROS (Verniquet et al. 1991).

Ultraviolet B (UV-B) radiation induces the activation of mitogen-activated protein kinase phosphatase 1 (MKP1) and its targets MPK3 and MPK6. UV-B-induced stomatal closure in *Arabidopsis* (*Arabidopsis thaliana*) is antagonistically regulated by MKP1 and MPK6 via modulating hydrogen peroxide (H_2O_2)-induced nitric oxide (NO) production in guard cells. It was confirmed by the use of mutants (Li et al. 2017). Diverse signaling pathway works in plants in response to UV-B radiation involving GPA1 (the Gα-subunit of heterotrimeric G proteins)-dependent activation of H_2O_2 production and subsequent Nia1-dependent NO accumulation in stomatal closure (He et al. 2013). UV RESISTANCE LOCUS 8 (UVR8) signaling involves constitutively photomorphogenic 1, the ELONGATED HYPOCOTYL5 (HY5) transcription factor, and the closely related HY5 HOMOLOG.

Stomata on abaxial epidermal strips of *Arabidopsis* ecotype Landsberg erecta got closed in response to increasing UV-B fluence rates that showed parallel increase in hydrogen peroxide (H_2O_2) and NO. The transgenic plants containing bacterial NO dioxygenase showed no NO accumulation, and the stomata did not close in response to UV-B, although H_2O_2 levels got increased. But when the uvr8-1 null mutant was exposed to UV-B radiation, stomata remained open, irrespective of the fluence rate. Neither NO nor H_2O_2 got increased in stomata of mutant. However, NO donor S-nitrosoglutathione induced closure of uvr8-1 stomata to the same extent that showed the UVR8 pathway regulates stomatal closure in *Arabidopsis* by a mechanism involving both H_2O_2 and NO generation in response to UV-B exposure (Tossi et al. 2014).

The UV-B irradiation led to a dose-dependent root growth inhibition and various morphological alterations of primary roots, such as swelling and excessive root hair formation, thus affecting plant microtubules (MTs) in both epidermal and cortical cells. The enhanced NO levels by exogenous NO donor application in plant cells act as protecting agent of MT organization as well as MT-related processes of root growth and development against disrupting effects of UV-B in *Arabidopsis thaliana* seedlings (Krasylenko et al. 2012). Hence, it can be seen that NO plays an important role in plant adaptation under chilling/cold stress condition.

15.3.6 Heavy Metal Stress

Heavy metals' (HMs) presence in the soil causes toxicity and hampers plant growth and development. Nitric oxide can counteract HM-induced ROS, either by direct scavenging or by stimulating antioxidant defense mechanism acting as secondary antioxidant. The imbalance or cross talk of/between NO and ROS concentration along with antioxidant system leads to nitrosative and oxidative stress or combination of both, i.e., nitroso-oxidative stress. ROS and RNS parameters suggest that the oxidative components are predominant compared with the nitrosative components in the root system of both species (Feigl et al. 2015).

During HM stress, the different organelles of plant cells can biosynthesize NO in parallel to the ROS, such as in mitochondria, chloroplasts, peroxisomes, cytoplasm, endoplasmic reticulum, and apoplast (Sahay and Gupta 2017), which can be correlated with application of NO donor (SNP) that successfully ameliorates adverse impact of AgNps on pea seedlings by regulating the Ag uptake, antioxidant system, oxidative stress, and anatomical structures of root and shoot (Tripathi et al. 2017).

Cadmium (Cd) is one of the most toxic heavy metals that inhibit physiological processes of plants. Nitric oxide (NO) is involved in the plant response to cadmium (Cd) stress (Gill et al. 2013). NO counteracts cadmium-induced cytotoxic processes mediated by reactive oxygen species (ROS) through cross talk between ROS, NO, and antioxidant responses and modulates protein changes in the plasma membrane in *Brassica juncea* (Verma et al. 2013). A quantitative proteomics approach showed 66 differentially expressed proteins, among which phospholipase D (PLD) was altered substantially after the treatment of Cd or Cd and NO in rice (Yang et al. 2016). Cd inhibits the growth of the primary root by inhibiting root meristem growth involving nitric oxide-mediated repression of auxin accumulation and signaling in *Arabidopsis*. Cd exposure causes decrease in auxin levels which is associated with reduced PIN1/3/7 protein accumulation, but not with reduced PIN1/3/7 transcript levels. Additionally, Cd stabilized AXR3/IAA17 protein to repress auxin signaling in Cd-mediated process which was confirmed by the use of NO-specific scavenger 2-(4-carboxyphenyl)-4,4,5,5-tetramethylimidazoline-1-oxyl-3-oxide (cPTIO) or NO synthase inhibitor $N(\omega)$-nitro-l-Arg-methylester (l-NAME). It suggests that NO participates in Cd-mediated inhibition of root meristem growth (Yuan and Huang 2016).

Cd causes reduced biomass, NO production, and chlorophyll (Chl a, Chl b, and total Chl) concentration but stimulated reactive oxygen species (ROS) and their own accumulation in plants. SNP (50 μM) substantially attenuated growth inhibition, reduced hydrogen peroxide (H_2O_2) and malondialdehyde (MDA) levels, stimulated ROS-scavenging enzymes/agents, and mitigated the $H^{(+)}$-ATPase inhibition in proton pumps in *Trifolium repens* L. plants. *S*-Nitrosylation process is involved in the ameliorating effect of SNP against Cd toxicity in leaves of *Boehmeria nivea* (L.) Gaud. This involvement exhibited a concentration-dependent property (Wang et al. 2015a, b). SNP also considerably upregulated the levels of jasmonic acid (JA) and proline in plant tissues but downregulated the levels of ethylene (ET) in both shoots and roots and the level of salicylic acid (SA) in roots only, which might be related to the elevated NO synthesis. Involvement of salicylic acid (SA) and nitric oxide was observed in protective reactions of wheat under the influence of heavy metals (Gil'vanova et al. 2012). Additionally, SNP (25–200 μM) regulated mineral absorption and, particularly at 50 μM, significantly enhanced the uptake of shoot magnesium (Mg) and copper (Cu) and of root calcium (Ca) and iron (Fe) which indicated that NO depleted Cd toxicity by eliminating oxidative damage, enhancing mineral absorption, regulating proton pumps, and maintaining hormone equilibrium (Liu et al. 2015). Such exogenously applied NO could promote the scavenging of reactive oxygen, keep the mineral nutrition in balance, and alleviate the damage of Cd stress to the leaf photosynthetic apparatus, making the tomato seedlings preserve their photosynthetic efficiency (Zhang et al. 2010).

Exogenous NO depletes Cd-induced toxicity by eliminating oxidative damage, reestablishing ATPase activity, and maintaining stress-related hormone equilibrium in white clover plants (Liu et al. 2015) and exerted an advantageous effect on alleviating the inhibitory effect of Cd on rice seed germination and seedling growth, which might interact with NO in rice (*Oryza sativa* L.) (He et al. 2014). Cd toxicity causes reduction in root number was reduction in plants but it was overcome by NaCl that led to reduction of NO accumulation in plants (Zhang et al. 2014). Nitric oxide-activated hydrogen sulfide is essential for cadmium stress response in Bermuda grass (*Cynodon dactylon* (L). Pers.) (Shi et al. 2014).

Cd exposure caused a significant decrease in total phenolic, GSH, and nitric oxide (NO) levels in maize plants (Akinyemi et al. 2017). GSH biosynthesis inhibition by buthionine sulfoximine (BSO) aggravated Cd toxicity. Chelation, sequestration promotion, and stimulating NO, SNO, and the antioxidant system through a redox-dependent mechanism are used by GSH to enhance tolerance to Cd stress (Hasan et al. 2016). It has been observed that HY1 confers cadmium tolerance by decreasing nitric oxide production and improving iron homeostasis in *Arabidopsis* (Han et al. 2014). Nitric oxide along with hydrogen sulfide alleviates cadmium-induced oxidative damage in alfalfa seedling roots (Li et al. 2012) and promotes $Cd(^{2+})$-induced *Arabidopsis* programmed cell death (PCD) by promoting MPK6-mediated caspase-3-like activation (Ye et al. 2012, 2013) and cadmium-induced PCD in roots and signaling response of yellow lupine plants, accompanied by the NADPH-oxidase-dependent superoxide anion $(O_{(2)}^{(\cdot-)})$ production (Arasimowicz-Jelonek et al. 2012).

GA-alleviated Cd toxicity is mediated through the reduction of the Cd-dependent NO accumulation and expression of Cd^{2+} uptake-related gene IRT1 in *Arabidopsis* (Zhu et al. 2012) or via calcium and change in endogenous NO with levels of non-protein thiol, protein thiol, and matrix polysaccharides (Zhang et al. 2012). Nitrogen (N) management is found to be a promising agronomic strategy to minimize cadmium (Cd) contamination in crops.

Heavy metals such as Zn can also induce oxidative stress and generate reactive oxygen and nitrogen species (ROS and RNS) and show effects such as reduced growth and yield in crop plants. Zn-induced NO production promoted an increase in reactive oxygen species accumulation in *Solanum nigrum* primary root tips by modulating the expression and activity of antioxidative enzymes and programmed cell death (PCD) (Xu et al. 2010). NO donor (SNP) also ameliorates zinc oxide nanoparticle-induced phytotoxicity in rice seedlings (Chen et al. 2015). It has been observed that GSNO plays role in Zn uptake modulation in hydroponically grown wheat plant (Buet et al. 2014).

Wild-type and iron uptake-inefficient tomato (*Solanum lycopersicum*) mutant (T3238fer) plants showed the upregulation of the Fe uptake system was responsible for NO_3^--facilitated Cd accumulation in plants (Luo et al. 2012). Iron (Fe) deficiency triggered significant accumulation of NO in the root system predominantly in the outer cortical and epidermal cells of the elongation zone in *Malus xiaojinensis* (Zhai et al. 2016), and various dicots increase their root branching which contributes to the enhancement of ferric-chelate reductase activity (Jin et al. 2011).

The deficiency of P, Fe, or both increased the cluster root number and cluster zones along with enhanced NO accumulation in pericycle cells and rootlet primordia at various stages of cluster root development with the expression of LaSCR1 and LaSCR2 which is crucial in cluster root formation suggesting the role of NO in the shared signaling pathway of the P- and Fe-deficiency-induced formation of cluster roots in white lupin (Meng et al. 2012). It leads to an increase in the expression of ferritin during the senescence of *Lotus japonicus* nodules (Chungopast et al. 2017). Glutathione has a role in nitric oxide-mediated iron-deficiency signaling and tolerance in *Arabidopsis* (Shanmugam et al. 2015).

Auxin, ethylene, and nitric oxide show interplay in the regulation of Fe-deficiency responses by strategy I plants (non-leguminous). Auxin acts upstream of ethylene and NO (Romera et al. 2011) supported by the report that NOS-generated rather than NR-generated NO acts downstream of auxin in regulating this Fe-deficiency-induced response (Jin et al. 2011). Iron uptake and transport-related gene expression gets impact by nitric oxide and glutathione in *A. thaliana* plants when grown under iron deficiency (Koen et al. 2012). GSH threshold requirement is observed for NO-mediated expression of the *Arabidopsis* AtFer1 ferritin gene in response to iron (Touraine et al. 2012).

Cell wall is the major component of root apoplast and main reservoir for iron in roots. Nitric oxide (NO) is involved in cell wall synthesis regulation. Iron deficiency elevated NO level in tomato (*Solanum lycopersicum*) roots. S-nitrosoglutathione, a NO donor, enhanced iron retention in root apoplast of iron-deficient plants, accompanied with a decrease of iron level in xylem sap (Ye et al. 2015).

Lipid peroxidation got induced via decreasing activation of antioxidant enzymes were observed with exogenous NO in tomato seedlings under Cu and Cd stress (Wang et al. 2016b). An NR-mediated early NO production in the shoots of hulless barley plays positive protective role from Cu toxicity through enhanced antioxidant enzyme activities and antioxidant pools (Hu et al. 2015) and by influencing ROS metabolism in *Arabidopsis* (Pető et al. 2013). Exogenous NO-mediated GSH-PC synthesis pathway in tomato under copper stress differently affected the subcellular distribution of Cu (Wang et al. 2014). Auxin and NO negatively regulate each other's level, and later NO intensifies the metal-induced cotyledon expansion, but mitigates elongation processes under $Cu^{(2+)}$ exposure (Pető et al. 2011). Copper-induced cross talk was observed among calcium, H_2O_2, NO, and a calcium-dependent activation of gene expression involving calmodulins and calcium-dependent protein kinases in *Ulva compressa* (González et al. 2012).

Pb exposure significantly decreased in vivo NO level, and exogenous NO partially overcome toxicity, but unable to restore the plant growth on prolonged Pb exposure (Kaur et al. 2015). Pb-triggered NO burst is linked with NR that has a role in Pb uptake by *P. crinitum* root cells (Yu et al. 2012).

The differential toxic effects of AgNPs and $AgNO_3$ are seen on *Brassica nigra* seed germination. Higher concentration of both suppressed HO-1 expression and is relieved by the HO-1 inducer, hematin, or NO donor, sodium nitroprusside (SNP) (Amooaghaie et al. 2015). NO also participates in tolerance to Mn excess, and negative effects of the highest SNP dose were also observed in *Matricaria chamomilla* (Kováčik et al. 2014). Information about NO in alleviating aluminum (Al) toxicity is quite limited, but evidences show that the exogenous NO treatments improve Al tolerance by activating antioxidative capacity to eliminate reactive oxygen species in plants (He et al. 2012).

Thus, it can be seen that NO is involved in countering the adverse effects induced by various heavy metals and contributes in maintaining the cellular homeostasis.

15.3.7 Mineral Deficiency Stress

To cope with mineral deficiency, plants have evolved many adaptive mechanisms from changes in morphology to altered physiological responses. Nitric oxide (NO) is found to be involved in the managing deficiency.

Iron (Fe) deficiency is a general agricultural problem that affects both the productivity and nutritional quality of plants (Shanmugam et al. 2015). Cell wall is the major component of root apoplast and acts as reservoir for iron in roots (Ye et al. 2015). Iron deficiencies impair chlorophyll biosynthesis and chloroplast development in plants that in turn affect plant productivity. NO is involved in iron homeostasis in plants. Most of the leaf iron (80%) is located in chloroplast. Under iron deficiency, NO treatment increases the chlorophyll content of leaves. It is accompanied by the accumulation of transcript encoding both the D1 protein of photo system II and the Rubisco large subunit (LSU) (Giardi et al. 1997). D1 protein is associated with the thylakoid membranes (Summers 1980). Oxidative stress causes cross-linking of Rubisco large subunit and protein degradation (Mehta et al. 1992). In addition, Rubisco LSU is directly fragmented by superoxide in 37 and 16 kDa polypeptides (Ishida et al. 1999). NO donors completely prevented leaf interveinal chlorosis in maize plants. But NO treatment did not increase iron content in plant organs indicating that root iron uptake was not enhanced.

Nitric oxide (NO), Fe-deficiency response, and hormonal signaling pathways are interconnected and involve morphological changes to the regulation of physiological processes and gene expression (chlorophyll a dioxygenase gene (MxCAO) in *Malus xiaojinensis* (Zhai et al. 2016), and the induction of Fe uptake-related genes during Fe deficiency, which is dependent on glutathione supply especially in the glutathione-deficient mutant (zir1), accumulated lower levels of Fe than the wild type suggesting its role in Fe-deficiency tolerance and NO-mediated Fe-deficiency signaling in *Arabidopsis* (Shanmugam et al. 2015).

S-Nitrosoglutathione, a NO donor, also significantly enhanced iron retention in root apoplast of iron-deficient plants, accompanied with a decrease of iron level in xylem sap in tomato (*Solanum lycopersicum*) roots (Ye et al. 2015). Plants grown under iron deficiency show that nitric oxide and glutathione both impact the expression of iron uptake- and iron transport-related genes as well as the content of metals in *A. thaliana* (Koen et al. 2012). The P and Fe deficiency induces formation of cluster roots in white lupin (*Lupinus albus*) which was mitigated by exogenously NO donor (Meng et al. 2012). A role for auxin, ethylene, and, recently, nitric oxide is found for the upregulation of the Fe acquisition genes which does not depend solely on these hormones and would act as activators. Signals, probably phloem Fe, may act as an inhibitor in alleviating Fe deficiency in strategy I (non-graminaceous) plants (Romera et al. 2011).

Iron-storing protein known as ferritin is found in bacteria, plants, and animal cells, and its level is regulated by nitric oxide in *Arabidopsis*. It has been observed that *Arabidopsis* cell suspension culture treated with NO donors showed the accumulation of ferritin both at mRNA and protein level. Iron is not necessary for this NO-mediated ferritin transcript accumulation. The pathway is serine/threonine phosphatase dependent and necessitates protein synthesis. Ferritin regulation through the IRDS sequence of the *Atfer*1 promotor was responsible for transcriptional repression under low iron supply, suggesting that NO, by acting downstream of iron in the induction of ferritin transcript accumulation, is therefore a key signaling molecule for regulation of iron homeostasis in plants (Murgia et al. 2002). Enzyme aconitase (cytosolic) in animals was found to be liked with the iron regulatory protein (IRP), a protein that controls iron homeostasis (Kaptain et al. 1991; Klausner et al. 1993). IRP binds to mRNAs containing a specific iron-responsive element (IRE) consensus sequences and regulates their translatability and/or stability. NO converts the cytosolic aconitase into an IRP by promoting the loss of the iron-sulfur cluster which otherwise prevents IRE binding; thus aconitase lays a dual role as a key NO sensor and regulator of iron homeostasis. The tobacco aconitase like animal counterparts was inhibited by NO donors, and aconitase gene NtACO1 shows similarity and conservation. This shows similarities between the plant and animal NO signaling mechanisms (Navarre et al. 2000).

Melatonin increases the tolerance of plants to Fe deficiency in a process dependent on the polyamine-induced NO production under Fe-deficient conditions (Zhou et al. 2016). Boron deficiency accelerates generation of reactive oxygen species (ROS), specifically OH-1 and induced cellular oxidative injury in watermelon seedlings, but counteracted by exogenously applied SNP that promoted leaf chlorophyll, photosynthesis, and consequently biomass production by reducing Boron accumulation, lipid membrane peroxidation, and ROS generation along with activating antioxidant enzymes (SOD, POD, and APX) that protects from ROS-induced cellular burst (Farag et al. 2017) (Figs. 15.4 and 15.5). NO cannot supply the deficient mineral but augments the plant to meet their needs and rescue from stress condition.

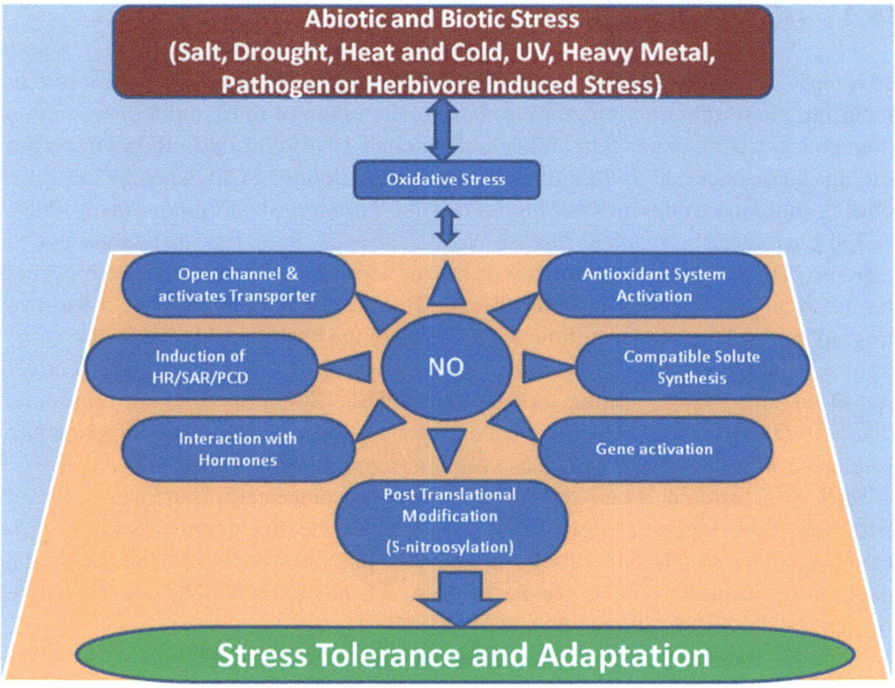

Fig. 15.4 NO and adaptation of abiotic and biotic stress

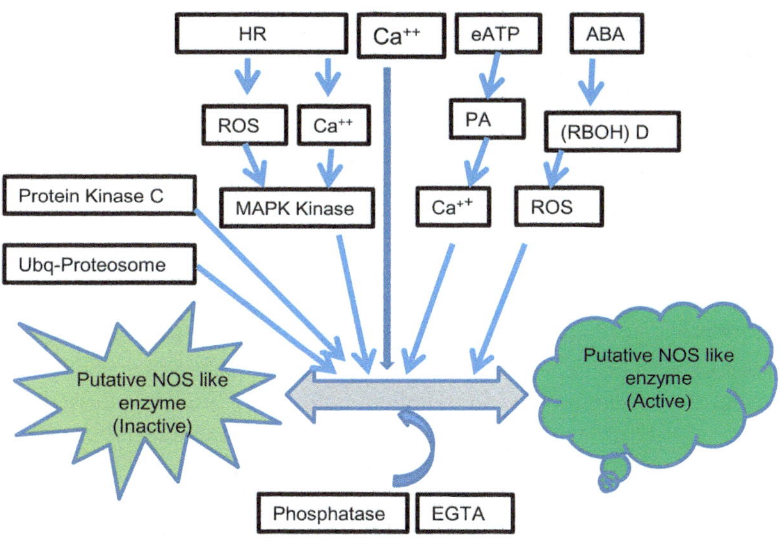

Fig. 15.5 Possible regulatory mechanism of putative plant NOS-like enzyme

15.4 NO and Biotic Stress

Both nitric oxide (NO) and reactive oxygen species (ROS) play an important role in imparting plant immunity (Asai et al. 2010). The effect of nitric oxide in system is often concentration dependent (Wink and Mitchell 1998). Similarly ROS effects are also dose dependent, high dose of ROS triggers HR-related PCD, whereas low dose induces antioxidant enzymes and blocks cell cycle progression (Vranova et al. 2002).

The characteristic of plant disease resistance is the induction of hypersensitive response (HR), where the formation of necrotic lesions at the infection site occurs that restricts pathogen growth and spread. This way it is thought to be a form of programmed cell death. One of the early events of the hypersensitive response is the rapid accumulation of reactive oxygen species and nitric oxide through the activation of enzyme system similar to NADPH oxidase (Keller et al. 1998; Lamb and Dixon 1997). After this local resistance response, tissue distal to the infection site develops a wider response known as systemic acquired resistance (SAR).

SAR is a form of broad-spectrum disease resistance and associated with an increase in those chemical signals that operate in a collective manner to confer protection against secondary infections, namely, the phytohormone salicylic acid (SA), glycerol-3-phosphate (G3P), azelaic acid (AzA), nitric oxide (NO), and reactive oxygen species (ROS) (El-Shetehy et al. 2015).

NO and ROS provide disease resistance against necrotrophic pathogens (Asai et al. 2010), to root-knot nematodes (Leonetti et al. 2011), and also play a crucial role in uredinial germination of the wheat stripe rust pathogen (*Puccinia striiformis* Westend f. sp. Tritici) (Pst) (Yin et al. 2016). Different putative NO-responsive genes have been identified by transcriptomic analyses performed with several NO donors in different plant species as well as tissues, and results are suggesting the implication of NO-responsive genes with plant adaptive responses to biotic stress processes (Mata-Pérez et al. 2016). Nitric oxide induces the hypersensitive cell death in the soybean cells by reactive oxygen intermediates and also functions independently of such intermediates to induce genes for the synthesis of protective natural products (Delledonne et al. 1998). *Arabidopsis thaliana* suspension culture generates elevated level of NO in response to avirulent bacteria, where elevated level of NO was sufficient to induce cell death that involves caspase-like activity (Clarke et al. 2000).

A defensive mechanism of mammalian phagocytes in response to organism involves the oxidative burst, releasing reactive oxygen species, which are ultimately toxic to the pathogen. The key enzyme of this respiratory burst is NADPH oxidase catalyzing the reaction, and similar enzyme bound to plasma membrane generates reactive oxygen in elicited cells of *Arabidopsis thaliana* (Desikan et al. 1996). Similarly, there is an alternate pathway for nitric oxide generation, that is, plasma membrane-bound enzyme, which catalyzes the formation of nitric oxide from nitrite detected in tobacco roots (Stohr et al. 2001). Hydrogen peroxide from the oxidative burst also plays a key role in the modulation of a localized hypersensitive response during the expression of plant disease resistance (Levine et al. 1994). It was observed that hydrogen peroxide alone is not sufficient for an efficient response (Delledonne et al. 1998). Nitric oxide is also involved in the HR (hypersensitive disease resistance response) as found by treatment of cell suspension by NO donor. Hydrogen peroxide and superoxide potentiated the cell death induced by endogenous ROS (Delledonne et al. 1998). Activation of the oxidative burst following pathogen recognition involves a protein kinase cascade (Levine et al. 1994; Chandra et al. 1996).

Plants respond to pathogen attack by activating diverse countering mechanism ranging from prevention of pathogen replication and spreading that causes rapid and localized cell death (hypersensitive response, HR) and the accumulation of antimicrobial compounds known as phytoalexins observed in soybean plants against attack by the fungus *Diaporthe phaseolorum* f. sp. *meridionalis* (Dpm), the causal agent of stem canker disease (Modolo et al. 2002).

Tomato plants accumulated high levels of H_2O_2 in response to wounding and elicitors without being toxic to plant. NO donors strongly inhibited the expression of wound-inducible proteinase inhibitor but did not inhibit the activation of octadecanoid pathway genes reflecting the NO role in downregulating the expression of wound-inducible defense gene during pathogenesis (Orozco-Cardenas and Ryan 2002). High levels of NO have been related to wounding after removing the primary root (Pagnussat et al. 2002). Potato tuber treated with a NO donor (NOC-18) induced the accumulation of the potato phytoalexin (reshitin) (Noritake et al. 1996).

NO involvement is found against fungus *Botrytis cinerea* and the expression of the salicylic acid (SA)-responsive gene PR-1 in *Nicotiana benthamiana*, but ROS function negatively in resistance during *B. cinerea-N. benthamiana* interaction (Asai et al. 2010). Nitric oxide-associated protein 1 (NOA1) is involved in various abiotic stress responses and required for plant resistance to pathogen infections. NOA1-silenced plants showed elevated levels of herbivory-induced jasmonic acid (JA), but decline in JA-isoleucine conjugate (JA-Ile) levels. It suggests that NOA1 mutant alters the allocation of carbon resources within the phenylpropanoid pathway and the involvement of NOA1 in defense (Wunsche et al. 2011).

NUBS-4190 (new bis-aryl-methanone compound) elicited NO production and defense responses and resistance in *Nicotiana benthamiana* against *Phytophthora infestans* without association of reactive oxygen generation and hypersensitive cell death. And, here, pathogenesis-related 1a (NbPR1a) and nitric oxide-associated 1 (NbNOA1) genes were involved in providing induced resistance that gives the constitutive accumulation of PR1a transcripts and NO-associated salicylic acid which was not seen in mutants (Monjil et al. 2013).

There exists an alternate signaling pathway that involves local respiratory burst oxidase homolog B (RBOHB)-dependent H_2O_2 production, and subsequent systemic NR-dependent NO generation leads to BR-mediated systemic virus resistance (Deng et al. 2016).

Analyzing the variation of resistance and susceptibility in *Arabidopsis* against *S. sclerotiorum* led to two conclusions: first, signaling of inducible defenses (especially by NO) is predominantly mediated by jasmonic acid and abscisic acid, which is influenced by ethylene and is independent of salicylic acid, and, second, by nitric oxide (NO) and reactive oxygen species, which are important signals required for providing resistance (Perchepied et al. 2010).

Nitric oxide is involved in the interaction of *Arabidopsis thaliana* with the biotrophic fungi, such as *Golovinomyces orontii* and *Erysiphe pisi*, where it was observed that NO accumulated rapidly and transiently at infection sites. *Arabidopsis* mutants with defective immune response accumulated lower NO levels as compared to wild type (Schlicht and Kombrink 2013). Thus, nitric oxide (NO) is known for its role in the activation of plant defense responses. The transgenic plants have calmodulin-dependent mammalian neuronal nitric oxide synthase (nNOS), into wild-type and NahG tobacco plants, and exhibited enhanced resistance toward diverse pathogenic range such as bacteria, fungi, and viruses because of high amounts of H_2O_2, with lower catalase activity as compared to the wild type. Such transgenic plants contained high levels of salicylic acid (SA) and induced an array of salicylic acid (SA), jasmonic acid (JA), and/or ethylene (ET)-related genes (Chun et al. 2012).

Lipopolysaccharide (LPS) induces nitric oxide (NO) production and defense gene expression that boost immunity of *Arabidopsis thaliana* (Sun and Li 2013). It has been found that Chitosan-induced immunity in *Camellia sinensis* (L.) O. Kuntze against blister blight disease is mediated by nitric oxide (Chandra et al. 2017). Hypersensitive response (HR) cell death requires the simultaneous and balanced production of NO and ROS. The model pathosystem *Arabidopsis*/*Pseudomonas* confirms the production of NO during the hypersensitive response that triggers hypersensitive cell death as well as mediates defense gene expression (Chen et al. 2014). Flavin biosynthesis participates in regulating NO and ROS production as well as involved in HR cell death. By silencing NbRibA, a gene that encodes a bifunctional enzyme, i.e., guanosine triphosphate cyclohydrolase II/3,4-dihydroxy-2-butanone-4-phosphate synthase, which participates in the biosynthesis of flavin, in *Nicotiana benthamiana*, compromised not only HR cell death as well as the NO and ROS production induced by INF1 elicitin and a constitutively active form of NbMEK2 (NbMEK2DD). It also induced high susceptibility to an oomycete (*Phytophthora infestans*) and ascomycete (*Colletotrichum orbiculare*). Such conditions were rescued by adding riboflavin, FMN, or FAD (Asai et al. 2010). An arginine-dependent enzyme activity was probably the main source of NO in tomato tissues, which shows resistance to tomato powdery mildew inhibited by animal NO synthase, but not by a plant nitrate reductase inhibitor. In resistant tomato genotypes, increased NO production was localized in infected tissues (Piterková et al. 2009).

NO exerts its biochemical function through the *S*-nitrosylation of reactive cysteine thiols on target proteins (Skelly and Loake 2013). Cellular levels of protein S-nitrosylation are regulated by S-nitrosoglutathionereductase1 (GSNOR1). GSNO reductase (GSNOR) is the enzyme responsible for the in vivo control of intracellular levels of GSNO. GSNO accumulation activates the jasmonic acid (JA)-dependent wound response and acts synergistically with salicylic acid. Thus, GSNOR appears to be a key regulator of systemic defense responses, in both wounding and pathogenesis in *Arabidopsis* (Espunya et al. 2012). Glutathione promoted the nuclear accumulation of nonexpressor of pathogenesis-related genes 1 (NPR1) protein accompanied by an elevated SA concentration and the activation of pathogenesis-related (PR) genes that showed induced resistance of *A. thaliana* against *Pseudomonas* infection (Kovacs et al. 2015).

In the absence of GSNOR1 function, GSNO accumulates, leading to dysregulation of total cellular *S*-nitrosylation. On the other hand, the endogenous NO accumulation in *Arabidopsis*, observed due to loss of function by mutation in NO overexpression 1 (NOX1), led to disabled resistance (R) gene-mediated protection, basal resistance, and defense against non-adapted pathogens (Yun et al. 2016).

Calcium signaling works as an important signal transduction pathway of higher plants in response stresses. Ca^{2+}-bound calmodulin (CaM) has a key role in decoding and transducing stress signals by activating specific targets. The pepper calmodulin gene CaCaM1 is involved in reactive oxygen species and nitric oxide generation required for cell death and the defense response (Choi et al. 2009). Endogenous plant elicitor peptides (Peps) can act to facilitate immune signaling and pathogen defense responses via Ca^{2+} elevation as an early event in a signaling cascade that activates plant immune responses. Nitric oxide and reduced nicotinamide adenine dinucleotide phosphate oxidase-dependent reactive oxygen species generation are also involved downstream from the Ca^{2+} signal in the Pep immune defense signal transduction cascade (Ma et al. 2013).

Nitric oxide (NO) and mitogen-activated protein kinases (MAPKs) are signaling molecules involved in the disease resistance of plants. NO and the nitric oxide synthase activity got increased during resistance of the tomato fruits against *Botrytis cinerea* invasion, chitinase, β-1,3-glucanase, polyphenol oxidase, and phenylalanine ammonia lyase, but such effects are weakened by the MAPK inhibitor (Zheng et al. 2014).

NO donor application immediately before inoculation of tobacco leaves with potato virus X and tobacco mosaic virus has shown declination in the accumulation of virus, indicating that NO can induce resistance rapidly which was reversed by use of NO scavenger and transgenic tobacco plants expressing increased alternative respiratory pathway capacity due to constitutive expression of the plant mitochondrial enzyme, alternative oxidase (AOX) (Li et al. 2014). Exogenous nitric oxide application induces disease resistance against *Monilinia fructicola* through activating the phenylpropanoid pathway in peach fruit (Li et al. 2016a). Exogenous application of NO is a potential method for inducing the disease resistance in fruit against fungal pathogens, and it also helps in extending the postharvest life of citrus fruit (Zhou et al. 2016).

The use of NO_3^- or NH_4^+ fertilizers influences the plant-pathogen interactions. NO_3^- feeding augments hypersensitive response (HR)-mediated resistance and enhances the production of polyamines such as spermine and spermidine. Ammonium nutrition can compromise defense and NH_4^+ nutrition leading to increased γ-aminobutyric acid (GABA) levels which may be a nutrient source for the pathogen. NO is mostly generated from NO_2^- by nitrate reductase and is elicited by both pathogen-associated microbial patterns and gene-for-gene-mediated defenses. Thus, it seems that nitric oxide (NO) production and associated defenses are NO_3^- dependent and are compromised by NH_4^+ (Mur et al. 2017).

Melatonin (*N*-acetyl-5-methoxytryptamine) is a naturally occurring small molecule, serving as important secondary messenger in the response of plants to various biotic and abiotic stresses. Both melatonin and nitric oxide (NO) levels in

Arabidopsis leaves were significantly induced by bacterial pathogen (Pst DC3000) infection. NO scavenger is significantly suppressed suggesting the expression of salicylic acid (SA)-related genes and bacterial disease resistance (Shi et al. 2015).

Both ROS and NO have emerged as important players not only in the defense responses after pathogenic or herbivore attacks but also found to be involved in symbiotic interactions between plants and microorganisms. A role for NO during symbiosis has been suggested in bacteria-legume interaction. NO was detected in soybean nodule using EPR spectroscopy, and there is also evidence of the presence of NO in young alfalfa nodules by using confocal microscopy (Herouart et al. 2002). NOS immunoreactivity has been demonstrated in *Lupinus albus* nodules (Cueto et al. 1996). There are some reports that NO act as negative regulator of nitrogen fixation due to its interaction with leghemoglobin, and some data indicate that modulation in NO level results in alteration of nodule numbers (Herouart et al. 2002). Thus, biotic stress is overcome by NO through various mechanisms to deter the pathogen and save the plant.

15.5 Conclusions

NO is now a well-established signal molecule in biological kingdom. In animals, it has dedicated enzyme system collectively known as nitric oxide synthase (NOS). In plants, no such enzyme has been purified till date. But plants have evolved enzyme diversity which generates nitric oxide. Such diversity provides flexibility to plants to adapt in changing environmental and intercellular conditions. These enzymes are putative NOS-like enzyme, nitrite reductase (NR), plasma membrane-bound Ni-NOR (PM-NiNOR), xanthine oxidoreductase (XOR), and polyamine oxidase. These enzymes provide the NO generation and availability as and when need arises.

Under both natural and agriculture conditions, plants are frequently exposed to stress. Such multilocational presence of enzyme helps in coping up the abiotic and biotic stress faced by plants. There are various abiotic stresses such as mineral deficiency, drought, salinity, low and high temperature, heavy metal, UV, and ozone exposure, and biotic stress includes herbivory and disease caused by pathogens. All these abiotic and biotic stress conditions induce the generation of reactive oxygen species (ROS) and reactive nitrogen species (RNS) that includes NO. These reactive species initiate several oxidatively destructive processes, trigger various signaling pathways, and induce many antioxidant enzymes (SOD, POD, catalase, etc.) and synthesis of protectant molecules like flavonoid, GST, or glutathione. These signaling pathways and protective molecules cope up the stress and help in adaptation of plants.

References

Abat JK, Deswal R (2009) Differential modulation of S-nitrosoproteome of *Brassica juncea* by low temperature: change in S-nitrosylation of Rubisco is responsible for the inactivation of its carboxylase activity. Proteomics 9(18):4368–4380

Adak S, Bilwes AM, Panda K, Hosfield D et al (2002) Cloning, expression, and characterization of a nitric oxide synthase protein from *Deinococcus radiodurans*. Proc Natl Acad Sci U S A 99:107–112

Ahmad P, Abdel Latef AA, Hashem A et al (2016) Nitric oxide mitigates salt stress by regulating levels of osmolytes and antioxidant enzymes in chickpea. Front Plant Sci 7:347

Akinyemi AJ, Faboya OL, Olayide I et al (2017) Effect of cadmium stress on non-enzymatic antioxidant and nitric oxide levels in two varieties of maize (*Zea mays*). Bull Environ Contam Toxicol 98(6):845–849

Amooaghaie R, Tabatabaei F, Ahadi AM (2015) Role of hematin and sodium nitroprusside in regulating Brassica nigra seed germination under nanosilver and silver nitrate stresses. Ecotoxicol Environ Saf 113:259–270

An L, Liu Y, Zhang M et al (2005) Effects of nitric oxide on growth of maize seedling leaves in the presence or absence of ultraviolet-B radiation. J Plant Physiol 162(3):317–326

Antoniou C, Chatzimichail G, Xenofontos R et al (2017) Melatonin systemically ameliorates drought stress-induced damage in Medicago sativa plants by modulating nitro-oxidative homeostasis and proline metabolism. J Pineal Res 62(4). https://doi.org/10.1111/jpi.12401

Arasimowicz-Jelonek M, Floryszak-Wieczorek J, Deckert J et al (2012) Nitric oxide implication in cadmium-induced programmed cell death in roots and signaling response of yellow lupine plants. Plant Physiol Biochem 58:124–134

Arc E, Galland M, Godin B et al (2013) Nitric oxide implication in the control of seed dormancy and germination. Front Plant Sci 4:346. https://doi.org/10.3389/fpls.2013.00346

Arora D, Jain P, Singh N et al (2016) Mechanisms of nitric oxide crosstalk with reactive oxygen species scavenging enzymes during abiotic stress tolerance in plants. Free Radic Res 50(3):291–303

Asai S, Mase K, Yoshioka H (2010) Role of nitric oxide and reactive oxygen [corrected] species in disease resistance to necrotrophic pathogens. Plant Signal Behav 5(7):872–874

Asgher M, Per TS, Masood A et al (2017) Nitric oxide signaling and its crosstalk with other plant growth regulators in plant responses to abiotic stress. Environ Sci Pollut Res Int 24(3):2273–2285

Astier J, Lindermayr C (2012) Nitric oxide-dependent posttranslational modification in plants: an update. Int J Mol Sci 13(11):15193–15208

Bai X-G, Chen J-H, Kong X-x, Todd CD (2012) Carbon monoxide enhances the chilling tolerance of recalcitrant Baccaurea ramiflora seeds via nitric oxide-mediated glutathione homeostasis. Free Radic Biol Med 53:710–720. https://doi.org/10.1016/j.freeradbiomed.2012.05.042

Barroso JB, Corpas IJ, Carreras A et al (1999) Localization of nitric oxide synthase in plant peroxisomes. J Biol Chem 274:36729–36733

Bavita A, Shashi B, Navtej SB (2012) Nitric oxide alleviates oxidative damage induced by high temperature stress in wheat. Indian J Exp Biol 50(5):372–378

Beard RA, Anderson DJ, Bufford JL et al (2012) Heat reduces nitric oxide production required for auxin-mediated gene expression and fate determination in tree tobacco guard cell protoplasts. Plant Physiol 159(4):1608–1623

Beligni MV, Lamattina L (1999) Is nitric oxide toxic or protective? Trends Plant Sci 4:299

Beligni MV, Lamattina L (2000) Nitric oxide stimulates seed germination and de-etiolation, and inhibits hypocotyl elongation, three light-inducible responses in plant. Planta 210:215–221

Bird LE, Ren J, Zhang J et al (2002) Crystal structure of SANOS, a bacterial nitric oxide synthase oxygenase protein from *Staphylococcus aureus*. Structure 10(12):1687–1696

Buet A, Moriconi JI, Santa-María GE et al (2014) An exogenous source of nitric oxide modulates zinc nutritional status in wheat plants. Plant Physiol Biochem 83:337–345

Cai W, Liu W, Wang WS et al (2015) Overexpression of rat neurons nitric oxide synthase in rice enhances drought and salt tolerance. PLoS One 10(6):e0131599. https://doi.org/10.1371/journal.pone.0131599

Camejo D, Romero-Puertas Mdel C, Rodríguez-Serrano M et al (2013) Salinity-induced changes in S-nitrosylation of pea mitochondrial proteins. J Proteome 79:87–99

Campbell WH (1996) Nitrate reductase, biochemistry comes of age. Plant Physiol 111:355–361

Cantrel C, Vazquez T, Puyaubert J et al (2011) Nitric oxide participates in cold-responsive phosphosphingolipid formation and gene expression in *Arabidopsis thaliana*. New Phytol 189(2):415–427. https://doi.org/10.1111/j.1469-8137.2010.03500.x

Castro L, Rodriguez M, Radi R (1994) Aconitase is readily inactivated by peroxynitrite, but not by its precursor nitric oxide. J Biol Chem 269:29498–29415

Chaki M, Carreras A, López-Jaramillo J et al (2013) Tyrosine nitration provokes inhibition of sunflower carbonic anhydrase (β-CA) activity under high temperature stress. Nitric Oxide 29:30–33

Chandra S, Martion GB, Low PS (1996) The pto kinase mediates a signaling pathway leading to the oxidative burst in tomato. Proc Natl Acad Sci U S A 93:13393–13397

Chandra S, Chakraborty N, Panda K et al (2017) Chitosan-induced immunity in *Camellia sinensis* (L.) O. Kuntze against blister blight disease is mediated by nitric-oxide. Plant Physiol Biochem 115:298–307

Chen K, Feng H, Zhang M et al (2003) Nitric oxide alleviates oxidative damage in the green alga *Chlorella pyrenoidosa* caused by UV-B radiation. Folia Microbiol (Praha) 48(3):389–393

Chen J, Xiao Q, Wu F et al (2010) Nitric oxide enhances salt secretion and Na(+) sequestration in a mangrove plant, *Avicennia marina*, through increasing the expression of $H^{(+)}$-ATPase and $Na^{(+)}/H^{(+)}$ antiporter under high salinity. Tree Physiol 30(12):1570–1585

Chen J, Xiong DY, Wang WH (2013) Nitric oxide mediates root K+/Na+ balance in a mangrove plant, kandelia obovata, by enhancing the expression of AKT1-Type K+ channel and Na+/H+ antiporter under high salinity. PLoS ONE 8(8):e71543. https://doi.org/10.1371/journal.pone.0071543

Chen J, Vandelle E, Bellin D et al (2014) Detection and function of nitric oxide during the hypersensitive response in *Arabidopsis thaliana*: where there's a will there's a way. Nitric Oxide 43:81–88

Chen J, Liu X, Wang C et al (2015) Nitric oxide ameliorates zinc oxide nanoparticles-induced phytotoxicity in rice seedlings. J Hazard Mater 297:173–182

Chen X, Tian D, Kong X et al (2016) The role of nitric oxide signalling in response to salt stress in *Chlamydomonas reinhardtii*. Planta 244(3):651–669

Choi HW, Lee DH, Hwang BK (2009) The pepper calmodulin gene CaCaM1 is involved in reactive oxygen species and nitric oxide generation required for cell death and the defense response. Mol Plant-Microbe Interact 22(11):1389–1400

Chun HJ, Park HC, Koo SC et al (2012) Constitutive expression of mammalian nitric oxide synthase in tobacco plants triggers disease resistance to pathogens. Mol Cell 34(5):463–471

Chungopast S, Duangkhet M, Tajima S et al (2017) Iron-induced nitric oxide leads to an increase in the expression of ferritin during the senescence of *Lotus japonicus* nodules. J Plant Physiol 208:40–46

Clarke A, Desikan R, Hurst RD et al (2000) NO way back: nitric oxide and programmed cell death in *Arabidopsis thaliana* suspension cultures. Plant J 24(5):667–677

Corpas FJ, Barroso JB, Esteban FJ et al (2002) Peroxisome as a source of nitric oxide in plant cells. Free Radic Biol Med 33:S73

Corpas FJ, Chaki M, Fernández-Ocaña A et al (2008) Metabolism of reactive nitrogen species in pea plants under abiotic stress conditions. Plant Cell Physiol 49(11):1711–1722

Corpas FJ, Hayashi M, Mano S et al (2009) Peroxisomes are required for in vivo nitric oxide accumulation in the cytosol following salinity stress of *Arabidopsis* plants. Plant Physiol 151:2083–2094

Correa-Aragunde N, Graziano M, Lamattina L (2004) Nitric oxide plays a central role in determining lateral root development in tomato. Planta 218:900–905. https://doi.org/10.1007/s00425-003-1172-7

Cueto M, Herandez-Perea O, Martin R et al (1996) Presence of nitric oxide synthase in roots and nodules of *Lupinus albus*. FEBS Lett 398:159–164

Dean JV, Harper JE (1986) Nitric oxide and nitrous oxide production by soybean and winged bean during the in vivo nitrate reductase assay. Plant Physiol 82:718–723

Delledonne M, Xia Y, Dixon RA et al (1998) Nitric oxide functions as a signal in plant disease resistance. Nature 394:585–588

Deng M, Moureaux T, Caboche M (1989) Tungstate, a molybdate analog inactivating nitrate reductase, deregulates the expression of the nitrate reductase structural gene. Plant Physiol 91:304–309

Deng XG, Zhu T, Zou LJ et al (2016) Orchestration of hydrogen peroxide and nitric oxide in brassinosteroid-mediated systemic virus resistance in *Nicotiana benthamiana*. Plant J 85(4):478–493

Desikan R, Hancock JT, Coffey MJ et al (1996) Generation of active oxygen in elicited cells of *Arabidopsis thaliana* is mediated by a NADPH oxidase like enzyme. FEBS Lett 383:213–217

Diao QN, Song YJ, Shi DM et al (2016) Nitric oxide induced by polyamines involves antioxidant systems against chilling stress in tomato (*Lycopersicon esculentum* Mill.) seedling. J Zhejiang Univ Sci B 17(12):916–930

Diao Q, Song Y, Shi D et al (2017) Interaction of polyamines, abscisic acid, nitric oxide, and hydrogen peroxide under chilling stress in tomato (*Lycopersicon esculentum* Mill.) seedlings. Front Plant Sci 14(8):203

Dinler BS, Antoniou C, Fotopoulos V (2014) Interplay between GST and nitric oxide in the early response of soybean (*Glycine max* L.) plants to salinity stress. J Plant Physiol 171(18):1740–1747

Drapier JC (1997) Interplay between NO and [Fe-S] clusters: relevance to biological systems. Comp Methods Enzymol 11(3):319–329

Du X, Zhang C, Guo W et al (2015) Nitric oxide plays a central role in water stress-induced tanshinone production in *Salvia miltiorrhiza* hairy roots. Molecules 20(5):7574–7585

Durner J, Wendehenne D, Klessig DE (1998) Defence gene induction in tobacco by nitric oxide, cyclic GMP and cyclic ADP ribose. Proc Natl Acad Sci U S A 95:10328–10333

El-Shetehy M, Wang C, Shine MB et al (2015) Nitric oxide and reactive oxygen species are required for systemic acquired resistance in plants. Plant Signal Behav 10(9):e998544. https://doi.org/10.1080/15592324.2014.998544

Esim N, Atici O, Mutlu S (2014) Effects of exogenous nitric oxide in wheat seedlings under chilling stress. Toxicol Ind Health 30(3):268–274

Espunya MC, De Michele R, Gómez-Cadenas A et al (2012) S-Nitrosoglutathione is a component of wound- and salicylic acid-induced systemic responses in *Arabidopsis thaliana*. J Exp Bot 63(8):3219–3227

Fan QJ, Liu JH (2012) Nitric oxide is involved in dehydration/drought tolerance in Poncirus trifoliata seedlings through regulation of antioxidant systems and stomatal response. Plant Cell Rep 31(1):145–154. https://doi.org/10.1007/s00299-011-1148-1

Fancy NN, Bahlmann AK, Loake GJ (2017) Nitric oxide functions in plant abiotic stress. Plant Cell Environ 40(4):462–472

Farag M, Najeeb U, Yang J et al (2017) Nitric oxide protects carbon assimilation process of watermelon from boron-induced oxidative injury. Plant Physiol Biochem 111:166–173

Feigl G, Lehotai N, Molnár Á et al (2015) Zinc induces distinct changes in the metabolism of reactive oxygen and nitrogen species (ROS and RNS) in the roots of two *Brassica* species with different sensitivity to zinc stress. Ann Bot 116(4):613–625

Flores-Perez U, Sauret-Gueto S, Gas E et al (2008) A mutant impaired in the production of plastome-encoded proteins uncovers a mechanism for the homeostasis of isoprenoid biosynthetic enzymes in *Arabidopsis* plastids. Plant Cell 20:1303–1315

Foissner I, Wendehenne D, Langebartels C et al (2000) In vivo imaging of an elicitor-induced nitric oxide burst in tobacco. Plant J 23:817–824

Foresi N, Correa-Aragunde N, Parisi G et al (2010) Characterization of a nitric oxide synthase from the plant kingdom: no generation from the green alga *Ostreococcus tauri* is light irradiance and growth phase dependant. Plant Cell 22(11):3816–3383

Fröhlich A, Durner J (2011) The hunt for plant nitric oxide synthase (NOS): is one really needed? Plant Sci 181(4):401–404. https://doi.org/10.1016/j.plantsci.2011.07.014

Fu J, Chu X, Sun Y et al (2015) Nitric oxide mediates 5-aminolevulinic acid-induced antioxidant defense in leaves of *Elymus nutans griseb.* exposed to chilling stress. PLoS One 10(7):e0130367

Gadelha CG, Miranda RS, Alencar NLM et al (2017) Exogenous nitric oxide improves salt tolerance during establishment of *Jatropha curcas* seedlings by ameliorating oxidative damage and toxic ion accumulation. J Plant Physiol 212:69–79

Garces H, Durzan D, Pedroso MC (2001) Mechanical stress elicits nitric oxide formation and DNA fragmentation in *Arabidopsis thaliana*. Ann Bot 87:567–574

Garcia-Mata C, Lamattina L (2001) Nitric oxide induces stomatal closure and enhances the adaptive plant responses against drought stress. Plant Physiol 126:1196–1204

Gardner PR, Gardner AM, Martin LA et al (1998) Nitric oxide dioxygenase; an enzymic function for flavohemoglobin. Proc Natl Acad Sci U S A 95:10378–10383

Gas E, Flores-Pèrez U, Sauret-Gűeto S et al (2009) Hunting for plant nitric oxide synthase provides new evidence of a central role for plastids in nitric oxide metabolism. Plant Cell 21:18–23

Ghadakchiasl A, Mozafari AA, Ghaderi N (2017) Mitigation by sodium nitroprusside of the effects of salinity on the morpho-physiological and biochemical characteristics of *Rubus idaeus* under in vitro conditions. Physiol Mol Biol Plants 23(1):73–83

Ghosh DK, Stuehr DJ (1995) Macrophage NO synthase: characterization of isolated oxygenase and reductase domains reveals a head to head subunit interaction. Biochemistry 34:801–807

Giardi MT, Masojidek J, Godde D (1997) Effect of abiotic stress on the turnover of the D_1 reaction center H protein. Physiol Plant 101:636–642

Gil'vanova IR, Enikeev AR, Stepanov SI et al (2012) Involvement of salicylic acid and nitric oxide in protective reactions of wheat under the influence of heavy metals. Prikl Biokhim Mikrobiol 48(1):103–108

Gill SS, Hasanuzzaman M, Nahar K et al (2013) Importance of nitric oxide in cadmium stress tolerance in crop plants. Plant Physiol Biochem 63:254–261

Gniazdowska A, Krasuska U, Debska K et al (2010) The beneficial effect of small toxic molecules on dormancy alleviation and germination of apple embryos is due to NO formation. Planta 232:999–1005. https://doi.org/10.1007/s00425-010-1214-x

Gong YW, Yuan YJ (2006) Nitric oxide mediates inactivation of glutathione S transferase in suspension culture of *Taxus cuspidata* during shear stress. J Biotechnol 123(2):185–192

González A, Cabrera Mde L, Henríquez MJ et al (2012) Cross talk among calcium, hydrogen peroxide, and nitric oxide and activation of gene expression involving calmodulins and calcium-dependent protein kinases in *Ulva compressa* exposed to copper excess. Plant Physiol 158(3):1451–1462

Guillas I, Zachowski A, Baudouin E (2011) A matter of fat: interaction between nitric oxide and sphingolipid signaling in plant cold response. Plant Signal Behav 6(1):140–142

Guo FQ, Okamoto M, Crawford NM (2003) Identification of a plant nitric oxide synthase gene involved in hormonal signaling. Science 302:100–103

Guo Z, Tan J, Zhuo C et al (2014) Abscisic acid, H_2O_2 and nitric oxide interactions mediated cold-induced S-adenosylmethionine synthetase in *Medicago sativa subsp. falcata* that confers cold tolerance through up-regulating polyamine oxidation. Plant Biotechnol J 12(5):601–612

Gupta KJ, Fernie AR, Kaiser WM et al (2011) On the origins of nitric oxide. Trends Plant Sci 16(3):160–168. https://doi.org/10.1016/j.tplants.2010.11.007

Gupta KJ, Shah JK, Brotman Y et al (2012) Inhibition of aconitase by nitric oxide leads to induction of the alternative oxidase and to a shift of metabolism towards biosynthesis of amino acids. J Exp Bot 63(4):1773–1784

Hall AV, Antoniou H, Wang Y (1994) Structural organization of human neuronal nitric oxide synthase gene (*NOS1*). J Biochem 269:33082–33090

Han B, Yang Z, Xie Y et al (2014) *Arabidopsis* HY1 confers cadmium tolerance by decreasing nitric oxide production and improving iron homeostasis. Mol Plant 7(2):388–403

Hao GP, Xing Y, Zhang JH (2008) Role of nitric oxide dependence on nitric oxide synthase-like activity in the water stress signaling of maize seedling. J Integr Plant Biol 50(4):435–442

Harper JE (1981) Evolution of nitrogen oxide(s) during in vivo nitrate reductase assay of soybean leaves. Plant Physiol 68:1488–1493

Hasan MK, Liu C, Wang F et al (2016) Glutathione-mediated regulation of nitric oxide, S-nitrosothiol and redox homeostasis confers cadmium tolerance by inducing transcription factors and stress response genes in tomato. Chemosphere 161:536–545

He JM, Bai XL, Wang RB et al (2007) The involvement of nitric oxide in ultraviolet-B-inhibited pollen germination and tube growth of *Paulownia tomentosa* in vitro. Physiol Plant 131(2):273–282. https://doi.org/10.1111/j.1399-3054.2007.00955.x

He H, Zhan J, He L et al (2012) Nitric oxide signaling in aluminum stress in plants. Protoplasma 249(3):483–492
He JM, Ma XG, Zhang Y et al (2013) Role and interrelationship of Gα protein, hydrogen peroxide, and nitric oxide in ultraviolet B-induced stomatal closure in *Arabidopsis* leaves. Plant Physiol 161(3):1570–1583
He J, Ren Y, Chen X et al (2014) Protective roles of nitric oxide on seed germination and seedling growth of rice (*Oryza sativa* L.) under cadmium stress. Ecotoxicol Environ Saf 108:114–119
Herouart D, Baudouin E, Frendo P (2002) Reactive oxygen species, nitric oxide and glutahione: a key role in the establishment of the legume-*Rhizobium* symbiosis? Plant Physiol Biochem 40:619–624
Hope BT, Michael GJ, Knigge KM (1991) Neuronal NADPH diaphorase is a nitric oxide synthase. Proc Nati Acad Sci USA 88:2811–2814
Hu Y, You J, Liang X (2015) Nitrate reductase-mediated nitric oxide production is involved in copper tolerance in shoots of hulless barley. Plant Cell Rep 34(3):367–379
Ignarro LJ, Buga GM, Wood KS et al (1987) Endothelium-derived relaxing factor produced and released from artery and vein is nitric oxide (endothelium-dependent relaxation/vascular smooth muscle/cyclic GMP). Proc Natl Acad Sci U S A 84:9265–9269
Ishida H, Makino A, Mae T (1999) Fragmentation of the large subunit of ribulose-1-5, biphosphate carboxylase by reactive oxygen species occurs near Gly-329. J Biol Chem 274:5222–5226
Jasid S, Simontacchi M, Bartoli CG (2006) Chloroplasts as a nitric oxide cellular source. Effect of reactive nitrogen species on chloroplastic lipids and proteins. Plant Physiol 142:1246–1255
Jiao C, Yang R, Zhou Y (2016) Nitric oxide mediates isoflavone accumulation and the antioxidant system enhancement in soybean sprouts. Food Chem 1(204):373–380. https://doi.org/10.1016/j.foodchem.2016.02.147
Jin CW, Du ST, Shamsi IH et al (2011) NO synthase-generated NO acts downstream of auxin in regulating Fe-deficiency-induced root branching that enhances Fe-deficiency tolerance in tomato plants. J Exp Bot 62(11):3875–3884
Kaptain S, Dewney WE, Tang L et al (1991) A regulated RNA finding protein also possesses aconitase activity. Proc Natl Acad Sci U S A 88:10109–10113
Katsuki S, Arnold WP, Murad F (1977) Effects of sodium nitroprusside, nitroglycerin, and sodium azide on levels of cyclic nucleotides and mechanical activity of various tissues. J Cyclic Nucleotide Res 3:239–247
Kaur G, Singh HP, Batish DR et al (2015) Exogenous nitric oxide (NO) interferes with lead (Pb)-induced toxicity by detoxifying reactive oxygen species in hydroponically grown wheat (*Triticum aestivum*) roots. PLoS One 10(9):e0138713. https://doi.org/10.1371/journal.pone.0138713
Keller T, Damude HG, Werner D et al (1998) A plant homolog of the neutrophil NADPH oxidase of 91-phox subunit genes encodes a plasma membrane protein with Ca^{++} binding motifs 47. Plant Cell 10:255–266
Keyster M, Klein A, Ludidi N (2012) Caspase-like enzymatic activity and the ascorbate-glutathione cycle participate in salt stress tolerance of maize conferred by exogenously applied nitric oxide. Plant Signal Behav 7(3):349–360
Klausner RD, Rauault TA, Herford JB (1993) Regulating the fate of m-RNA: the control of cellular iron metabolism. Cell 72:190–128
Klepper LA (1979) Nitric oxide (NO) and nitrogen dioxide (NO2) emissions from herbicide-treated soybean plants. Atmos Environ 13:537–542
Knowles RG, Moncada S (1994) Nitric oxide synthase in mammals. Biochemistry 298:249–258
Koen E, Szymańska K, Klinguer A et al (2012) Nitric oxide and glutathione impact the expression of iron uptake- and iron transport-related genes as well as the content of metals in *A. thaliana* plants grown under iron deficiency. Plant Signal Behav 7(10):1246–1250
Kong W, Huang C, Chen Q et al (2012) Nitric oxide alleviates heat stress-induced oxidative damage in *Pleurotus eryngii var. tuoliensis*. Fungal Genet Biol 49(1):15–20
Kováčik J, Babula P, Klejdus B et al (2014) Unexpected behavior of some nitric oxide modulators under cadmium excess in plant tissue. PLoS One 9(3):e91685

Kovacs I, Durner J, Lindermayr C (2015) Crosstalk between nitric oxide and glutathione is required for Nonexpressor of Pathogenesis-Related genes 1 (NPR1)-dependent defense signaling in *Arabidopsis thaliana*. New Phytol 208(3):860–872. https://doi.org/10.1111/nph.13502

Krasylenko YA, Yemets AI, Sheremet YA et al (2012) Nitric oxide as a critical factor for perception of UV-B irradiation by microtubules in *Arabidopsis*. Physiol Plant 145(4):505–515

Kuo WN, Kuo TW, Jones DL et al (1995) Nitric oxide synthase immune reactivity in baker's yeast lobster and wheat germ. Biochem Arch 11:73–78

Lamattina L, Beligni MV, Garcia-mata C et al (2001) Method of enhancing the metabolic function and the growing conditions of plants and seeds. US Patent, US 624384 B1

Lamb CJ, Dixon RA (1997) The oxidative burst in plant disease resistance. Annu Rev Plant Physiol Plant Mol Biol 76:419–422

Lea PJ (1999) Nitrate assimilation. In: Lea PJ (ed) Plant biochemistry and molecular biology. Wiley, London, pp 163–192

Leonetti P, Melillo MT, Bleve-Zacheo T (2011) Nitric oxide and hydrogen peroxide: two players in the defence response of tomato plants to root-knot nematodes. Commun Agric Appl Biol Sci 76(3):371–381

Leshem Y (2001) Nitric oxide in plants. Kluwer Academic Publishers, London

Leshem YY, Haramaty E (1996) The characterization and contrasting effects of the nitric oxide free radical in vegetative stress and senescence of *Pisum sativum* Linn. foliage. J Plant Physiol 148:258–263

Leshem YY, Haramaty E, Iluz D et al (1997) Effect of stress nitric oxide (NO): interaction between chlorophyll fluorescence, galactolipid fluidity and lipoxygenase activity. Plant Physiol Biochem 35:573–579

Leshem YY, Wills RBS, Veng-Va Ku V (1998) Evidences for the function of the free radical gas-nitric oxide (NO) – as an endogenous maturation and senescence-regulating factor in higher plants. Plant Physiol Biochem 36:825–833

Levine A, Tenhapen R, Dixon RA et al (1994) H_2O_2 from the oxidative burst orchestrates the plant hypersensitive response. Cell 79:583–593

Li F-C, Wang J, M-M W (2017) Mitogen-activated protein kinase phosphatases affect UV-B-induced stomatal closure via controlling NO in guard cells. Plant Physiol 175. https://doi.org/10.1104/pp.16.01656

Li H, Raman CS, Glaser CB et al (1999) Crystal structures of zinc-free and bound heme domain of human inducible nitric oxide synthase implication for dimer stability and comparison with endothelial nitric oxide synthase. J Biol Chem 274:21276–21284

Li W, Liu X, Ajmal Khan M et al (2005) The effect of plant growth regulators, nitric oxide, nitrate, nitrite and light on the germination of dimorphic seeds of *Suaeda salsa* under saline conditions. J Plant Res 118(3):207–214

Li L, Wang Y, Shen W (2012) Roles of hydrogen sulfide and nitric oxide in the alleviation of cadmium-induced oxidative damage in alfalfa seedling roots. Biometals 25(3):617–631

Li Z, Liang WS, Carr JP (2014) Effects of modifying alternative respiration on nitric oxide-induced virus resistance and PR1 protein accumulation. J Gen Virol 95(Pt 9):2075–2081. https://doi.org/10.1099/vir.0.066662-0

Li G, Zhu S, Wu W et al (2016a) Exogenous nitric oxide induces disease resistance against *Monilinia fructicola* through activating the phenylpropanoid pathway in peach fruit. J Sci Food Agric. 2016 Nov 18. https://doi.org/10.1002/jsfa.8146

Li X, Pan Y, Chang B et al (2016b) NO promotes seed germination and seedling growth under high salt may depend on EIN3 protein in Arabidopsis. Front Plant Sci 6:1203

Li Y, Gao L, Han R (2016c) A combination of He-Ne laser irradiation and exogenous NO application efficiently protect wheat seedling from oxidative stress caused by elevated UV-B stress. Environ Sci Pollut Res Int 23(23):23675–23682

Liao WB, Huang GB, Yu JH et al (2012) Nitric oxide and hydrogen peroxide alleviate drought stress in marigold explants and promote its adventitious root development. Plant Physiol Biochem 58:6–15

Lin Y, Yang L, Paul M et al (2013) Ethylene promotes germination of *Arabidopsis* seed under salinity by decreasing reactive oxygen species: evidence for the involvement of nitric oxide simulated by sodium nitroprusside. Plant Physiol Biochem 73:211–218

Liu Y, Jiang H, Zhao Z (2010) An, nitric oxide synthase like activity-dependent nitric oxide production protects against chilling-induced oxidative damage in *Chorispora bungeana* suspension cultured cells. Plant Physiol Biochem 48:936–944

Liu SL, Yang RJ, Pan YZ et al (2015) Exogenous NO depletes Cd-induced toxicity by eliminating oxidative damage, re-establishing ATPase activity, and maintaining stress-related hormone equilibrium in white clover plants. Environ Sci Pollut Res Int 22(21):16843–16856

Liu WC, Li YH, Yuan HM et al (2017) WD40-REPEAT 5a functions in drought stress tolerance by regulating nitric oxide accumulation in *Arabidopsis*. Plant Cell Environ 40(4):543–552

Luo BF, Du ST, Lu KX et al (2012) Iron uptake system mediates nitrate-facilitated cadmium accumulation in tomato (*Solanum lycopersicum*) plants. J Exp Bot 63(8):3127–3136

Ma Y, Zhao Y, Walker RK et al (2013) Molecular steps in the immune signaling pathway evoked by plant elicitor peptides: Ca^{2+}-dependent protein kinases, nitric oxide, and reactive oxygen species are downstream from the early Ca^{2+} signal. Plant Physiol 163(3):1459–1471

Mackerness AHS, John CF, Jordan B et al (2001) Early signaling components in ultraviolet-B responses: distinct roles for different reactive oxygen species and nitric oxide. FEBS Lett 489(2–3):237–242

Maier J, Hecker R, Rockel P et al (2001) Role of nitric oxide synthase in the light induced development of sporengiophores in *Phycomyces blackesleeanus*. Plant Physiol 126:1323–1330

Mallick N, Mohn FH, Soeder CJ (2000) Evidence supporting nitrite-dependent NO release by the green microalga *Scenedesmus obliquus*. J Plant Physiol 157:40–46

Manai J, Gouia H, Corpas FJ (2014) Redox and nitric oxide homeostasis are affected in tomato (*Solanum lycopersicum*) roots under salinity-induced oxidative stress. J Plant Physiol 171(12):1028–1035

Mata-Pérez C, Begara-Morales JC, Luque F et al (2016) Transcriptomic analyses on the role of nitric oxide in plant disease resistance. Curr Issues Mol Biol 19:121–128

Mehta RA, Fowcett TW, Porath D et al (1992) Oxidation stress causes rapid membrane translocation and in vivo degradation of ribulose-1,5-biphosphate corboxylase\oxygenase. J Biol Chem 267:2810–2816

Mendel RR, Stellmeyer (1995) Molybdenum cofactor (nitrate reductase) biosynthesis in plants: first molecular analysis. In: MRC T, Falavigna (eds) Current plant science and biotechnology in agriculture vol 22: current issues in plant molecular and cellular biology. Kluwer, Dordrecht, pp 577–582

Meng ZB, Chen LQ, Suo D et al (2012) Nitric oxide is the shared signalling molecule in phosphorus- and iron-deficiency-induced formation of cluster roots in white lupin (*Lupinus albus*). Ann Bot 109(6):1055–1064

Mengel A, Chaki M, Shekariesfahlan A et al (2013) Effect of nitric oxide on gene transcription-S-nitrosylation of nuclear proteins. Front Plant Sci 4:293. https://doi.org/10.3389/fpls.2013.00293. PMCID: PMC3729996

Messner S, Leitner S, Bommassar C (2009) *Physarum* nitric oxide synthases: genomic structures and enzymology of recombinant proteins. Biochem J 418(3):691–700

Modolo LV, Cunha FQ, Braga MR et al (2002) Nitric oxide synthase mediated phytoalexin accumulation in soybean cotyledons in response to the *Diaporthe phaseolorum* f. sp. *Meridionalis* elicitor. Plant Physiol 130:1288–1297

Monjil MS, Shibata Y, Takemoto D et al (2013) Bis-aryl methanone compound is a candidate of nitric oxide producing elicitor and induces resistance in *Nicotiana benthamiana* against *Phytophthora infestans*. Nitric Oxide 29:34–45

Monreal JA, Arias-Baldrich C, Tossi V et al (2013) Nitric oxide regulation of leaf phosphoenolpyruvate carboxylase-kinase activity: implication in sorghum responses to salinity. Planta 238(5):859–869

Moreau M, Lee GI, Wang Y et al (2008) AtNOS/A1 is a functional *Arabidopsis thaliana* cGTPase and not a nitric oxide synthase. J Biol Chem 283:32957–32967

Mott HR, Carpenter JW, Campbell SL (1997) Structural and functional analysis of a mutant ras protein, that is insensitive to nitric oxide activation. Biochemistry 36:3540–3644

Mur LAJ, Simpson C, Kumari A et al (2017) Moving nitrogen to the centre of plant defence against pathogens. Ann Bot 119(5):703–709

Murgia I, Delledonne M, Soave C (2002) Nitric oxide mediates iron-induced ferritin accumulation in *Arabidopsis*. Plant J 20:521–528

Navarre DA, Wendehenne D, Durner J et al (2000) Nitric oxide modulates the activity of tobacco aconitase. Plant Physiol 122:573–582

Negi S, Santisree P, Kharshiing E et al (2010) Inhibition of the ubiquitin-proteasome pathway alters cellular levels of nitric oxide in tomato seedlings. Mol Plant 3(5):854–869

Neill SJ, Desikan R, Clarke A et al (2002) Hydrogen peroxide and nitric oxide as signaling molecules in plants. J Exp Bot 53:1237–1242

Neill S, Barros R, Bright J et al (2008) Nitric oxide, stomatal closure, and abiotic stress. J Exp Bot 59(2):165–176. https://doi.org/10.1093/jxb/erm293

Ninnemann H, Maier J (1996) Indication for the occurrence of nitric oxide synthase in fungi and plants and the involvement in photoconidiation of *Neurospora crasa*. Photochem Photobiol 64:393–398

Noritake T, Kawakita K, Doke N (1996) Nitric oxide induces phytoalexin accumulation in potato tuber tissue. Plant Cell Physiol 37:113–116

Notton BA, Hewitt EJ (1971) The role of tungsten in the inhibition of nitrate reductase activity in spinach (*Spinacia oleracea* L.) leaves. Biochem Biophys Res Commun 44:702–710

Oaks A (1994) Primary nitrogen assimilation in higher plants and its regulation. Can J Bot 72:869–885

Orozco-Cardenas ML, Ryan CA (2002) Nitric oxide negatively modulates wound signaling in tomato plants. Plant Physiol 130:487–493

Pagnussat GC, Simontacchi M, Puntarulo S et al (2002) Nitric oxide is required for root organogenesis. Plant Physiol 129:954–956

Palma JM, Corpas FJ, del Rio LA (2009) Proteome of plant peroxisomes: new perspectives on the role of these organelles in cell biology. Proteomics 9:2301–2312

Palmer RMJ, Ferrige AG, Moncada S (1987) Nitric oxide release accounts for the biological activity of endothelium-derived relaxing factor. Nature 327:524–526. https://doi.org/10.1038/327524a0

Pan J, Wang W, Li D et al (2016) Gene expression profile indicates involvement of NO in *Camellia sinensis* pollen tube growth at low temperature. BMC Genomics 17(1):809

Pedroso MC, Magalhaes JR, Durzan D (2000) A nitric oxide burst precedes apoptosis in angiosperm and gymnosperm callus cells and foliar tissues. J Exp Bot 51:1027–1036

Peng D, Wang X, Li Z et al (2016) NO is involved in spermidine-induced drought tolerance in white clover via activation of antioxidant enzymes and genes. Protoplasma 253(5):1243–1254

Perchepied L, Balagué C, Riou C et al (2010) Nitric oxide participates in the complex interplay of defense-related signaling pathways controlling disease resistance to *Sclerotinia sclerotiorum* in *Arabidopsis thaliana*. Mol Plant-Microbe Interact 23(7):846–860

Pető A, Lehotai N, Lozano-Juste J et al (2011) Involvement of nitric oxide and auxin in signal transduction of copper-induced morphological responses in *Arabidopsis* seedlings. Ann Bot 108(3):449–457

Pető A, Lehotai N, Feigl G et al (2013) Nitric oxide contributes to copper tolerance by influencing ROS metabolism in *Arabidopsis*. Plant Cell Rep 32(12):1913–1923

Piterková J, Petrivalský M, Luhová L et al (2009) Local and systemic production of nitric oxide in tomato responses to powdery mildew infection. Mol Plant Pathol 10(4):501–513

Planchet E, Gupta JK, Sonoda M et al (2005) Nitric oxide emission from tobacco leaves and cell suspensions: rate limiting factors and evidence for the involvement of mitochondrial electron transport. Plant J 41:732–743

Planchet E, Verdu I, Delahaie J et al (2014) Abscisic acid-induced nitric oxide and proline accumulation in independent pathway under water-deficit stress during seedling establishment in *Medicago truncatula*. J Exp Bot 65(8):2161–2170

Poór P, Kovács J, Szopkó D et al (2013) Ethylene signaling in salt stress- and salicylic acid-induced programmed cell death in tomato suspension cells. Protoplasma 250(1):273–284

Raman CS, Li H, Martasek P et al (1998) Crystal structure of constitutive endothelial nitric oxide synthase: a paradigm for protein function involving a novel metal center. Cell 95:939–950

Ribeiro EA, Cunha FQ, Tamashiro WMSC et al (1999) Growth phase-dependent subcellular localization of nitric oxide synthase in maize cells. FEBS Lett 44:282–286

Rockel P, Strube F, Rockel A et al (2002) Regulation of nitric oxide (NO) production by plant nitrate reductase in vivo and in vitro. J Exp Bot 53(366):103–110

Romera FJ, García MJ, Alcántara E et al (2011) Latest findings about the interplay of auxin, ethylene and nitric oxide in the regulation of Fe deficiency responses by strategy I plants. Plant Signal Behav 6(1):167–170

Sahay S, Gupta M (2017) An update on nitric oxide and its benign role in plant responses under metal stress. Nitric Oxide 67:39–52

Sakihama Y, Kamura SN, Yamasaki H (2002) Nitric oxide production mediated by nitrite reductase in green algae *Chlamydomonas reinhardtii*: an alternative NO production pathway in photosynthetic organism. Plant Cell Physiol 43(3):290–297

Santa-Cruz DM, Pacienza NA, Polizio AH et al (2010) Nitric oxide synthase-like dependent NO production enhances heme oxygenase up-regulation in ultraviolet-B-irradiated soybean plants. Phytochemistry 71(14–15):1700–1707

Santa-Cruz DM, Pacienza NA, Zilli CG et al (2014) Nitric oxide induces specific isoforms of antioxidant enzymes in soybean leaves subjected to enhanced ultraviolet-B radiation. J Photochem Photobiol B 141:202–209

Schlicht M, Kombrink E (2013) The role of nitric oxide in the interaction of *Arabidopsis thaliana* with the biotrophic fungi, *Golovinomyces orontii* and *Erysiphe pisi*. Front Plant Sci 4:351

Sehrawat A, Deswal R (2014) S-nitrosylation analysis in *Brassica juncea* apoplast highlights the importance of nitric oxide in cold-stress signaling. J Proteome Res 13(5):2599–2619

Sen S, Chema IR (1995) Nitric oxide synthase and calmodulin immunoreactivity in plant embryonic tissue. Biochem Arch 11:221–227

Shanmugam V, Wang YW, Tsednee M et al (2015) Glutathione plays an essential role in nitric oxide-mediated iron-deficiency signaling and iron-deficiency tolerance in *Arabidopsis*. Plant J 84(3):464–477

Shi S, Wang G, Wang Y et al (2005) Protective effect of nitric oxide against oxidative stress under ultraviolet-B radiation. Nitric Oxide 13(1):1–9

Shi H, Ye T, Chan Z (2014) Nitric oxide-activated hydrogen sulfide is essential for cadmium stress response in bermudagrass (*Cynodon dactylon* (L). Pers.) Plant Physiol Biochem 74:99–107

Shi H, Chen Y, Tan DX et al (2015) Melatonin induces nitric oxide and the potential mechanisms relate to innate immunity against bacterial pathogen infection in *Arabidopsis*. J Pineal Res 59(1):102–108

Silveira NM, Hancock JT, Frungillo L et al (2017) Evidence towards the involvement of nitric oxide in drought tolerance of sugarcane. Plant Physiol Biochem 115:354–359

Simon-Sarkadi L, Ludidi N, Kocsy G (2014) Modification of cadaverine content by NO in salt-stressed maize. Plant Signal Behav 9(1):e27598. https://doi.org/10.4161/psb.27598

Sirova J, Sedlarova M, Piterkova J et al (2011) The role of nitric oxide in the germination of plant seeds and pollen. Plant Sci 181:560–572

Skelly MJ, Loake GJ (2013) Synthesis of redox-active molecules and their signaling functions during the expression of plant disease resistance. Antioxid Redox Signal 19(9):990–997

Song NK, Jeong CS, Choi HS (2000) Identification of nitric oxide synthase in *Flammulina velutipes*. Mycologia 92(6):1027–1032

Stohr C, Strube F, Marx G et al (2001) A plasma membrane-bound enzyme of tobacco roots catalyses the formation of nitric oxide from nitrite. Planta 212:835–841

Stuehr DJ (1996) Purification and properties of nitric oxide synthases. Methods Enzymol 268:324–333

Stuehr DJ, Cho HJ, Known NS et al (1991) Purification and characterization of the cytokine-induced macrophage nitric oxide synthase: an FAD- and FMN- containing flavoprotein. Proc Natl Acad Sci U S A 88:7773–7777

Summers LA (1980) The bipyridinium herbicides. Academic, New York

Sun A, Li Z (2013) Regulatory role of nitric oxide in lipopolysaccharides-triggered plant innate immunity. Plant Signal Behav 8(1):e22554. https://doi.org/10.4161/psb.22554

Talwar PS, Gupta R, Maurya AK et al (2012) *Brassica juncea* nitric oxide synthase like activity is stimulated by PKC activators and calcium suggesting modulation by PKC-like kinase. Plant Physiol Biochem 60:157–164

Tanou G, Filippou P, Belghazi M et al (2012) Oxidative and nitrosative-based signaling and associated post-translational modifications orchestrate the acclimation of citrus plants to salinity stress. Plant J 72(4):585–599

Tossi V, Lamattina L, Cassia R (2009) An increase in the concentration of abscisic acid is critical for nitric oxide-mediated plant adaptive responses to UV-B irradiation. New Phytol 181(4):871–879. https://doi.org/10.1111/j.1469-8137.2008.02722.x

Tossi V, Amenta M, Lamattina L et al (2011) Nitric oxide enhances plant ultraviolet-B protection up-regulating gene expression of the phenylpropanoid biosynthetic pathway. Plant Cell Environ 34(6):909–921

Tossi V, Lombardo C, Cassia R et al (2012) Nitric oxide and flavonoids are systemically induced by UV-B in maize leaves. Plant Sci 193–194:103–109

Tossi V, Lamattina L, Jenkins GI et al (2014) Ultraviolet-B-induced stomatal closure in *Arabidopsis* is regulated by the UV RESISTANCE LOCUS8 photoreceptor in a nitric oxide-dependent mechanism. Plant Physiol 164(4):2220–2230

Touraine B, Briat JF, Gaymard F (2012) GSH threshold requirement for NO-mediated expression of the *Arabidopsis* AtFer1 ferritin gene in response to iron. FEBS Lett 586(6):880–883

Tripathi DK, Singh S, Singh S et al (2017) Nitric oxide alleviates silver nanoparticles (AgNps)-induced phytotoxicity in *Pisum sativum* seedlings. Plant Physiol Biochem 110:167–177

Tun NN, Santa-Catarina C, Begum T et al (2006) Polyamines induce rapid biosynthesis of nitric oxide (NO) in *Arabidopsis thaliana* seedlings. Plant Cell Physiol 47:346–354

Valentovicova K, Haluskova L, Huttova J (2010) Effect of cadmium on diaphorase activity and nitric oxide production in barley root tips. J Plant Physiol 167:10–14

Verma K, Mehta SK, Shekhawat GS (2013) Nitric oxide (NO) counteracts cadmium induced cytotoxic processes mediated by reactive oxygen species (ROS) in *Brassica juncea*: cross-talk between ROS, NO and antioxidant responses. Biometals 26(2):255–269

Verniquet F, Gaillar J, Neuburger M et al (1991) Rapid inactivation of plant aconitases by hydrogen peroxide. Biochemistry 276:643–648

Vranova E, Inze D, Van Breusegem F (2002) Signal transduction during oxidative stress. J Exp Bot 53:1227–1236

Wang YH, Li XC, Zhu-Ge Q (2012) Nitric oxide participates in cold-inhibited *Camellia sinensis* pollen germination and tube growth partly via cGMP in vitro. 10.1371/journal.pone.0052436

Wang Y, Luo Z, Du R et al (2013) Effect of nitric oxide on antioxidative response and proline metabolism in banana during cold storage. J Agric Food Chem 61(37):8880–8887

Wang J, Yu SX, Zhang M et al (2014) Exogenous NO mediated GSH-PCs synthesis pathway in tomato under copper stress. Ying Yong Sheng Tai Xue Bao 25(9):2629–2636

Wang P, Du Y, Hou YJ et al (2015a) Nitric oxide negatively regulates abscisic acid signaling in guard cells by S-nitrosylation of OST1. Proc Natl Acad Sci U S A 2:613–618

Wang YW, Tsednee M, Karunakaran K et al (2015b) Glutathione plays an essential role in nitric oxide-mediated iron-deficiency signaling and iron-deficiency tolerance in *Arabidopsis*. Plant J 84(3):464–477

Wang YJ, Dong YX, Wang J et al (2016a) Alleviating effects of exogenous NO on tomato seedlings under combined Cu and Cd stress. Environ Sci Pollut Res Int 23(5):4826–4836

Wang H, Huang J, Li Y et al (2016b) Involvement of nitric oxide-mediated alternative pathway in tolerance of wheat to drought stress by optimizing photosynthesis. Plant Cell Rep 35(10):2033–2044

Wang W, Sheng X, Shu Z et al (2016c) Combined cytological and transcriptomic analysis reveals a nitric oxide signaling pathway involved in cold-inhibited Camellia sinensis pollen tube growth. Front Plant Sci 7:456

Warner RL, Kleinhofs A (1992) Genetics and molecular biology of nitrate metabolism in higher plants. Physiol Plant 85:245–252

Wei HR, Meng YL, Sun Y et al (2010) Effects of exogenous nitric oxide on high bush blueberry PSII photochemical activity and antioxidant system under high temperature stress. Ying Yong Sheng Tai Xue Bao 21(10):2529–2535

Wilkinson JQ, Crawford NM (1993) Identification and characterization of a chlorate resistant mutant of Arabidopsis with mutations in both *Nia1* and *Nia2* nitrate reductase structural genes. Mol Gen Genet 239:289–297

Wink DA, Mitchell JB (1998) Chemical biology of nitric oxide insight into regulatory, cytotoxic and cytoprotective mechanism in nitric oxide. Free Radic Biol Med 25:434–456

Wu JC, Chen JQ, Liang J et al (2009) Effects of exogenous NO on ascorbate-glutathione cycle in loquat leaves under low temperature stress. Chin J Appl Ecol 20:1395–1400

Wünsche H, Baldwin IT, Wu J (2011) Silencing NOA1 elevates herbivory-induced jasmonic acid accumulation and compromises most of the carbon-based defense metabolites in *Nicotiana attenuata* (F). J Integr Plant Biol 53(8):619–631

Xie Y, Ling T, Han Y et al (2008) Carbon monoxide enhances salt tolerance by nitric oxide-mediated maintenance of ion homeostasis and up-regulation of antioxidant defence in wheat seedling roots. Plant Cell Environ 31(12):1864–1881

Xie Y, Mao Y, Lai D et al (2013) Roles of NIA/NR/NOA1-dependent nitric oxide production and HY1 expression in the modulation of *Arabidopsis* salt tolerance. J Exp Bot 64(10):3045–3060

Xin L, Shuqiu Z, Chenghou L (2003) Involvement of nitric oxide in the signal transduction of salicylic acid regulating stomatal movement. Chin Sci Bull 48:449–452

Xin L, Wuliang S, Shuqiu Z et al (2005) Nitric oxide involved in signal transduction of jasmonic acid-induced stomatal closure of *Vicia faba* L. Chin Sci Bull 50:520–525

Xiong J, Zhang L, Fu G et al (2012) Drought-induced proline accumulation is uninvolved with increased nitric oxide, which alleviates drought stress by decreasing transpiration in rice. J Plant Res 125(1):155–164

Xu J, Yin H, Li Y et al (2010) Nitric oxide is associated with long-term zinc tolerance in *Solanum nigrum*. Plant Physiol 154(3):1319–1334

Xue L, Li S, Sheng H et al (2007) Nitric oxide alleviates oxidative damage induced by enhanced ultraviolet-B radiation in *Cyanobacterium*. Curr Microbiol 55(4):294–301

Yadu S, Dewangan TL, Chandrakar V et al (2017) Imperative roles of salicylic acid and nitric oxide in improving salinity tolerance in *Pisum sativum* L. Physiol Mol Biol Plants 23(1):43–58

Yamasaki H, Sakihama Y (2000) Simultanious production of nitric oxide and peroxynitrite by plant nitrate reductase: in vitro evidence for the NR-dependent formation of active nitrogen species. FEBS Lett 468:89–92

Yamasaki H, Sakihama Y, Takahashi S (1999) An alternative pathway for nitric oxide production in plants: new features of an old enzyme. Trends Plant Sci 4:128–129

Yan LJ, Levine RI, Sohal RS (1997) Oxidative damage during aging targets mitochondrial aconitase. Proc Natl Acad Sci U S A 54:11168–11172

Yang Q, He H, Li H et al (2011) NOA1 functions in a temperature-dependent manner to regulate chlorophyll biosynthesis and rubisco formation in rice. PLoS One 6(5):20015. https://doi.org/10.1371/journal.pone.0020015. PMCID: PMC3100308

Yang L, Ji J, Harris-Shultz KR (2016) The dynamic changes of the plasma membrane proteins and the protective roles of nitric oxide in rice subjected to heavy metal cadmium stress. Front Plant Sci 7:190

Ye Y, Li Z, Xing D (2012) Sorting out the role of nitric oxide in cadmium-induced Arabidopsis thaliana programmed cell death. Plant Signal Behav 7(11):1493–1494

Ye Y, Li Z, Xing D (2013) Nitric oxide promotes MPK6-mediated caspase-3-like activation in cadmium-induced *Arabidopsis thaliana* programmed cell death. Plant Cell Environ 36(1):1–15

Ye YQ, Jin CW, Fan SK et al (2015) Elevation of NO production increases Fe immobilization in the Fe-deficiency roots apoplast by decreasing pectin methylation of cell wall. Sci Rep 5:10746

Yin S, Gao Z, Wang C et al (2016) Nitric oxide and reactive oxygen species coordinately regulate the germination of *Puccinia striiformis f. sp. tritici* urediniospores. Front Microbiol 7:178

Yu Q, Sun L, Jin H et al (2012) Lead-induced nitric oxide generation plays a critical role in lead uptake by *Pogonatherum crinitum* root cells. Plant Cell Physiol 53(10):1728–1736

Yuan HM, Huang X (2016) Inhibition of root meristem growth by cadmium involves nitric oxide-mediated repression of auxin accumulation and signalling in Arabidopsis. Plant Cell Environ J 39(1):120–135. https://doi.org/10.1111/pce.12597

Yun BW, Skelly MJ, Yin M et al (2016) Nitric oxide and S-nitrosoglutathione function additively during plant immunity. New Phytol 211(2):516–526

Zhai L, Xiao D, Sun C et al (2016) Nitric oxide signaling is involved in the response to iron deficiency in the woody plant *Malus xiaojinensis*. Plant Physiol Biochem 109:515–524

Zhang M, An L, Feng H et al (2003) The cascade mechanisms of nitric oxide as a second messenger of ultraviolet B in inhibiting mesocotyl elongations. Photochem Photobiol 77(2):219–225

Zhang YK, Cui XM, Yang SX et al (2010) Effects of exogenous nitric oxide on active oxygen metabolism and photosynthetic characteristics of tomato seedlings under cadmium stress. Ying Yong Sheng Tai Xue Bao 21(6):1432–1438

Zhang M, Dong JF, Jin HH et al (2011) Ultraviolet-B-induced flavonoid accumulation in *Betula pendula* leaves is dependent upon nitrate reductase-mediated nitric oxide signaling. Tree Physiol 31(8):798–807

Zhang L, Chen Z, Zhu C (2012) Endogenous nitric oxide mediates alleviation of cadmium toxicity induced by calcium in rice seedlings. J Environ Sci 24(5):940–948

Zhang B, Shang S, Jabben Z et al (2014) Sodium chloride alleviates cadmium toxicity by reducing nitric oxide accumulation in tobacco. Ecotoxicol Environ Saf 110:56–60

Zhao Z, Chen G, Zhang C (2001) Interaction between reactive oxygen species and nitric oxide in drought-induced abscisic acid synthesis in root tips of wheat-seedling. Aust J Plant Physiol 28:1050–1061

Zhao L, Zhang F, Guo J et al (2004) Nitric oxide functions as a signal in salt resistance in the calluses from two ecotypes of reed. Plant Physiol 134(2):849–857

Zhao MG, Tian QY, Zhang WH (2007) Nitric oxide synthase dependent nitric oxide production is associated with salt tolerance in *Arabidopsis*. Plant Physiol 144:206–217

Zheng Y, Hong H, Chen L et al (2014) LeMAPK1, LeMAPK2, and LeMAPK3 are associated with nitric oxide-induced defense response against *Botrytis cinerea* in the *Lycopersicon esculentum* fruit. J Agric Food Chem 62(6):1390–1396

Zhou Y, Li S, Zeng K (2016) Exogenous nitric oxide-induced postharvest disease resistance in citrus fruit to *Colletotrichum gloeosporioides*. J Sci Food Agric 96(2):505–512

Zhu XF, Jiang T, Wang ZW et al (2012) Gibberellic acid alleviates cadmium toxicity by reducing nitric oxide accumulation and expression of IRT1 in *Arabidopsis thaliana*. J Hazard Mater 239–240:302–307

Zimmer-Prados LM, Moreira AS, Magalhaes JR et al (2014) Nitric oxide increases tolerance responses to moderate water deficit in leaves of *Phaseolus vulgaris* and *Vigna unguiculata* bean species. Physiol Mol Biol Plants 20(3):295–301

Zou T, Zheng LP, Yuan HY et al (2012) The nitric oxide production and NADPH-diaphorase activity in root tips of *Vicia faba* L. under copper toxicity. Plant Omics J 5(2):115–121

Printed by Printforce, the Netherlands